MBL Lectures in Biology
Volume 9

# THE BIOLOGY OF PARASITISM

A Molecular and Immunological Approach

# MBL LECTURES IN BIOLOGY

# THE BIOLOGY OF PARASITISM
## A Molecular and Immunological Approach

### Editors

## Paul T. Englund
Department of Biological Chemistry
Johns Hopkins School of Medicine
Baltimore, Maryland

## Alan Sher
Head, Immunology and Cell Biology Section
Laboratory of Parasitic Disease
National Institute of Allergies and Infectious Diseases
Bethesda, Maryland

Alan R. Liss, Inc. • New York

**Address all Inquiries to the Publisher**
**Alan R. Liss, Inc., 41 East 11th Street, New York, NY 10003**

**Printed in the United States of America**

**Library of Congress Cataloging in Publication Data**

The Biology of parasitism.

(MBL lectures in biology ; v. 9)
Includes bibliographies and index.
1. Parasitism. I. Englund, Paul T. II. Sher,
Alan. III. Marine Biological Laboratory (Woods
Hole, Mass.) IV. Series. [DNLM: 1. Parasites.
W1 MB999 v.9 / QX 4 B6145]
QL757. B57   1988        591.5'249        88-8947
ISBN 0-8451-2208-8
ISBN 0-8451-2209-6 (Paperback)

# Cover:

**Upper Left:** Stage specificity of antiparasite vaccines. See Sher, page 171.

**Upper Right:** Scanning electron micrograph of a *T. brucei brucei* among red blood cells. ×5500. See Donelson, page 372.

**Lower Left:** Separation of *Plasmodium falciparum* chromosomal DNA molecules by pulsed field gradient gel electrophoresis. Such separations provide a powerful means of mapping the parasite genome and analyzing the structure of individual chromosomes. Figure courtesy of Thomas E. Wellems, M.D., Ph.D., Malaria Section, Laboratory of Parasitic Diseases, National Institute of Allergy and Infectious Diseases, National Institutes of Health, Bethesda, Maryland.

**Lower Center:** Schematic diagram showing trafficking of membrane proteins from the intracellular *P. falciparum* parasite to the RBCM. See Howard, page 137.

**Lower Right:** Sketch of the structure of a monomeric subunit of the MITat 1.2 VSG, based on crystallographic evidence. See Turner, page 361.

# Contents

## PARASITE IMMUNOLOGY

## PARASITE MOLECULAR BIOLOGY, BIOCHEMISTRY, AND GENETICS

# Contributors

**R.F. Anders,** The Walter and Eliza Hall Institute of Medical Research, Melbourne, Australia **[201]**

**Stephen M. Beverley,** Department of Biological Chemistry and Molecular Pharmacology, Harvard Medical School, Boston, MA 02115 **[431]**

**Barry R. Bloom,** Department of Microbiology and Immunology, Albert Einstein College of Medicine, Bronx, NY 10461 **[265]**

**A.E. Butterworth,** Medical Research Council and Department of Pathology, University of Cambridge, Cambridge CB2 1QP, England **[43]**

**Richard Carter,** Medical Research Council External Scientific Staff, Institute of Animal Genetics, University of Edinburgh, Edinburgh, Scotland **[225]**

**Tamara L. Doering,** Department of Biological Chemistry, Johns Hopkins School of Medicine, Baltimore, MD 21205 **[401]**

**John E. Donelson,** Department of Biochemistry, University of Iowa, Iowa City, IA 52442 **[371]**

**Dennis M. Dwyer,** Cell Biology and Immunology Section, Laboratory of Parasitic Diseases, National Institutes of Health, Bethesda, MD 20892 **[449]**

**Thomas E. Ellenberger,** Department of Biological Chemistry and Molecular Pharmacology, Harvard Medical School, Boston, MA 02115 **[431]**

**Paul T. Englund,** Department of Biological Chemistry, Johns Hopkins School of Medicine, Baltimore, MD 21205 **[401]**

**Michael Gottlieb,** Department of Immunology and Infectious Diseases, The Johns Hopkins University School of Hygiene and Public Health, Baltimore, MD 21205 **[449]**

**Gerald W. Hart,** Department of Biological Chemistry, Johns Hopkins School of Medicine, Baltimore, MD 21205 **[401]**

**Dale Hereld,** Department of Biological Chemistry, Johns Hopkins School of Medicine, Baltimore, MD 21205 **[401]**

**Russell J. Howard,** Malaria Section, Laboratory of Parasitic Diseases, National Institute of Allergy and Infectious Diseases, National Institutes of Health, Bethesda, MD 20892; present address: DNAX Research Institute of Molecular and Cellular Biology, Palo Alto, CA 94304-1104 **[111]**

**David M. Iovannisci,** Department of Biological Chemistry and Molecular Pharmacology, Harvard Medical School, Boston, MA 02115 **[431]**

The numbers in brackets are the opening page numbers of the contributors' articles.

**Stephanie L. James,** Departments of Medicine and Microbiology, The George Washington University Medical Center, Washington, DC 20037 **[249]**

**Keith A. Joiner,** Laboratory of Clinical Investigation, National Institute of Allergy and Infectious Diseases, National Institutes of Health, Bethesda, MD 20892 **[309]**

**Geoffrey M. Kapler,** Department of Biological Chemistry and Molecular Pharmacology, Harvard Medical School, Boston, MA 02115 **[431]**

**Jessica L. Krakow,** Department of Biological Chemistry, Johns Hopkins School of Medicine, Baltimore, MD 21205 **[401]**

**Carole A. Long,** Malaria Research Group, Department of Microbiology and Immunology, Hahnemann University, Philadelphia, PA 19102 **[233]**

**R.M. Maizels,** Departments of Pure and Applied Biology, Imperial College of Science and Technology, London SW7 2BB, England **[285]**

**Philip D. Marsden,** Núcleo de Medicina Tropical, University of Brasília, 70910 Brasília, DF, Brazil **[77]**

**Adolfo Martínez-Palomo,** Department of Experimental Pathology, Center for Research and Advanced Studies, National Polytechnical Institute, 07000 Mexico, D.F. Mexico **[61]**

**Wayne J. Masterson,** Department of Biological Chemistry, Johns Hopkins School of Medicine, Baltimore, MD 21205 **[401]**

**G.S. Nelson,** Department of Parasitology, Liverpool School of Tropical Medicine, Liverpool L3 5QA, England **[13]**

**Ruth S. Nussenzweig,** Department of Medical and Molecular Parasitology, New York University Medical Center, New York, NY 10016 **[183]**

**Victor Nussenzweig,** Department of Pathology, New York University Medical Center, New York, NY 10016 **[183]**

**Miercio E.A. Pereira,** Tufts University School of Medicine, New England Medical Center Hospitals, Division of Geographic Medicine and Infectious Diseases, Boston, MA 02111 **[105]**

**Maria Petrillo-Peixoto,** Department of Biological Chemistry and Molecular Pharmacology, Harvard Medical School, Boston, MA 02115 **[431]**

**E.R. Pfefferkorn,** Department of Microbiology, Dartmouth Medical School, Hanover, NH 03756 **[479]**

**David L. Sacks,** Laboratory of Parasitic Diseases, National Institute of Allergy and Infectious Diseases, National Institutes of Health, Bethesda, MD 20892 **[93]**

**Phillip Scott,** Laboratory of Parasitic Diseases, National Institute of Allergy and Infectious Diseases, National Institutes of Health, Bethesda, MD 20892 **[249]**

**M.E. Selkirk,** Department of Biochemistry, Imperial College of Science and Technology, London SW7 2BB, England **[285]**

**Alan Sher,** Immunology and Cell Biology Section, Laboratory of Parasitic Diseases, National Institute of Allergy and Infectious Diseases, National Institutes of Health, Bethesda, MD 20892 **[169]**

**Barbara J. Sina,** Department of Biological Chemistry and Molecular Pharmacology, Harvard Medical School, Boston, MA 02115 **[431]**

**Andrew Spielman,** Department of Tropical Public Health, Harvard School of Public Health, Boston, MA 02115 **[147]**

**Mervyn J. Turner,** Biochemical Parasitology, Merck Sharp & Dohme Research Laboratories, Rahway, NJ 07065 **[349]**

**David Walliker,** Department of Genetics, University of Edinburgh, Edinburgh EH9 3JN, Scotland **[467]**

**Ching C. Wang,** Department of Pharmaceutical Chemistry, University of California San Francisco, San Francisco, CA 94143 **[413]**

**Samuel Ward,** Department of Embryology, Carnegie Institution of Washington, Baltimore, MD 21210; present address: Department of Molecular and Cellular Biology, University of Arizona, Tucson, AZ 85721 **[503]**

**Kenneth S. Warren,** Director, Health Sciences Division, The Rockefeller Foundation, New York, NY 10036 **[3]**

**Donald L. Wassom,** Department of Pathobiological Sciences, School of Veterinary Medicine, University of Wisconsin, Madison, WI 53706 **[329]**

# Preface

In July 1888, the *Falmouth Enterprise* (a weekly newspaper published in Falmouth, Massachusetts) announced a new scientific establishment in "Woods Hole." The *Enterprise* article noted that the new Marine Biological Laboratory (MBL) "seeks to give advanced workers, investigators, experts, etc., opportunities for pursuing their researches in life science, in the expectation that they will thereby increase the sum of human knowledge and solve some of the great problems, which still exist about marine animals and plants."

In the one hundred years since its founding, the MBL has had notable success at offering generations of investigators the opportunity to increase the sum of human knowledge and to solve some of the great problems of life sciences. The laboratory has had an especially good track record at mounting state-of-the-art summer courses in many areas of basic biology. Internationally renowned among biologists, the courses are distinguished by their constantly evolving curricula, with course directorships turning over every five years.

Almost from the beginning, MBL summer courses have played a major role in biological research, helping to shape the disciplines of cell biology, developmental biology, and neurobiology. The nine-year-old Biology of Parasitism course, one of the youngest MBL summer courses, was founded in the hope it would make a similar contribution to parasitology.

In the late 1970s, when the Biology of Parasitism course was conceived, it was clear that the powerful new tools of molecular biology were not being applied in any serious and systematic way to the study of parasites. In part this was because of a historical (and largely gratuitous) schism between parasitology and microbiology. Additionally, the field was not well-funded compared with other areas of biomedical research.

Around 1980, research on cancer, a disease that affects 10 million people worldwide, was supported by $900 million in the United States alone. At the same time, research on schistosomiasis, which affects 200 million people and results in 700,000 deaths annually, was supported by less than $3 million worldwide.

Recognizing that the benefits of modern biology were bypassing parasitology, a number of foundations and agencies turned their attention—and their resources—to the problem. This was a wise and well-timed move. Unlike many other biomedical problems, parasitic diseases have well-known causes. With a relatively small investment, there was hope of making a significant improvement in the lives of many hundreds of millions of people.

The MBL was chosen to host the training camp for the campaign to modernize parasitology. True, the field of parasitology is not marine in any sense, but the MBL was the appropriate place nonetheless, for a number of reasons. Perhaps most important, the laboratory has a long history of multi-disciplinary programs—in education and in research. As the MBL has neither a parasitology department nor a microbiology department (nor, for that matter, *any* formal department), there was no chance that any territorial struggles would limit the scope or success of a new course. At the MBL it is relatively easy to bring together an international faculty of parasitologists, microbiologists, immunologists, molecular biologists, pharmacologists, vector biologists, entomologists, and cell biologists.

In the eight years it has been offered, the Biology of Parasitism course has fulfilled all expectations. Through 1987, the course has trained 128 students, most of whom have continued research careers in parasitology or closely related fields. A literature search in the fall of 1987 turned up more than 500 papers written by former students. In 1986, the course was described in the *ASM News* (American Society for Microbiology) as "an instrumental force in modernizing this field of research (parasitology)." Photographs of participants in the Biology of Parasitism course, 1980–1987, appear in the Appendix, pages 519–522.

We are indebted to the energetic leadership of course co-directors Alan Sher and Paul T. Englund for bringing together this volume, which makes the course lectures available to investigators, educators, and students of parasitology beyond the classrooms and laboratories of the MBL.

**Harlyn O. Halvorson**
**President and Director**
**Marine Biological Laboratory**
**Woods Hole**
**February 1988**

# Introduction

This book is based on the Biology of Parasitism course at the Marine Biological Laboratory (MBL) in Woods Hole, Massachusetts. During the past eight years the course has been a major factor in the renaissance of the field of parasitology. An important reason for its influence has been its emphasis on the application of molecular biology and modern immunology to the investigation of parasites responsible for human disease. The faculty and lecturers teach not only the most important current research on parasites, but also relevant areas of basic biochemistry, immunology, and molecular biology. For nine or ten weeks each summer, sixteen outstanding students, from every part of the world, participate in this course. It serves to inspire a new group of scientists each year, and it also contributes to the field by serving as an international forum for exchange of ideas and for establishment of research collaborations. Because of the Biology of Parasitism course, the MBL is now a major intellectual center for modern parasitology.

The concept of the Biology of Parasitism course was developed in the late 1970s by Drs. Joshua Lederberg, Kenneth Warren, and Anthony Cerami. Because of the MBL's great tradition in biological sciences and summer education programs, it was considered the logical institution to host the course. Dr. Paul Gross, then Director of the MBL, gave enthusiastic support. With financial assistance from the Edna McConnell Clark and Rockefeller foundations, the MBL provided space for the course laboratory. Dr. John David of Harvard Medical School was invited to be the first Course Director.

Dr. David launched the course in the summer of 1980. Because of the enthusiasm of the faculty, the multidisciplinary nature of its curriculum, and the novelty of parasites as biological systems, the course was an instant success. Modern molecular and immunological approaches were applied, in many cases for the first time, to the study of parasites of medical importance. In subsequent years the course became widely known and the number of applications grew substantially. One measure of success of the course is that many former students have proceeded to make their own contributions to research in parasitology.

We became Co-directors of the Biology of Parasitism course in 1985. Although we were initially anxious about taking over such a highly successful enterprise, we were fortunate each summer to have a truly outstanding and exuberant faculty. Our task was also eased by the generous financial support of the Edna McConnell Clark Foundation, the MacArthur Foundation, the Burroughs-Wellcome Fund, the Wellcome Trust, and New England Biolabs. The course is currently organized in two sections. During the first 4½ weeks the students conduct organized laboratory exercises in molecular biology, biochemistry, immunology, and membrane biology and immunochemistry of parasites. During this period they also study the biology of the major parasites and vectors. The second 4½ weeks are devoted to independent research projects that reflect the individual interests of the students. Throughout the course there are morning lectures on parasite biology and on basic molecular biology and immunology. There are also several lectures on nonparasitic infectious diseases such as leprosy and AIDS.

In 1989 the course leadership will again change hands. The new Directors, Drs. John Donelson and Carole Long, will continue to emphasize molecular biology and immunology, without neglecting parasite biology and epidemiology. With the strong support of the MBL and its new Director, Dr. Harlyn Halvorson, with continued financial commitments from the MacArthur Foundation and other funding sources, and with expanding interest in the field of parasitology, the future of the course appears to be secure.

This book, a collection of essays by current and former faculty, conveys the intellectual spirit of the Biology of Parasitism Course. We have asked each contributor to summarize major concepts in an area rather than to comprehensively review the literature or present current data. Like the Woods Hole course, the book covers a wide range of subjects and expresses many different, and sometimes conflicting, points of view. We have organized the chapters within the topics of parasite biology and disease, parasite immunology, and parasite molecular biology, biochemistry, and genetics. However, because of the multidisciplinary nature of the field, some articles are not easily categorized.

All of the contributors to this volume share with us an enormous enthusiasm for the field of parasitology and for the philosophy of the Biology of Parasitism Course. We express to them our deepest appreciation for their important contributions to this book and to their continuing support of the MBL teaching program. We hope that the book will serve as an important summary not only to present, former, and future participants in the course but also to all students of parasitology. Finally, we hope that this volume will testify to the dynamic growth and keen intellectual excitement of contemporary research on parasites.

**Alan Sher**
**Bethesda, Maryland**
**Paul T. Englund**
**Baltimore, Maryland**

# THE BIOLOGY OF PARASITES
# AND PARASITIC DISEASE

The Biology of Parasitism, pages 3–12
© 1988 Alan R. Liss, Inc.

# The Global Impact of Parasitic Diseases

**Kenneth S. Warren**

*Director, Health Sciences Division, The Rockefeller Foundation, New York, New York 10036*

The impact of parasitic diseases on the peoples of this world is truly enormous. The protozoan and helminth organisms which cause these diseases are ubiquitous and are particularly prevalent in the vast tropical regions. Particularly hard hit by these infectious agents are children: 75% of the global population is in the developing world and approximately 50% of these more than 3 billion people are under 15 years of age; 15% are in the age group 0–4. Ninety-seven percent of all infant and child deaths occur in this latter group, and most of these fatalities are due to infectious diseases (Grant, 1983). Table 1—which ranks the bacterial, viral, protozoan, and helminthic diseases of Africa, Asia, and Latin America on the basis of prevalence, mortality, and morbidity—gives some idea of the overwhelming nature of this problem (Walsh and Warren, 1979). While diarrheal and respiratory diseases rank one and two, they are each a composite of many different infections, both viral and bacterial in etiology. Malaria ranks third and continues to increase in prevalence, mortality and morbidity because of resistance of the mosquito vectors to insecticides and the parasites to antimalarial drugs. Several of the protozoan diseases—such as African sleeping sickness caused by *Trypanosoma gambiese* and *rhodesiense*, and Kala Azar, caused by *Leishmania donovani*—are almost universally fatal unless treated. Therapy, however, is inadequate, often requiring the arsenic and antimony-containing drugs which were introduced 70–80 years ago. While often not fatal, many of the parasitic diseases are grossly disfiguring such as espundia (mucocutaneous leishmaniasis) of Latin America in which the face is gradually destroyed, and elephantiasis (bancroftian and Malayan filariasis) which causes severe distortion of the limbs and genitalia. Blinding is the major complication of onchocerciasis also known as river blindness. The vast numbers of individuals infected with hookworm, ascaris and trichuris, each accounting for almost a billion victims, suffer from often subtle forms of malnutrition, including iron deficiency. For many of these diseases diagnosis is complex and difficult and treatment is inadequate; vaccines are not now available for any human parasitic disease.

TABLE 1. Prevalence, Mortality, and Morbidity of the Major Infectious Diseases of Africa, Asia, and Latin America, 1977-1978[a]

| Infection | Infections (thousands/yr) | Deaths (thousands/yr) | Disease (thousands of cases/yr) | Average No. of days of life lost (per case) | Relative Personal Disability[b] |
|---|---|---|---|---|---|
| Diarrheas | 3-5,000,000 | 5-10,000 | 3-5,000,000 | 3-5 | 2 |
| Respiratory infections | | 4-5,000 | | 5-7 | 2-3 |
| Malaria | 800,000 | 1,200 | 150,000 | 3-5 | 2 |
| Measles | 85,000 | 900 | 80,000 | 10-14 | 2 |
| Schistosomiasis | 200,000 | 500-1,000 | 20,000 | 600-1,000 | 3-4 |
| Whooping cough | 70,000 | 250-450 | 20,000 | 21-28 | 2 |
| Tuberculosis | 1,000,000 | 400 | 7,000 | 200-400 | 3 |
| Neonatal tetanus | 120-180 | 100-150 | 120-180 | 7-10 | 1 |
| Diphtheria | 40,000 | 50-60 | 700-900 | 7-10 | 3 |
| Hookworm | 7-900,000 | 50-60 | 1,500 | 100 | 4 |
| South American trypan- somiasis | 12,000 | 60 | 1,200 | 600 | 2 |
| Onchocerciasis | | | | | |
| Skin disease | 30,000 | Low | 2-5,000 | 3,000 | 3 |
| River blindness | | 20-50 | 200-500 | 3,000 | 1-2 |
| Meningitis | 150 | 30 | 150 | 7-10 | 1 |
| Amebiasis | 400,000 | 30 | 1,500 | 7-10 | 3 |
| Ascariasis | 800,000- 1,000,000 | 20 | 1,000 | 7-10 | 3 |
| Poliomyelitis | 80,000 | 10-20 | 2,000 | 3,000+ | 2 |
| Typhoid | 1,000 | 25 | 500 | 14-28 | 2 |
| Leishmaniasis | 12,000 | 5 | 12,000 | 100-200 | 3 |
| African trypanosomiasis | 1,000 | 5 | 10 | 150 | 1 |
| Leprosy | | Very low | 12,000 | 500-3,000 | 2-3 |
| Trichuriasis | 500,000 | Low | 100 | 7-10 | 3 |
| Filariasis | 250,000 | Low | 2-3,000 | 1,000 | 3 |
| Giardiasis | 200,000 | Very low | 500 | 5-7 | 3 |
| Dengue | 3-4,000 | 0.1 | 1-2,000 | 5-7 | 2 |
| Malnutrition | 5-800,000 | 2,000 | | | |

[a]Based on estimates from the World Health Organization and its Special Programme for Research and Training in Tropical Diseases, confirmed or modified by extrapolations from published epidemiologic studies performed in well defined populaitons (see references). Figures do not always match those officially reported, because underreporting is great.

[b]1 denotes bedridden; 2, able to function on own to some extent; 3, ambulatory; 4, minor.

While these parasites tend to be found in greatest numbers in the tropics, their distribution may be cosmopolitan. In far northern climates, trichinosis, echinococcosis (hydatid disease) and fish tapeworm (*Diphyllobothrium latum*) infection are common. Toxoplasmosis is found throughout the world with more than 30% of the population in the United States and Europe infected. The geographic distribution of the more important of these parasitic infections is shown in Table 2 (Warren and Mahmoud, 1985).

In addition to the challenge posed by the prevalence of these organisms and the morbidity and mortality caused by them, scientists have become fascinated by their unique host-parasite interrelationships in terms of ecology, epidemiology and pathogenesis. The two groups of parasites, protozoa and helminths, are very different: the former are unicellular, multiply within the definitive human host, and are frequently intracellular in habitat; the latter are large multicellular organisms which don't multiply within humans and undergo complex metamorphoses and migrations within the host.

Protozoa as defined by the population biologists Anderson and May (1979) are "microparasites (also viruses and bacteria) characterized by small size, short generation times, extremely high rates of direct reproduction within the host, and a tendency to induce immunity to reinfection in those hosts that survive the initial onslaught. The duration of infection is typically short in relation to the expected life span of the host, and therefore is of a transient nature." These organisms have developed elaborate mechanisms for the evasion of host immune responses. Many of them are intracellular in habitat. *Leishmania* are able to survive within host cell phagolysosomes, *Toxoplasma* prevent lysosomal fusion with phagosomes, and *Trypanosoma cruzi* moves

**TABLE 2. Major Areas of Distribution of Parasitic Disease**

| Worldwide | Africa | Asia | Latin America |
|---|---|---|---|
| Amebiasis | Leishmaniasis | Leishmaniasis | Leishmaniasis |
| Giardiasis | Malaria | Malaria | Malaria |
| Toxoplasmosis | Trypanosomiasis, Afr. | Filariasis | Trypanosomiasis, Am. |
| Ascariasis | Filariasis | Clonorchiasis | Filariasis |
| Toxocariasis | Onchocerciasis | Opisthorchiasis | Onchocerciasis |
| Echinococcosis | Paragonimiasis | Fasciolopsiasis | Schistosomiasis |
| Enterobiasis | Schistosomiasis | Paragonimiasis | |
| Fascioliasis | | Schistosomiasis | |
| Hookworm | | | |
| Strongyloidiasis | | | |
| Tapeworms | | | |
| Trichinosis | | | |
| Trichuriasis | | | |

out of the phagosome into a cytoplasmic habitat. The African trypanosomes have developed an elaborate system of antigenic variation which may involve as many as several hundred different variant surface antigens. Furthermore, different strains of the parasite have different repertoires of antigens. Other evasive mechanisms include shedding and renewal of surface antigens, and modification of host immune responsiveness including immunosuppression and polyclonal lymphocyte activation.

Helminths as defined by Anderson and May (1979) are "macroparasites which tend to have much longer generation times than the microparasites. Direct multiplication within the host is either absent or occurs at a low rate. The immune responses elicited by these metazoans generally depend on the number of parasites present in a given host, and tend to be of relatively short duration. Macroparasitic infections, therefore, tend to be of a persistent nature with hosts being continually reinfected." Distribution of parasites within hosts is overdispersed with only a small proportion of individuals carrying large numbers of worms. In most cases, disease occurs only in those with heavy infections. Helminths, too, have many ways of evading host immune responses.

Of the two principal parasite groups, helminths, all of which are visible to the naked eye, have been known since ancient times and among primitive peoples. Protozoa were not seen until 1681 when Leeuwenhoek, who invented the microscope, first observed "animalcules" in his own feces. The first parasitic protozoan, *Entamoeba histolytica*, was reported by Losch in 1875 (Foster, 1965). Parasitology as a scientific discipline didn't begin, however, until the period 1900–1918, when the impetus provided by colonialism throughout the developing world led to a concerted effort to study the infectious diseases of the tropics. As described by Worboys (1983) it "emerged to provide the zoological underpinning for tropical medicine. Its subject matter was defined on the one hand by the tropical medicine curriculum and on the other by the botany-zoology division and the way tropical parasites fitted neatly into the type-based zoological curriculum." While all infectious agents are parasites, one of the early parasitologists, Nuttall, called bacteria and viruses "vegetable parasites." Protozoa and helminths were termed "animal parasites," not because they infect animals but because they, themselves, are animals.

*Parasitology*, the first journal in this area, began publication in the United Kingdom in 1908 and defined its subject matter as "disease-transmitting insects, malaria, trypanosomiasis, spirochaetoses, piroplasmosis, and plague, as well as parasitic worms." The *Journal of Parasitology*, which appeared in the United States in 1914, was more specific, stating that its area of interest was "animal parasites," and noting that "emphasis will be laid on the morphology, life history and biology of zooparasites, and the relations of animals to disease" (Warren, 1981). Parasitology did not become established until the period 1914–1940, "when scientists began to call themselves parasitologists, when parasitological institutes and associations were founded, when parasitological education, usually postgraduate, became available, and the journals began to proliferate and were founded in the major scientific nations" (Worboys, 1983).

The separation of parasitology from the study of bacteria and viruses, which has been characterized as the field of microbiology, has had negative consequences. While departments of microbiology have flourished in medical schools and in research institutes throughout the world, parasitology became immured in the relatively few schools of public health and tropical medicine. It is striking to note that microbiology has provided the underpinning for the two great fields of modern biology—immunology and molecular biology— and also provided effective means of treatment (antibiotics) and control (vaccines). In contrast, parasitology produced no breakthroughs in basic science and is still plagued by the lack of effective and nontoxic drugs; to repeat, vaccines are not available for any human parasitic disease.

With respect to the education of parasitologists, a survey by the American Society of Parasitologists in 1978 revealed that of 141 educational institutions in the United States only 4.7% of graduate courses in parasitology were devoted to immunology, 3.6% to physiology and biochemistry, 0.7% to ultrastructure and none to molecular biology (Weinstein, 1981). A study of the literature of parasitology for the year 1979 revealed that among 19 English-language journals in the areas of general parasitology, specialty parasitology, tropical medicine, general medicine, specialty medicine, veterinary medicine, general science and specialty sciences a total of only 972 papers were published. In that year, the *Journal of Immunology* alone published 960 papers (Warren, 1981.) A comparative study of journal citation analysis in parasitology, virology and bacteriology for 1982 found that the former is relatively isolated and out of the mainstream of modern biological research. Examination of the subject matter in English language textbooks of parasitology corroborated concerns that these textbooks have been "dominated by the traditional subject areas of morphology, biology and life cycles, pathogenicity and symptomatology, diagnosis and treatment, and epidemiology and control" (Warren et al., 1983).

A study of the funding available for parasitological research in 1977 was startling indeed. Global support for research on malaria, schistosomiasis, filariasis, amebiasis, and ascariasis—infections involving 2.2 billion people—

was not much more than $10 million, while U.S. government support for research on cancer—an affliction of 10 million people—was $815 million (Rockefeller Foundation, unpublished data). A later report on support by U.S. government agencies for all human parasitic diseases revealed a maximum of $40 million (Cook, 1981). Almost $20 million of this funding was provided by the Department of Defense and the Agency for International Development for malaria alone. In that year, the National Science Foundation contribution to research on parasitology was $33,000. The National Institute of Allergy and Infectious Diseases Tropical Medicine Program has gradually increased its support from approximately $20 million in 1978 to about $30 million in recent years. It is interesting to note that the localized and relatively rare disease African sleeping sickness has received more NIH support than any other parasitic disease over recent years. This does not relate to the importance of the disease in human terms, but bears on the quality of work being done largely by molecular biologists on the mechanisms controlling antigenic variation. Also of interest is that in spite of the considerable increase in NIH funding, and the enormous expansion in basic scientific research on parasites, only one study section deals with all aspects of parasitic and tropical diseases from molecular biology to snail morphology.

In the last decade, great changes have occurred in the quality and quantity of research on parasitic diseases. At a meeting held in New Orleans in 1980 entitled "The Current Status and Future of Parasitology," the following prescient remarks were made by Joshua Lederberg (1981), President of Rockefeller University. "The germ theory, and the basic techniques for the recognition and isolation of various species of pathogenic microorganisms in pure culture established by Louis Pasteur, Robert Koch, and Ferdinand Cohn, gave us a sweeping scientific principle looking for the appropriate question to which it would be a solution. As far as I am aware, this principle has had more important consequences for the improvement of public health than any other concept in the history of mankind. But today we are uneasy because we cannot so easily replicate that kind of comprehensive advance for our remaining health problems such as heart disease, cancer, and schizophrenia. . . . The introduction of the germ theory of disease led to a rapid penetration of one large set of public health problems. I hold the view that it is precisely in parasitic infection that we have the nearest analogue to that kind of opportunity. My major premise is that when our understanding of the eukaryotic agents of infection can be brought to comparable levels of depth and insight—complicated by the fact that parasites are indeed eukaryons, often intracellular in habitat, and resemble the metabolism of the host more closely than do bacterial parasites—we will see advances as sudden and as spectacular as were achieved for most of the bacterial infections. . . . The assemblage of scientists at this conference is testimony to the capacity to mobilize a diverse set of experiences and intellectual resources. If the material resources could be made available, with the kind of nucleation represented here, very rapid progress, a veritable new wave of research would at this point be inevitable."

New sources of support began to appear in the latter 1970s. The largest of these was the Tropical Diseases Research Programme (TDR) of the World Health Organization. Funding for this program, most of which was provided by the great bilateral aid agencies, reached a peak of $25 million about 5 years ago and has declined since to an annual level of about $20–24 million. Five groups of parasitic diseases are included in the program—malaria, trypanosomiasis, leishmaniasis, schistosomiasis, and filariasis; leprosy is the sixth disease. The subject areas of particular interest, and the grants provided, are decided by Steering Committees for each disease; grants average about $25,000 per annum. At approximately the same time that the TDR program began, the Edna McConnell Clark Foundation started a unique research program devoted only to schistosomiasis. Groups of investigators from all over the world were gathered together to develop a strategic plan, which was updated approximately every 3 years. Funding averaged about $2.5 million per year. In the last few years, support for the schistosomiasis program has been declining while a similar program devoted to the blinding diseases, trachoma and onchocerciasis, has been developed. In 1978, the Rockfeller Foundation began a program called the Great Neglected Diseases (GND) of Mankind. Research units were established in the mainstream of modern biomedical investigations within medical schools and research institutes in Australia, Thailand, Israel, Egypt, England, Sweden, the United States and Mexico. The tenets of this undertaking were of investigator-initiated research on any of the GND from the basic to the field level, the latter being done collaboratively with scientists in the developing world. Support was pledged for 8 years, and investigators were gathered together in a network through an annual meeting to foster communication and collaboration. At least 70% of the work in this program has been on parasitic diseases. In the early 1980s, the John D. and Catherine T. MacArthur Foundation began a program on the Biology of Parasitic Diseases in which ten outstanding units in major universities and research institutes from Australia to Mexico are being supported for a period of 5 years to do state-of-the-art research on the great parasitic diseases. They are linked by a computer network and meet annually.

The effect of all of these programs on the field of parasitology has been enormous. Not only has training gone on in many of the research-oriented programs, but special courses have been set up to bring outstanding young scientists into the field. The first of these was begun, at the suggestion of Joshua Lederburg, by the Edna McConnell Clark and Rockefeller Foundations at the Marine Biological Laboratories in Woods Hole, Massachusetts. This course has trained more than 100 young investigators to work on parasitic diseases using the most advanced techniques of cellular biology, immunology, and molecular biology. Major meetings have been devoted to parasitic diseases, including the prestigious Gordon Conferences, the recent addition of molecular parasitism to the UCLA conferences, and a FASEB conference on molecular biology and infectious diseases planned for the summer of 1988.

Even textbooks, those great flagships of research and teaching, are slowly beginning to change direction. In 1982 *Modern Parasitology* (Cox, 1982) appeared, as did the second edition of *Immunology of Parasitic Infections* (Cohen and Warren, 1982). In 1984, *Tropical and Geographical Medicine* (Warren, Mahmoud, 1984) was published. This book contains introductory sections on genetics and the biochemistry and immunology of parasitic infections. The chapters covering specific diseases begin with state-of-the-art science on each parasitic organism.

Over the last decade, superb new drugs have been developed by industry for schistosomiasis and filariasis and field tested by the TDR program of WHO. The Clark Foundation's program on schistosomiasis and the Rockefeller Foundation's 16-year longitudinal study on the island of St. Lucia have resulted in new and cost-effective strategies for control. Many donors have supported investigations of the remarkably complex means of evasion of host defense mechanisms by the African trypanosomes. In fact, these parasitic organisms have now become research models for molecular biologists. Fundamental discoveries have been made, such as a macrophage-released mediator, cachectin, circulating in the blood of emaciated rabbits with trypanosomiasis. Cachectin, which is virtually identical with tumor-necrosis factor, appears, among numerous other activities, to be a crucial mediator of irreversible shock. A remarkable toxin released by *Entamoeba histolytica* was discovered by a student in the Woods Hole course. And in the particularly neglected area of vaccines, protective antigens have been tested in laboratory animals for malaria (both sporozoite and blood stream stages), schistosomiasis (at least eight in as many different laboratories), filariasis, and, perhaps, hookworm. There are many other discoveries, and even more important, an air of excitement among investigators that bodes well for the future.

In conclusion it should be noted that the negative impact of parasitic diseases has undergone little amelioration up to the present time. In spite of the billions of people still infected with intestinal helminths; the fact that malaria continues its depredations having been only momentarily deflected by the great and failed attempt at eradication; and the recrudescence of African sleeping sickness in Africa, there is now great hope for the future. With the application of modern science, particularly biotechnology, vaccines are now in the offing against many of the great parasitic diseases of mankind. John Maddox, editor of *Nature,* has recently pointed out "that the greatest promise of biotechnology, at the present time, is in the area of vaccines," and particularly noted the progress of vaccine development for malaria and schistosomiasis (Maddox, 1986). Studies of the basic biology of parasitic organisms, and of the host-parasite relationship, is undergoing a virtual renaissance. Large numbers of bright young investigators trained by various programs and courses are devoting their talents to this problem. Parasitology has become fashionable not only in the great institutes of basic science, but in the clinical departments and geographic medicine divisions of medical schools. Thus, while the impact of parasitic diseases on people remains relatively undiminished, the impact of modern science on parasitic diseases should soon enable us drastically to decrease the predations of these parasites on mankind.

## REFERENCES

Anderson RM, May RM (1979): Population biology of infectious diseases: Part I. Nature 280:361.

Cohen S, Warren KS (1982): Immunology of Parasitic Infections. Oxford: Blackwell Scientific Publications.

Cook JA (1981): Sources of funding for training and research in parasitology. In K.S. Warren and E.F. Purcell (eds): The Current Status and Future of Parasitology. New York: Josiah Macy, Jr. Foundation, pp 121–135.

Cox FEG (1982): Modern Parasitology. Oxford: Blackwell Scientific Publications.

Foster WD (1965): A History of Parasitology. Edinburgh: E&S Livingstone Ltd.

Grant JP (1983): The State of the World's Children 1984. Oxford: Oxford University Press.

Lederberg J (1981): The future of parasitology: An overview. In K.S. Warren and E.F. Purcell (eds): The Current Status and Future of Parasitology. New York: Josiah Macy, Jr. Foundation, pp 157–165.

Maddox J (1986): The new technology of medicine. Nature 321:807.

Walsh JA, Warren KS (1979): Selective primary health care: An interim strategy for disease control in developing countries. N Engl J Med 301:967–974.

Warren KS (1981): The present status of the parasitology literature. In K.S. Warren and E.F. Purcell (eds): The Current Status and Future of Parasitology. New York: Josiah Macy, Jr. Foundation, pp 142–154.

Warren KS, Goffman W, Chernin E (1983): The status of the parasitology literature: Linkages to modern biology. In K.S. Warren and J.Z. Bowers (eds): Parasitology: A Global Perspective. New York: Springer-Verlag, ch 15, pp 178–190.

Warren KS, Mahmoud AAF (1984): Tropical and Geographical Medicine. New York: Mc-Graw-Hill.

Warren KS, Mahmoud AAF (1985): Geographic Medicine for the Practitioner, Second Edition. New York: Springer-Verlag.

Weinstein PP (1981): Teaching parasitology. In K.S. Warren and E.F. Purcell (eds): The Current Status and Future of Parasitology. New York: Josiah Macy, Jr. Foundation, pp 51–62.

Worboys M (1983): The emergence and early development of parasitology. In K.S. Warren and J.Z. Bowers (eds): Parasitology: A Global Perspective. New York: Springer-Verlag, Introduction, pp 1–18.

The Biology of Parasitism, pages 13–41
© 1988 Alan R. Liss, Inc.

# Parasitic Zoonoses

## G.S. Nelson

*Department of Parasitology, Liverpool School of Tropical Medicine, Liverpool L3 5QA, England*

## INTRODUCTION

The purpose of this chapter is to define what is meant by the terms *"zoonoses"* and *"zooprophylaxis,"* to summarise the zoonotic aspect of the parasitic diseases of relevance to the UNDP/World Bank/World Health Organization (WHO) Special Programme for Research and Training in Tropical Diseases, and to discuss in more detail two cosmopolitan zoonoses, namely trichinosis and hydatid disease. These two diseases have been selected because of the author's personal involvement with their epidemiology in Africa and also because they demonstrate a) the great diversity of domestic and wild animals involved in the maintenance of parasites in nature; b) the intrinsic variability of parasites in different geographical areas and the significance of this variability on the epidemiology of diseases and c) the need to study human behaviour when investigating the transmission of parasites and in devising measures for their control.

## WHAT ARE THE ZOONOSES?

The first reference in the literature to zoonoses is by Moses! (Deuteronomy, Chapter 14). He says, "And the swine is unclean unto you, ye shall not eat of their flesh nor touch their dead carcasses, ye shall not eat anything that dieth of itself." Here we have the first clear recognition of the possible public health importance of infections such as trichinosis and *Taenia solium* and many other infections which man acquires directly from pigs. Moses was also aware of other types of zoonoses, such as anthrax, which is acquired from dead carcasses. However, by modern standards he was not particularly ethical because he goes on to say "Thou canst give it to the stranger that is within thy gate that he may eat it or sell it to an alien"! Most Christians and Jews no longer observe this Mosaic law, but it was accepted by the prophet Mohammed, and he was much more successful in persuading his Muslim followers that pork and pork products were anathema to all true believers.

The word *zoonosis* was first popularised more than 100 years ago in Germany by Virchow, the father of modern pathology, who was interested in diseases of animals and particularly those transmitted from animals to man. But the term led to much confusion and it was only after many meetings and a great deal of controversy that a joint Expert Committee of the World Health Organisation and the Food and Agricultural Organisation (FAO) defined the zoonoses as "Those diseases and infections naturally transmitted between vertebrate animals and man" (WHO, 1959). The term is not a biological principle, and it is etymologically unsound; but as Schwabe (1969) has emphasised, "although it is in a sense pre-Copernican, the zoonosis concept is fully justified by its great practical usefulness in public health."

The more we know about infectious diseases, the more we realise that there are very few diseases where man is the sole natural host of the infective agent and many infections which were generally thought to be restricted to animals are now known to be capable of transmission to man. In the past, too much emphasis has been given to the role of domestic animals as reservoir hosts without realising the significance of maintenance cycles in wild animals, and this failure to recognise wild animal hosts has led to strategies for control and for eradication which have inevitably led to disillusionment.

Public health workers dealing with the zoonoses have been mainly concerned with the direction of transmission between man and animals, so that rational control strategies could be developed. Their approach has been unashamedly anthropocentric, with man in the centre of the universe (Fig. 1). As a result, the public health classification of zoonoses consists of labels based on medical rather than zoological concepts. In contrast, veterinarians and biologists studying the zoonoses have been more interested in the ecology of the organisms and of their animal hosts and they often use the term *zoonoses* to describe diseases of animals where there is no evidence of human infection with the same organism. This has resulted not only in a conflict of ideas but also in a confusing and often contradictory terminology. For example, Kaegel (1951) introduced the term *anthropozoonoses* to include all infections which man acquires from animals, and this term, which was popularised by Pavlowsky (1966), is still used by WHO and FAO and by most public health and veterinary workers in Eastern Europe. Wagener (1957) emphasised the two-way movement of infections between man and animals but confused the issue by using the term *anthropozoonoses* in the opposite sense of infections animals acquire from man. Garnham (1958) introduced the term "euzoonoses" for *Taenia saginata* and *T. solium,* where there is a unique obligatory link betwen man and animals for the maintenance

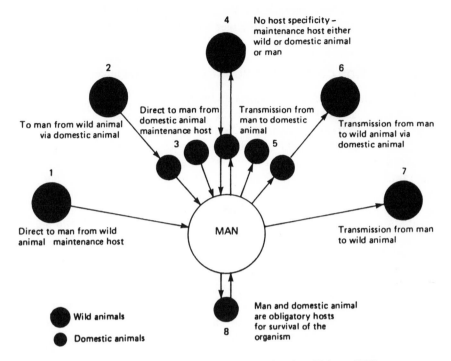

Fig. 1.  Zoonoses, epidemiological categories (from Nelson, 1979).

of the organism (category 8, Fig. 1). He also introduced the term *parazoo-nosis* to replace *anthropozoonoses* but this term has not been generally accepted. Nelson (1960) made matters worse by introducing the term *am-phixenosis* to categorise infections such as *Schistosoma japonicum* and *Try-panosoma cruzi* where there was very little host specificity and the infections went in both directions from man to animals and animals to man (category 4, Fig. 1). MacArthur (1960) took exception to all of these terms and called them Humpty-Dumptyisms, but despite MacArthur's denunciation this con-fusing terminology was perpetuated by Hoare (1962). In trying to simplify the classification, Schwabe (1969) introduced a new set of labels, including *metazoonoses, saprozoonoses,* and *direct zoonoses*; this is probably the best method of categorising the zoonoses but there is still no generally acceptable classification. One of the main problems with these attempts to categorise the zoonoses is that they generally fail to take into account the dynamics of transmission both in place and time so that any particular zoonosis will be characterised in different categories in different ecological situations. Also they often fail to take into account the stage of evolution of the infection in relation to man's association with either wild or domestic animals.

## THE EVOLUTION OF PARASITIC ZOONOSES

More than 40 species of protozoa and more than 100 species of helminths are recorded as infecting man (Beaver et al., 1984). Only a few of these parasites are restricted to man and very few have been acquired phylogenetically from his prehominid or anthropoid ancestors. Parasites which man inherited from apes and monkeys and which are now maintained by interhuman transmission include malaria; most of the nonpathogenic intestinal protozoa, the pinworm *Enterobius vermicularis*; and some of the filarial parasites, including *Brugia malayi*, *Loa loa*, *Mansonella perstans*, and *M. streptocerca*. At the present time, apes and monkeys, especially those in captivity and in nature reserves, are more likely to be affected by zoonoses in reverse, such as pulmonary tuberculosis or amoebiasis which they acquire from man (category 7, Fig. 1).

The earliest human communities of hunter gatherers acquired their parasitic infections directly from wild animals (category 1, Fig. 1) and many remote rural communities in the Arctic and the tropical rain forest still acquire their infections in this way. For example, in the Arctic, Eskimos acquire trichinosis from eating the flesh of the walrus or polar bear and they acquire the malignant hydatid, *Echinococcus multilocularis*, from the faeces of the arctic fox. In the rain forest of Africa, pygmies are frequently infected with *Strongyloides fuelleborni* from monkeys, and in South America the Amerindians in the Amazon region are infected with several species of *Leishmania* and with *Trypansoma cruzi* from a wide range of rodents, marsupials, and edentates (Lainson, 1982; Molyneux and Ashford, 1983). But even in advanced industrial societies transmission of infections from wild animals to man still occurs. For example, directly from exotic foods, such as trichinosis in New York from eating bear meat (Roselle et al., 1965), or food poisoning from exposure to the excreta of rodents and other animals which are maintenance hosts of *Salmonella*, still the commonest cause of food poisoning in the world.

The zoonoses became much more prevalent following the domestication of wild animals. This began with the domestication of the dog more than 10,000 years ago, probably by hunting tribes who domesticated the wolf which shared the feast when an excess of meat was available. The cat was a much later acquisition, following the development of settled agriculture where its virtues as a slayer of rodents was so greatly appreciated that it became venerated. The first records from Egypt date from the 16th century B.C. When a cat died the owners went into mourning and the animal was mummified and buried in consecrated ground. According to Davis and Dent (1968) this was "on such a scale that not long ago an enterprising merchant brought a whole shipload of mummified cats to Manchester, England to sell as manure." Edward Jenner, who developed his interest in natural history and comparative pathology as an assistant to John Hunter, was particularly interested in the transmission of infections from animals to man, and he can be justly acclaimed as the first scientific "zoonosologist." He gained notoriety for his experiments on the beneficial effect of cross protection between cowpox and smallpox, but he was aware that other infections which man acquired from animals might be more serious. Jenner (1798) says, "the deviation of man from the state in which he is placed by nature seems to have proven to him a prolific source of disease. From the love of splendour, from the indulgence of luxury and from his fondness for amusement, he has familiarised himself with a great number of animals. The wolf disarmed of his ferocity is now pillowed in the lady's lap; the cat the little tiger of our island, whose natural home is the forest is equally domesticated and caressed." We now know that dogs share at least 50 organisms with man, including some of the more important zoonoses, such as rabies, leishmaniasis, hydatid disease, and toxocariasis; and the domestic cat is the primary source of one of the commonest parasitic infections of man and livestock, namely, toxoplasmosis. In some situations the dog and cat are passing on an infection to man where the organism is still maintained in wild animals, e.g., hydatid disease in the Arctic (category 2, Fig. 1) or the organism may have evolved as a "domestic" zoonosis with maintenance in dogs and sheep as with hydatid disease in Europe (category 3, Fig. 1).

The domestication of herbivores such as the goat, sheep, reindeer, cattle, camels, and horses for food, milk, and clothing, and later for riding and as draft animals, brought man into closer contact with many other infections, notably anthrax, brucellosis and tuberculosis, but the animal which was the worst offender for parasitic zoonoses was the domestic pig which is a source of infections with *T. solium*, trichinosis, balantidiasis, fascioliasis and gnathostomiasis. It is also the original source of ascariasis and probably the whipworm *Trichocephalus trichurus* and the hookworm *Necator americanus*.

The more recent domestication of poultry, wild fowl, game birds, together with parrots and other cage birds has resulted in another crop of zoonoses, including ornithosis and a variety of arboviruses. It has been suggested that influenza has its origin in birds and it is now considered that the most prevalent food poisoning in man is caused by varieties of *Salmonella* which are maintained in poultry. Fortunately birds are very rarely incriminated as hosts of human parasitic disease except for abortive infections of trematodes causing cercarial dermatitis and rare infections with nematodes such as *Syngamus*.

## ZOONOTIC ASPECTS OF THE PARASITIC DISEASES OF INTEREST TO THE WHO SPECIAL PROGRAMME FOR RESEARCH AND TRAINING IN TROPICAL DISEASES

If we accept the WHO/FAO definition of the zoonoses, which includes infections going in either direction from animals to man or from man to animals, then all of the diseases of interest to the Special Programme are zoonoses, and they are all rapidly evolving so that in some situations they have escaped from the zoonotic cycle and become anthroponoses adapted for interhuman transmission.

Although malaria was acquired by man from his prehominid ancestors (and there are very similar if not identical species in the anthropoid apes of Africa; e.g., *P. rodhaini* is identical with *P. malariae* [Garnham, 1973; McWilson Warren, 1975]), for all practical purposes malaria is maintained in man by interhuman transmission. The situation in South America is particularly interesting because here *P. malariae* (*P. brasilianum*) and *P. vivax* (*P. simium*) have been transmitted from man to monkeys, but because the vectors of the monkey and human parasites are distinct there is no continuing transmission back to man. The malaria zoonotic situation is summarised in Figure 2.

*Trypanosoma brucei rhodesiense* is a parasite of game animals in southern Africa, and here man is only an incidental host, but in East Africa the parasite occurs in cattle and game animals and the present epidemic in Uganda is due to interhuman transmission. A similar evolution has occurred with *T. b. gambiense*, the cause of sleeping sickness in West Africa; here there is increasing evidence of an important animal reservoir (Molyneux, 1986). However, the successful control of the disease by mass chemotherapy suggests that in most areas animals are of no significance in transmission to man (Fig. 3). In contrast, trypanosomiasis in South America is found in an enormous range of wild and domestic animals, and although man is a relatively recent host of *T. cruzi*, there is evidence of interhuman transmission and of genetic variation of the organism account for geographical variations in the epidemiology of the disease (Molyneux and Ashford, 1983; Garcia-Zapata and Marsden, 1986)(Fig. 3).

All types of cutaneous leishmaniasis in South America are zoonoses with no evidence of interhuman transmission (Fig. 4). The same is true of *Leish-*

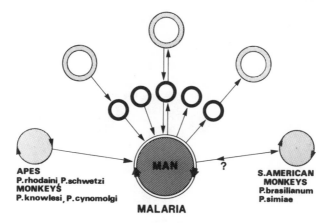

Fig. 2.   Malaria as a zoonosis (shaded circles = positive; open circles = negative).

*mania major* in the Old World, but *L. tropica* has escaped from the zoonotic cycle and has became a major urban problem (Ashford, 1986) (Fig. 4). The same is true of *L. donovani* in India, where there has been a complete break in the link between man and animals. The complexity of the situation with regard to the leishmanias is summarised in Figure 4.

Schistosomiasis is a major zoonosis in South East Asia, where *S. japonicum* is found in 29 species of mammals (Mao Shou Pai, 1962). The situation with regard to *S. mansoni* and *S. haematobium* is less clear (Nelson, 1975). In some parts of tropical Africa human infections are acquired from baboons and possibly rodents (see reviews by Nelson, 1960, 1983; Fenwick, 1969; Kawashima et al., 1978). There is no evidence of an animal reservoir in Egypt. An enormous range of rodents and marsupials are infected with *S. mansoni* in South America, representing a vast zoonosis in reverse (category 7, Fig. 1). It is not known if any of these animals have become true reservoir hosts (Fig. 5).

The situation with regard to filarial infections in man is complex. *Mansonella perstans* and *M. streptocerca* are parasites of anthropoid apes and man and *Loa loa papionis* is widespread in monkeys. Only a single gorilla has been found infected with *Onchocerca volvulus*, and there is no evidence that animals are involved in transmission to man; but this parasite has evolved from *Onchocerca* species in cattle and wild antelopes, and it is not known if a reservoir still exists in nature (Nelson, 1965). Man seems to be the only host of *Wuchereria bancrofti,* and control campaigns based on mass chemotherapy have been successful; but *Brugia malayi* is found in leaf-eating

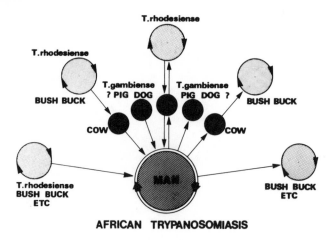

AFRICAN TRYPANOSOMIASIS

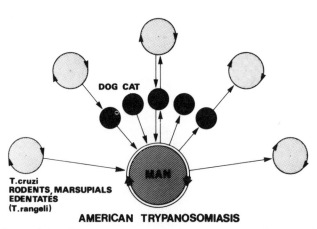

AMERICAN TRYPANOSOMIASIS

Fig. 3. African and American trypanosome infections as zoonoses (shaded circles = positive; open circles = negative).

Fig. 4. Leishmanial infections as zoonoses (shaded circles = positive; open circles = negative).

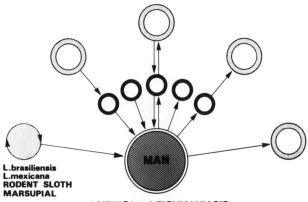

AMERICAN LEISHMANIASIS

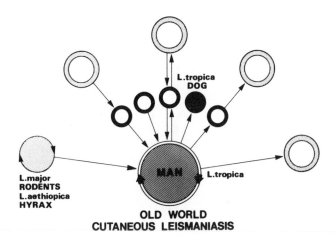

OLD WORLD
CUTANEOUS LEISMANIASIS

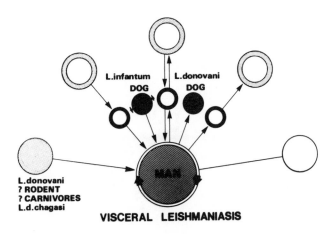

VISCERAL LEISHMANIASIS

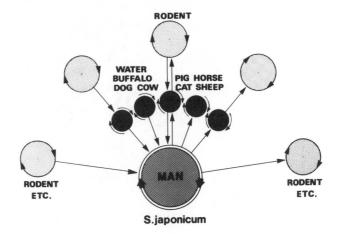

S.japonicum

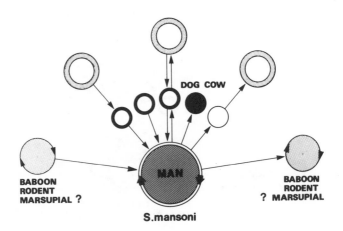

S.mansoni

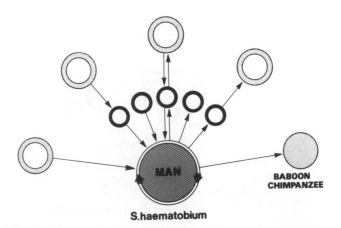

S.haematobium

monkeys and reinfection is known to occur after mass treatment in areas where zoonotic infections are prevalent (Fig. 6) (Mak, 1987).

Leprosy is the only nonparasitic disease of concern to the Special Programme, and it is generally considered to be restricted to man; but this is not true—natural infections have been recorded in the mangabey in West Africa, and there have been several well-documented cases of leprosy in the United States in people handling wild armadillos (Lumpkin et al., 1983, Smith et al., 1983). These observations suggest that much more attention should be given to possible reservoir hosts of leprosy.

## NEW OR EMERGING ZOONOSES

With new developments in the identification of infective agents, together with man's intrusion into almost all the remote areas of the world, it has been inevitable that there have been outbreaks of hitherto extremely rare or unknown diseases. This has been compounded by the acquisition of wild animals as pets or for zoological exhibits or for their use as laboratory animals.

The most dramatic example of a new zoonosis was the outbreak of Marburg disease in laboratory workers in Germany and Yugoslavia. This haemorrhagic fever was acquired from vervet monkeys imported from Uganda so that their kidneys could be used as cell lines for the development of the poliomyelitis vaccine (Martini and Siegert, 1971). The disease was superficially similar to yellow fever, and there was a high mortality but the virus was completely new to science. Fortunately the epidemic was contained by strict quarantine measures, but there was a subsequent case several weeks later when the wife of one of the laboratory workers developed Marburg disease and the virus was isolated from her husband's semen. An even more severe outbreak occurred in Zaire and the southern Sudan with the related Ebola virus with more than 40 deaths among the staff of Meridi Hospital. Transmission was by direct contamination with blood from infected individuals, but there was also evidence of transmission from infected syringes and by sexual intercourse (Pattyn, 1978). The sudden appearance of these viral zoonoses suggest that insufficient attention is being given to the possible

Fig. 5. Schistosome infections as zoonoses (shaded circles = positive; open circles = negative).

zoonotic origin of the present epidemic of AIDS. In a recent review entitled "The epidemiology of AIDS world wide," by Lifson et al. (1986), there is no mention of the possible origin of the virus from primates, and yet monkeys are infected with closely related viruses, and the present epidemic in Africa which was first recognised in Zaire may well have its origin in monkeys in the rain forest.

A parasitic zoonosis which was recognised for the first time in recent years is capillariasis, which was first seen as a devastating epidemic of a

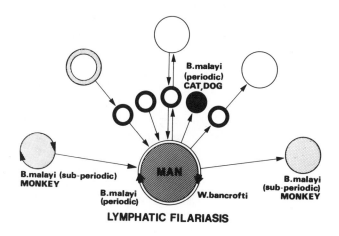

**LYMPHATIC FILARIASIS**

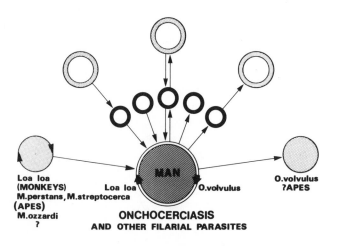

**ONCHOCERCIASIS**
**AND OTHER FILARIAL PARASITES**

Fig. 6.   Filarial infections as zoonoses (shaded circles = positive; open circles = negative).

malabsorption syndrome in the Philippines. The parasite was identified as *Capillaria philippinensis* with transmission to man through the larval stage from fish, but the natural definitive host, which may be a bird, is still not known (Cross et al., 1980). Another remarkable "new" zoonosis traced to eating uncooked fish is anisakiasis or terranoviasis, where man develops helminthomas of the gut which are often mistaken for tumours. The genus *Terranova* was named after the Dundee whaler used by Scott in his last expedition to the Antarctic and was described by Lt. E.L. Atkinson, who was parasitologist on the expedition (Nelson, 1977). These parasites are maintained in nature in a cycle between marine mammals and fish, and the increased prevalence, especially in Japan, is due to the custom of eating raw fish and the increased affluence of the country which has required access to distant fishing grounds in the Antarctic.

One of the more recent parasitic zoonoses which has caused considerable alarm is babesiosis. Infections were first seen in Europe, where they were always fatal, and it was suggested that man only became infected if immunity was impaired due to previous splenectomy (Garnham, 1973). More recent observations from New England and the islands of Nantucket and Martha's Vineyard in the United States indicate that the infection need not be fatal and that man is susceptible to the roden parasite *B. microti* even with an intact spleen (Ruebush et al., 1977). An interesting feature of this zoonosis is that the conservation of deer has led to an increase in the tick vectors, and man is infected as a result of leisure activities such as hiking and golf. The same is true of Lyme disease, tularaemia and even bubonic plague in the United States.

Other protozoal zoonoses which are becoming increasingly recognised include sarcocystis, cryptosporidial and coccidial infections. There has also been concern over the increase in parasitic zoonoses due to the use of immunosuppressive drugs but more particularly as a result of AIDS, where both toxoplasmosis and pneumocystosis are important and often fatal complications in adults. Toxoplasmosis is a zoonosis acquired from cats and from eating meat, but very little is known about the role of animals in the transmission of *Pneumocystis*.

### Zooprophylaxis, or the Jennerian Principle of Cross-Protection

The WHO/FAO definition of the zoonoses includes the concept of disease and infections naturally transmitted between vertebrate animals and man. The most neglected group of zoonoses are those infections which man acquires from animals but which produce either no disease or only a low degree of pathogenicity. Yet these infections by stimulating man's immune mechanisms are probably of much more biological importance in terms of our evolution and survival than the more dramatic diseases which are acquired from animals. Again, this aspect of the zoonoses was first observed and tested by Jenner in his studies on cowpox versus smallpox. Although this phenomenon of heterologous immunity and cross-protection stimulated by organisms of animal origin was first recognised more than 200 years ago, its epidemiological significance has received only scant attention. To draw attention to this aspect of the zoonoses, in relation to schistosomiasis, Nelson et al. (1962) used the term *zooprophylaxis*, which had been used by entomologists to describe the phenomenon of the deviation of mosquitoes from biting man by providing cattle for them to feed on. However, the concept is much broader and includes the idea of an immunological barrier preventing the spread of more pathogenic organisms. Zooprophylaxis was defined as "the prevention or amelioration of disease in man as a result of previous exposure to heterologous infection of animal origin" (Nelson, 1974). But, like all hybrid nomenclatures, it has not been generally accepted and in recognition of the origin of the concept, it might be best to refer to this phenomenon as the "Jennerian principle of cross-protection."

Cross-protection between different organisms has now been demonstrated through the whole range of infective agents from viruses to helminths. A few examples of cross-protective immunity are shown in Table 1. They are all based on good laboratory or epidemiological evidence; but despite Jenner's observation, the extent of natural cross-immunity between low-grade pathogens and more virulent organisms has not been clearly established in the field. One reason is the lack of species-specific diagnostic tests, particularly in parasitic infections. On the other hand, in the laboratory, there is clear

TABLE I. Examples of Zooprophylaxis or the "Jennerian Principle of Cross-Protection"

| Low-grade pathogen | More virulent organism |
| --- | --- |
| Compox and other pox viruses | Smallpox (Jenner, 1798) |
| Non-pathogenic arboviruses (e.g., Wesselsbrons) | Yellow fever (Henderson et al., 1970) |
| Murray Valley fever | Japanese encephalitis (Andrews, 1967) |
| Bovine tuberculosis | Leprosy and Buruli ulcer (Nelson, 1958; Uganda Buruli Group, 1969) |
| *Toxoplasma gondii* | Viruses, protozoa, and fungi (Krahenbuhl and Remington, 1982) |
| Dog hookworm | Human hookworm (Miller, 1978) |
| Bovine schistosomiasis | Human schistosomiasis (Nelson, 1974) |
| *Babesia* | Malaria (Cox, 1978) |

evidence of cross-protection at the generic level between most groups of helminths. For example, it has been observed that schistosomiasis is usually less severe in areas where there are concomitant infections with schistosomiasis of animals, and cross-protection has been demonstrated in the laboratory with a wide range of different species. This can work in both directions, with animal parasites protecting against the human parasites and the human parasites acting against parasites in livestock (Nelson, 1974).

The most potent stimulus for protective immunity in schistosomiasis comes from the early stages of development, and this is the basis of the attenuated larval vaccine used against *S. bovis* in cattle in the Sudan (Majid et al., 1980). In the same way, heterologous infections which are rejected at an early stage of development in man, such as *S. bovis* and *S. rodhaini* or even the zoonotic *S. japonicum* in Taiwan, are probably responsible for stimulating a protective immune response which helps to ameliorate subsequent infection with the human parasites. The same phenomenon probably occurs with hookworms, where man is constantly exposed to the infective larvae of the very prevalent hookworms from cats and dogs such as *Ancylostoma caninum*. The studies by Miller (1978) who developed the first commercial vaccine for the control of canine hookworms, showed that attentuated *A. caninum* larvae were strongly protective against not only homogolous *A. caninum* infections but also *A. braziliense* and *Uncinaria stenocephalum*; and he suggested that the vaccine would also act against *A. duodenale* and *Necator americanus* in man. The cross-protective immunity in schistosomes and hookworms develops against parasites in the same or closely related genera, but *Toxoplasma* acts in a much more generalised way in that infections with this protozoa, which is probably the most widespread of all parasitic zoonoses, is a potent stimulus to the cell-mediated responses of the host and previous toxoplasmosis infections can limit subsequent infection with not only other protozoa, but even viruses and fungi (Cox, 1978; Krahenbuhl and Remington, 1982). With the recent development of species-specific diagnostic reagents, it should now be possible to study the epidemiological consequences of cross-protective immunity through a wide range of organisms. It is probably no exaggeration to say that this phenomenon accounts for man's survival in an environment where he is constantly exposed to an enormous variety of pathogens.

## HYDATID DISEASE AND TRICHINOSIS

The life cycle of all species of *Echinococcus* is similar, with the egg containing the infective embryo passed in the faeces of carnivores. Following ingestion by man or a herbivore, the embryo escapes from the egg and migrates throughout the body, eventually developing into the larval hydatid stage, which grows to produce numerous young tapeworms (protoscoleces), which in turn will grow into adult worms in the intestine of the carnivore host when the cyst is eaten.

The first evidence that the cysts were the intermediate stages of the *Echinococcus* tapeworm of dogs was obtained by Von Siebold (1853) when he successfully infected dogs with hydatid material from sheep. Similar results were obtained by Naunyn (1863), who infected dogs with material from a large hepatic cyst from man. It was assumed that *Echinococcus granulosus* was the only species and that transmission was confined to dogs and domestic herbivores, and for many years very little attention was paid to the possible role of wild animals as true maintenance hosts of the parasite.

The renaissance in research on the biology of hydatids began with the studies by Cameron (1926) in England and Canada followed by Rausch (1952) and Rausch and Schiller (1956) in Alaska, by Vogel (1957) in Germany, by Schwabe and Abou-Daoud (1961) in the Middle East, by Rausch and Nelson (1963) and Nelson and Rausch (1963) in East Africa, by Verster (1965) in South Africa, and by Lukashenko (1971) in the USSR. They all recognised that there were different species or strains of *Echinococcus* in wild herbivores and wild carnivores, and it became clear that the cycle of *E. granulosus* between the dog and domestic herbivores was a recent evolutionary development. These observations were summarised in the FAO/UNEP/WHO "Guidelines for Surveillance, Prevention and Control of Echinococcosis/Hydatidosis" (WHO, 1981) and the subject has been extensively reviewed in *The Biology of Echinococcus and Hydatid Disease*, edited by Thompson (1986).

The first clear account of subspecific variation was produced by Williams and Sweatman (1963), who recognised that the parasite in horses in the United Kingdom was distinct from the sheep parasite. They designated this form *E. granulosus equinus*, and this was soon followed by several other subspecies—for example, *E.g. borealis* for the parasite maintained in wolves and reindeer, *E. g. canadensis* in wolves and moose, and *E. g. felidis* in the lion and warthog; but these designations have not been accepted (Smyth, 1979). The parasite from the horse has received the most attention, with increasing evidence that although it is widespread in England and Ireland it is not a threat to man as is the dog-sheep parasite. Comparative studies have revealed minor morphological differences and quite distinct characteristics in culture. They also differ in their infectivity to different intermediate hosts including monkeys and recent observations on their biochemistry and molecular genetics also reveal differences (McManus and Bryant, 1986).

Schwabe (1969) first emphasised the significance of human behaviour in relation to transmission of hydatid disease. He was particularly interested in the way in which man ingests the eggs of *Echinococcus* from dog faeces and he demonstrated that the use of dog faeces for tanning leather was responsible for the unusually high prevalence of hydatid cysts in shoemakers in Lebanon. This aspect of the life cycle was of concern to Leuckart (1863). In his observations on the high prevalence in Iceland he comments, "The treatment of the quacks is exactly suited to keep up the epidemic, for amongst their remedies dog urine and fresh dog excrement play a conspicuous part." Cobbold (1864) refers to this observation of Leuckhart's and says it explains why more women than men were infected in Iceland: "the women, probably, obey more implicitly the dictates of the 'quacks' who supply them with the filthy medicines above mentioned"! He goes on to explain why the upper classes in England are rarely infected—this he attributes to their consumption of beer in preference to water and he concludes that "beer drinking on the whole, is preferable to water drinking; if however the so called Adam's ale is filtered and pure no evil can possibly result from its moderate imbibition"!

The use of faeces as medicaments is an ancient custom (Hoeppli, 1959). Martin Luther was treated with garlic and horse faeces to relieve his ague, and in Africa witch doctors still use ground-up tapeworms from human faeces to treat tapeworm infections (Nelson, 1972). Unfortunately the use of *Taenia solium* as a medicament can be fatal, and the patient may develop epilepsy due to generalised cysticercosis (Heinz and McNab, 1965).

In most parts of the world man is only an incidental or accidental host of *Echinococcus*, and he plays no part in the life cycle of the organism or its maintenance in nature except through his agricultural pursuits in relation to the livestock industry and his use of dogs for herding or as pets. Turkana in Kenya is exceptional. Here there is the highest prevalence of the disease in man in the world, and dogs and jackals have access to human corpses because of the absence of burial customs, so that man plays an active role as an intermediate host (French et al., 1982; Macpherson, 1983). This may account for the evolution of a strain of the parasite with exceptionally fertile cysts in man, but the unusually high prevalence rate is more likely to be related to the very intimate association of the people with their dogs. The dogs in Turkana are used as "nurses" to clean babies when they vomit or defaecate; to act as guards to protect children when their mothers are working; and in the desert region where there is a great shortage of water, the dogs may be used by the women to clean their household utensils and to clean themselves when they are menstruating. The women also prepare a medicament from dogs' faeces to treat burns and wounds and as a lubricant to prevent their necklaces from abrading their skin (Nelson and Rausch, 1963; French and Nelson, 1982; French et al., 1982; Nelson, 1986).

With the recognition by the people that they might benefit from surgery there was a demand for treatment by the Flying Doctor Service of the African Medical and Research Foundation, and over a 10-year period up to 1977 they operated on more than 1,500 people out of an estimated total population of only 100,000. The incidence of new infections remained fairly constant, and it was decided that a detailed study was necessary to discover more about the life cycle and to devise methods for control. It was particularly important to know if there was a wildlife cycle. Our studies have now shown that although there is a wildlife cycle of *Echinococcus* in Masailand involving wild carnivores and wild herbivores, there is no wildlife cycle in Turkana (Macpherson et al., 1983; Nelson, 1986). The Masai live in harmony with the wildlife, whereas the Turkana will kill and eat almost all mammals. In order to defend themselves from cattle raiders from neighbouring tribes and from Uganda and Ethiopia, they have acquired rifles which they have used to kill off most of the wild herbivores. This has eliminated potential wild animal hosts, and the absence of a wildlife cycle in Turkana makes a practicable control programme realistic. A major effort is now in progress. This is based on a) case detection with ultrasound and serology, b) treatment with albendazole and surgery, and c) the elimination of stray dogs and registration and treatment of the remaining dogs with praziquantel (Macpherson et al., 1986).

Case detection has not been easy because of the very high false-negative antibody levels in people with large viable cysts, but this has been overcome by detecting circulating antigens and immune complexes (Craig et al., 1986).

One of the problems in the epidemiological studies has been the identification of *Echinococcus* eggs in dog faeces and in the environment, because the eggs are identical morphologically with *Taenia* eggs from man and the numerous species of *Taenia* in dogs and jackals. This problem has been resolved by the development of a test based on a monoclonal antibody, which is specific for *Echinococcus* (Craig et al., 1986) and the test is now being developed as an essential tool for monitoring the success of the control programme.

Treatment of the disease with albendazole has produced remarkable results even with very large cysts, and in inoperable cases where there has been extensive involvement of the bone or multiple cysts in the abdomen (Okelo, 1986). The benefits of surgery and chemotherapy are now widely recognised by the people, and the health education campaign in Turkana is now spearheaded by patients who have benefitted from treatment. They have no difficulty in understanding the complexity of the life cycle and they have developed a repertoire of more than 20 songs in the local idiom which explain the way in which the people become infected and how the disease can be controlled. There is at long last some hope that transmission will be reduced and that the people who are already infected will be cured, but without the preliminary scientific study of the epidemiology of the disease and of the life cycle of the parasites, this major public health programme would have been based on false premises with inevitable disillusionment of the public health authorities and the people of Turkana.

## TRICHINOSIS WITH SPECIAL REFERENCE TO AFRICA

Trichinosis was the first generalised infection of man in which the aetiological agent was recognised and where a knowledge of the life cycle led to practical methods of control. It is also a parasitic disease where recent advances in knowledge of the diversity of the parasites and their hosts has revolutionised ideas on how the parasites are maintained in nature and how the disease can be controlled. *Trichinella* has a simple life cycle with the cystic stage in the muscles of man and animals. Transmission occurs when the cysts are ingested by a meat-eating animal, including man. The larvae develop directly in the intestine into very small ephemeral adult worms which mature and produce larvae, which then migrate through the muscles, where they persist for many years in a viable state waiting to be ingested by a new host.

*Trichinella* was discovered by James Paget in the muscles of a cadaver in the dissecting rooms in St. Bartholomews Hospital in London on February 2, 1835. It is no wonder that he later attained fame and a knighthood because in 1835 he was still a medical student and he had the courage to deliver an account of his observations to the Abernethian Society. In his *Memoirs and Letters of Sir James Paget*, Stephen Paget (1901) quotes Sir James as saying, "All the men in the dissecting room, teachers included, saw the little specks in the muscles, but I believe that I alone looked at them and observed them; no one trained in natural history could have failed to do so." In 1835 Paget's discovery was regarded as zoologically interesting but of no clinical importance, and it was not until 1859 that Virchow and Zenker unravelled the life cycle and demonstrated that trichinosis was a prevalent and often fatal disease of man.

The disease soon became recognised in almost all parts of the world where pork products were consumed, and although every effort was made to control trichinosis, as late as 1944 a national autopsy survey in the United States revealed an infection rate in man of 16% (Wright et al., 1944). In recent years there has been a decline in the prevalence of pig-transmitted trichinosis as an incidental result of the enforcement of cooking garbage and pigswill for the control of hog cholera and swine vesicular exanthema. Improvements in food processing, storage, and cooking have also had their effect, but pig-transmitted trichinosis is still prevalent and epidemics continue to occur. In the largest outbreak on record, 1,122 persons were infected in Poland (Kozar, 1970) and 146 cases occurred in an outbreak in Germany in 1982 (Feldmieier et al., 1984). A recent unusual epidemic in France was traced to eating horsemeat imported from the United States (Ancelle et al., 1985).

The obsession with the "domestic" cycle and with pigs as the main source of infection for man has had a similar inhibiting effect on ideas on the epidemiology of trichinosis as the obsession with dogs and domestic herbivores had in relation to hydatid disease. Textbooks continue to produce misleading life-cycle diagrams for both of these parasitic diseases, and public health workers are often unaware of the significance of wild animals as reservoir hosts. It is only in recent years that it has been realised that the true maintenance hosts of *Trichinella* are carrion feeding or cannibalistic wild carnivores. Domestic pigs are artificial hosts forced into cannibalism either by feeding them garbage-containing pork scraps or by overcrowding where dead pigs are often consumed by their companions. Peridomestic rats are also "artificial" hosts. They have access to pork scraps on dumps, in sewers, and around farms and abattoirs. Under conditions of overcrowding and food shortage they may be cannibalistic, but like domestic pigs they are secondary hosts of no significance in the natural cycle of transmission.

The following account is restricted to observations on *Trichinella* in the Arctic and in Africa because it is here that variations in the parasite which determine the epidemiology of the infection in man and animals were first observed.

### *TRICHINELLA* IN AFRICA

Trichinosis was unknown in Africa south of the Sahara until 1961, when exceptionally heavy infections were seen in 11 boys on the lower slopes of Mount Kenya (Forrester et al., 1961). Because there were no previous reports in either man or animals, it was assumed by both the veterinary and medical authorities that the parasite had been introduced in pork products for feeding the British troops involved in operations against the Mau Mau. However, the infected individuals gave a clear account of killing and eating a wild pig and eventually *Trichinella* cysts were found in both the bushpig (*Potamochoerus porcus*) and a leopard near the area where the boys were found infected. The authorities were still not convinced that the infection was enzootic, and it was only when high infection rates were found in hyaenas and other carnivores in the Rift Valley area that it was accepted that the infection was indigenous to Kenya (Nelson et al., 1963). This led to an increased awareness of the problem elsewhere in Africa and further outbreaks of the disease in man were subsequently reported by Gretillat and Vassilaides (1967) from as far away as Senegal and also by Bura and Willett (1977) from Tanzania, where Sachs and Taylor (1966) had confirmed infections in hyaenas and jackals and also in warthogs. The widespread distribution of the parasite in carnivores was confirmed by Young and Kruger (1968) in South Africa where they suggested that trichinosis might be responsible for limiting the population density of lions. The importance of carnivores as hosts of *Trichinella* in Africa has been discussed in detail by Nelson (1970, 1982).

The absence of *Trichinella* in wild rodents and the failure to infect laboratory rats and domestic pigs suggested that the parasite in Kenya differed from *T. spiralis* as seen in Europe and the United States. Although the African parasite was morphologically identical with *T. spiralis*, and although it produced the same clinical manifestations in man, it was obviously quite different in its epidemiology and mode of transmission. This was not easily accepted by parasitologists who had been brought up with the idea that the hosts par excellence of *Trichinella* were rats and pigs. It was necessary to examine further isolates and to compare the parasite from Kenya with parasites from other parts of the world and particularly from other wild carnivores. This led to comparative studies of isolates from Kenya with isolates from the Arctic and temperate regions. The first indication that parasites not only from Africa but also from the Arctic differed quite markedly from *T. spiralis* from temperate regions was produced by Nelson et al. (1966). They found that strains from the hyaena in Kenya and a brown bear from Alaska were of very low infectivity to rats and domestic pigs. Similar differences were subsequently observed with isolates from other animals and from other localities in the Arctic and, as a result of genetic studies, Britov and Boev (1972) named the Arctic parasite *Trichinella nativa*. The Arctic parasite differs from *T. spiralis* not only in its low infectivity to domestic pigs but also in its tolerance to low temperatures and in its infectivity and virulence in a wide range of hosts Dick (1983). Differences have been seen in the isoenzymes of the different isolates by Flockhart et al. (1982), and in their DNA by Rollinson et al. (1986), but there is still some controversy as to whether they should be designated as distinct species. Their life cycles are illustrated in Figure 7.

All these studies give support to the idea that the Arctic and African parasites are of ancient origin and that they are distinct from *T. spiralis* as

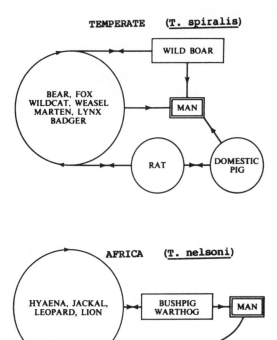

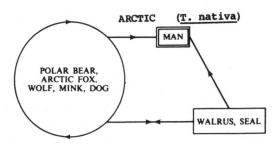

Fig. 7. Maintenance cycles and transmission to man of *Trichinella spiralis*, *T. nelsoni*, and *T. nativa*.

seen in domestic pigs. The survival of *T. nativa* at low temperature is also of great public health significance, because one of the established methods of rendering infected meat suitable for human consumption is freezing at a temperature of $-15°C$ for 20 days or $-29°C$ for 6 days (Steele, 1970). Fortunately the low infectivity for the domestic pig suggests that the cycle in wildlife in the Arctic is not such a threat to the pig industry as the parasite in temperate regions.

Although *Trichinella* is widespread in wild carnivores in Africa, they are not the usual source of infection to man because eating carnivore flesh is usually taboo. There are some exceptions. For example, some Nilohamite groups in East Africa will eat animals such as lion or leopard, and some tribes in West Africa and the Sudan will eat dogs, but the main host of *Trichinella* in Africa, the hyaena, is very rarely eaten by man. Human infections are almost always from eating bushpigs or warthogs, which are secondary hosts. It is not a major public health problem, but the parasite is widespread wherever there are carrion feeding and cannibalistic carnivores. These are the true maintenance hosts of *Trichinella*. They are increasing as a result of conservation, so they will ensure the persistence of a reservoir of *Trichinella* as long as man ensures their survival.

It may be of some significance in relation to the geographical distribution of the different species of *Trichinella* that both *T. nelsoni* and *T. nativa* stimulate a marked degree of protective immunity against *T. spiralis* from Europe and North America (Nelson et al., 1966). This cross-immunity might explain the absence of "domestic" strains of *T. spiralis* in the Arctic and in Africa. The main danger for Africa is the possibility of introducing *T. spiralis* in domestic pigs from Europe, especially as standards of meat hygiene and pig husbandry in Africa are often very low and pigs may be fed on uncooked garbage.

These studies on *Trichinella* and hydatid disease in Kenya illustrate the danger of accepting established dogma and of generalising about parasitic diseases. The more we study their zoonotic aspects and their biological properties, the more we realise the complexity of their epidemiology.

In conclusion it can be said that in all public health programmes concerned with parasites and vector-borne diseases, it is essential, if rational control measures are to be introduced, that physicians and health workers cooperate with zoologists and basic scientists in understanding the variability of the parasites and their host range, and that a knowledge of the epidemiology of transmission requires the expertise of not only parasitologists but also of anthropologists who have an intimate knowledge of human behaviour. Chairman Mao once said, "Knowledge begins with practice and theoretical knowledge acquired through practice must then return to practice." This has always been my guiding principle, and it is an appropriate motto for the Course on Biology of Parasitism at Woods Hole, where I have been privileged to share my experience in Africa with so many enthusiastic research workers who are beginning their careers in this field.

## ACKNOWLEDGMENTS

I am grateful to Professor Henry Ford, Editor of the *Journal of Comparative Pathology,* for permission to include in this chapter material from a paper entitled "More than a hundred years of parasitic zoonoses with special reference to trichinosis and hydatid disease," which was produced for the centenary volume of the *Journal of Comparative Pathology.* The illustrations of reservoir hosts of relevance to the WHO Special Programme for Research Training in Tropical Diseases were presented at the British/Scandinavian Joint Meeting on Tropical Medicine and Parasitology, and I am grateful to Dr. Harald Fuglsang for encouraging me to publish them. I am also grateful to Dr. Ralph Muller for permission to reproduce a modified version of the life cycles of the *Trichinella* species which first appeared in his *Worms and Disease.* It has been a pleasure and an honour to participate in the Biology of Parasitism Course at Woods Hole, and for this I owe a great deal to John and Roberta David, who first invited me to take part in the course, and to Alan Sher and Paul Englund, who suggested that my contribution was worthy of a more permanent record.

## REFERENCES

Ancelle T, Dupony-Camet J, Heyer F, Faurant C, Lapierre J (1985): An outbreak of trichinosis in the Paris area due to horsemeat. Lancet 2:660.

Andrews CH (1967): The Natural History of Viruses. London: Widenfeld and Nicolson.

Ashford RW (1986): The leishmaniases. In H.M. Gilles (ed): Epidemiology and Control of Tropical Diseases, Clinics in Tropical Medicine and Communicable Diseases. London: W.B. Saunders Co., pp 513–533.

Beaver PC, Jung RC, Cupp EW (1984): Clinical Parasitology, 9th edition. Philadelphia: Lea & Febiger.

Britov VA, Boev SN (1972): Taxonomic status of different strains of *Trichinella* and the nature of their focal occurrence. Vsesoyuznaya Konferentisya (VIII) po prirodnoi ochagovosti boleznei zhivetnkich i ohrane ikh chrislennosti 1:83–84.

Bura MWT, Willett WC (1977): An outbreak of trichinosis in Tanzania. East Afr Med J 54:185–193.

Cameron TWM (1926): Observations on the genus *Echinococcus* Rudolphi 1801. J Helminthol 4:13–22.

Cobbold TS (1864): Entozoa, An Introduction to Helminthology. London: Groombridge.

Craig PS, Zeyhle E, Romig T (1986): Hydatid disease: Research and control in Turkana, II. The role of immunological techniques for the diagnosis of hydatid disease. Trans R Soc Trop Med Hyg 80:183–192.

Cross JH, Singson CN, Battad S, Basaca-Sevilla V (1980): Intestinal capillariasis: Epidemiology, parasitology and treatment. In Health Policies in Developing Countries. Royal Society of Medicine International Congress and Symposium No. 24. London: Academic Press.

Davis PDC, Dent AA (1968): Animals That Changed the World. New York: Crowell-Collier Press.

Dick TE (1983): Species and infraspecific variation. In W.E. Campbell (ed): Trichinella and Trichinosis. New York: Plenum Press.

Feldmieir H, Blaumeiser G, Fischer H, Stein HA (1984): Epidemiology of *Trichinella spiralis* infections in the Eifel region, 1982. Dtsch Med Wochenschr 109:205–210.

Fenwick A (1969): Baboons as reservoir hosts of *Schistosoma mansoni*. Trans R Soc Trop Med Hyg 63:557–567.

Flockhart HA, Harrison SE, Dobinson AR, James ER (1982): Enzyme polymorphism in *Trichinella*. Trans R Soc Trop Med Hyg 76:541–545.

Forrester ATT, Nelson GS, Sander G (1961): The first record of an outbreak of trichinosis in Africa south of the Sahara. Trans R Soc Trop Med Hyg 55:503–513.

French CM, Nelson GS (1982): Hydatid disease in the Turkana District of Kenya, II. A study in medical geography. Ann Trop Med Parasitol 76:439–457.

French CM, Nelson GS, Wood M (1982): Hydatid disease in the Turkana District of Kenya, I. The background to the problem with hypotheses to account for the remarkably high prevalence of the disease in man. Ann Trop Med Parasitol 76:425–437.

Garnham PCC (1958): Zoonoses or infections common to man and animals. J Trop Med Hyg 61:92–94.

Garnham PCC (1973): Distribution of malaria parasites in primates, insectivores and bats. Symp Zool Soc Lond 33:377–404.

Garcia-Zapata MTA, Marsden PD (1986): Chagas disease. In H.M. Gilles (ed): Epidemiology and Control of Tropical Diseases, Clinics in Tropical Medicine and Communicable Diseases. London: W.B. Saunders Co., pp 557–585.

Gretillat S, Vassilaides G (1967): Presence de *Trichinella spiralis* (Owen, 1835) chez les carnivores et suides sauvages de la region du delta du fleuve Senegal. C R Acad Sci 264:1297–1300.

Heinz HJ, McNab GM (1965): Cysticercosis in the Bantu of Southern Africa. S Afr J Med Sci 30:19–31.

Henderson BE, Cheshire PP, Kirva GB, Lule M (1970): Immunological studies with yellow fever and selected African Group B arboviruses in rhesus and vervet monkeys. Am J Trop Med Hyg 19:110–118.

Hoare CA (1962): Reservoir hosts and natural foci of human protozoal infections. Acta Trop (Basel) 19:281–317.

Hoeppli R (1959): Parasites and Parasitic Infections in Early Medicine and Sciences. Singapore: University of Malaya Press.

Jenner E (1798): An Enquiry Into the Causes and Effects of the Variolae Vaccinae. London: Sampson Low.

Kaegel A (1951): Zoonoses (Anthropozoonoses). Basel: Reinhardt.

Kawashima K, Datamine D, Sakamoto M, Shimada M, Nojima H, Miyahara M (1978): Investigations on the role of wild rodents as reservoirs of human schistosomiasis in the Taveta area of Kenya. Jap J Trop Med Hyg 6:195–203.

Kozar Z (1970): Trichinosis in Europe. In S.E. Gould SE (ed): Trichinosis in Man and Animals. Springfield: Charles C. Thomas.

Krahenbuhl JL, Remington JS (1982): The immunology of *Toxoplasma* and toxoplasmosis. In S. Cohen and K.S. Warren (eds): Immunology of Parasitic Infections. Oxford: Blackwell.

Lainson R (1982): Leishmanial parasites of mammals in relation to human disease. In M.A. Edwards and U. McDonnel (eds): Animal Disease in Relation to Animal Conservation. London: Academic Press.

Lifson AR, Ancelle RA, Brunet JB, Curran JW (1986): The epidemiology of AIDS worldwide. In A.J. Pinching (ed): AIDS and HIV Infection. Clinics in Immunology and Allergy, 6. London: W.B. Saunders Co.

Leuckart RF (1863): Die Menschlichen Parasiten und die von Ihnen Herruhrenden Krahkheiten. Leipzig: C.F. Wintersche.

Lukashenko NP (1971): Problems of epidemiology and prophylaxis of alveococcosis (multilocular echinococcosis): A general review with particular reference to the USSR. Int J Parasitol 1:125–134.

Lumpkin LR, Cox GF, Wolf JE, Jr (1983): Leprosy in five armadillo handlers. J Am Acad Dermatol 9:899–903.

MacArthur WM (1960): Discussion, pp 320–321. In Nelson GS (1960): Schistosome infections as zoonoses in Africa. Trans R Soc Trop Med Hyg 54:301–324.

Macpherson CNL (1983): An active intermediate host role for man in the life cycle of *Echinococcus granulosus* in Turkana, Kenya. Am J Trop Med Hyg 32:397–404.

Macpherson CNL, Karstad L, Stevenson P, Arundel JH (1983): Hydatid disease in the Turkana District of Kenya, III. The significance of wild animals in the transmission of *Echinococcus granulosus* with particular reference to Turkana and Masailand in Kenya. Ann Trop Med Parasitol 77:61–73.

Macpherson CNL, Wachira TM, Zeyhle E, Romig T, Macpherson C (1986): Hydatid disease: Research and control in Turkana, IV. The pilot control programme. Trans R Soc Trop Med Hyg 80:196–200.

Majid AA, Bushara HO, Saad AM, Hussein MF, Taylor MG, Dargie JD, Marshall TF de C, Nelson GS (1980): Observations on cattle schistosomiasis in the Sudan, a study in comparative medicine. III. Field testing of an irradiated *Schistosoma bovis* vaccine. Am J Trop Med Hyg 29:452–455.

Mak JW (1987): Epidemiology of lymphatic filariasis. In D. Evered and S. Clark (eds): Filariasis. Ciba Foundation Symposium 127:5–14, Chichester: John Wiley & Sons.

Máo Shou Pai (1962): Important achievements in the control of bilharziasis in China. In G.E.W. Wolstenholme and M. O'Connor (eds): Ciba Foundation Symposium on Bilharziasis. London: J.A. Churchill Ltd.

Martini GA, Siegert R (1971): Marburg Virus Disease. New York: Springer Verlag.

McManus DP, Bryant C (1986): Biochemistry and physiology of echinococcosis. In R.C.A. Thompson (ed): The Biology of *Echinococcus* and Hydatid Disease. London: Allen and Unwin.

McWilson Warren (1975): Malaria. In W.T. Hubbert, W.F. McCulloch, and P.R. Schnurrenberger (eds): Diseases Transmitted From Animals to Man, 6th Edition. Illinois: Charles C. Thomas, pp 789–798.

Miller TA (1978): Industrial development and field use of the canine hookworm vaccine. Adv Parasitol 16:333–342.

Molyneux DH (1986): African trypanosomiasis. In H.M. Gilles (ed): Epidemiology and Control of Tropical Diseases, Clinics in Tropical Medicine and Communicable Diseases. London: W.B. Saunders Co., pp 513–533.

Molyneux DH, Ashford RW (1983): The Biology of *Trypanosoma* and *Leishmania* Parasites of Man and Domestic Animals. London: Taylor and Francis.

Naunyn B (1863): Ueber die zu *Echinococcus hominis* geborige tanie. Arch Anat Physiol 412–416.

Nelson GS (1958): Leprosy in the West Nile District of Uganda: An epidemiological study with special reference to the distribution of leprosy in Africa. Trans R Soc Trop Med Hyg 52:176–185.

Nelson GS (1960): Schistosome infections as zoonoses in Africa. Trans R Soc Trop Med Hyg 54:301–324.

Nelson GS (1965): Filarial infections as zoonoses. J Helminthol 39:229–250.

Nelson GS (1970): Trichinosis in Africa. In S.E. Gould (ed): Trichinosis in Man and Animals. Springfield: Charles C. Thomas.

Nelson GS (1972): Human behaviour in the transmission of parasitic diseases. In E. Canning and C.A. Wright (eds): Behavioural Aspects of Parasite Transmission. London: Linnean Society, Academic Press.

Nelson GS (1974): Zooprophylaxis with special reference to schistosomiasis and filariasis. In E.J.L. Soulsby (ed): Parasitic Zoonoses. Clinical and Experimental Studies. London: Academic Press.

Nelson GS (1975): Schistosomiasis. In W.T. Hubbert, W.F. McCulloch, and P.R. Schnurrenberger (eds): Diseases Transmitted From Animals to Man, 6th Edition. Illinois: Charles C. Thomas, pp 620–640.

Nelson GS (1977): A milestone on the road to the discovery of the life cycles of the human schistosomes. Am J Trop Med Hyg 26:1093–1100.

Nelson GS (1979): The parasite and the host. In R.J. Donaldson (ed): Parasites and Western Man. Lancaster: MTP Press Ltd., pp 1–10.

Nelson GS (1982): Carrion-feeding cannibalistic carnivores and human disease in Africa with special reference to trichinosis and hydatid disease in Kenya. Symp Zool Soc Lond 50:181–198.

Nelson GS (1983): Wild animals as reservoir hosts of parasitic diseases of man in Kenya. In J.D. Dunsmore (ed): Tropical Parasitoses and Parasitic Zoonoses. Perth: World Association of the Advancement of Veterinary Parasitology.

Nelson GS (1986): Hydatid disease: Research and control in Turkana, Kenya I. Epidemiological observations. Trans R Soc Trop Med Hyg 80:177–182.

Nelson GS, Rausch RL (1963): *Echinococcus* infections in man and animals in Kenya. Ann Trop Med Parasitol 57:136–149.

Nelson GS, Blackie EJ, Mukundi J (1966): Comparative studies on geographical strains of *Trichinella spiralis*. Trans R Soc Trop Med Hyg 60:471–480.

Nelson GS, Guggisberg CWA, Mukundi J (1963): Animal hosts of *Trichinella spiralis* in East Africa. Ann Trop Med Parasitol 57:332–346.

Nelson GS, Teesdale C, Highton RB (1962): The role of animals as reservoirs of bilharaziasis in Africa. In G.E.W. Wolstenholme and M. O'Connor (eds): Ciba Foundation Symposium on Bilharziasis. London. J.A. Churchill Ltd.

Okelo GBA (1986): Hydatid disease: Research and control in Turkana, III. Albendazole in the treatment of inoperable hydatid disease in Kenya—a report on 12 cases. Trans R Soc Trop Med Hyg 80:193–195.

Paget S (1901): Memoirs and Letters of Sir James Paget. London: Longmans Green.

Pattyn SR (1978): Ebola virus haemorrhagic fever. In Proceedings of International Colloquium on Ebola Virus Infection and Other Haemorrhagic Fevers. Amsterdam: Elsevier/North Holland Biomedical Press.

Pavlowsky EM (1966): In N.D. Levine (ed): Natural Nidality of Transmissible Diseases. Urbana: University of Illinois Press.

Rausch RL (1952): Hydatid disease in boreal regions. Arctic 5:157–174.

Rausch RL, Nelson GS (1963): A review of the genus *Echinococcus* Rudolphi 1801. Ann Trop Med Parasitol 57:127–135.

Rausch RL, Schiller EL (1956): Studies on the helminth fauna of Alaska, XXV. The ecology and public health significance of *Echinococcus sibiricensis* on St. Lawrence Island. Parasitology 46:395–419.

Rollinson D, Walker TK, Simpson AJG (1986): The application of recombinant DNA technology to problems of helminth identification. Parasitology [Suppl] 91:53–71.

Roselle HA, Schwartz DT, Geer FG (1965): Trichinosis from New England bear meat. Report of an epidemic. N Engl J Med 272:304–305.

Ruebush TK, Cassadau PB, Marsh HJ, Lisker SA, Voorhees DB, Mahoney EB, Healy GR (1977): Human babesiosis on Nantucket Island. Clinical features. Ann Int Med 86:6–9.

Sachs R, Taylor AS (1966): Trichinosis in a spotted hyaena (*Crocuta crocuta*) of the Serengeti. Vet Res 78:704.

Schwabe CW (1969): Veterinary Medicine and Human Health, 2nd Edition. London: Bailliere Tindall & Cassel.

Schwabe CW, Abou-Daoud K (1961): Epidemiology of echinococcosis in the Middle East. I. Human infection in Lebanon, 1949 to 1959. Am J Trop Med Hyg 10:374–381.

Smith JH, Folse DS, Long EC et al. (1983): Leprosy in wild armadillos (*Dasypus novemcinctus*) of the Texas Gulf Coast: Epidemiology and mycobacteriology. J Reticuloenthothial Soc 34:75–88.

Smyth JD (1979): An *in vitro* approach to taxonomic problems in trematodes and cestodes, especially *Echinococcus*. Symp Br Soc Parasitol 17:75–101.

Steele JH (1970): Epidemiology and control of trichinosis. In: S.E. Gould (ed): Trichinosis in Man and Animals. Springfield: Charles C. Thomas.

Thompson RCA (1986): The Biology of Echinococcus and Hydatid Disease. London: Allen and Unwin.

Uganda Buruli Group (1969): BCG vaccinations against *Mycobacterium ulcerans* infection (Buruli ulcer). First results of a trial in Uganda. Lancet 1:111–115.

Verster AJM (1965): Review of *Echinococcus* species in South Africa. Onderstepoort J Vet Res 32:7–118.

Vogel H (1957): Uber den *Echinococcus multilocularis* Suddeutschlands. I. Das Bandwurm-Stadium von Stammen menschlicher und tierischer Herkunft. Z Tropenmed Parasitol 8:404–454.

Von Siebold CT (1853): Ueber die verwandlung der *Echinococcus* brut in taenien. Z Wissen Zool 4:409–425.

Wagener K (1957): Zoonoses. Berl Munch Tierarztl Wochenschr 70:12.

Williams RJ, Sweatman GK (1963): On the transmission, biology and morphology of *Echinococcus granulosus equinus*, a new subspecies of hydatid tapeworm in horses in Great Britain. Parasitology 53:391–407.

World Health Organization 1959): Zoonoses. Technical Report Series No. 169. Geneva: WHO.

World Health Organization (1981): FAO/UNEP/WHO Guidelines for Surveillance, Prevention and Control of Echinococcosis/Hydatidosis. Geneva: WHO.

Wright WH, Jacobs L, Walton AC (1944): Studies in trichinosis, XVI. Public Health Rep 59:669–681.

Young E, Kruger SP (1968): *Trichinella spiralis* (Owen, 1835) Railliet, 1895. Infestation of wild carnivores and rodents in South Africa. S Afr Med Assoc 38:441–443.

The Biology of Parasitism, pages 43–59

# Control of Schistosomiasis in Man

## A.E. Butterworth

*Medical Research Council and Department of Pathology, University of Cambridge, Cambridge CB2 1QP, England*

## INTRODUCTION

This chapter is not intended as a comprehensive review of schistosomiasis, but instead focuses on two aspects of the disease in man. First, what methods are currently available for its control, and what are their advantages and disadvantages? And secondly, how may recent advances, both in understanding the immunology of human infections and in applying recombinant DNA techniques to the cloning of schistosome antigens, lead to improved methods of control through the development of vaccines?

### The Need for Control

The three main species of schistosome that are pathogenic for man, *Schistosoma mansoni*, *haematobium* and *japonicum*, are currently estimated to affect over 200 million people, mainly in South America and the Caribbean, Africa and to a lesser extent the Near East; and China, Southeast Asia and the Philippines. It is extremely difficult to obtain accurate estimates of morbidity and mortality: but a rough approximation might be that, in areas of high transmission, 5–10% of heavily infected people will eventually die as a result of their infection, while many others will suffer more-or-less severe, chronic ill-health, especially during childhood and adolescence. However, this is only an estimate, and it should be stressed that there are marked differences in the severity of disease, not only between the different species but also within the same species but in different areas. In irrigation schemes, for example—notoriously, the Gezira scheme on the Blue Nile in the Sudan—*S. mansoni* infection is associated with a high level of morbidity and mortality. In contrast, in more natural foci of transmission in rural areas of sub-Saharan Africa, mortality from *S. mansoni* infections may be low, in spite of high prevalence of infection. It is not yet known whether this difference is solely dependent on intensity of infection, as has previously been supposed, or whether other factors may also be involved, of which the most likely are genetic variations in either parasite or host and the status of the host with respect to nutrition and to other infectious diseases. However, the important point is that disease attributable to schistosome infection is still extensive and even increasing in some areas, in spite of attempts at control. This is largely due to the widespread development of both large and small irrigation schemes, as rapid population growth and extensive drought puts progressively increasing pressure on more effective land and water usage. In Africa particularly, the effective harnessing of water for agricultural purposes must be a major long-term goal; and such harnessing, among many other problems, will undoubtedly lead to a massive increase in schistosomiasis unless control measures can be developed that are effective, cheap, and long-lasting.

### Aspects of the Schistosome Life Cycle

A short description of the schistosome life cycle may help in appreciating the aims of control (Fig. 1). The adult worms, about 1 cm long, live in pairs in the small mesenteric or pelvic veins, depending on the species, and the female of the pair lays eggs at the rate of about 300 a day. These eggs, containing a larval stage, the miracidium, surrounded by a hard, chitinous shell, can now do one of two things.

First, they can pass up the hepatic portal vein (in the case of *S. mansoni*, the species dealt with in detail in this chapter) and lodge in the liver; here they elicit the development of an intense, cell-mediated granulomatous reaction which, with its associated fibrosis, is responsible for the clinical manifestations of hepatosplenomegaly, portal hypertension and, in severe cases, death following rupture of oesophageal varices.

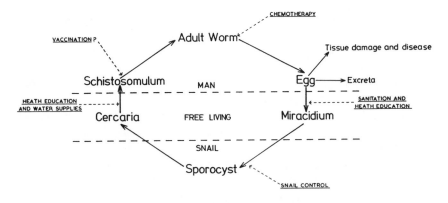

Fig. 1. Life cycle of *Schistosoma* and points of possible intervention for control.

Alternatively, the eggs can pass out through the intestinal wall and into the faeces, subsequently hatching in fresh water to release the free-living miracidia, which go on to infect aquatic snails of certain restricted species. Within the snail, they undergo a series of asexual replication steps, eventually emerging as a second free-living stage, the cercaria, which swims actively in the water for several hours. When people come into contact with such water, the cercaria penetrates directly through the skin, shedding its tail as it goes in and transforming into the young schistosome or schistosomulum. This now migrates via the bloodstream and lungs to the hepatic portal system, where it matures, forms pairs, and starts to lay eggs about 5 weeks after infection.

Several general points about this life cycle should be borne in mind. First, it has been shown in several experimental hosts that the adult worm survives for long periods of time, even in hosts that have acquired the capacity to resist reinfection: and, in man, the mean life span of the adult worm is of the order of 2–5 years, although individual worms can live much longer. The phenomenon of an acquired resistance to reinfection in the continued presence of adult worms from a primary infection is referred to as concomitant immunity and has two important consequences from the point of view of control. First, for an immune response to be effective, it must be able to attack the young schistosomulum at some stage during its development: by the time that it has matured into the adult worm, it has acquired the capacity to evade immune attack. Secondly, once an individual is infected, he will retain his worm burden, and, in the absence of treatment, continue to contaminate the environment with eggs for many years.

In addition, the adult worms do not replicate within the mammalian host. This is in complete contrast to infections with most protozoa, for example, in which infection with a single organism is theoretically capable of leading to fulminant disease. In schistosomiasis, in contrast, the extent of disease depends directly on the number of eggs deposited in the tissues; and this in turn depends on the numbers of worms harboured by the host and the length of time that they have been present. In other words, it is the *intensity* of infection that determines disease, not simply the presence or absence of infection. Examination of intensities of infection within endemic communities, as determined by measuring egg outputs in the faeces, commonly reveals a negative binomial distribution: a few patients harbour most of the worms and have high intensity infections, and these are the ones who will go on to develop severe disease. The implication of this is that, in order to reduce disease, a control measure does not have to be 100% effective; any measure such as vaccination that even partly reduces the establishment of the adult worms may have a great impact on the prevalence of *disease*, by reducing the numbers of patients in the high-intensity category.

## CURRENT METHODS OF CONTROL

The cycle of infection described above might theoretically be interrupted or reduced by any one of a number of means (Fig. 1):

1. Reduction either of the level of contamination of the waterbodies with excreta, or of the extent of contact with contaminated water, both of which require improved sanitation, health education, and water engineering;
2. Removal of the snail intermediate host;
3. Treatment of infected individuals within a community, thereby killing the adult worms and preventing the output of eggs in the excreta; and
4. Vaccination, leading to a reduction in the establishment of adult worms and hence a reduction both in pathology and in egg excretion.

Each of the first three methods has been applied, and has both advantages and disadvantages.

### Water Supplies, Sanitation, and Health Education

The desirable ideal for the prevention of schistosomiasis would be the introduction of piped water and drainage to each household. Unfortunately, this is completely impracticable in cost terms in developing countries, not only in subsistence rural areas but also in large irrigation schemes and in urban areas of high population density. Instead, a cheaper compromise must be accepted, which might consist of the construction of pit latrines, the digging of boreholes or wells, and the pumping of water to communal standpipes or shower and laundry units. These measures can be quite effective. Their main advantages are that they invariably have strong community support, that their effectiveness is not diminished by local movement of people into or out of an area, and that they will also have an impact on other water-borne diseases. The disadvantage is that they do depend on the availability of a supply of clean water, which is difficult to obtain in arid areas, and that the construction and maintenance of such a supply may be unmanageably expensive. Local habits must also be taken into account. In irrigation schemes, for example, both defaecation and the majority of water contact may take place outside the villages, so that the installation of clean water supplies within a village may have little impact. In other areas, it may be found that children are banned by their elders from using pit latrines; and, since it is the children who have the highest egg output, and who contribute most to contamination of the environment, the use of pit latrines by adults may have relatively little effect on transmission.

Tied in with sanitation and water engineering, therefore, is health education, with the general aim of conveying the idea of the nature of the disease and the life cycle of the parasite, in order to reduce both contamination of the waterbodies and unnecessary contact with contaminated water. In this context, it is the children who are the main targets for such education: and, as primary schooling becomes increasingly widely available, this becomes more and more feasible. In the coastal areas of Kenya, for example, it used to be considered normal for an adolescent boy to be passing red urine, as a result of *S. haematobium* infection. Any boy who did not was considered to be sickly; and this was probably true, since the connotation was that for some reason he had less water contact than his peers. Now that primary education has become more widely available, it is generally recognised that passing red urine is abnormal, and may lead to later severe disease. One has, in effect, a large captive audience—the primary school population—and an increasing desire on the part of the community as a whole to be informed about the nature of disease and the steps that they themselves can take to prevent it.

## Snail Control

Although alternatives exist, the most widespread approach to snail control is by the use of molluscicides, which can be applied in several ways.

At one extreme, large irrigation schemes that are fed by a single source can be treated simply by the application of large amounts of molluscicide at a single point on the main water course. This approach has been used, for example, in the Fayoum irrigation scheme in Egypt, which is fed by a single offshoot from the Nile. The molluscicide has been found to be still active over 100 km from the point of application, and treatment of a single point affects over 40,000 km of canals and drainage ditches; this approach is stated to have resulted in a reduction in prevalence in this area from 46% before control to 7% after seven years of control (Mobarak, 1982; Fenwick, 1987).

At the other extreme, it is possible to treat in a focal manner those pools or watercourses in dispersed rural areas that actually contain large numbers of infected snails, at the appropriate time of year (Klumpp and Chu, 1987). This approach has been particularly successfully applied in the study in St. Lucia, described below, in which focal mollusciciding in one valley over a period of several years was associated both with a progressive reduction in the numbers of total and infected snails and with a progressive decrease in prevalence of infection within the human population (Prentice et al., 1981).

The main advantage of mollusciciding in general is that it is independent both of community support and of population movement. The disadvantages, however, are that it is extremely expensive, even when applied focally; it requires extensive and skilled precontrol studies and virtually indefinite continuation of control; and molluscicide-resistant strains of snails may eventually be selected.

Other methods of snail control may also be considered. Biological control by predators and competitors has not been routinely used, although there are some candidate organisms; but environmental control (together with chemotherapy) has been widely applied in the case of *S. japonicum* in China. This species is more difficult to control than *S. mansoni* or *S. haematobium*, since it is a zoonosis. Cattle are infected, as are a variety of rodents, and there is always the danger that control measures targeted at human populations will leave behind an animal reservoir of infection. In China, the main method of snail control is to allow, at periodic intervals, the drainage of water from the irrigation ditches. Trenches are then dug at the bottom, and the amphibious snail hosts are shovelled into these trenches and suffocated with earth before the ditch is refilled with water. This approach, in combination with chemotherapy, has proved remarkably successful in reducing the distribution of *S. japonicum* in many areas of China (Shou-Pai and Bao-Ruo, 1982). However, the technique is particularly labour-intensive, and requires a strong element of subordination of individual and family responsibilities to those of the community as a whole, and this may well be difficult to achieve in different types of society.

## Chemotherapy

Chemotherapy is currently the favoured method, either alone or in combination with other measures, for the control of schistosomiasis, and is particularly strongly supported by the World Health Organization. Safe and effective drugs, in particular praziquantel, now exist for the treatment of all three species of human schistosome; and, in addition to reducing transmission, chemotherapy has the obvious added advantage of removing the worm burden from the human host, and therefore of reducing the likelihood of the development of subsequent morbidity. Chemotherapy may be applied either by mass treatment of all infected individuals within a community or, in an attempt to reduce costs, by treatment targeted at certain groups. These may either be the most heavily infected individuals, since they are the ones who will eventually develop disease, or the infected children, since they are the ones most likely to be contaminating the environment.

A good example of the relative merits of chemotherapy, in comparison with other methods of control, comes from the Caribbean island of St. Lucia, on which the Rockefeller Foundation established in 1965 a programme designed to compare the different methods that were then available (Jordan, 1977, 1986). The island was favourable for such a study, in that it was highly endemic for schistosomiasis and is extremely mountainous; most of the population live in the valleys, and there is relatively little movement from one valley to the next. This meant that it was possible to take three geographically distinct control areas and apply different measures in each.

In the first, the approach was to provide a cheap supply of clean water; in some communities an individual tap in each household, but in most a communal shower and laundry unit. After the introduction of the water supplies, a period of intensive health education was carried out. In the second valley, molluscicides were applied at repeated and frequent intervals to the whole of the valley's river system; and, in the third, a majority of infected individuals were treated, with retreatment subsequently being carried out at intervals over the following three years.

The effects on transmission were then measured by examining the incidence of new infections, the changes in overall prevalence, and the reduction in contamination of the environment as estimated by exposure of "sentinel" snails. All three measures had a reasonably marked effect on transmission, as reflected both by a reduction in incidence of new infections and by a reduction in infection in sentinel snails. However, chemotherapy was by far the most effective, and had the added advantage of reducing prevalence more extensively. In addition, it was considerably cheaper than the other methods, although this effect would be less marked if the study were repeated today; the newly available drugs, although less toxic, are more expensive. The early stages of this extensive experimental study can therefore be summarised quite simply as showing that, when the various methods of control were applied in isolation, chemotherapy was the most effective and least expensive.

The next stage was to determine the duration of the effect of chemotherapy. Treatment was administered in each of the 4 years between 1973 and 1976; no further treatment was then given, but the communities were followed for a further 4 years. As was expected, the incidence of new infections progressively rose with time, especially in the high transmission villages in the valley. The conclusion was that, although transmission had not been broken, it had been usefully reduced, and that retreatment was not required for a period of 3–4 years.

Since that time, chemotherapy-based control programmes, with or without other methods as ancillary measures, have been applied on a wide scale. Notable successes have been reported, in particular in Brazil, with a long-term reduction not only in parasitological indices of infection but also, more importantly, in morbidity (Sleigh et al., 1986). However, in spite of its current popularity, chemotherapy has serious potential disadvantages. First, the newly available generation of schistosomicidal drugs such as praziquantel, although safe and effective, are extremely expensive. Secondly, and tied in with this problem, is the fact that, in areas of high transmission, reinfection may occur rapidly: the younger, nonimmune subjects in particular may regain up to 40% of their pretreatment egg counts within 1 year. Retreatment must therefore be carried out frequently and for an indefinite period, and this requires secure and long-term funding on a large scale. Finally, the possibility must be considered of the eventual emergence of drug-resistant strains of parasite. The finding that resistance to a commonly used drug, oxamniquine, has developed under both laboratory and field conditions suggests that schistosomiasis is no exception to the general rule for infectious diseases that, in circumstances of heavy drug pressure, selection of drug-resistant strains will eventually occur.

## FUTURE CONTROL METHODS: THE DEVELOPMENT OF A VACCINE

Considerable advances have been made in the control of schistosomiasis over the last 20 years, in particular through the development and application of the new generation of safe and effective drugs. However, this must not lead to complacency, or to the view that the problem of schistosomiasis is now solved in principle, and that there is no room for further improvement. Specific problems concerning both chemotherapy and other control measures have been mentioned in the preceding section, and these may be viewed in a more general perspective. In developed nations, major reductions in the prevalence of various infectious diseases and in their associated morbidity and mortality have previously been brought about by socioeconomic changes, which have included health education, improved living conditions and public health programmes, improved nutrition, and improved delivery of health services. However, one cannot adopt the laissez-faire attitude of sitting back and waiting for socioeconomic conditions to change in Third World countries.

With respect to chemotherapy, I know of no infectious disease in which adequate and long-term control has been achieved solely through drug delivery. Effective drugs may lead to a false sense of security. In the case of malaria, for example, the attitude in the 1960s was very similar to that now adopted in some quarters for schistosomiasis: that chemotherapy and the use of residual insecticides was sufficient for control. Twenty years later, the very high prevalence of malaria, now associated with major problems of drug resistance, has made the search for alternative control measures, especially vaccines, of high priority. In more general terms, it is difficult to accept the argument that it is in any way satisfactory, as a method for long-term control, to allow an individual patient or group of patients to become infected with a severe and potentially lethal, albeit chronic, disease, and then to treat them once they have become infected. The cliché that "prevention is better than cure" is trite only because it is sufficiently true as to be self-evident.

A vaccine against schistosomiasis, were one to become available, would have two major advantages, other than the general value of being a preventive rather than a curative measure. First, through the auspices of the Expanded Programme of Immunisation (EPI) of the World Health Organization, effective immunisation programmes are now established in many developing countries. These programmes are served by dedicated and enthusiastic staff who see the practical yields of their efforts; they commonly receive, in contrast to other health programmes, secure and long-term funding: and their effectiveness is increasing annually. Secondly, the development of a schistosome vaccine and its inclusion within EPI schemes would mean that there would be less need for *separate* hierarchical structures specifically for the control of schistosomiasis: it could be treated more loosely as a disease to be covered within the EPI framework, and would therefore compete to a lesser extent with other medical or other problems of equal or greater priority for restricted national funds.

Immunologists have been promising a schistosome vaccine for sufficiently long that an amused or angry scepticism among administrators of practical control programmes is entirely understandable. However, the last 5 years have seen developments of a new order of magnitude, which justify a considerable degree of optimism that a useful vaccine could be produced in the near future. This section summarises briefly the background work that has led to this sense of optimism, and outlines the future problems that remain to be overcome. Aspects of this problem are also covered in the article by Sher (see Sher, pp. 169–182).

### Immunity in Experimental Animals

**Nature and mechanisms of immunity.** The possibility that a schistosome vaccine might eventually be developed began to emerge during the 1960s, when it became clear that a variety of experimental animals, in particular the mouse and the rhesus monkey, acquired the capacity to resist reinfection following a primary infection. This observation led to the concept of concomitant immunity, and posed the question of why the adult worms of a primary infection continue to survive, apparently in the face of an immune response that is capable of destroying the young larvae of a new, challenge infection. Various mechanisms have been proposed, including the acquisition by the older worm of a coating of host-derived molecules that mask parasite antigens, the simple loss of expression from the outer of the two parasite tegumental membranes of parasite antigens, and the development of a tegument that is refractory to immune attack. Evidence for each of these mechanisms has been obtained, and it is likely that more than one may operate. From the point of view of vaccine development, however, the importance of this early work was in directing the focus of attention on the developing schistosomulum as the primary target for immune attack.

Subsequently, experiments in particular in the mouse model indicated that at least part of the resistance observed in chronically infected animals was nonspecific in nature and attributable to an attrition of schistosomula as a consequence of egg-induced pathology. The relevance of this phenomenon to human disease is not clear at present, largely because the difference in weight between the two hosts makes it impossible to test in mice the consequence of infections with low worm burdens per unit of body weight of the level that might be anticipated in man. The problem of resistance associated with egg-induced pathology is not seen in the rat; in this species, however, the fact that it is a nonpermissive host must be borne in mind, although the fact that a host fails to permit the full development of adult worms to the egg-laying stage does not necessarily indicate that it will show unusual immune responses against the schistosomula.

In the face of these problems, much attention has been focussed in recent years on the immunity that can be induced, in both mice and rats as well as primates and cattle, by immunisation with irradiation-attenuated larvae (cercariae or schistosomula). Provided that the dose of irradiation is optimal—sufficient to ensure a transient survival of the larva and initial migration to the lungs, without development through to egg-laying adult worms—a consistent and high degree of immunity can be induced in which there is no evidence for nonspecific effects. This immunity is species specific and can be reproducibly transferred with serum, although a role for cell-mediated effector mechanisms is not excluded by this observation. Current interest in the biology of the immunity elicited by irradiated larvae is focussed on two main areas:

1. The site of attrition of the young larvae of a challenge infection, with the skin, the lung, and the post-lung phases of migration being considered to be important by different groups; and
2. The relative roles in mediating immunity of antibody-dependent, cell-mediated effector mechanisms directed against schistosomulum surface anti-

gens, and of cell-mediated mechanisms, including macrophage activation, which need not have specificity for surface antigens.

There is no reason to assume that either the various sites or the different mechanisms of attrition are mutually exclusive. It would seem reasonable that immunity against a multicellular parasite with a complex migration pattern in its definitive host might well be multifactorial. From the point of view of vaccine development, the relevant point is that different effector mechanisms, or responses acting at different stages of migration, might well have specificity for different target antigens; and, therefore, that a cocktail of antigens might show an additive or even a synergistic effect in eliciting protection, in comparison with its individual components.

**Vaccine antigens.** In this context, much attention has recently been paid to the identification and molecular cloning of protective antigens (see Simpson and Cioli, 1987). This field is discussed in more detail by Sher (pp. 169–182), but some points are worth re-emphasising from the point of view of this discussion. The finding that irradiated larvae elicit high levels of specific immunity was itself of considerable interest, and has led to the development of a vaccine for *S. bovis* that has been successfully tested in the field and that could be developed commercially (Taylor and Bickle, 1986). For the human schistosomes, however, three problems preclude the use of irradiated organisms: the ethics of their administration; the cost of their preparation, when weighed against the available funds for schistosomiasis control in man; and the problems in producing sufficient material from a naturally maintained life cycle for large-scale production. The advent of molecular cloning techniques offers a way of bypassing these problems, by producing large amounts of individual antigens as expressed products in *E. coli*, yeast, or other expression vectors.

After many years of failure, an increasing number of laboratories have now reported the successful immunisation of both mice and rats with a range of crude and purified antigens, in protocols designed to elicit either cell-mediated responses (James, 1986) or humoral antibodies (Smith and Clegg, 1985). In particular, the production of a considerable number of monoclonal antibodies that confer passive protection in vivo has demonstrated that antibodies against a single antigen can mediate protection (e.g., Grzych et al., 1982); and such monoclonals have been used for affinity purification of native antigens for protection studies. These experiments have generated considerable optimism about the production of individual vaccine antigens by recombinant DNA techniques, and work in this area is now in progress in several laboratories. One recombinant antigen, p28, has been shown to elicit protection in both rats and hamsters, and is now being tested in baboons (Balloul et al., 1987). A second, Sm97, has been shown to be effective in mice when administered with BCG to elicit a cell-mediated response; this antigen is the schistosome equivalent of paramyosin (Lanar et al., 1986). Other recombinant molecules are also reaching the stage of being tested for their protective capacity. The next major question to be tackled, therefore, is whether it will be better to use a single antigen or a cocktail of several molecules; this is discussed in more detail below.

### Immunity in Man

Animal models of schistosome infection and immunity are of tremendous value in demonstrating the potential range of protective immune responses, and the antigens against which such responses may be directed. However, the various model systems differ quite markedly from each other, and arguments that any one model is a more accurate reflection of the situation in man than any other are usually both hypothetical and tenuous. An understanding of the extent and mechanisms of human immunity is necessary for obtaining the maximum benefit from a vaccine and, although appropriate hypotheses for testing may be derived from animal models, the actual testing of such hypotheses for their relevance to human immunity can only be carried out with human subjects observed under conditions of field exposure.

Such studies are expensive, time-consuming, and difficult to execute, to such an extent that there has until recently been considerable controversy about the actual existence of acquired immunity in man. Early arguments were based on the shape of the age-specific prevalence and intensity curves of human schistosome infection, with a rise during the first 10 or 15 years of life, followed by a decline. More refined studies have compared changes of infection in children followed longitudinally with time; in adult immigrants to endemic areas; in people (such as canal cleaners) assumed to be heavily exposed to infection; in individuals living in areas of high and low transmission, respectively; and in individuals living in communities in which transmission has or has not been reduced by snail control. All of these studies have supported the idea of the slow development of an age-dependent resistance to reinfection, but have suffered from the problem that the lack of direct observation of water-contact patterns have made it difficult to estimate the relative contribution of acquired immunity and reduced exposure to infection as factors in the reduced levels of infection among older age groups.

During recent years, therefore, several studies have been undertaken, for *S. haematobium* in the Gambia and for *S. mansoni* in Kenya, Egypt, and Brazil, to examine intensities of reinfection after treatment of individuals whose levels of contact with contaminated water can subsequently be observed. These "treatment and reinfection" studies offer several theoretical advantages:

1. The initial treatment, even though it may clearly affect the immune response, makes it possible to exclude variations in intensity of infection due to a variable loss of adult worms from established infections; any eggs that are detected during the post-treatment period, in individuals whose initial response to treatment is satisfactory, will be attributable to new infections acquired during that period.

2. Estimates of intensity of reinfection after treatment, rather than simply of the presence or absence of reinfection, should make it possible to detect an incomplete immunity that limits but does not prevent the establishment of new worms. Such incomplete immunity—similar to that observed in most experimental systems—may be of great practical value, not only in limiting transmission but also in reducing the numbers of patients who bear heavy infections, and who therefore go on to develop severe disease.

3. Direct observations of water contact are naturally subject to a high degree of variation both between and within individuals. However, observations may be made sufficiently precise that they allow the demonstration of statistically significant relationships between observed exposure and actual reinfection. Given that it is possible to obtain meaningful estimates of exposure, then it is also possible to examine other changes that may limit infection, including the development of an age-dependent resistance that is independent of exposure.

The various treatment and reinfection studies that have now been carried out in different parts of the world have yielded essentially similar conclusions. First, both the extent and nature of water contact changes with age, such that older individuals show reduced levels of exposure. On top of this, however, there is clear evidence, both at the individual level and for age groups of patients, of the slow development of an additional age-dependent resistance to reinfection. In other words, the observed reduction in water contact is insufficient to account for the very marked reduction in reinfection with age. This is particularly marked in women, who may continue to show high levels of exposure associated with a range of household-related water contact activities. In addition, preliminary evidence from *S. mansoni* infection in Kenya has indicated that the observed age-dependent resistance (superimposed on changes attributable to water contact patterns) is *also* dependent on recent experience of infection, in that it is not seen among individuals who are not detectably infected and who are subsequently followed for new infections.

The most reasonable interpretation of an age-dependent resistance that is also dependent on previous experience of infection is the slow development of an acquired immunity—rather than, for example, age-dependent physiological changes such as skin thickness. A range of studies have therefore been carried out to test for immunological correlates of the observed resistance. Studies on *S. haematobium* in the Gambia have shown a relationship between resistance to reinfection and both high eosinophil counts and the levels of antibodies against adult worm antigens (Hagan et al., 1985, 1987). Studies on *S. mansoni* infections among a group of Kenyan schoolchildren have revealed a different and somewhat more complicated picture (Butterworth et al., 1987).

We originally worked on the hypothesis that the development of resistance would be associated with the development of one or more potentially protective immune responses. This hypothesis, however, was oversimplistic and incorrect: *all* of these heavily exposed children, both the older, resistant individuals and the younger children who remained susceptible to reinfection after treatment, showed a range of potentially protective immune responses, including, for example, high levels of antibodies mediating eosinophil-dependent killing of schistosomula in vitro. The distinguishing feature of the younger children is that they *also* showed high levels of lgM antibodies against egg carbohydrate antigens. This led to the hypothesis that these antibodies might block antischistosomulum effector mechanisms, and prevent the expression of immunity, a phenomenon for which there was already a precedent in the rat model (Grzych et al., 1984). Subsequent experiments showed directly that lgM antibodies could block the eosinophil-dependent killing of schistosomula that was mediated by IgG antibodies from the same infection sera (Khalife et al., 1986); and more recently, a role for IgG2 blocking anticarbohydrate antibodies has also been demonstrated (Dunne et al., in preparation).

As a result of these findings, we propose that the very *slow* development of immunity in children can be explained by:

1. The early development of a range of protective responses directed against the schistosomulum surface;

2. The simultaneous early development of antibodies that block these protective responses. These blocking antibodies may be elicited by egg polysaccharides (a major immunogenic stimulus during natural infection), cross-reacting with schistosomulum surface carbohydrate epitopes, and may be of either IgM or IgG2 isotypes; and

3. The slow decline with time in the levels of blocking antibodies, thus permitting the expression of the protective responses and hence of immunity.

Clearly, this interpretation must be treated with caution. Immunity in human schistosomiasis, as in the various animal models, will almost certainly turn out to be multifactorial; and the findings described above, being correlative in nature, do not exclude a possible role for other, unidentified immunological factors. However, insofar as they represent one aspect that may influence the slow acquisition of immunity during natural infection, they bode well for the development of a vaccine. If the continued susceptibility of younger children to reinfection after treatment is governed by the presence of blocking antibodies elicited in response to egg antigens, then the early administration *before* natural infection of an effective vaccine—for example, a recombinant peptide of the type described above—would prevent establishment of the initial infection, and would therefore prevent the development of blocking antibodies in response to egg antigens. It should therefore be possible to convert the young child into the equivalent of the immune adult, with high levels of protective responses but low levels of blocking antibodies, and showing a strong and long-lasting immunity.

## CONCLUSIONS AND FUTURE WORK

Three developments summarised in the preceding section offer grounds for considerable optimism about the rapid development of a schistosome vaccine: the reproducible demonstration of acquired immunity in experimental animals both with irradiated larvae and with isolated antigens; the molecular cloning of various candidate protective antigens; and the demonstration and partial analysis of immunity in man. However, much remains to be done. At the research level, each cloned antigen must be tested, and its protective effects optimised, initially in rodent models but also in primates; and a preliminary structure for coordinated primate experiments has now been established. At an early stage, it will also be necessary to decide whether a single antigen will be sufficient or whether a cocktail vaccine would be preferable. There are considerable difficulties in the commercial manufacture of a cocktail vaccine, but two points will have to be taken into account before the idea is discarded. First, a combination of antigens, each eliciting different effector responses and possibly directed against different stages in schistosomulum, may be more effective than a single product. In no experimental system has complete immunity yet been achieved; and this may not be necessary, since the aim is to reduce the numbers of patients with heavy infections rather than to prevent infection entirely. However, a maximum efficacy is clearly desirable. In addition, there may be a genetically determined heterogeneity of individual responses to single cloned antigens, which would result in a vaccine of variable efficacy: such variation might be reduced with a cocktail preparation.

At that stage, the problem becomes one of commercial manufacture of an ethically acceptable preparation, with new elements of both cost and quality control. It is unlikely that commercial companies will see this as a profit-making venture; extensive support from external agencies will therefore be required, and this must be generated in the near future. Finally, attention must be paid to the design of vaccine trials under field conditions, after initial demonstrations of safety and efficacy in man. These present, perhaps, less of a problem than might be imagined; they could be similar in approach to the "treatment and reinfection" studies that have already yielded invaluable results.

The problem of vaccine development is not yet solved, but it is now a far more realistic proposition than it was 10 or even 5 years ago. The *need* for such a vaccine, although disputed in some quarters, is to my mind undeniable. This need may be more generally recognised in a few years time; in the meantime, it is important that those who are now riding a crest of a wave in vaccine development continue their impetus, in order that they may have a usable material to deliver when the time is ripe.

## REFERENCES

Balloul JM, Sondermeyer P, Dreyer D, Capron M, Grzych JM, Pierce RJ, Carvallo D, Lecocq JP, Capron A (1987): Molecular cloning of a protective antigen of schistosomes. Nature 326:149–153.

Butterworth AE, Bensted-Smith R, Capron A, Capron M, Dalton PR, Dunne DW, Grzych JM, Kariuki HC, Khalife J, Koech D, Mugambi M, Ouma JH, Arap Siongok TK, Sturrock RF (1987): Immunity in human schistosomiasis mansoni: Prevention by blocking antibodies of the expression of immunity in young children. Parasitology 94:281–300.

Fenwick A (1987): The role of molluscicides in schistosomiasis control. Parasitol Today 3: 70–73.

Grzych JM, Capron M, Bazin H, Capron A (1982): *In vitro* and *in vivo* effector function of rat IgG2a monoclonal anti-*S. mansoni* antibodies. J Immunol 129:2739–2743.

Grzych JM, Capron M, Dissous C, Capron A (1984): Blocking activity of rat monoclonal antibodies in experimental schistosomiasis. J Immunol 133:998–1003.

Hagan P, Wilkins HA, Blumenthal UJ, Hayes RJ, Greenwood BM (1985): Eosinophils and resistance to *Schistosoma haematobium* in man. Parasite Immunol 7:625–632.

Hagan P, Blumenthal UJ, Chaudri M, Greenwood BM, Hayes RJ, Hodgson J, Kelly C, Knight M, Simpson AJG, Smithers SR, Wilkins HA (1987): Resistance to reinfection with

*Schistosoma haematobium* in Gambian children: Analysis of their immune responses. Trans R Soc Trop Med Hyg (in press).

James SL (1986): Induction of protective immunity against *Schistosoma mansoni* by a nonliving vaccine. III. Correlation of resistance with induction of activated larvacidal macrophages. J Immunol 136:3872–3877.

Jordan P (1977): Schistosomiasis—research and control. Am J Trop Med Hyg 26:877–886.

Jordan P (1986): Schistosomiasis: Research to Control. Cambridge: Cambridge University Press.

Khalife J, Capron M, Capron A, Grzych JM, Butterworth AE, Dunne DW, Ouma JH (1986): Immunity in human schistosomiasis mansoni. Regulation of protective immune mechanisms by IgM blocking antibodies. J Exp Med 164:1626–1640.

Klumpp RK, Chu KY (1987): Focal mollusciciding: An effective way to augment chemotherapy of schistosomiasis. Parasitol Today 3:74–76.

Lanar DE, Pearce EJ, James SL, Sher A (1986): Identification of paramyosin as schistosome antigen recognized by intradermally vaccinated mice. Science 234:593–596.

Mobarak AM (1982): The schistosomiasis problem in Egypt. Am J Trop Med Hyg 31:87–91.

Prentice MA, Jordan P, Bartholomew RK, Grist E (1981): Reduction in transmission of *Schistosoma mansoni* by a four-year focal mollusciciding programme against *Biomphalaria glabrata* in Saint Lucia. Trans R Soc Trop Med Hyg 75:789–798.

Shou-Pai M, Bao-Ruo S (1982): Schistosomiasis control in the People's Republic of China. Am J Trop Med Hyg 31:92–99.

Simpson AJG, Cioli D (1987): Progress towards a molecularly defined vaccine for schistosomiasis. Parasitol Today 3:26–28.

Sleigh AC, Hoff R, Mott KE, Maguire JH, Da Franca Silva JT (1986): Manson's schistosomiasis in Brazil: 11-year evaluation of successful disease control with oxamniquine. Lancet i:635–637.

Smith M, Clegg JA (1985): Vaccination against *Schistosoma mansoni* with purified surface antigens. Science 227:535–538.

Taylor MG, Bickle QD (1986): Towards a schistosomiasis vaccine. Parasitol Today 2:132–134.

The Biology of Parasitism, pages 61–76
© 1988 Alan R. Liss, Inc.

# Biology of Amebiasis: Progress and Perspectives

Adolfo Martínez-Palomo

*Department of Experimental Pathology, Center for Research and Advanced Studies, National Polytechnic Institute, 07000 Mexico, D.F. Mexico*

## INTRODUCTION

Amebiasis is the infection of humans with the protozoan parasite *Entamoeba histolytica*, which has a cosmopolitan distribution. The motile form of the parasite, the trophozoite, usually lives as a harmless commensal in the lumen of the large intestine, where it multiplies and differentiates into cysts, the resistance form, responsible for the transmission of the infection. In most cases, particularly in developed countries, *E. histolytica* induces no signs of symptoms in the condition known as lumenal amebiasis. The parasite, however, can also act as a pathogen; this is seen mainly in certain underdeveloped countries, where it causes invasive amebiasis. The most common forms of symptomatic infections are dysentery and diarrhea, which occur when virulent strains invade the intestinal mucosa. These are generally self-limited in course but occasionally may give rise to potentially fatal complications. Extraintestinal lesions, mainly liver abscesses, occur when trophozoites spread through blood-borne dissemination. Amebic liver abscess may be lethal unless promptly diagnosed and properly treated (Sepúlveda and Martínez-Palomo, 1984).

Renewed interest in the study of amebiasis over the past 15 years has resulted in a wealth of new knowledge concerning the biological, clinical, and epidemiological aspects of the disease. As a result, the infection can no longer be considered one of the neglected parasitic diseases; but, on the other hand, it is clear that the basic biological problems that are relevant to the understanding and the control of amebiasis remain to be solved.

## GLOBAL DISTRIBUTION OF AMEBIASIS

Although estimates of the worldwide distribution of amebiasis suggest that only a small fraction of the total number of infected individuals have invasive intestinal or extraintestinal amebiasis, in countries where invasive forms are a public health problem, a significant percentage of those with infections present clinical evidence of the disease. Even in temperate zones, where affluent societies generally flourish, and where the number of cases of invasive amebiasis is much lower, familiarity with this disease is important since the failure to identify an amebic infection may result in a lethal outcome: for example, intestinal amebiasis may be treated as ulcerative colitis. In addition, high rates of infections may exist among certain immigrant groups, and epidemic outbreaks can occur in institutions such as schools or mental hospitals (Martínez-Palomo and Martínez-Báez, 1983).

The most common modes of transmission are by food contaminated with cysts or direct fecal-oral transmission from person to person. The greatest risk is associated with cyst passers, especially when they are engaged in the preparation and handling of food. Amebiasis can also be a sexually transmitted disease, especially among homosexual men. In certain urban homosexual populations lumenal amebiasis has reached hyperendemic levels during recent years. So far, almost all cases reported in homosexuals have been produced by nonpathogenic amoebas, but the introduction of pathogenic strains is an ever-present possibility.

Invasive amebiasis is a major health and social problem in areas of Africa, Asia, and Latin America where inadequate sanitary conditions and the presence of highly virulent strains of E. histolytica may combine to sustain a high incidence of symptomatic infections. However, the wide array of imperfect tests for diagnosing amebiasis severely restricts understanding of its magnitude and epidemiology. The limitations of the available epidemiological data include the study of biased population samples, inability to distinguish easily between pathogenic and nonpathogenic strains, variation in case definitions, and methodological differences in the examination of stool and serological specimens (Walsh, 1986a). The proportion of the population infected with amoebas throughout the world appears to have remained approximately the same; however, as a result of the population growth, the prevalence of infection has increased (Walsh, 1986b). Estimates by Walsh (1986a,b) suggest that in 1984 probably 500 million people were infected with the parasite and 38 million people developed disabling colitis or amebic liver abscesses. Approximately 40,000–100,000 deaths were attributable to amebiasis and, on a global scale, amebiasis probably comes third among parasitic causes of death, behind only malaria and schistosomiasis. Better knowledge of amebiasis frequency, transmission, and epidemiology will require improved diagnostic tools and a clear understanding of the biological and epidemiological significance of strain differences.

## THE PARASITE

*E. histolytica* is one of the most primitive eukaryotes. The motile activity and pleomorphism of the trophozoites is based on a simple cytoplasmic layout which lacks a number of organelles usually found in most eukaryotes: a structured cytoskeleton and cytoplasmic microtubules, a membranous system equivalent to the Golgi complex and endoplasmic reticulum of higher eukaryotes, mitochondria, and a system of primary and secondary lysosomes (McLaughlin and Aley, 1985; Martínez-Palomo, 1982, 1986). This small (10–40 $\mu$m), fragile, and temperature-sensitive protozoan is, however, capable of colonizing the large intestine of a sizeable percentage of the world's population. Furthermore, under unknown circumstances it can invade the intestinal lining and eventually destroy practically any tissue in the human body, while, at the same time, successfully evading the immune responses of the human host. Unless checked by adequate treatment, this parasite may continue its destructive activity in cases of invasive amebiasis until the host dies—along with the parasite; thus, an invasive infection can often be biologically suicidal for the amoebas.

The life cycle of *E. histolytica* consists of three consecutive stages, namely, trophozoite, cyst, and metacyst. Trophozoites dwell in the colon, where they multiply by binary fission and encyst, producing typical four-nucleated cysts after two successive nuclear divisions. Details on the mechanism of nuclear division, including the determination of the number of chromosomes, are unknown. Cysts are found in the formed stools of carriers as hyaline bodies 8–20 $\mu$m in diameter, with a rigid wall, probably made up by chitin and glycoproteins, that protects the amoebas outside the human body. Cysts do not develop within tissues. Trophozoites are of no importance in the transmission of the disease, since they are short-lived in the environment and do not survive exposure to hydrochloric acid and digestive enzymes in the gastrointestinal tract.

Some of the problems that have hindered a more thorough understanding of the biology of the parasite are the apparent simplicity of the structural organization of the cytoplasm, which renders almost useless the large body of information concerning the cell biology of more developed eukaryotes; the presence of numerous and potent proteases and nucleases; the fragility of amoebas, which hinders manipulation; the exquisite sensitivity of axenic cultures to minor variations in the components of the culture medium, and the lack of a culture medium that would promote encystation of axenic amoebas. There is not even certainty about the molecular and cellular bases of such fundamental processes as the differentiation of trophozoites into cysts, the switch from harmless commensal to harmful invader, the mechanisms of evasion of the host immune response, and, as previously mentioned, the cellular changes that take place during division. It is thus obvious that much more research on the biology of the parasite is needed.

During the last decade, progress in the knowledge of the parasite has been achieved mainly at the cellular level, while work on the molecular biology of the parasite just started with the characterization of the actin gene of pathogenic amoebas (Edman et al., 1987).

As shown in Figure 1 the pleomorphism of amoebas is best revealed by scanning electron microscopy. The surface of trophozoites may show lobopodia, endocytic stomata, filopodia, and a posterior uroid (González-Robles and Martínez-Palomo, 1983). The regions of attachment involved in contact cytolysis show no unusual features. Much of the cytoplasm consists of a vacuolar system mainly involved in endocytosis (Fig. 2); other functions of the vacuolar system remain to be explored. A lattice of tubules and vesicles superficially resembling smooth endoplasmic reticulum may also be found with transmission electron microscopy, but it remains to be determined whether this membranous system is involved in the channeling of toxins and enzymes related to the destructive activity of the parasite. Many inclusions of unknown nature have also been described, including filamentous and polyhedral viruses, but efforts to relate the presence of the viruses to amebic virulence have so far failed (Mattern et al., 1979). In spite of the presence of large amounts of cytoplasmic actin, microfilaments are usually not evident (Meza et al., 1983). Microtubules have not been identified in the cytoplasm, and only the dividing nuclei show microtubular bundles (Martínez-Palomo, 1982). Further studies on the cytoskeleton should be encouraged in view of the involvement of cytoskeletal components in motile processes related to the cytopathic effect of virulent amoebas, such as adhesion, phagocytosis, and possibly, the export of toxic substances by means of exocytosis (Martínez-Palomo, 1987).

Trophozoites are facultative aerobes with peculiar glycolytic enzymes also found in certain bacteria (Weinbach, 1981). It remains to be demonstrated whether this protean metabolism relates to the apparent ability of pathogenic forms to go from the environment of the intestinal lumen at low oxygen pressure to that encountered upon invasion of solid organs with abundant blood supply. In many other respects, amoebas resemble anaerobic and microaerophilic bacteria more than typical eukaryotes; e.g., they lack glutathione metabolism (Fahey et al., 1984). In spite of much recent work on the biochemistry of the parasite, the detailed characterization of surface antigens has not been achieved yet. The membrane lipid composition is unusual (Cerbón and Flores, 1981) and at least a dozen glycoproteins are found in

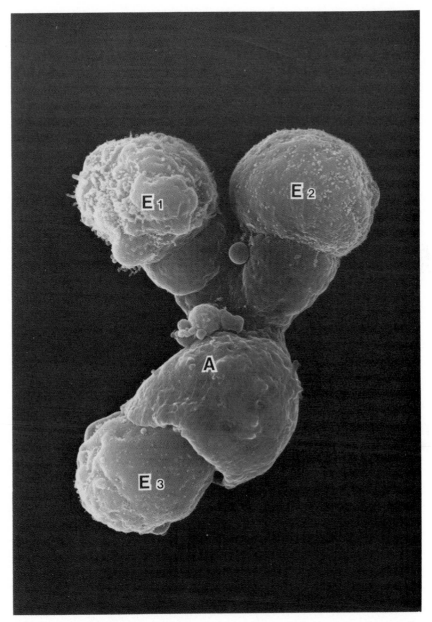

Fig. 1. Scanning electron micrograph of a trophozoite of *E. histolytica* (A) in culture ingesting simultaneously three epithelial cells (E1–E3). The surface of the amoeba is smooth in comparison to the surface of epithelial cells. ×21,000.

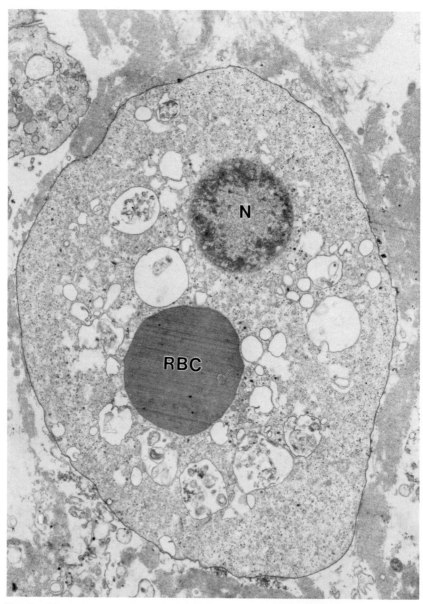

Fig. 2. Transmission electron micrograph of an amoeba localized at the border of an intestinal ulceration in a guinea pig. The cytoplasm is made up largely of a vacuolar system. The larger vacuoles contain cellular debris, and one is occupied by remnants of a red blood cell (RBC). Erythrophagocytosis is the hallmark of invasive amoebas in the large intestine. N = nucleus. ×41,000.

purified plasma membranes (Aley et al., 1980). Antibodies interacting with surface antigen elicit a striking mobilization of the complexes towards the uroid (Pinto da Silva et al., 1975). This capping phenomenon, as well as resistance to complement (Calderón and Tovar, 1986; Reed et al., 1986) and masking with host proteins are hypothetical means to evade that the parasite may use to evade the host humoral (Trissl, 1982; Kretschmer, 1986). In spite of these and other advances, biochemists and immunologists still have a long way to go in the exploration of this bizarre protozoan.

## STRAIN DIFFERENCES

Two puzzling aspects of the biology of E. histolytica are the variable expression of its pathogenic potential and the general restriction of human invasive amebiasis to certain geographical areas (although epidemic outbreaks may occur anywhere), despite the worldwide distribution of the parasite (Sepúlveda and Martínez-Palomo, 1984; Martínez-Palomo, 1987).

A major advance in the understanding of amebiasis has been the demonstration of consistent biochemical differences between pathogenic and nonpathogenic strains of E. histolytica. Nonpathogenic strains were first shown to differ from those isolated from cases of human invasive amebiasis in certain surface properties, thus reinforcing the hypothesis of Brumpt (1925), who stated that one species of amoeba was responsible for lumenal amebiasis, while invasive amebiasis was caused by a morphologically similar, but different species. In amebiasis, as is the case with the study of other parasitic diseases, much is to be learned from a historical review of the development of present knowledge, because many of the crucial problems that still remain unsolved have been amply analyzed by our predecessors, some of whom proposed interesting leads that can now be explored with modern biotechnological tools.

The initial suggestions of the existence of strain differences included the high susceptibility of pathogenic strains to agglutinate in the presence of concanavalin A (Martínez-Palomo et al., 1973), their lack of negative surface charge at neutral pH (Trissl et al., 1977), their high erythrophagocytic capacity (Trissl et al., 1978), and their ability to destroy monolayers of epithelial cells in culture. Furthermore, trophozoites defective in phagocytosis were found simultaneously to lose their virulence and their phagocytic activity (Orozco et al., 1983).

A difference in metabolic markers between pathogenic and nonpathogenic strains of E. histolytica has further strengthened the hypothesis of strain diversity. Sargeaunt et al. (1982) have shown, with the use of the isoenzyme technique applied to recent isolates of amoebas obtained from several continents, that parasites from well characterized cases of invasive amebiasis can be clustered into at least eight different patterns, distinguished by the presence of a $\beta$-band and the absence of an $\alpha$-band in the enzyme phosphoglucomutase, and by fast running bands in hexoquinase in most patterns. All remaining zymodemes were found in amoebas isolated from probable carriers. As expected, pathogenic patterns of isoenzymes are also present in some isolates from well characterized asymptomatic cyst passers (Meza et al., 1986).

What appears to be a fairly reasonably tested hypothesis, i.e., that pathogenic strains are distinct from the nonpathogenic ones has, however, been recently challenged by observations that suggest that the isoenzyme patterns may not be stable, at least under in vitro conditions. It has been reported that the axenization of a strain with a nonpathogenic isoenzyme pattern results in reversion to a pathogenic zymodeme, with the acquisition of virulence. Futhermore, the reverse may also be found; i.e., a pathogenic strain, upon culture with certain bacteria, can switch to a nonpathogenic isoenzyme pattern (Mirelman et al., 1986a,b). Initially it was thought that these observations could be induced by selective pressure on heterogeneous populations, but the same results have been obtained with cloned cultures. As a consequence, a lively and important debate has been raised on the interpretation of isoenzyme differences (Mirelman, 1987; Sargeaunt, 1987). The general epidemiological features of amebiasis and the fact that the most common extraintestinal lesion (liver abscess) generally occurs without amoebas being associated with bacteria tend to support Sargeaunt's hypothesis (Martínez-Palomo, 1987). However, the field clearly needs more research, in order to reconcile the observations obtained in cultured amoebas with the real situation in the human host.

## BIOLOGICAL BASIS OF LYTIC ACTIVITY

Probably the most striking aspect of the host-parasite interplay that takes place in invasive amebiasis is the powerful lytic activity of E. histolytica, for which the parasite was named by Schaudinn, the discoverer of the syphilis agent, who died at the age of 35 from self-inflicted amebiasis. Several recent reviews address specifically the question of the cellular basis of the cytopathic effect of the parasite. For descriptive purposes, the cytolytic action of E. histolytica on cultured mammalian cells (Gitler and Mirelman, 1986; Martínez-Palomo, 1986, 1987; Pérez-Tamayo, 1986; Ravdin, 1986) can be divided into four stages: adhesion, cytolysis following contact, phagocytosis, and

intracellular degradation. These stages are based on time-lapse microcine-
matographical and electron microscopic observations (Martínez-Palomo et
al., 1985).

Adhesion of amoebas to target cells (Fig. 3) and extracellular matrix
components seems to be an absolute prerequisite for lysis to occur, at least
under in vitro conditions. Sonicates of *E. histolytica* contain a lectin that
appears to have N-acetylglucosamine-containing glycoconjugates as receptors (Kobiler and Mirelman, 1980, 1981). Adhesion may also be mediated by

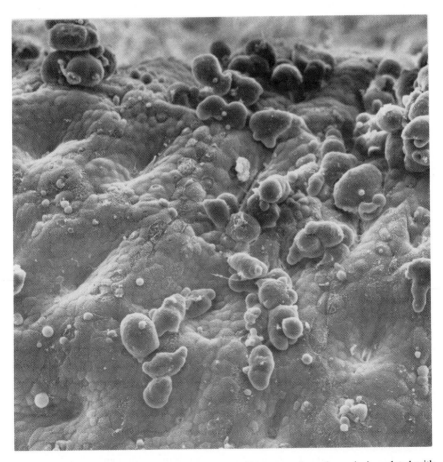

Fig. 3.  Scanning electron micrograph of the cecal mucosa of a guinea pig inoculated with
trophozoites of *E. histolytica*. During the first hours of interaction, no damage is seen in
epithelial cells. A large number of amoebas are seen attached to the mucosal surface. ×500.

an amebic adhesion inhibitable by N-acetyl-D-galactosamine (Ravdin et al., 1985). Blockage by specific carbohydrates and monoclonal antibodies may inhibit adhesion and reduce the cytopathic effect in vitro (Arroyo and Orozco, 1987).

Whether these lectins have any role in adhesion of amoebas to human intestinal epithelial cells has yet to be determined. It should be remembered that the adhesion of the parasites to the lumenal surface of intestinal cells is the exception rather than the rule. The small proportion of amoebas that attach directly to the apical aspect of epithelial cells are limited to interglandular regions where old cells slough continuously as part of the renewal process of the intestinal mucosa. Furthermore, strains of differing virulence show the same amount of lectin per amoeba and similar degrees of attachment to human intestinal epithelial cells in culture (Kobiler and Mirelan, 1981). A lack of correlation between virulence and attachment of amoebas to target cells has also been found (Orozco et al., 1982). Thus, in contrast to several enteropathogenic bacteria in which molecules involved in attachment determine the virulence of a given strain, both pathogenic and nonpathogenic amoebas seem to have equally effective attachment mechanisms. Of course, this does not rule out the potential interest of developing agents that would specifically interfere with attachment, thereby preventing the invasion of the intestine by the parasite.

So far, the crucial switch from a harmless commensal—in luminal amebiasis—to a powerful tissue destroyer—in invasive intestinal amebiasis—has remained an unsolved mystery. A clue to the understanding of this change may lie in the local conditions at the interface between the mucous lining of the colon and the bacterial flora associated with the amoebas. Anaerobic conditions and bacteria ingested by the trophozoites apparently favor the lowering of the redox potential in the parasite, thus facilitating its electron transport system, the first by virtue of oxygen deprival, bacteria acting as broad-range scavengers for oxygen molecules (Bracha and Mirelman, 1984). This in vitro observation indicates that virulence of a given strain of E. histolytica may depend to a considerable extent on the activity of the electron transport system, or the reducing power of the parasites. The local intestinal microenvironment would, at any rate, act only by enhancing the invasive capacity of the amoebas, because colonic ulcerations may be induced in rodents with axenic amoebas, provided the intestinal transit and the intestinal contents are eliminated previous to the inoculation (Anaya Velázquez et al., 1985). Otherwise, amoebas fail to colonize and invade the colon of rodents, one of the limiting factors being the mucous blanket that provides a barrier to the direct access of trophozoites to the intestinal epithelium (Leitch et al., 1985).

After contact, the lysis of target cells requires intact amebic actin microfilament function, as evidenced by inhibition with cytochalasins B and D (Guerrant et al., 1981). Lysis is also blocked by calcium-channel blockers, suggesting that calcium fluxes in the amoeba and entry of calcium into the target cell are involved in the cytopathic effect (Ravdin and Guerrant, 1982). A phospholipase may be involved in the damage to the plasma membrane of host cells. Saíd Fernández and López Revilla (1983) found haemolytic activity from homogenates of *E. histolytica* and, more recently, Long Krug et al. (1985) have demonstrated a calcium-dependent phospholipase, inhibitable by various pharmacological agents known to block phospholipase activity. Fractionation has revealed that the enzyme is associated with the plasma membrane of the parasite, an indication of its possible role in contact-dependent amebic cytolytic action.

*E. histolytica* in culture contains and under certain in vitro conditions releases the following substances: proteases that round up cultured cells (Lushbaugh et al., 1984; Scholze et al., 1986); pore-forming proteins that insert into natural or artificial membranes, creating an ionic imbalance (Lynch et al., 1982; Young et al., 1982); enzymes that degrade collagen and oligosaccharides in the extracellular matrix (Muñoz et al., 1982; Gadasi and Kessler, 1983; Trissl, 1983); and even neurotransmitter-like compounds that induce water secretion in intestinal samples (McGowan et al., 1983). Despite these results and the correlation between virulence and some of the above-mentioned factors, none of them have proved so far to be instrumental in the pathogenesis of amebiasis.

After contact-dependent cytolisis of target cells occurs, pathogenic amoebas ingest the lysed cells, although they can also engulf living cells. In general, pathogenic strains show a high rate of erythrophagocytosis; in contrast, nonpathogenic amoebas and those of attenuated virulence ingest few red blood cells (Trissl et al., 1978). Further exploration of the relationship between high phagocytic rates and virulence has shown that trophozoites defective in phagocytosis simultaneously lose their virulence and erythrophagocytic activity. This correlation was demonstrated with a clone deficient in phagocytosis selected through irradiation of trophozoites of a wild-type pathogenic strain fed on 5-bromo-2'-deoxyuridine-loaded bacteria. In addition, virulence revertants obtained through successive liver passages simultaneously recover both the ability to produce liver abscess and a high erythrophagocytic rate (Orozco et al., 1983). These results show that phagocytosis is related to the aggresive mechanisms of the parasite.

In summary, the tissue-destroying capacity of amoebas cannot be ascribed to a single protein, toxin, enzyme, organelle, or cell function, despite many attempts to identify a single causative factor. The striking lytic and scavenging activity of the parasite appears to be the result of a combination of liberation of toxins, enzymes, the active motility of the parasite, a particularly avid phagocytic activity, and an efficient cytoplasmic machinery that rapidly degrades ingested cellular and extracellular components (Martínez-Palomo, 1986).

It should be noted that the above-mentioned cytolytic mechanisms have yet to be demonstrated in vivo. In experimental intestinal and liver amebic lesions, necrosis of host tissues is also the result of death and disintegration of inflammatory cells upon contact with the amoebas. In animal models, *E. histolytica* trophozoites do not directly damage liver cells when inoculated intraportally. Rather, tissue destruction is the result of the accumulation and subsequent lysis of leukocytes and macrophages surrounding the amoebas (Tsutsumi et al., 1984). In this respect, in vitro studies have indicated that normal human polymorphonuclear neutrophils, peripheral blood mononuclear cells, monocytes, and nonactivated macrophages are lysed by pathogenic amoebas; only activated macrophages are capable of killing virulent amoebas (Salata et al., 1986).

Both humoral and cell-mediated immune responses are induced in invasive amebiasis. Clinical observations suggest that protective immunity prevents recurrent hepatic abscesses in patients treated with chemotherapy. One should bear in mind that cases of diarrhea due to *E. histolytica* may spontaneously resolve; in contrast, almost all cases of liver abscess progress towards a fatal evolution, unless adequate antiamebic chemotherapy is used. It is thus evident that in many intestinal invasive infections the defense response of the host is effective, whereas in most cases of liver abscess—and of other forms of extraintestinal amebiasis as well—immunological mechanisms fail to cope with the amebic infection.

The participation of immunological mechanisms in the pathogenesis of invasive amoebiasis has not been fully elucidated. Circulating antiamebic antibodies in humans are of interest mainly for diagnostic and epidemiological consideration, but they seem to be of relatively little relevance in controlling established invasive infections. On the other hand, there are no clear-cut histopathological, clinical, or immunological indications to suggest that pathogenic immune complexes play a role in invasive amebiasis (Trissl, 1982; Kretschmer, 1986). Even though complement has been shown to be amebicidal, virulent strains may be resistant to it. The major role in combating invasive infection due to *E. histolytica* appears to be due to cell-mediated immune mechanisms. Despite a lower T4:T8 ratio and a heterogeneous lymphocyte proliferation in response to mitogens, after treatment for hepatic abscess sensitization to amebic antigens develops and specific effector mechanisms are present (Salata et al., 1986).

## DIAGNOSTIC METHODS

One of the most important research priorities, and one that has received scant attention, is the development of simplified and reliable diagnostic and survey techniques for the diagnosis of amebiasis. The diagnosis of lumenal amebiasis relies on the identification of cysts in feces; and that of invasive intestinal forms, on the microscopic finding of hematophagous amoebas. The procedure is unreliable, time-consuming, and plagued with false-negative and false-positive results. For this reason, the detection of the parasite with new biotechnological methods using monoclonal antibodies or DNA probes should be encouraged. The potential of these new techniques in terms of increasing specificity, sensitivity, and ease of use has already been demonstrated in certain parasitic diseases, such as malaria and leishmaniasis (Wirth et al., 1986).

## REFERENCES

Aley SB, Scott WA, Cohn ZA (1980): Plasma membrane of *Entamoeba histolytica*. J Exp Med 152:391–404.

Anaya-Velázquez F, Martínez-Palomo A, Tsutsumi V, González-Robles A (1985): Intestinal invasive amebiasis: An experimental model in rodents using axenic or monoxenic strains of *E. histolytica*. Am J Trop Med Hyg 34:723–730.

Arroyo R, Orozco E (1987): Localization and identification of an *Entamoeba histolytica* adhesin. Mol Biochem Parasitol 23:151–158.

Bracha R, Mirelman D (1984): Virulence of *Entamoeba histolytica* trophozoites: Effects of bacteria, microaerobic conditions, and metronidazole. J Exp Med 160:353–368.

Brumpt E (1925): Etude sommaire de l'"Entamoeba dispar" n.sp. Amibe à quistes quadrinucléés, parasite de l'homme. Bull Acad Med (Paris) 94:943–952.

Calderón J, Tovar R (1986): Loss of susceptibility to complement lysis in *Entamoeba histolytica* HML by treatment with human serum. Immunology 58:467–471.

Cerbón J, Flores J (1981): Phospholipid composition and turnover of pathogenic amoebas. Comp Biochem Physiol [B] 69:487–492.

Edman U, Meza I, Agabian N (1987): Genomic and cDNA actin sequences from a virulent strain of *Entamoeba histolytica*. Proc Natl Acad Sci USA 84:3024–3028.

Fahey RC, Newton GL, Arrick B, Overdank-Bogart T, Aley SB (1984): *Entamoeba histolytica*: A eukaryote without glutathione metabolism. Science 224:70–72.

Gadasi H, Kessler E (1983): Correlation of virulence and collagenolytic activity in *Entamoeba histolytica*. Infect Immun 39:528–531.

Gitler C, Mirelman D (1986): Factors contributing to the pathogenic behavior of *Entamoeba histolytica*. Annu Rev Microbiol 40:237–261.

González-Robles A, Martínez-Palomo A (1983): Scanning electron microscopy of attached trophozoites of pathogenic *Entamoeba histolytica*. J Protozool 30:692–700.

74 / Martínez-Palomo

2alGuerrant RL, Brush J, Ravdin JI, Sullivan HA, Mandell GL (1981): Interaction between *Entamoeba histolytica* and human polymorphonuclear neutrophils. J Infect Dis 143:83–93.
Kobiler D, Mirelman D (1980): Lectin activity in *Entamoeba histolytica* trophozoites. Infect Immun 29:221–225.
Kobiler D, Mirelman D (1981): Adhesion of *Entamoeba histolytica* trophozoites to monolayers of human cells. J Infect Dis 144:539–546.
Kretschmer R (1986): Immunology of amebiasis. In A. Martínez-Palomo (ed): Amebiasis. Amsterdam: Elsevier Biomedical, pp 95–168.
Leitch CJ, Dickey AV, Udezulu IA, Bailey GB (1985): *Entamoeba histolytica* trophozoites in the lumen and mucus blanket of rat colon studied in vivo. Infect Immun 47:68–73.
Long Krug SA, Fischer HJ, Hysmith RM, Ravdin JA (1985): Phospholipase A enzymes of *Entamoeba histolytica* description and subcellular localization. J Infect Dis 152:536–541.
Lushbaugh WB, Hofbauer AF, Pittman FE (1984): Proteinase activities of *Entamoeba histolytica* cytotoxin. Gastroenterology 87:17–27.
Lynch EC, Rosenberg IM, Gitler C (1982): An ion-channel forming protein produced by *Entamoeba histolytica*. EMBO J 1:801–804.
Martínez-Palomo A (1982): The Biology of *Entamoeba histolytica*. Chichester: Research Studies Press.
Martínez-Palomo A (1986): Biology of *Entamoeba histolytica*. In A. Martínez-Palomo (ed): Amebiasis. Amsterdam: Elsevier Biomedical, pp 11–43.
Martínez-Palomo A (1987): The pathogenesis of amoebiasis. Parasitol Today 3:111–118.
Martínez-Palomo A, Martínez-Báez M (1983): Amebiasis. Rev Infect Dis 5:1093–1102.
Martínez-Palomo A, González-Robles A, de la Torre M (1973): Selective agglutination of pathogenic strains of *Entamoeba histolytica* induced by Con A. Nature N Biol 245:186–187.
Martínez-Palomo A, González-Robles A, Chávez B, Orozco E, Fernández-Castelo S, Cervantes A (1985): Structural bases of the cytolytic mechanisms of *Entamoeba histolytica*. J Protozool 32:166–175.
Mattern CFT, Kester DB, Diamond LS (1979): Experimental amebiasis. IV. Amebal viruses and the virulence of *Entamoeba histolytica*. Am J Trop Med Hyg 28:653–657.
McGowan K, Kane A, Asarkof N, Wicks J, Guerina V, Xellum J, Baron S, Gintzler AR, Donowitz M (1983): *Entamoeba histolytica* causes intestinal secretion: Role of serotonin. Science 221:762–764.
McLaughlin J, Aley S (1985): The biochemistry and functional morphology of *Entamoeba*. J Protozool 32:221–240.
Meza I, Sabanero M, Cázares F, Bryan J (1983): Isolation and characterization of actin from *Entamoeba histolytica*. J Biol Chem 258:3936–3941.
Meza I, de la Garza MA, Meraz B, Gallegos M, de la Torre M, Tanimoto M, Martínez-Palomo A (1986): Isoenzyme patterns of *Entamoeba histolytica* isolates from asymptomatic carriers: Use of gradient acrylamide gels. Am J Trop Med Hyg 35:1134–1139.
Mirelman D (1987): Effect of culture conditions and bacterial associates on the zymodemes of *Entamoeba histolytica*. Parasitol Today 3:37–40.
Mirelman D, Bracha R, Chayen A, Aust-Kettis A, Diamond LS (1986a): *Entamoeba histolytica*: Effect of growth conditions and bacterial associates on isoenzyme patterns and virulence. Exp Parasitol 62:142–148.

Mirelman D, Bracha R, Wexler A, Chayen A (1986b): Alterations of isoenzyme patterns of a cloned culture of nonpathogenic *Entamoeba histolytica* upon changes in growth conditions. Arch Invest Med (Mex) [Suppl] 17:187–193.

Muñoz ML, Calderón J, Rojkind M (1982): The collagenase of *Entamoeba histolytica*. J Exp Med 155:42–51.

Orozco ME, Martínez-Palomo A, González-Robles A, Guarneros G, Mora-Galindo J (1982): Las interacciones entre lectina y receptor median la adherencia de *E. histolytica* a células epiteliales. Arch Invest Med (Mex) 13(Suppl 3):153–157.

Orozco ME, Guarneros G, Martínez-Palomo A, Sánchez T (1983): *Entamoeba histolytica*. Phagocytosis as a virulence factor. J Exp Med 158:1511–1521.

Pérez-Tamayo R (1986): Pathology of amebiasis. In: A. Martínez-Palomo (ed): Amebiasis. Amsterdam: Elsevier Biomedical, pp.45–94.

Pinto da Silva P, Martínez-Palomo A, González-Robles A (1975): Membrane structure and surface coat of *Entamoeba histolytica*. Topochemistry and dynamics of the cell surface: Cap formation and microexudate. J Cell Biol 64:538–550.

Ravdin JI (1986): Pathogenesis of disease caused by *Entamoeba histolytica*: Studies of adherence, secreted toxins, and contact-dependent cytolisis. Rev Infect Dis 8:247–260.

Ravdin JI, Guerrant RL (1982): A review of the parasite cellular mechanisms involved in the pathogenesis of amebiasis. Rev Infect Dis 4:1185–1207.

Ravdin JI, Murphy CF, Salata RA, Guerrant RL, Hewlett EL (1985): N-acetyl-D-galactosamine-inhibitable adherence lectin of *Entamoeba histolytica*. I. Partial purification and relation to amebic virulence in vitro. J Infect Dis 151:804–815.

Reed SL, Curd JG, Gigli I, Guillin FD, Braude AI (1986): Activation of complement by pathogenic and nonpathogenic *Entamoeba histolytica*. J Immunol 136:2265–2270.

Saíd-Fernández S, López-Revilla R (1983): Latency and heterogeneity of *Entamoeba histolytica* hemolysins. Z Parasitenkd 69:435–438.

Salata RA, Martínez-Palomo A, Murray HW, Canales L, Trevino N, Segovia E, Murphy CF, Ravdin JI (1986): Patients treated for amebic liver abscess develop cell-mediated immune responses effective in vitro against *Entamoeba histolytica*. J Immunol 136:2633–2639.

Sargeaunt PG (1987): The reliability of *Entamoeba histolytica* zymodemes in clinical diagnosis. Parasitol Today 3:40–43.

Sargeaunt PG, Jackson TFHG, Simjee AE (1982): Biochemical homogeneity of *Entamoeba histolytica* isolates, especially those from liver abscess. Lancet i:1386–1388.

Scholze H, Otte J, Werries E (1986): Cysteine proteinase of *Entamoeba histolytica*. II. Identification of the major split position in bovine insulin B-chain. Mol Biochem Parasitol 18:113–121.

Sepúlveda B, Martínez-Palomo A (1984): Amebiasis. In K.S. Warren and A.A.F. Mahmoud (eds): Tropical and Geographical Medicine. New York: McGraw-Hill, pp. 305–318.

Trissl D (1983): Glycosidases of *Entamoeba histolytica*. Z Parasitenkd 69:291–298.

Trissl D (1982): Immunology of *Entamoeba histolytica* in human and animal hosts. Rev Infect Dis 4:1154–1184.

Trissl D, Martínez-Palomo A, Arguello C, de la Torre M, de la Hoz R (1977): Surface properties related to concanavalin A-induced agglutination. A comparative study of several Entamoeba strains. J Exp Med 145:652–655.

Trissl D, Martínez-Palomo A, de la Torre M, de la Hoz R, Pérez de Suárez E (1978): Surface properties of Entamoeba: Increased rates of human erythrocyte phagocytosis in pathogenic strains. J Exp Med 148:1137–1145.

Tsutsumi V, Mena-López R, Anaya-Velázquez F, Martínez-Palomo A (1984): Cellular bases of experimental amebic liver abscess formation. Am J Pathol 117:81–91.
Walsh J (1986a): Problems in recognition and diagnosis of amebiasis. Estimation of the global magnitude of morbidity and mortality. Rev Infect Dis 8:228–238.
Walsh J (1986b): Amebiasis in the world. Arch Invest Med (Mex) [Suppl]17:385–389.
Weinbach EC (1981): Biochemistry of enteric parasitic protozoa. Trends Biochem Sci 6:254–257.
Wirth D, Rogers WO, Barker R, Dourado H, Suesebang L, Albuquerque B (1986): Leishmaniasis and malaria: New tools for epidemiologic analysis. Science 234:975–979.
Young JD-E, Young TM, Lu LP, Unkeless JC, Cohn ZA (1982): Characterization of a membrane pore-forming protein from *Entamoeba histolytica*. J Exp Med 156:1677–1690.

The Biology of Parasitism, pages 77–92
© 1988 Alan R. Liss, Inc.

# South American Trypanosomiasis and Leishmaniasis: Endemic Diseases of Continental Dimensions Affecting Poor, Neglected, and Underfunded People

**Philip D. Marsden**

*Núcleo de Medicina Tropical, University of Brasilia, 70910 Brasilia, DF, Brazil*

## INTRODUCTION

At present trypanosomiasis and leishmaniasis are endemic infectious diseases chiefly affecting the families of subsistence farmers in rural areas. With the current trend to urbanisation of the populations of South American countries this situation could change. Triatomine bugs, which transmit Chagas' disease, have been found in *favelas,* and epidemics of kala azar have been reported among children of poor families in several cities of northeastern Brazil.

It is assumed that the reader has a grasp of the biomedical aspects of these infections, which have been reviewed elsewhere (Marsden, 1984a; Marsden and Jones, 1986).

## SOUTH AMERICAN TRYPANOSOMIASIS (*TRYPANOSOMA CRUZI* INFECTION, CHAGAS' DISEASE)

Calculations based on available data of the number of people infected for each South American country have been possible in the case of South American trypanosomiasis (Chagas' disease) (Table 1). However, such statistics are often inaccurate for developing countries due to variations in prevalence within the country itself and underreporting. The figures in Table 1 represent infections with *T. cruzi* and not lesions of the heart or gut that Chagas described and that are known by his name. These can occur in 25–30% of affected individuals but are usually limited to a much smaller (again a geographical difference) number of seroreactors. Still, in many South

**TABLE 1. Chagas' Disease in Latin America[a]**

| Country | Total population (millions) | % Rural | Estimated cases (millions) | Main vector |
|---|---|---|---|---|
| Argentina | 26.393 | 21 | 2.640 | T. infestans |
| Belize | 0.145 | ND | 0.003 | T. dimidiata |
| Bolivia | 4.647 | 77 | 1.858 | T. infestans |
| Brazil | 119.024 | 36 | 6.340 | T. infestans |
| Chile | 10.857 | 20 | 0.367 | T. infestans |
| Colombia | 26 | 29 | 1.217 | R. prolixus |
| Costa Rica | 2.110 | 59 | 0.130 | T. dimidiata |
| Ecuador | 6.521 | 58 | 0.180 | T. dimidiata |
| El Salvador | 4.300 | 60 | 0.322 | T. dimidiata |
| Fr. Guiana | 0.080 | ND | 0.021 | R. pictipes? |
| Guatemala | 7.110 | 64 | 0.730 | T. dimidiata |
| Guyana | 0.835 | 73 | 0.208 | R. prolixus? |
| Honduras | 3.400 | 68 | 0.213 | R. prolixus |
| Mexico | 69.950 | 34 | 3.798 | T. barberi |
| Nicaragua | 2.400 | 52 | 0.114 | T. dimidiata |
| Panama | 1.630 | 49 | 0.226 | R. pallescens |
| Paraguay | 2.880 | 56 | 0.397 | T. infestans |
| Peru | 16.800 | 42 | 0.643 | T. infestans |
| Suriname | 0.352 | ND | 0.147 | R. pictipes? |
| Uruguay | 2.886 | 17 | 0.278 | T. infestans |
| Venezuela | 13.913 | 30 | 4.865 | R. prolixus |
| Total | 322.233 | 36 | 24.697 | |

[a]ND: no data. From Schofield (1985).

American countries, Chagas' disease is a common cause of hospital admission. Fifty percent of all patients with heart failure due to chronic chagasic cardiomyopathy have a life expectancy under 2 years and can only be treated symptomatically. In 1983 we calculated for our hospital that pacemaker implantation for an individual with cardiac arrhythmia due to Chagas' disease cost U.S. $6,118, and surgical treatment of mega oesophagus and megacolon cost $3,529 and $4,412, respectively. A patient with heart muscle failure will have several hospital admissions and may incur costs many times greater than these in the course of a year. This represents a considerable drain on hospital services and expenditures from any national health insurance scheme. Such considerations have prompted several South American countries (e.g., Brazil, Argentina, Venezuela) to mount national control programmes (Marsden, 1984c). Since this is the most pertinent aspect of Chagas' disease research, the Brazilian programme (the best documented) is considered below in some detail. The Venezuelan programme, fuelled with petrodollars, appears to have had reasonable success in spite of the fact that the principal vector, *Rhodnius prolixus*, proved resistant to benzene hexachloride (BHC), and Dieldrin had to be used. Up-to-date information on the status of the Argentinan programme is difficult to obtain, although one area continues to report large numbers of acute cases.

## THE BRAZILIAN CHAGAS' DISEASE CONTROL PROGRAMME

Poorly constructed houses with good hiding places for vector triatomine bugs result in transmission to sleeping children of what is a lifelong infection. There has been a lag phase of some 40 years in Brazil since the discovery of the importance of residual insecticide spraying in controlling domiciliated bugs and the application of this finding to all affected areas. The failure to implement this vital discovery was due to the low priority given to Chagas' disease in the past. However, as a result of research demonstrating the importance of spraying, in 1983 funds became available though the Brazilian National Health Insurance Scheme and the last 3 years have seen a remarkable improvement in the situation. Plans for Chagas' disease control in 1986 have recently been published (Dias, 1986).

Now all municipalities in rural Brazil where domiciliated triatomine bugs transmit *T. cruzi* infections to man have been sprayed with residual insecticide. Evidently highly domiciliated species such as *Triatoma infestans*, the principal vector, are more vulnerable to such insecticide application since ecotopes are rare outside the house fabric and no genetic resistance is known.

The greater part of the attack phase of insecticide spraying has been done with BHC, a nondegradable chlorinated hydrocarbon. This was applied in a water-miscible solution and killed triatomines passing over the sprayed surface for 3 months; and some anti-triatominal activity could be detected for up to 6 months. Reasons for BHC's continued use despite the hazard of environmental pollution were its relatively low cost and easy availability. The Brazilian Ministry of Health bought BHC from various parts of the world (South Africa, Spain, etc.) as well as from Brazilian manufacturers. Recently BHC has been replaced in the Brazilian control programme by Deltamethrin, a synthetic pyrethroid. This is not manufactured in Brazil and is much more expensive than BHC. However, because its effects last twice as long and the dose required is half as much, the price of application per house is little different from that of BHC. Only a single application per year is necessary. Spraying houses is a costly process, and calculations for our field area suggest that wide oscillations occur from year to year, depending on the price of materials and labour—$5–25 per house (Garcia-Zapata et al., 1986).

While it is admirable that Brazil with her current economic difficulties should enter 1987 with at least the attack phase of the Chagas' disease control programme completed, the phase of vigilance is the current research worry. In the old regulations the vigilance phase consisted principally of ministry personnel attempting to capture bugs manually in suspected houses over a limited time period (30 min). Our work in central Brazil has shown that such manual captures are inaccurate in houses after spraying, since the small numbers of surviving bugs are unlikely to be captured by such a crude technique (Garcia-Zapata and Marsden, 1986). Better are techniques involving householder participation, where vigilance is continuous. Thus, throughout the yearly observation period the householder is provided with a suitable container (self-sealing plastic bag) and instructed to put any bug captured into the bag and send it to a central collecting point. Then his house is visited and if necessary resprayed. A second simple technique involves applying a thick paper sheet to a section of wall over the chief blood-meal source (the nuptial bed). The ministry is now distributing an informative poster for this purpose. Over the year any feeding bugs are likely to take shelter during the day behind the paper and frequently leave characteristic defaecation marks when the sheet is examined.

Such longitudinal vigilance techniques provide more reliable information on how a community is faring regarding the persistence of vector triatomines in houses after a spraying programme. We thought our good initial results with community participation could have been the result of our work in the area for many years, but we have shown that our methods work equally well in a virgin area recently subjected to the attack phase. The success of community collaboration could vary depending on the social situation. For example, one doubts if the rapid success achieved in rural Brazil could be achieved in a remote Bolivian Indian community, without a prior extensive education programme.

Another important aspect of a Chagas' disease control programme is the hidden benefits that it usually brings to affected rural communities. For example, an initial requirement is access by a communication line—usually a road, which inevitably improves with use. Along this road come not only the public health personnel to implement the programme but also eventually schoolteachers to collaborate in health-education programmes. In our own area there has been a dramatic increase in the number of schoolteachers; and at the time of writing there are four resident doctors in the village, and a small hospital has opened, all due to the initiative of the current local mayor. In our experience when rural schoolteachers are present, a bug-vigilance scheme utilising schoolchildren to report bug presence through the schoolteacher to the reference center functions well.

Our system of vigilance providing householders with plastic bags to collect bugs and tacking paper sheets to walls as indicators for bug faeces was designed for remote rural farms visited yearly. If a responsible, trainable person can be identified in such a farm, he can be taught the spraying procedure; and if he is provided with protective clothing and insecticide, he can apply insecticide if bugs are noted. In carefully selected farms this method has been employed with good results in our field studies.

The tendency with effective vigilance schemes is to implant them at a local level. The concept of having a vigilance team which sequentially follows the spraying team in an area is still hardly implemented in practice. Such a vigilance group would study community attitudes, implant triatomine information posts through schoolteachers, and in remote areas use the longitudinal vigilance methods described.

Six years after the initial attack phase which reached all localities infested with *T. infestans*, we still are registering a low incidence of this species in our study area of Mambaí, Goiás, Brazil. Vigilance and selective spraying when bugs are found continues in this area. With the recent use of Deltamethrin in place of BHC the tendency is for *T. infestans* to diminish more rapidly. A problem arises however, when other triatomines invade the peridomicile after control of *T. infestans*. Should they be treated as seriously? Usually only a single winged adult is captured, and since the whole house is resprayed, we can make no observations regarding colonisation potential. Of the five species captured in houses, four, at least, have been known to colonise with varying frequency. Probably, however, with all of them the risk of transmission to man is much less than that posed by *T. infestans*.

Since Mambaí is only one ecological situation in Chagas' disease control and a multitude of different vectors exist in Latin America, the reader can imagine the amount of applied research that remains to be done.

It should be emphasised that planning of the Brazilian control programme was greatly assisted by the results of a national serological survey and a entomological survey which mapped where communities with high infection rates and suitable vectors were to be found. For countries such as Bolivia, where Chagas' disease is a real problem, this would be an essential first step to planning control measures. Unfortunately, as in many South American countries, financial constraints do not even allow this first step.

The last step in control programmes is house improvement. This should take place in selected houses after the attack and vigilance phases. An unpublished study from our group shows that even new houses proximal to bug-infested ones will be constantly reinvaded by flying adults. Wall plastering is probably the best alternative to eliminate bug resting places. Preferably this is done by the householder himself. The last house plastered by our group cost $250, but the owner was incapable of helping and difficulties in access to his home raised the price.

A real worry with the current Brazilian programme is whether it will continue to command a high priority in government circles. Other endemic disease problems are serious enough to contemplate reducing the expenditure on Chagas' disease. These are epidemic malaria in new agricultural settlements in Amazonia, the transmission of dengue by *Aedes aegypti* and *A. albopictus*, and the threat of yellow fever transmission from the same mosquitoes. Ironically, these serve as powerful arguments in favour of devolving the responsibilities of vigilance in Chagas' disease-affected communities on local administrations.

## OTHER CONSIDERATIONS IN SOUTH AMERICAN TRYPANOSOMIASIS

The national serological survey in Brazil utilised eluates from blood dried on filter paper examined by the indirect anti-IgG immunofluorescent test (Camargo et al., 1984). Fourteen laboratories nationwide participated in this study, and more than 1½ million specimens were examined from representative samples of rural populations. In spite of continuous quality control provided by the central laboratory, false positives have been detected, which is probably inevitable in such a large survey. An antigen specific for *T. cruzi* that will not cross-react with leishmanial infections has been developed but is not widely applied. However, the best serological techniques, especially in postcontrol surveys, still have to be determined. Although entomological parameters such as the number of bug-infested houses, density of those populations, and frequency of infection with *T. cruzi* in captured bugs are valuable indicators, the final, most significant parameter is to show that, in the control area in question, positive serology in children born after the attack phase has fallen significantly. This requires a reliable test that can be done on a small volume of blood, preferably from a finger prick. Elisa test (ELISA) is a recommended procedure (Hoff et al., 1985), but surprisingly there is still no standard recommendation. Like syphilis, in chronic infections with *T. cruzi* the clinical investigator is dependent on reliable serology. Few investigative immunologists are interested in serology in spite of its immediate practical applications to important human problems.

Of course even if triatomine transmission of *T. cruzi* were controlled tomorrow we would still be left with millions of people who are already infected. The slow evolution of chronic chagasic cardiomyopathy means that little change will be noted in the frequency of this common cause of hospital admission in endemic areas during the next few decades. In the early, acute phase of the infection when parasites are easily detected in the peripheral blood, there is a general consensus that such patients should receive specific treatment. Doubts arise in patients with chronic infections regarding such treatment, but certainly a target group would be those incubating a progressive heart muscle inflammation. Many believe this process to have an autoimmune basis. Despite repeated attempts estimating autoantibodies of various types and complement-mediated trypanosome lysis, no reliable method of detecting active cardiomyopathy in chronic infections has been possible. By the time electrocardiographic changes or heart failure have developed, specific therapy is of little use.

In actual practice many physicians in South America do use specific therapy in the chronic phase in spite of the lack of evidence that it is beneficial. The two drugs with activity against both the tissue amastigote and the trypomastigote are Nifurtimox (a nitrofuran) and Benznidazole (a nitroimidazole) (McGreevy and Marsden, 1986). Both are not licensed in the United States due to mutagenicity and frequent side effects such as dermatoses, psychosis, and peripheral neuritis which occur at the end of the long courses of oral therapy. It appears that *T. cruzi* strains show a geographical variation in susceptibility to these drugs, being more sensitive in Chile and Argentina than in central Brazil. Even with lengthy evaluation of the effect of these drugs, serology rarely converts to negative and evaluation relies heavily on repeated xenodiagnosis to detect a minimal circulating parasitaemia. The significance of a positive xenodiagnosis in chronic human infections is unclear. However, acute cases are described where seroconversion has occurred after chemotherapy and observations over many years have shown no sequelae.

Purine analogues (allopurinol, allopurinol riboside) are a possible alternative chemotherapeutic approach. Marr and Do Campo (1987, Reviews of Infectious Disease) have recently reviewed chemotherapy of Chagas' disease using these compounds. Progress towards clinical trials has been slow as both allopurinol and allopurinol riboside fail to eliminate all parasites from experimental animals. Since there are such difficulties in the chemotherapy of Chagas' disease, it is not surprising that the emphasis is on control of house-dwelling vector bugs.

Since infection can also be acquired by transfusion of infected blood, adequate screening of blood banks is necessary, using reliable serology and preferably rejecting seroreactive donors. Congenital Chagas' disease also occurs in seroreactive mothers, but it is rare ($<2\%$). All the evidence indicates that attacking important domiciliated vector bugs such as *T. infes-*

*tans, Panstrongylus megistus, Rhodnius prolixus,* and *Triatoma dimidiata* is the correct solution.

## LEISHMANIASIS

In South America the epidemiological situation regarding the transmission of leishmaniasis to man is complex. In terms of infant mortality visceral leishmaniasis is most important, with many foci being identified in the dry northeast of Brazil as well as other Latin American countries such as Venezuela, Colombia, Paraguay, and Honduras. The causative organism is usually characterised as *Leishmania donovani chagasi* (Ldc). Risk factors include young age and malnutrition (Badaró et al., 1986). Without appropriate therapy the reticuloendothelial system is progressively invaded by the parasite, immunosuppression is marked, and the child often dies of a secondary infection of the lung or intestine.

Other parasites have been isolated from skin granulomas in all countries of the Continental Americas with the exception of Chile, Uruguay, and Canada. These parasites fall into two groups—namely, the *Leishmania mexicana* complex and the *Leishmania braziliensis* complex, each consisting of several subspecies. Mapping of the distribution of these various subspecies using taxonomy employing specific monoclonal antibodies and isoenzyme electrophoresis of cultural flagellate extracts is still in progress, but sufficient information exists to make some generalisations in Brazil (Grimaldi et al., 1987). First the subspecies *Leishmania braziliensis braziliensis* (Lbb) is the most prevalent, which is unfortunate since it is the organism most frequently associated with nasal metastases or espundia (Marsden, 1986). In our 5-year prospective study in one endemic area where this is virtually the sole parasite in transmission, the annual incidence of disease was 8.1/1,000 inhabitants and a prevalence of 14.9% (Jones et al., in press). However, we have recent experience of a severe epidemic in a region nearby where the attack rate in 1984 was 82 per 1,000.

Fortunately we estimate mucosal involvement to be much rarer than previously supposed ($>5\%$ of patients with a skin infection with Lbb).

Lbb is said to produce more skin destruction than other cutaneous species, although no good comparative study exists. Certainly skin ulceration is rapid deep and slow to heal. Another subspecies commonly isolated in Brazil is *Leishmania mexicana amazonensis* (Lma), which, although it produces a smaller skin granuloma, can give rise to the rare anergic hansenoid leishmaniasis (Convit's syndrome), which is virtually incurable. *Leishmania braziliensis guyanensis* (Lbg) is restricted to an area north of the Amazon river, often producing multiple skin lesions with lymphatic spread. *Leishmania braziliensis panamensis* (Lbp) is identified in Panama and Colombia from leishmanial ulcers and can also be associated with espundia. A number of other mexicana complex subspecies exist in the Americas.

Obviously, since each subspecies has a different epidemiological cycle with different vector phlebotomines, animal reservoirs, and clinical manifestations, space will not permit a discussion of some fifteen entities. In some cases little is known. It appears clear, however, that visceral leishmaniasis (Ldc) and mucocutaneous leishmaniasis caused by Lbb are the most important human infections in the New World. The rest of our discussion of New World leishmaniasis will be restricted to these diseases.

Only *Lutzomyia longipalpis* sandflies are known to transmit Ldc, and when epidemics occur the Brazilian health authorities use peridomiciliary dichloro-diphenyl-trichloro-ethane (DDT) spraying to control this vector as well as the identification and destruction of infected dogs. There is no similar national programme for tegumentary leishmaniasis, as doubt still exists as to the value of these measures. Often human infections are endemic and sporadic, and treatment of individual cases is preferred. For Lbb infections, this is also the case. Frequently patients will use topical treatment with folk remedies. Since the tendency is for the skin ulcers to heal with time, a large number of medicaments have their advocates as a cure for cutaneous leishmaniasis.

Studies in both visceral and cutaneous leishmaniasis show that the most reliable drugs are still the pentavalent antimonials. Two of the these appeared on the market shortly after the Second World War, namely, sodium stibogluconate (Pentostam) and N-methyl glucamine (Glucantime). Only the latter is available in South America, and it is bought by the ministry of health in Brazil for distribution to physicians needing to treat patients. Without this assistance the situation would be more difficult since patients can rarely afford to buy more than a few ampoules, and high doses may be required for cure. The problems of giving daily parenteral therapy over a period of weeks of potentially toxic drugs under field conditions in a health post cannot be exaggerated.

There is no immediate prospect that these old toxic heavy metal drugs will be replaced by simpler oral therapy. In fact the chief current suggestion is to combine allopurinol with antimonials in the hope of reducing treatment time and achieving better results with this combination in intractable forms. (Our recent results with this combination in mucosal leishmaniasis have been poor.) Allopurinol alone has been disappointing. While some workers are optimistic that alternative therapy will be found (Gutteridge, 1987), it is the author's view that nothing new will be ready for Lbb in the next decade. As a result therapeutic aspects of pentavalent antimonials have been recently updated (Marsden, 1985). The reason for this view is that even if a strong drug candidate for all forms of leishmaniasis were available, it would take about a decade to establish its position by human trials. First trials have to be done in visceral leishmaniasis, where a whole series of parameters can be used to evaluate the drug's effect. Next, it has to be tried on a nonmetastasising form of cutaneous leishmaniasis. Then in a controlled hospital trial it must be evaluated in Lbb cutaneous and mucosal cases. Finally, if it is promising, a field trial can be undertaken. Each one of these trials takes years to complete. Also, there are many problems in the design of these trials. A major problem in evaluating tegumentary leishmaniasis is that it must be a form where parasites are relatively easy to recover (e.g., Lbg) so that a parasitological criterion of cure can be used, since clinical criteria tend to be subjective. Also, a double-blind trial is preferred. Further, since virtually all types of tegumentary leishmaniasis show a tendency to heal spontaneously over a variable time frame, the numbers in trial groups will have to be relatively large. This latter observation suggests a focus for immunologists: Why do leishmanial granulomata pursue this variable course, and why is the immunological control of such events as necrosis, granuloma formation, and fibrosis so unpredictable? It is strange that the immunological response in Lbb is usually so marked and yet healing is so slow with or without chemotherapy. Other unsolved problems include an explanation of the relatively rare occurrence of mucosal metastases and why immunological paralysis producing anergic hansenoid leishmaniasis (Convit's syndrome) is only associated with members of the Mexicana group.

As Gutteridge (1987) points out, there are two other possibilities other than allopurinol—namely, an eight amino quinoline (which could have toxicity problems) and liposome-linked antimony, which still has a whole series of developmental problems (Alving, 1986). Both of these are likely to be effective only in visceral leishmaniasis. Patients with espundia continue to take massive doses of antimonials in the hope of a cure before the clinician resorts to amphotericin B or pentamidine, the second-line drugs. The literature continues to be filled with uncontrolled trials of any drug used for parasitic disease. Not even Pentostam and Glucantime have been adequately compared in a human trial. The manufacturers have difficulties in standardizing these drugs and would probably drop them if they were not under pressure from the medical community to continue producing them.

We have recently been involved in an epidemic of human Lbb infections that quite swamped our small field clinic facility. It occurred in an area where no forest had been cleared and affected man, dogs, and equine species. It has been suggested that the latter could have brought the infection into the area. House spraying with DDT appeared to have little effect, although the peak of the epidemic had past. The principal vector of Lbb in the area, *Lutzomyia whitmani*, dominates in the peridomicile. The epidemic disappeared as mysteriously as it came. Glucantime was in such short supply that patients in general received much lower doses than we recommend, but eventually their ulcers closed. They are being followed-up to see if relapses occur. The Brazilian Ministry of Health has no procedure to control epidemics of cutaneous leishmaniasis apart from the use of Glucantime.

Children with kala azar are usually debilitated and immunosuppressed, and the use of antimonial drugs has to be carefully monitored. There is a case for not admitting the child to hospital, where the risk of secondary infections is great, but, rather, trying to treat the child at home. Fortunately, Brazilian kala azar usually responds well to antimony therapy, unlike its Kenyan equivalent.

## OTHER CONSIDERATIONS IN SOUTH AMERICAN LEISHMANIASIS

As regards kala azar a group of infected children without classical disease have been recognised in an endemic area by means of ELISA serology. Follow-up suggested that those going on to develop classical kala azar were those with some degree of malnutrition. Serology is also of diagnostic value in Lbb infections using either immunofluorescence (IFA) or ELISA. The prognostic value of serology in both Ldc and Lbb is still under investigation.

The leishmanin skin test is usually positive in Lbb infections, but negative in kala azar, only turning positive after treatment. The reason for the emphasis on these immunological diagnostic aids is that, in contrast to Ldc, Lbb is very difficult to isolate and characterise. Indeed, this must influence attempts to map the distribution of various leishmanial species since Lbb will tend to be underrepresented. Attempts continue to develop a more practical test using monoclonal antibodies to identify Lbb parasites in primary biopsy material. DNA hybridisation of tissue smears does not work for Lbb due to the low number of organisms, but more sensitive techniques are being developed. At present, by the time the clinician receives the parasite-identification result the patient has been treated and has gone home. The clinician must therefore establish the common parasites in his area and treat by inference. Since we believe that mexicana group lesions and even braziliensis group infections other than Lbb require much less antimony, it would be an advantage to know the causative organism at the time of consultation. Thus, although new molecular taxonomic methods are helping to unravel the complexities of subspecies geographical distribution, they only help the individual patient by providing clues as to frequency in the area of certain organisms. In actual practice an experienced clinician can detect leishmaniasis in the great majority of cases without any complex tests.

With such unsatisfactory chemotherapy results, forms of control other than the treatment of human cases must be briefly considered (Marsden, 1984b). Outbreaks of kala azar have followed cessation of house spraying with residual insecticides for malaria, but this is of doubtful value in Lbb. Here identification of infected domestic animals and their destruction or treatment could have some effect. Vaccination with live promastigote vaccines has been used in Israel, Russia, and Iran against Old World leishmaniasis. An ill-defined dead polyvalent vaccine has been tried in man in Brazil, but the results are inconclusive. It is probable that animal studies will be funded to try and define cross-protection between various leishmanial species in the New World and determine whether live or dead immunogens induce good protection against Ldc or Lbb. If such vaccines were developed, however, there might be real practical problems in their use, since the sporadic, endemic nature of these human infections would make it difficult to define the population that should be vaccinated.

## FINAL CONSIDERATIONS

These two great endemic disease problems are good examples of poor man's medicine. The poor, with their low socioeconomic status and malnutrition, are particularly prone to infections. There is also poverty in terms of the number of biomedical personnel involved in trying to find solutions. Note that our brief review regarding these diseases is only concerned with diagnosis, treatment, and control, since these problems require urgent solutions. The primary message is also clear: concentrate on Chagas' disease vector control and find a cheap oral drug to treat leishmaniasis. A vaccine is not the top priority in either case. The principal conclusion from this brief discussion is that such protozoal diseases as those discussed here deserve more attention. Help has come from an unexpected quarter, namely, the scientific community of immunologists and molecular biologists. For various reasons trypanosomes and leishmania are attractive models, and many grant requests today begin with general assertions of the importance of these diseases in global terms before passing on to the specific proposal. Some of these proposals in the future must be geared to more practical considerations such as those discussed here if solutions are to be forthcoming for these endemic diseases.

There is no doubt that the scientific community will respond, as is illustrated by the course at Woods Hole, Massachusetts. Often it is beneficial for the worker to see the disease at first hand. For example, our field projects in the University of Brasília are constantly being visited by scientists for this purpose.

Certainly, in both diseases there have been failures by the people responsible for the distribution of the very limited funds available; otherwise, we would not be in the situation we are in today. One of these failures relates simply to lack of understanding of the problem. (Administrators of funding agencies rarely leave urban hotels and, perhaps most important, rarely speak to practicing clinicians concerned with day-to-day problems; indeed, clinicians concerned with endemic tropical disease are a dying breed.) For example, people with a good clinical grasp of the problem would never have let the research on a possible schistosomal vaccine get so out of hand. Scientifically it is a fascinating problem, but Grodzinski (1987) is right—we are never going to use all those gene libraries. Also, the drugs available today are excellent. Targeted mass chemotherapy aimed at the risk groups who develop serious complications is the correct policy. The history of schistosomiasis research in the last 25 years deserves to be recorded. It involves the decision by the Rockefeller Foundation not to withdraw from a disease linked to irrigation at a time when they practically withdrew support from all the other great endemic disease research and also the successful trajectory of several medical popularisers who got the ear of certain granting agencies. The actual answers came from research on the old hit-and-miss basis by pharmaceutical companies: a fungus got into a vat containing Lucanthone, and produced Hycanthone, from which came Oxamniquine. An Egyptian thought the organophosphorus insecticide protecting his fruit trees might work against *Schistosoma haematobium*, and it did, leading to metrifonate. Praziquantel was used in veterinary medicine for years before the powers-that-be would prepare it for human use. This last drug is a good example of how veterinary drugs aid chemotherapy of human helminthic disease; Ivermectin (in filariasis) is another. There are no such parallels in protozoal infections.

Most drug development to date has been along rather empirical lines, based on either clinical observations of drug effect or behavior in animal screens. While chemists are good at modifying molecules, there are still few examples where a knowledge of parasite metabolism led to effective drug synthesis, although there is hope for the future in this regard. A few American soldiers got leishmaniasis, which led to a rescreening of part of the immense library of antimalarial compounds for antileishmanial activity with some success. However, the animal screen tested mainly for visceral leish-maniasis, although cutaneous disease is much more likely to infect soldiers in the field in Latin America.

The author must plead guilty to recommending unqualified support for malaria vaccine some years ago. It seemed at the time the test case for a vaccine for protozoal disease in a field where control options were diminishing yearly. In spite of spectacular research the actual use of such a vaccine for needy African families seems remote (Bruce Chwatt, 1987). Everybody knew it wasn't being developed for such people anyway. What is already happening in applied malaria research is a return to reevaluating old control methods, such as bed nets and screens and attack of larval breeding sites (Anonymous, 1984). After all, it was these old methods that eradicated *Anopheles gambiae* from northeast Brazil and enabled the building of the Panama canal. Also, the use of chemoprophylaxis is again commanding attention. Incredibly, there are doctors in Amazonas who recommend nothing when the value of a weekly dose of chloroquine is undisputed in an area where much vivax transmission occurs and not *every* strain of *Plasmodium falciparum* is drug resistant. A certain slackness has crept into malaria control, born of the concept that the problem can be solved by mass insecticide spraying and/or mass chemotherapy. A more pragmatic approach is necessary with an adequate educational component to avoid what is often preventable mortality.

With both diseases considered here attempts to stimulate interest in developing drugs among pharmaceutical companies have been unsuccessful. Good reasons exist for this lack of interest (Hoekenga, 1983). Developing countries with these problems don't have the scientific muscle to promote drug development. Considering the panorama of successes in the chemotherapy of infectious disease, we evidently need a larger number of biomedical research workers looking at the problem. For any such worker interested in this field it is a race against time to try and prevent the young generations growing up in endemic areas from becoming infected and developing serious complications.

The resources necessary to develop a drug for human use are quite beyond universities and research institutes. Neither should drug companies be criticised, since economic considerations must come into their deliberations and there is little profit in developing a drug either for *T. cruzi* or *Leishmania*. For instance, the author has it on good authority that Bayer never recovered the development costs of Nifurtimox, and global sales of a pentavalent antimonial hardly justify its manufacturing problems. Perhaps a solution could be found in tax relief. Certainly, a solution has to be found, since better chemotherapy is the principal need in relation to the infections discussed.

Brazil's current example of a Chagas' disease control programme has to be copied by other South American countries. It is an investment for the future as well as a source of national pride. Much nationalism is associated with research on endemic disease, most of it misguided. For me it doesn't matter if a Hottentot research worker discovers the treatment answer for Bolivian children infected with *T. cruzi*. Unfortunately, a united global scientific community telling the politicians what to do will always remain a pipe dream. Meanwhile countries of Latin America struggle under a burden of endemic protozoal disease that the United States has never known.

## REFERENCES

Alving CR (1986): Liposomes as drug carriers in leishmaniasis and malaria. Parasitol Today 2:101–107.

Anonymous (1984): Malaria Control as Part of Primary Health Care. WHO Tech Report Series 712.

Badaró R, Jones TC, Lorenço R, Cerf BJ, Sampaio D, Carvalho EM, Rocha H, Teixeira R, Johnson WD (1986): A prospective study of visceral leishmaniasis in an endemic area of Brazil. J Infect Dis 154:639–649.

Bruce Chwatt LJ (1987): The challenge of malaria vaccine. Trials and tribulations. Lancet 1:371–373.

Camargo ME, Silva GR, Castilho EA, Silveira AC (1984): Inquérito sorológico da prevalência de infecção chagásica no Brasil 1975/1980. Rev Inst Med Trop São Paulo 26:192–204.

Dias JCP (1986): The programme of Chagas' disease control in Brazil for 1986. Rev Soc Bras Med Trop 19:129–133.

Garcia-Zapata MTA, Marsden PD (1986): Chagas' disease. In H. Gillies (ed): Clinics in Tropical Medicine and Communicable Disease. London: Saunders Co., Vol 1, no 3, pp 557–585.

Garcia-Zapata MT, Marsden PD, Virgens D, Penna R, Soares V, Brasil IA, Castro CN, Prata A, Macêdo V (1986): O controle da transmissão da doença de Chagas em Mambaí, Goiás-Brasil (1982–1984). Rev Soc Bras Med Trop 19:219–225.

Grimaldi G, David JR, McMahon Pratt D (1987): Identification and distribution of new world Leishmania species characterised by serodeme analysis using monoclonal antibodies. Am J Trop Med Hyg 36:270–287.

Grodzinski L (1987): Too many gene libraries. Parasitol Today 3:120–121.

Gutteridge WE (1987): New anti-protozoal agents. Int Parasitol 17:121–129.

Hoekenga MT (1983): The role of pharmaceuticals in the total health care of developing countries. Am J Trop Med Hyg 32:437–446.

Hoff R, Todd CW, Maguire JH, Piesman J, Mott KE, Mota EE, Sleigh A, Sherlock IA, Weller TH (1985): Serologic surveillance of Chagas' disease. Ann Soc Belg Med Trop 65(Suppl l):187–196.

Jones TC, Johnson WE, Jr, Barretto AC, Lago E, Badaró R, Cerf B, Reed SG, Netto EM, Tada MS, França F, Weise K, Golightly L, Fikrig E, Costa JML, Cuba CC, Marsden PD: Epidemiology of American cutaneous leishmaniasis due to *Leishmania braziliensis braziliensis*. J Infect Dis (published).

Marr and Do Campo (1987): Rev Infect Dis.

Marsden PD (1984a): Chagas' disease clinical aspects. In H.M. Gilles (ed): Recent Advances in Tropical Medicine. Edinburgh: Churchill Livingstone, pp 63–78.

Marsden PD (1984b): Strategies for control of disease in the developing world. XIV. Leishmaniasis. Rev Infect Dis 6:736–744.

Marsden PD (1984c): Strategies for control of disease in the developing world. XVI. Chagas' disease. Rev Infect Dis 6:855–865.

Marsden PD (1985): Pentavalent antimonials: Old drugs for new diseases. Rev Soc Bras Med Trop 18:187–198.

Marsden PD (1986): Mucosal leishmaniasis (Espundia, Escomel 1911). Trans R Soc Trop Med Hyg 80:859–876.

Marsden PD, Jones TC (1986): Clinical manifestations, diagnosis and treatment of leishmaniasis. In K.P. Chang and P.S. Bray (eds): Leishmaniasis: Amsterdam: Elsevier, pp

McGreevy P, Marsden PD (1986): American trypanosomiasis and leishmaniasis. In W.C. Campbell and R.S. Rew (eds): Chemotherapy of Parasitic Diseases. Plenum Publishing Co., pp 115–127.

Schofield CJ (1985): Control of Chagas' disease vectors. Br Med Bull 41:187–194.

The Biology of Parasitism, pages 93–103
© 1988 Alan R. Liss, Inc.

# Developmental Biology of *Leishmania* Promastigotes

## David L. Sacks

*Laboratory of Parasitic Diseases, National Institute of Allergy and Infectious Diseases, National Institutes of Health, Bethesda, Maryland 20892*

## INTRODUCTION

*Leishmania* species are known to multiply as intracellular amastigotes in macrophages of their vertebrate hosts and as extracellular promastigotes in the midgut of their sandfly vector. An additional aspect of the developmental cycle of *Leishmania* parasites has only recently been emphasized. This concerns the sequential development of sandfly promastigotes from a noninfective to infective stage. This critical developmental stage was not generally appreciated before because, unlike the infective invertebrate stages of other hemoflagellates, it does not appear to have a readily distinguishable morphological identity. This paper reviews the evidence demonstrating the existence of infective-stage or metacyclic promastigotes and describes recent studies on the molecular characterization of these developmental forms.

## DEMONSTRATION OF INFECTIVE-STAGE PROMASTIGOTES WITHIN THE PHLEBOTOMINE MIDGUT

The first experimental evidence of transmission of leishmaniasis by the bite of sandflies was by Shortt et al. in 1931, who observed transmission of *L. donovani* to hamsters by the bite of *Phlebotomus argentipes*. The first transmission of leishmaniasis to man by sandflies was proven by Adler and Ber in 1941. Both investigators believed that there was one particular form of promastigote in the sandfly which is adapted for life in the vertebrate. Morphological differences between dividing midgut forms and those found anteriorly have been described (Killick-Kendrick, 1979), but until recently there has been no evidence that these changes reflect development of promastigotes into an infective stage. By comparing directly the infectivity of promastigotes as they developed temporally within the fly, we were able to demonstrate that promastigotes undergo sequential development from a noninfective to an infective stage during growth within the midgut (Sacks and Perkins, 1984; Sacks and Perkins, 1985). Clones of *L. major* and *L. mexicana* promastigotes that were recovered from midguts of *P. papatasi* and *L. longipalpis*, respectively, 3 days after fly infection were avirulent for BALB/c mice, whereas midgut promastigotes that were recovered on days 4–7 after infection were progressively more virulent. The generation of optimally infective parasite populations shortly after bloodmeal passage coincides with the time at which another meal is sought by the fly. The generation of infective promastigotes during, and indeed as a consequence of, their early growth within the midgut clearly means that they are infective prior to their invasion of the mouthparts. This means that the low numbers of highly characteristic forms found within the mouthparts are not necessarily required for transmission, and that if the more abundant parasites found posterior to the mouthparts can be regurgitated forward and deposited on the skin during feeding, then this would be a highly effective means of transmission, since these populations will contain infective stage promastigotes.

We (da Silva and Sacks, 1987) and others (Killick-Kendrick, 1979) have chosen to term this infective promastigote stage "metacyclic" promastigotes, by analogy with the term used to denote the infective invertebrate stages of other hemoflagellates.

## DEMONSTRATION OF METACYCLIC PROMASTIGOTES DEVELOPING WITHIN AXENIC CULTURE

There is considerable evidence that identical developmental events accompany the growth of promastigotes within axenic culture. In early studies, investigators found a correlation between the age of *L. donovani* in culture and their infectivity for experimental animals (Giannini, 1974; Keithly, 1976). We found more recently that cloned promastigotes of *L. major* taken from logarithmic (log) phase cultures were unable to establish intracellular growth within mouse peritoneal macrophages in vitro, and they were relatively avirulent for normally susceptible BALB/c mice. As cultures approached the stationary phase, the virulence behavior of the promastigotes progressively increased, both in vitro and in vivo (Sacks and Perkins, 1984; Sacks et al., 1985). The discovery that both noninfective and infective stages of *Leishmania* promastigotes could be found within cultured populations was an important step toward identifying those specific developmental changes that are associated with the ability of these parasites to adapt to and survive within the vertebrate host, since sufficient quantities of each stage could be obtained for antigenic, biochemical, and genetic analyses.

## IDENTIFICATION OF DEVELOPMENTALLY REGULATED MOLECULES EXPRESSED BY METACYCLIC *L. MAJOR* PROMASTIGOTES

While the metacyclogenesis of *Leishmania* promastigotes is not accompanied by an obvious morphological change, we have been able to identify a biochemical surface marker for metacyclic promastigotes of *L. major* (Sacks et al., 1985). These organisms fail to be agglutinated by the lectin peanut agglutinin (PNA) at concentrations which agglutinate all organisms from log phase cultures as well as all noninfective promastigotes remaining within stationary phase cultures (Fig. 1). These results indicate that metacyclogenesis involves changes in surface carbohydrates and the lectin, which has specificity for d-galactose-containing sugars, can be used to both enumerate and purify metacyclic *L. major* promastigotes from culture. Figure 2 shows a light micrograph of *L. major* promastigotes separated on the basis of PNA agglutination. Metacyclics are extremely uniform in shape and size, slender with elongated flagella approximately twice the length of the cell body, and highly motile.

By immunizing mice with purified metacyclics, we have raised a monoclonal antibody (3F12) recognizing a molecule which is expressed on the surface of, and released by, metacyclic promastigotes and not promastigotes from log phase cultures (Sacks and da Silva, 1987). The molecule can be metabolically labeled with [$^{14}$C]-glucose, [$^3$H]-mannose, [$^3$H]-galactose, and [$^3$H]-palmitic acid, but not with [$^{35}$S]-methionine or [$^3$H]-leucine. In addition, the molecule is the major species surface labeled by [$^3$H]-sodium borohydride

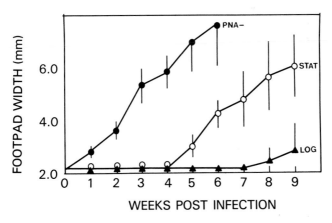

Fig. 1. Infectivity of log phase, stationary phase, and PNA$-$ purified *L. major* promastigotes for BALB/c mice.

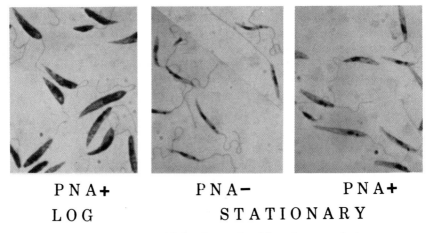

P N A +       P N A –       P N A +

L O G          S T A T I O N A R Y

Fig. 2.   Giemsa-stained light micrographs of *L. major* promastigotes.

following treatment with periodate or galactose oxidase. In all these respects, the molecule is similar to the previously described *L. major* glycolipid (Handman et al., 1984) also referred to as excretory factor (EF), which is the major surface and released glycoconjugate of these cells; however, the glycolipid recognized by 3F12 can be distinguished from the molecule previously described because its migration on SDS-polyacrylamide gel electrophoresis, while still heterodispersed, is of higher relative molecular weight. A close relationship between the two molecules is indicated by the finding that another monoclonal antibody, WIC-79.3 (de Ibarra et al., 1982; Handman et al., 1984) recognizes both forms of the glycolipid—one produced and released only by log phase or noninfective promastigotes, and one produced and released only by metacyclic promastigotes generated within stationary phase cultures. The two developmental forms of the glycolipid, surface labeled using [$^3$H]-borohydride, are shown in Figure 3. We have determined that PNA binds to the log form of the glycolipid, and that the loss of PNA binding during growth appears to be due to a glycosylation event, for which the 3F12 antibody has specificity, which modifies the log form of the glycolipid such that it is no longer able to bind the lectin. The possibility that the molecule becomes sialated is suggested by the sensitivity of the metacyclic form of the glycolipid to 1 mM periodate (Fig. 3) and to acid hydrolysis.

A survey of a number of virulent and avirulent *L. major* strains and clones established an absolute association between the ability of promastigotes to initiate infection in BALB/c mice and their expression and release of the

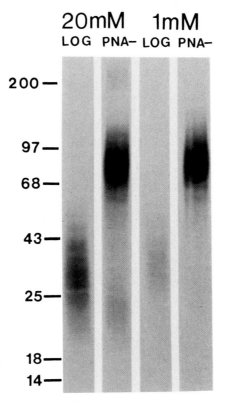

Fig. 3. SDS-PAGE analysis of [³H]-borohydride surface-labeled log phase and PNA−
promastigotes following treatment with 20 mM and 1 mM sodium periodate.

developmentally regulated, 3F12 binding form of the glycolipid. It will be of
interest to determine whether the lipid containing glycoconjugates which are
produced by other *Leishmania* species (Kaneshiro et al., 1982; Handman et
al., 1984) also display developmental polymorphisms. In this regard, it has
recently been reported by King et al. (1987) that the lipophosphoglycan of *L.
donovani* is expressed primarily by stationary as opposed to log phase
promastigotes. Metacyclogenesis is almost certainly associated with other
molecular events. Enhancement and modification of the leishmanial surface
protease, gp63, has already been described (Kweider et al., 1987). The
expression of a stage-specific glycosyl transferase would also be predicted.
Nonetheless, the glycolipid is the major surface and released glycoconjugate

of these cells, and its ability to influence infectivity has already been established (Handman and Goding, 1985; Handman et al., 1986). The demonstration that a unique form of the glycolipid is expressed during metacyclogenesis might provide a molecular clue as to the adaptive mechanisms of promastigote survival within the vertebrate host.

## CELLULAR AND MOLECULAR COMPONENTS OF METACYCLIC PROMASTIGOTE SURVIVAL

In order for the *Leishmania* promastigotes which are inoculated by the sandfly to initiate infection, they must survive microbicidal defense mechanisms to which they are exposed within even nonimmune vertebrate hosts. These defenses include the lytic effects of alternative complement pathway activation and the oxygen-dependent and -independent leishmanicidal activities of their host macrophages. Metacyclic promastigotes appear to be uniquely adapted to survive both of these killing mechanisms.

Increased resistance of stationary phase promastigotes of a number of *Leishmania* species to killing by fresh normal sera has been reported by Franke et al. (1985). Whereas 100% of log phase promastigotes of *L. b. panamensis* and *L. donovani* were killed by as low as 2% normal human serum, a subpopulation of stationary-phase promastigotes survived incubation in undiluted serum. The serum resistance of purified metacyclic promastigotes of *L. major* was not so absolute, but could be demonstrated in up to 20% human serum. The lysis of serum sensitive promastigotes was due to complement activation, because it was inhibited in heat-inactivated and EDTA-chelated serum. Surprisingly, recent studies by Puentes et al. (1988) suggest that serum resistance in metacyclic *L. major* promastigotes does not simply reflect inefficient complement activation, since both developmental forms bound radiolabeled C3 after incubation in serum, with approximately 80% of the C3 deposited as hemolytically active C3b in each case. An intriguing observation was that a major portion of deposited C3b on metacyclics was not covalently bound. This difference might be related to the developmental modification of the glycolipid, since this surface molecule was found to be the predominant acceptor site of C3 deposition on both log and metacyclic promastigotes. How these molecular differences might explain serum resistance is not clear, but remains an important area of study. Resistance to complement-mediated lysis in the face of efficient complement activation might seem rather paradoxical, but it begins to make sense in the light of the increasing evidence that complement activation and C3 deposition are critical to leishmanial intracellular infection and survival.

The ability of *L. major* metacyclic promastigotes to initiate intracellular infections within mouse peritoneal macrophages *in vitro* has already been mentioned. The inability of log phase promastigotes to do so is not a consequence of their failure to attach to and gain entrance into the macrophage, but rather their inability to survive those microbicidal events which are associated with the infection process. Toxic oxygen metabolites, including superoxide anion, hydrogen peroxide, and hydroxyl radicals generated by activated phagocytic cells, are able to inactivate *Leishmania* promastigotes (Murray, 1981; Reiner and Kazura, 1982). We have not found any difference in the sensitivity of metacyclic promastigotes or their log phase counterparts to killing by $H_2O_2$ generated in a cell-free system. We have found, however, that metacyclic promastigotes trigger a minimal respiratory burst upon infection, even at high parasite-to-macrophage ratios, whereas log phase parasites trigger a substantial burst (da Silva et al., manuscript in preparation). Similar differences were observed by Murray (1982) and Channon et al. (1984) when *L. donovani* promastigotes (presumably log phase) and amastigotes were compared. These differences in cellular activation may explain in part the basis of metacyclic promastigote persistence within certain macrophage populations.

Since cellular activation can be associated with receptor-mediated endocytosis (Yamamoto and Johnston, 1984; Berton and Gordon, 1983), we have been exploring the possibility that the two promastigote developmental stages use different receptors for their attachment and uptake and that the use of different receptors influences their subsequent intracellular fate. We have focused our attention on receptors for the third component of complement, since one of these receptors (CR3) has already been implicated as being important for attachment of *L. donovani* and *L. major* promastigotes to mouse macrophages (Blackwell et al., 1985; Mosser and Edelson, 1985). By plating mouse or human macrophages on surfaces which have been precoated with monoclonal antibodies directed against C3 receptors, these receptors can be modulated to the underside of the cell and made unavailable for binding of their specific ligands. In this way, we have determined that the human CR1 receptor, which has highest affinity for C3b, is the major receptor used by metacyclic *L. major* promastigotes, since modulation of this receptor inhibited attachment over a 40-min incubation by 50–80% (da Silva, manuscript in preparation). In contrast, the attachment of log phase promastigotes to human macrophages could not be inhibited by CR1 modulation except within the first 5 min, indicating use of alternative receptors. These results were the same even when parasites were opsonized with C3 by serum incubation prior to infection. While opsonization dramatically enhanced attachment and uptake of both developmental stages, CR1 modulation still only inhibited attachment and uptake of metacyclics. Furthermore, the enhanced uptake of metacyclics was still associated with only minimal cellular activation as measured by $H_2O_2$ production, consistent with previous reports that receptor mediated endocytosis via CRl fails to trigger activation pathways (Wright and Silverstein, 1983).

The unique association of metacyclic promastigotes with CR1 may itself be a consequence of the developmental modification of the lipid containing surface glycoconjugate since, as mentioned, this molecule is the major acceptor for covalent C3b attachment and in the case of metacyclics, noncovalent C3b attachment to the parasite surface. This molecule might, therefore, influence both serum resistance and intracellular survival.

## METACYCLOGENESIS AND *LEISHMANIA* VIRULENCE

The observation that promastigote populations are not necessarily uniform with respect to infectivity means not only that their heterogeneity must be taken into account in molecular and biochemical studies, as has been just described, but also in comparative studies of *Leishmania* virulence. Differences in virulence between leishmanial promastigote strains and clones have been repeatedly described (Dawidowicz et al., 1975; Ayesta et al., 1985; Handman et al., 1983). These differences become difficult to interpret, since until recently metacyclic promastigotes could not be distinguished from noninfective organisms and, therefore, the size of the "effective" inocula may not have been comparable. This point is clearly demonstrated in recent studies in which we have used the loss of agglutination with PNA as a phenotypic marker for metacyclic promastigotes of *L. major* in order to compare the degree to which metacyclogenesis occurred for different strains and clones during growth. We found that there was considerable variation in the efficiency of metacyclogenesis between different strains and clones, and even within the same clone when promastigotes propogated for different lengths of time in culture were compared (da Silva and Sacks, 1987). For example, a previously described virulent clone (Handman et al., 1983) generated 20–30% metacyclics during stationary growth, in contrast to a reportedly avirulent clone for which, on closer examination, we found metacyclogenesis to be extremely inefficient (less than 1%) and delayed until very late stationary cultures. Those metacyclics which could be recovered from these aging cultures were virulent for BALB/c mice. The loss of virulence associated with frequent subculture could also be attributed to a drastic diminution in metacyclogenesis potential over time. A cloned stock of *L.*

*major* which yielded approximately 50% metacyclics during growth within its first passage, by the 100th passage generated fewer than 10% metacyclics during growth (Fig. 4). Thus, metacyclogenesis does not appear to be stable for even cloned lines of *Leishmania* promastigotes, and virulence comparisons between different strains and clones can only be meaningfully made if the metacyclic populations contained within the respective inocula are determined.

## CONCLUDING REMARKS

*Leishmania* promastigotes can be shown to undergo sequential development from a noninfective to an infective stage during growth within the sandfly midgut and within axenic culture. Metacyclogenesis is, therefore, as fundamental a feature of the *Leishmania* life cycle as it is of the life cycle of other hemoflagellates, (e.g., African trypanosomes and *Trypanosoma cruzi*), for which differentiation of invertebrate developmental stages to infective

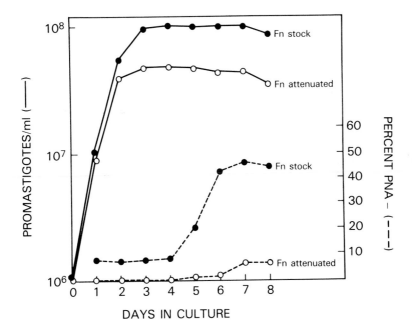

Fig. 4. Comparison of promastigote growth and the percent PNA− promastigotes generated during growth of a cloned virulent stock of *L. major* (Fn strain, NIH) prior to and after attenuation by repeated subculture.

forms is a readily distinguishable event. Metacyclic *Leishmania* promastigotes appear to be *preadapted* to vertebrate survival because, in the absence of vertebrate environmental influences, they demonstrate profoundly enhanced resistance to potent microbicidal mechanisms to which they will become exposed within even nonimmune vertebrate hosts. The signals for these adaptive changes are not understood and may be extremely complex. We have focused attention, instead, on the molecular nature of the changes themselves. We have identified developmentally regulated surface molecules and have begun to study their role in promoting successful parasitism.

## REFERENCES

Adler S, Ber M (1941): The transmission of *Leishmania tropica* by the bite of *Phlebotomus papatasii*. Indian J Med Res 29:803–909.

Ayesta C, Arguello C, Hernandez AG (1985): *Leishmania braziliensis*: Cell surface differences in promastigotes of pathogenic and nonpathogenic strains. Exp Parasitol 59:185–191.

Berton G, Gordon S (1983): Modulation of macrophage mannosyl-specific receptors by cultivation on immobilized zymosan. Effects on superoxide-anion release and phagocytosis. Immunology 4:705–715.

Blackwell JM, Ezekowitz RAB, Roberts MB, Channon JY, Sim RB, Gordon S (1985): Macrophage complement and lectin-like receptors bind *Leishmania* in the absence of serum. J Exp Med 162:324–335.

Channon JY, Roberts MB, Blackwell JM (1984): A study of the differential respiratory burst activity elicited by promastigotes and amastigotes of *Leishmania donovani* in murine resident peritoneal macrophages. Immunology 53:345–355.

da Silva R, Sacks DL (1987): Metacyclogenesis is a major determinant of *Leishmania* promastigote virulence and attenuation. Infect Immun 55:2802–2807.

Dawidowicz K, Hernandes AG, Infante RB, Convit J (1975): The surface membrane of *Leishmania*. 1. The effect of lectins on different stages of *Leishmania braziliensis*. J Parasitol 61:950–953.

de Ibarra AL, Howard JG, Snary D (1982): Monoclonal antibodies of *Leishmania major*: Specificities and antigen location. Parasitology 85:523–531.

Franke ED, McGreevy PB, Katz SP, Sacks DL (1985): Growth cycle dependent generation of complement resistant *Leishmania* promastigotes. J Immunol 134:2713–2718.

Giannini MS (1974): Effects of promastigote growth phase, frequency of subculture, and host age on promastigote-initiated infections in *Leishmania donovani* in the golden hamster. J Protozool 21:521–527.

Handman E, Goding JW (1985): The *Leishmania* receptor for macrophages is a lipid-containing glycoconjugate. EMBO J 4:329–338.

Handman E, Greenblatt CL, Goding JW (1984): An amphipathic sulphated glycoconjugate of *Leishmania*: Characterization with monoclonal antibodies. EMBO J 3:2301–2306.

Handman E, Hocking RE, Mitchell GF, Spithill TW (1983): Isolation and characterization of infective and non-infective clones of *Leishmania tropica*. Mol Biochem Parasitol 7:111–126.

Handman E, Schnur LF, Spithill TW, Mitchell GF (1986): Passive transfer of *Leishmania* lipopolysaccharide confers parasite survival in macrophages. J Immunol 137:3608–3615.

Kaneshiro ES, Gottlieb M, Dwyer DM (1982): Cell surface origin of antigens shed by *Leishmania donovani* during growth in axenic culture. Infect Immun 37:558–568.

Keithly JS (1976): Infectivity of *Leishmania donovani* amastigotes and promastigotes for golden hamsters. J Protozool 23:244–248.

Killick-Kendrick R (1979): Biology of *Leishmania* in phlebotomine sand flies. In W.H.R. Lumsden and D.A. Evans (eds): Biology of the Kinetoplastida, Vol 2. London: Academic Press, pp 395–460.

King DL, Chang Y, Turco SJ (1987): Cell surface lipophosphoglycan of *Leishmania donovani*. Mol Biochem Parasitol 24:47–54.

Kweider M, Lemesre J, Darcy F, Kusnierz JP, Capron A, Santoro F (1987): Infectivity of *Leishmania braziliensis* promastigotes is dependent on the increasing expression of a 65,000 dalton surface antigen. J Immunol 138:299–305.

Mosser DM, Edelson PJ (1985): The mouse macrophage receptor for C3bi (CR3) is a major mechanism in the phagocytosis of *Leishmania* promastigotes. J Immunol 135:2785–2789.

Murray HW (1981): Susceptibility of *Leishmania* to oxygen intermediates and killing by normal macrophages. J Exp Med 153:1302–1315.

Murray HW (1982): Cell-mediated immune response in experimental visceral leishmaniasis. II. Oxygen-dependent killing of intracellular *Leishmania donovani* amastigotes. J Immunol 129:351–359.

Puentes SM, Sacks DL, da Silva RP, Joiner KA (1988): Complement binding by two developmental stages of *Leishmania major* promastigotes varying in expression of a surface lipophosphoglycan. J Exp Med 167:887–902.

Reiner NE, Kazura JW (1982): Oxidant-mediated damage of *Leishmania donovani* promastigotes. Infect Immun 36:1023–1027.

Sacks DL, da Silva RP (1987): The generation of infective stage *Leishmania major* promastigotes is associated with the cell surface expression and release of a developmentally regulated glycolipid. J Immunol 139:3099–3106.

Sacks DL, Hieny S, Sher A (1985): Identification of cell surface carbohydrate and antigenic changes between noninfective and infective developmental stages of *Leishmania major* promastigotes. J Immunol 135:564–569.

Sacks DL, Perkins PV (1984): Identification of an infective stage of *Leishmania* promastigotes. Science 223:1417–1419.

Sacks DL, Perkins PV (1985): Development of infective stage *Leishmania* promastigotes within phlebotomine sandflies. Am J Trop Med Hyg 34:456–459.

Shortt HE, Smith ROA, Swaminath CS, Krishnan KV (1931): Transmission of Kala-azar by the bite of *Phlebotomus argentipes*. Indian J Med Res 18:1373–1375.

Wright SD, Silverstein SC (1983): Receptors for C3b and C3bi promote phagocytosis but not release of toxic oxygen from human phagocytes. J Exp Med 158:2016–2022.

Yamamoto K, Johnston RB (1984): Dissociation of phagocytosis from stimulation of the oxidative metabolic burst in macrophages. J Exp Med 159:405–416.

The Biology of Parasitism, pages 105–109
© 1988 Alan R. Liss, Inc.

# Does *Trypanosoma cruzi* Modulate Infection by Inherent Positive and Negative Control Mechanisms?

**Miercio E.A. Pereira**

*Tufts University School of Medicine, New England Medical Center Hospitals, Division of Geographic Medicine and Infectious Diseases, Boston, Massachusetts 02111*

One of the steps in the infection of mammalian hosts by several protozoa involves host cell–parasite binding as a prelude to parasite penetration into the host cell. Evidence has accumulated over the past decade or so to indicate that this initial contact of the protozoa with the respective host cell is mediated by specific recognition mechanisms. The indication is indirect and based largely on the inhibition of parasite binding and penetration by monosaccharides, glycoproteins, specific antibodies to parasite surface molecules, and enzymatic digestion of host and/or parasites. Thus, presently available evidence suggests that parasites developed mechanisms to facilitate infection by producing receptors, generally present on the outer membrane, that react with complementary ligands on the host cells. These parasite molecules therefore have a positive effect on infection, and they provide a rational basis for vaccine research, because blocking receptor binding activity should result in decreased infection. But while positive control of infection is attributed to the parasite, down regulation of infection is generally thought to be solely due to the host immune defense. Accordingly parasitism would be a consequence of an equilibrium between positive and negative influences from the parasite and the host, respectively (Bloom, 1979). In this paper, we put forward the hypothesis that the protozoa *Trypanosoma cruzi* evolved mechanisms to regulate infection at the cellular level through inherent positive and negative influences, and that the host may in turn disturb the influence by inhibiting the positive or negative parasite controls.

Inhibition of host cell infection by *T. cruzi* can be achieved with antibody to a surface glycoprotein of $M_r$ 85 kilodaltons (kDa) (GP-85) (Alves et al., 1986). This molecule is present exclusively on infective trypomastigotes and appears to be functionally similar to, if not identical with, the fibronectin receptor of *T. cruzi* which has also been found to be a surface molecule of $M_r$ 85 kD (Onaissi et al., 1986); antibody to the fibronectin receptor as well as the receptor itself inhibits infection of host cells in vitro (Onaissi et al., 1985). The gene encoding a trypomastigote-specific protein of *T. cruzi* of $M_r$ 85 kD has been cloned and found to contain a nonapeptide tandemly repeated several times (Peterson et al., 1986). Thus, since anti-GP-85 antibodies and GP-85 itself inhibit infection in vitro, one can say that this surface glycoprotein exerts a positive control on infection. Several parasite proteins that behave like *T. cruzi* GP-85 are known, the most notable of which being the circumsporozoite protein of malaria sporozoites (Yoshida et al., 1980).

We have discovered a developmentally regulated neuraminidase activity in *T. cruzi* (Pereira, 1983). The activity is predominantly found on infective trypomastigotes, is located on the outer membrane, and has an $M_r$ of 160K–200K. There are two lines of evidence suggesting that the trypomastigote neuraminidase controls infection by a negative mechanism. The first stems from the effect of antineuraminidase antibodies on the infection of host cells in vitro (Cavallesco and Pereira, 1988), in that such antibodies enhance, rather than inhibit, infection, as would be expected for an activity that has a negative control. The second piece of evidence is provided by the results with high-density lipoprotein (HDL), a specific inhibitor of the parasite neuraminidase (Prioli et al., 1987). HDL, at concentrations that inhibit the parasite enzyme, augments infection in a manner similar to the action of antineuraminidase antibodies. The enchancement of infection by the antibodies or by HDL was prevented by addition to the monolayers of exogenous neuraminidase, such as that from *Vibrio cholera*, whose activity is not inhibited by the two probes that affect the *T. cruzi* enzyme. Thus, although the *T. cruzi* neuraminidase is inhibited by specific antibodies or by HDL, infection is not augmented and remains at a basal level if neuraminidase activity (from *V. cholera* in the case of the experiment described above) is present during the infection. This result, therefore, suggests that sialylation determines the extent of infection and that the negative effect of neuraminidase on infection results from the ability of the enzyme to release free sialic acid from sialoglycoconjugates. If this is so, sialyltransferase, which has an effect opposite that of neuraminidase on sialylation should have a positive influence on infection. And recent results suggest an enhancement of *T. cruzi* infection by fetuin and other substrates of the sialyltransferase (Piras, 1987). The site of the functional sialic acid has not been determined, but it may be on the parasite itself, which is known to contain sialic acid (Pereira et al., 1980). It is interesting to note that not all trypomastigotes seem to express neuraminidase at a given time, as judged by the finding that the antineuraminidase antibodies recognize a subpopulation of 20–40% parasites as determined by immunofluorescence. Moreover, when trypomastigotes are treated with the antibodies and complement, the surviving parasites (60–80% of starting trypanosomes) have very low or no neuraminidase activity and exhibit an increased ability to infect cells in vitro (Cavallesco and Pereira, 1988). Thus, if the neuraminidase has a negative effect on infection, it must be exerted by a subset of trypomastigotes ($NA^+$). Whether or not $NA^+$ trypomastigotes are capable of infecting cells has not been determined, but they appear to decrease the infective property of the $NA^-$ parasites to culture cells in vitro. These findings predict that $NA^-$ trypomastigotes express GP-85 and other factors with a positive effect on infection.

Some circumstantial evidence strongly argues in favor of the hypothesis that the endogenous *T. cruzi* neuraminidase exerts a negative effect on infection. Thus, according to the hypothesis, strains with relatively high neuraminidase activity should be less virulent that strains with low enzyme activity; and this correlation has been confirmed experimentally (Pereira and Hoff, 1986). On the other hand, since HDL has been found to be a specific *T. cruzi* neuraminidase inhibitor, animals with relatively high concentrations of HDL should be more susceptible to infection than those with lower HDL content. Accordingly, $C_3H$, Balb/c, or other strains of mice which are highly susceptible to *T. cruzi* infection (Trischmann et al., 1978) have a higher HDL concentration than C57BL/6 (Paigen et al., 1985), a strain highly resistant to infection (Trischmann et al., 1978). Inhibition of neuraminidase activity paralleled the HDL concentrations in the plasma of the mice strain referred to above (Rosenberg et al., submitted). Furthermore, male rats are generally more susceptible to *T. cruzi* infection than their female counterparts, in accordance with their relative HDL concentration (Paigen et al., 1985).

The idea that *T. cruzi* and other parasites infect cells by inherent positive and negative mechanisms is just beginning to be exploited. For example, in *Leishmania major* infection in vitro, the parasite apparently enters the macrophages through complement receptors on the host cell which recognize the complement component C3 covalently bound to the parasite outer membrane (Wozengraft et al., 1986). It has been recently shown (Mosser and Edelson, 1987) that the success of *Leishmania* infection depends on the relative proportion of C3b and iC3b bound to the parasite surface. It appears that if *Leishmania* penetrates the macrophage through its C3b receptors, the parasite will thrive, whereas if it enters through the host iC3b receptors, it will stimulate the respiratory burst and inhibit parasite multiplication. It remains to be determined if the *Leishmania* promastigotes that activate C3 to C3b belong to a subset of cells distinct from those that activate C3 to iC3b. Thus, the activation of C3 by *L. tropica* may be functionally equivalent to the sialylation in *T. cruzi*. Other examples similar to the ones described above should be forthcoming, as it is reasonable to expect host cell–parasite interaction to be regulated by positive and negative mechanisms. A negative mechanism should be particularly prevalent in those parasites which produce a chronic infection, of which *T. cruzi* is an example par excellence, for it induces infection in man (Chagas' disease) that lasts for decades. And it is obvious that not only the molecules that exert a positive effect but also those that have a negative influence on infection are important in vaccination protocol experiments, as dramatically demonstrated with a *Leishmania* glycolipid which can produce protective immunity in experimental animals (Handman and Mitchell, 1985), while the sugar moiety of the glycoconjugate has a totally opposite effect (Mitchell and Handmann, 1986).

## REFERENCES

Alves MJM, Abuin G, Kuwijima VJ, and Colli W (1986): Partial inhibition of trypomastigote entry into cultured mammalian cells by monoclonal antibodies against a surface glycoprotein of *Trypanosoma cruzi*. Mol Biochem-Parasitol 21:75–82.

Bloom B (1979): Games parasites play: How parasites evade immune surveillance. Nature 279:21–26.

Cavallesco R and Pereira MEA (1988) Antibody to *Trypanosoma cruzi* neuraminidase enhances infection *in vitro* and identifies a subpopulation of trypomastigotes. J Immunol 140:617–625.

Handman E and Mitchell GF (1985): Immunization with *Leishmania* receptor for macrophages protects mice against cutaneous leishmaniasis. Proc Natl Acad Sci (USA) 82:5910–5941.

Mitchell GF and Handman E (1986): The glycoconjugate derived from a *Leishmania major* receptor for macrophages is a suppressogenic, disease-promoting antigen in murine cutaneous leishmaniasis. Parasite Immunol 8:255–263.

Mosser DM and Edelson PJ (1987): The third component of complement (C3) is responsible for the intracellular survival of *Leishmania major*. Nature 327:329–332.

Ouaissi MA, Cornette J, and Capron A (1985): *Trypanosoma cruzi*: Modulation of parasite-cell interaction by plasma fibronectin. Eur J Immunol 15:1096–1101.

Ouaissi MA, Cornette J, and Capron A (1986): Identification and isolation of *Trypanosoma cruzi* trypomastigote cell surface protein with properties expected of a fibronectin receptor. Mol Biochem Parasitol 19:210–211.

Paigen B, Morrow A, Brandon C, Mitchell D, and Holmes P (1985): Variation in susceptibility to atherosclerosis among inbred strains of mice. Atherosclerosis 57:65–73.

Pereira MEA (1983): A developmentally regulated neuraminidase activity in *Trypanosoma cruzi*. Science 219:1444–1446.

Pereira MEA and Hoff R (1986): Heterogenous distribution of neuraminidase activity in strains and clones of *Trypanosoma cruzi* and its possible association with parasite myotropism. Mol Biochem Parasitol 20:183–189.

Pereira MEA, Loures MA, Villaltey F, and Andrade AFB (1980): Lectin receptors as markers for *Trypanosoma cruzi* developmental stages and a study of the interaction of wheat germ agglutinin with sialic acid residues on epimastigote cells. J Exp Med 152:1375–1392.

Peterson DS, Wrightsman RA, and Manning JE (1986): Cloning of a major surface-antigen gene of *Trypanosoma cruzi* and identification of a monopeptide repeat. Nature 322:566–568.

Piras MM, Henriquez D, and Piras R (1987): The effect of fetuin and other sialoglycoproteins on the *in vitro* penetration of *Trypanosoma cruzi* trypomastigotes into fibroblastic cells. Molec Biochem Parasitol 22:135–143.

Prioli RP, Rosenberg I, and Pereira MEA (1987): Specific inhibition of *Trypanosoma cruzi* neuraminidase by the human plasma glycoprotein "cruzin". Proc Natl Acad Sci USA 84:3097–3101.

Rosenberg I, Prioli RP, and Pereira MEA: Levels of HDL, a *T. cruzi* neuraminidase inhibitor, correlate inversely with mouse strain susceptibility to *T. cruzi* infections (submitted).

Trischmann T, Tanowitz H, Wittner M, and Bloom B (1978): *Trypanosoma cruzi*: Role of the immune response in the natural resistance of inbred strains of mice. Exp Parasitol 45:160–168.

Wozengraft AO, Sayers G, and Blackwell JM (1986): Macrophage type 3 complement receptors mediate serum-independent binding of *Leishmania donovani*. J Exp Med 164:1332–1337.

Yoshida N, Nussenzweig RS, Potocnjak P, Nussenzweig V, and Aikawa M (1980): Hybridoma produces protective antibodies directed against the sporozoite stage of malaria parasite. Science 207:71–73.

The Biology of Parasitism, pages 111–145
© 1988 Alan R. Liss, Inc.

# *Plasmodium falciparum* Proteins at the Host Erythrocyte Membrane: Their Biological and Immunological Significance and Novel Parasite Organelles Which Deliver Them to the Cell Surface

**Russell J. Howard**

*Malaria Section, Laboratory of Parasitic Diseases, National Institute of Allergy and Infectious Diseases, National Institutes of Health, Bethesda, Maryland 20892*

## INTRODUCTION

The clinical symptoms of malaria are caused entirely by the asexual proliferation of this parasite in the bloodstream. Except for the brief period of seconds or minutes during which extracellular merozoites attach to and invade new host erythrocytes, the blood stages of malaria are intraerythrocytic. Thus, its 24–48 h cycle of growth in size, differentiation, and asexual multiplication is performed within the environment of the host erythrocyte cytoplasm. This intracellular environment would appear to afford important advantages for a parasitic organism: within the host erythrocyte the parasite plasmamembrane is inaccessible to the host immune system. Immune destruction mechanisms which operate against extracellular bacteria in the blood or against the plasmamembrane of free-living trypanosomes or worm parasites are therefore avoided. Furthermore, the host erythrocyte cytoplasm constitutes a readily available supply of protein (predominantly hemoglobin) that is utilized as a source of amino acids, nitrogen, and perhaps energy as well. A special feeding organelle at the parasite plasmamembrane, the cytostome, serves to internalize host erythrocyte cytoplasm into cytoplasmic food vacuoles within which the erythrocyte proteins are degraded proteolytically.

Russell J. Howard's present address is DNAX Research Institute of Molecular and Cellular Biology, Palo Alto, California 94304-1104.

In this article I will develop the concept that although this intraerythrocytic environment appears to be a safe (and edible!) haven for the parasite's asexual cycle, numerous disadvantages come with the territory, disadvantages the parasite had to surmount in order to survive. After describing the potential disadvantages of this intraerythrocytic existence, I will elaborate on specific ways by which parasite-induced alterations in the host erythrocyte membrane overcame some of these disadvantages. These alterations include the insertion of malarial proteins into the host erythrocyte outer membrane. Since the asexual malaria parasite is generally defined as the intracellular organism bounded by its plasmamembrane, parasite-induced alterations at the host membrane imply parasite manipulation of its environment (the host cell) at a considerable distance. The intracellular malaria parasite is separated from the host cell membrane by the parasitophorous vacuole membrane which ensheaths it (PVM; derived topologically from invagination and pinching off of the erythrocyte membrane during merozoite invasion). The host erythrocyte cytoplasm also separates the parasite and host cell membrane. Insertion of malarial proteins into the host cell membrane therefore implies mechanisms for protein transport and trafficking (i.e., appropriate "labeling" as to final destination) from parasite to infected erythrocyte surface membrane. I will present immuno-electron microscopic and transmission electron microscopic evidence for a novel system of diverse organelles involved in this subcellular trafficking. Given the properties of mature uninfected erythrocytes, the extent and nature of "extraparasitic" parasite-derived organelles and changes in the host erythrocyte blur the distinction between host cell and parasite. I will argue that the malaria parasite does not end at its plasmamembrane. This remarkable host-cell–parasite interaction demands that we consider the infected erythrocyte as considerably greater than the sum of its constituent parts, a new cell that is neither erythrocyte nor malaria parasite.

Since extensive reviews of parasite-induced changes in erythrocyte membrane properties (Howard and Barnwell, 1983; Sherman, 1985) and parasite proteins inserted into this membrane (Howard, 1987) have been published recently, this article will not attempt an exhaustive summary of recent progress nor a historical analysis of developments in this field. I will also assume that the reader is familiar with the basic biology, immunology, and biochemistry of malaria and the basic properties of the human erythrocyte. It is my intention to describe reasonable, testable hypotheses for this interaction of parasite and host cell, based particularly on data from our own work, and thereby describe the conceptual springboard for experiments my laboratory will pursue in the coming years.

## THE HOST CELL: MATURE CIRCULATING ERYTHROCYTES

The metabolic functions of the mature erythrocyte are homeostatic rather than biosynthetic. The oxygen-carrying function of erythrocyte hemoglobin demands control of intracellular pH, cation concentrations, energy balance, and redox potential. Glucose is transported into erythrocytes and energy generated by its conversion to lactate, which is then exported back through the erythrocyte membrane. Mature erythrocytes do not have metabolic demands for exogenous amino acids or nucleotides since there is no protein synthesis after the reticulocyte stage and no nucleic acid synthesis. There is also no demand for intracellular lipids or lipid precursors for organelle biosynthesis or turnover. The fluxes of amino acids, nucleotides, and lipids across the membrane of normal erythrocytes are therefore low.

Although mature erythrocytes are biconcave discoidal in shape in the peripheral circulation, this cell is amazingly deformable, especially during passage through the small fenestrations in basement membrane between spleen cords and sinuses wherein the membranes on either side of the erythrocyte are compressed tightly together (Schnitzer et al., 1972).

Finally, mature erythrocytes are devoid of intracellular membranes or organelles. They totally back the apparatus for organelle biogenesis (i.e., precursor organelles and membranes) and also lack the machinery involved in subcellular protein trafficking.

Three questions of relevance to survival of an organism *within* the host erythrocyte are immediately obvious from this facile consideration of some erythrocyte properties.

### Parasite Metabolic Demands Imply Alterations in the Erythrocyte Membrane

The first question concerns the modifications in intracellular metabolite pools and transport properties of the host cell membrane that are required to support the growth of an organism that requires not homeostasis, but explosive growth. One invading merozoite produces 10–40 daughter parasites, depending on the malaria species. Asexual replication therefore demands sufficient intracellular supplies of carbon and nitrogen sources, precursor amino acids, and nucleosides/nucleotides for protein synthesis and nucleic acid synthesis, respectively, and lipids/lipid precursors for membrane biosynthesis. All of these precursor compounds for energy generation and biosynthesis must pass across the membrane barrier of the host erythrocyte membrane. It is therefore not unexpected that the permeability and specific transport properties of endogenous transport systems in the host erythrocyte membrane will be insufficient (in specificity or capacity) for the demands of

zoites, schizonts, and segmenters are sequestered from the peripheral circulation since they attach specifically to the endothelial cells lining capillaries of numerous organs (Luse and Miller, 1971; MacPherson et al., 1985). In the small venous capillaries of brain, the attached infected erythrocytes can actually occlude blood flow, leading to local anoxia and tissue necrosis (see Oo et al., 1987). This pathology is evident as "cerebral malaria," characterized by nervous dysfunction, coma, and often death (see Marsden and Bruce-Chwatt, 1975; MacPherson et al., 1985). Attachment of infected erythrocytes to capillary endothelial cells is mediated by special morphological alterations induced on the infected cell surface by the intracellular parasite. These "knobs" measure roughly 100 nm in diameter and represent a convex protrusion of the host erythrocyte membrane, lipid bilayer, and underlying cytoskeleton (Trager et al., 1966; Luse and Miller, 1971; Langreth et al., 1979; Gruenberg et al., 1983; Leech et al., 1984a; Allred et al., 1986). The concavity of the knob under the host cell membrane contains a cup-shape of electron-dense material when viewed by standard transmission electron microscopy.

Two advantages to the malaria parasite may be conferred by knobs and infected-cell cytoadherence to endothelial cells. One advantage involves evasion of passage through the spleen, the other involves the requirement of *P. falciparum* for relatively low $O_2$ levels.

The spleen plays a major role in surveillance of the properties of erythrocytes and removal of cells with abnormal mechanical or cell-surface properties. Erythrocytes with reduced deformability are removed by the spleen (Sandza et al., 1974; Card et al., 1983; Driessen et al., 1982) as are erythrocytes containing inclusions (Schnitzer et al., 1972). *P. falciparum*-infected erythrocytes containing mature parasites have grossly impaired deformability compared with uninfected erythrocytes and immature-infected cells (Cranston et al., 1984). The large intracellular parasite also represents a large inclusion unable to pass through the narrow fenestrations in endothelial cell wall barriers involved in erythrocyte filtration. The spleen also possesses localized specific and nonspecific immune mechanisms able to recognize alterations in cell surface charge, carbohydrates, and antigen expression (see Wyler et al., 1979; Kreier and Green, 1980). Infected cells are altered in their expression and distribution of surface carbohydrates (Sherman and Greenan, 1986). New malarial antigens are also expressed on the surface of *P. falciparum*-infected cells (Hommel et al., 1982, 1983; Leech et al., 1984b; Marsh and Howard, 1986; Howard et al., 1988). Taken together, these mechanisms for removal of altered erythrocytes, all applicable to *P. falciparum*-infected cells, represent a powerful selective force for a mechanism, such as cytoadherence to endothelial cells, that would reduce the passage of mature infected cells through this organ.

Cytoadherence of *P. falciparum*-infected cells, particularly in venous capillaries, confers a second advantage to the parasite. The malaria parasite is microaerophilic (Scheibel et al., 1979) and is therefore protected from oxidant stress in the relatively low $O_2$ tension of venous capillaries.

## How Can the Parasite Alter the Host Erythrocyte Membrane?

Alterations in the cell surface membrane of infected erythrocytes are therefore necessary in order to allow the metabolite uptake and export properties necessary for an intracellular organism undergoing multiplication. Furthermore, a mechanism for evasion of localized splenic immunity has evolved in which the host cell membrane acquires the property of endothelial cell cytoadherence. These concepts lead to the question of how can the malaria parasite alter the cell surface membrane properties appropriately when it is distant from the host cell membrane? There is now evidence from two lines of research which begins to answer this question. Monoclonal antibodies specific for particular *P. falciparum* antigens have been used to study the distribution of antigens within the infected cell. From simple indirect immunofluorescence experiments at the microscopic level, it is apparent that some malarial antigens are localized external to the malaria parasite in the host cell cytoplasm and infected cell surface membrane. Immuno-electron microscopy studies, including cryo-thin-section immuno-electron microscopy, have localized much more precisely where these antigens are located and provide morphological evidence for their route of subcellular transport.

Second, there is now evidence from standard transmission electron microscopy for a much greater diversity of new, parasite-dependent membrane structures in the cytoplasm of the host erythrocyte. Maurer's clefts—elongated unit membrane vesicles—have long been known from electron microscopy studies of malaria-infected erythrocytes to represent a new structure absent from normal uninfected erythrocytes (Aikawa, 1966, 1971; Rudzinska and Trager, 1968). The morphological diversity of new structures in the host cell cytoplasm is now seen to be greater than had been previously appreciated, and, in some examples presented below, one can correlate the distribution of *P. falciparum* antigens defined by immuno-electron microscopy to novel organelles seen by transmission electron microscopy in the erythrocyte cytoplasm.

## MALARIAL PROTEINS TRANSPORTED TO THE HOST ERYTHROCYTE MEMBRANE

Numerous *P. falciparum* protein antigens have been localized to the host erythrocyte membrane (reviewed by Howard, 1987). The movement of some of these proteins from the parasite to the cell surface membrane will be discussed here. The proteins I will focus on are called PfEMP1, PfEMP2, PfHRP1, and PfHRP2. (PfEMP denotes "*P. falciparum* erythrocyte membrane protein," and PfHRP denotes "*P. falciparum* histidine-rich protein" (see Howard, 1987).)

### PfEMP1

PfEMP1 is a *P. falciparum* protein expressed on the external surface of erythrocytes containing mature parasites (trophozoites and schizonts). This protein possesses several properties consistent with an important role in the cytoadherence property acquired by *P. falciparum*-infected erythrocytes. PfEMP1 is either the infected cell surface receptor for the endothelial cell surface, or physically very closely associated with another molecule which acts as receptor. The properties of this malarial protein and evidence for its role in cytoadherence have been reviewed recently (see Howard, 1987). PfEMP1 appears to have two roles for the malaria parasite in evasion of host immunity: 1) By mediating infected cell cytoadherence to capillary endothelial cells the localized specific and nonspecific "immune" mechanisms of the spleen are avoided, as discussed above. 2) PfEMP1 exists in multiple antigenic forms, suggesting that antigenic diversity of this cell surface molecule may allow the parasite to evade parasiticidal immune responses developed in the host against antigenic forms of the protein seen in prior infections. The antigenic diversity of PfEMP1 (Leech et al., 1984b; Howard et al., 1988) is not incompatible with the suggested role of this protein in mediation of cytoadherence to endothelial cells if this very large molecule bears two types of domains: one domain involved in cytoadherence that is antigenically and structurally conserved but poorly immunogenic (only immune adults from West Africa make antibodies against a pan-specific conserved epitope on PfEMP1 [Marsh and Howard, 1986; R. Howard et al., unpublished data]) and a second domain that is antigenically diverse and highly immunogenic that functions in immune evasion of specific antibodies. This hypothesis has been discussed elsewhere in greater detail (Howard, 1987) and will be resolved once the PfEMP1 protein or gene has been sequenced and studied in greater structural detail.

As yet there are no monoclonal antibodies or monospecific polyclonal antibodies which react with PfEMP1. All the above-mentioned studies which elucidated the cell surface location of this protein, its association with knobs, its antigenic diversity and properties which link it to the cytoadherence property (Leech et al., 1984b; Aley et al., 1984, 1986; Howard et al., 1988) utilized complex monkey or human antisera shown to contain IgG antibodies which react differentially with different antigenic forms of this protein at the surface of intact cells. It is therefore impossible at this stage to study the route of movement of PfEMP1 from intracellular parasite to the surface of the infected erythrocyte. Later, in Figure 8, a scheme is presented which could account for this protein's movement and its topological distribution in the various membranes with which it would be associated.

## PfHRP1

At least three histidine-rich proteins (HRP) are expressed by asexual *P. falciparum* parasites (Howard, 1987). PfHRP1, or the knob-associated HRP, was the first *P. falciparum* protein to be correlated with the expression of knobs at the host erythrocyte membrane (Kilejian, 1984). All $K^+$ parasites studied to date express PfHRP1, regardless of their capacity to cytoadhere to endothelial cells (Kilejian, 1984; Hadley et al., 1983; Leech et al., 1984a; Vernot-Hernandez and Heidrich, 1985). In contrast, PfHRP1 is not expressed by any $K^-$ parasite. PfHRP1 is identified as a broad band at $M_r \sim 90$ kilodaltons (kD), strongly labeled by uptake of $[^3H]$-histidine. PfHRP1 is not extracted from infected cells by Triton X-100 and other detergents which fail to disrupt the erythrocyte cytoskeleton (Leech et al., 1984a). From analysis of the proteins present in Triton X-100–insoluble residues of $K^+$-infected cells and electron microscopy, it was suggested that PfHRP1 was localized in the submembrane electron-dense material at knobs (Leech et al., 1984a). Detergent treatments, such as SDS, which disrupted and solubilized the electron-dense material and host erythrocyte cytoskeleton also solubilized PfHRP1. Recent immuno-electron microscopy studies employing specific antibodies to PfHRP1 have confirmed this localization of PfHRP1 (Ardeshir et al., 1987; Pologe et al., 1987; Taylor et al., 1987; Culvenor et al., 1987).

We have studied the subcellular localization of PfHRP1 employing a specific rabbit antiserum raised against the $M_r$ 90.7, kD band of Malayan Camp Strain PfHRP1, and a mouse monoclonal antibody (McAb 89) which reacts with PfHRP1 but not the other *P. falciparum* HRP (Taylor et al., 1987). From immuno-electron microscopy studies in which antibodies were added to intact $K^+$-infected erythrocytes and identified with ferritin- or gold-conjugated secondary reagents, we concluded that none of the epitopes on PfHRP1 defined by our antibodies were surface exposed. Permeabilization of the erythrocyte outer membrane by saponin treatment did allow binding of anti-PfHRP1 antibodies. Immuno-electron microscopy then revealed antibodies clustered at the submembrane electron-dense material at knobs (Taylor et al., 1987). We concluded that PfHRP1 plays a structural and/or functional role under the knob protrusion. It is also possible that part of the molecule, not yet identified with antibodies, is exposed on the cell surface where it could participate in the cytoadherence of infected cells to endothelium.

The sequence of PfHRP1 has been obtained in several laboratories (Ardeshir et al., 1987; Ellis et al., 1987; Culvenor et al., 1987; Pologe et al., 1987; Sharma and Kilejian, 1987). As anticipated from the biosynthetic labeling studies, the molecule is extraordinarily rich in histidine, with stretches of six to nine contiguous histidine residues in one region. Other regions of the molecule are also of particular interest since they are very rich in lysine and contain extended repeat sequences. Future studies will presumably reveal which regions of PfHRP1 interact with other proteins, and in particular, which regions might interact with host cytoskeletal proteins to contribute to the knob deformation of the membrane and underlying electron-dense material.

The subcellular trafficking of PfHRP1 has been examined by cryo-thin-section immuno-electron microscopy using McAb 89 and protein A-gold (Fig. 1 and Taylor et al., 1987; Ardeshir et al., 1987; Culvenor et al., 1987; Pologe et al., 1987). Very few gold grains were seen within the parasite cytoplasm, consistent with the immunofluorescence microscopy results and suggesting that newly synthesized PfHRP1 is rapidly exported from the parasite. PfHRP1 was localized to electron-dense spheres (perhaps secretory granules?) in the host erythrocyte cytoplasm. It was also at the parasite plasmamembrane and parasitophorous vacuole membrane and with large membrane whorls in the erythrocyte cytoplasm and under the cell surface membrane (Fig. 1). The cryo-thin-section immuno-electron microscopy results suggest that the bulk of PfHRP1 is in the host erythrocyte cytoplasm with only a minor component in electron-dense material at knobs. Perhaps PfHRP1 has a role in ferrying other parasite proteins to and from the cell surface so that most of the molecule is in transit at any time. It can be noted that PfHRP1 has never been detected in culture supernatants, so it appears not to be secreted from intact infected cells.

### PfHRP2

PfHRP2 is expressed by $K^+$ and $K^-$ parasites (Leech et al., 1984a; Kilejian, 1984) and from comparison of its sequence (Wellems and Howard, 1986) with that of PfHRP1 does not appear to be structurally related to the knob-associated HRP. Like PfHRP1, PfHRP2 is strongly and preferentially labeled by uptake of $^3$H-histidine. PfHRP2 contains 34% histidine, 37%

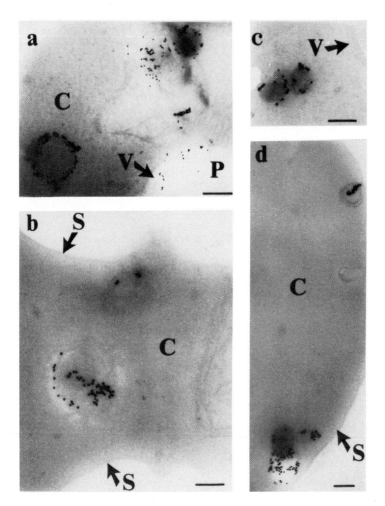

Fig. 1. Localization of PfHRP1 by cryo-thin-section immuno-electron microscopy using monoclonal antibody 89. Thin sections of mature trophozoite-infected cells of K⁺ Malayan Camp strain *P. falciparum* were treated sequentially with monoclonal antibody 89 and protein A-gold. **a:** PfHRP1 is localized within the parasite (P) and host cell cytoplasm (C). Spherical electron-dense organelles bear PfHRP1, predominantly on their periphery. Bar = 1 $\mu$m. **b:** PfHRP1 is also associated with a membrane whorl within the erythrocyte cytoplasm. Bar = 0.5 $\mu$m. **c:** Two electron-dense spherical organelles coated with PfHRP1 within the erythrocyte cytoplasm. Bar = 1 $\mu$m. **d:** Electron-dense spherical organelles about the erythrocyte surface membrane and a double membrane whorl containing PfHRP1 is seen fusing with the erythrocyte surface membrane. Bar = 0.5 $\mu$m. Additional immuno-electron microscopy experiments with membranes from saponin-lyzed erythrocytes identified PfHRP1 as a constituent of the electron dense material under knobs (Taylor et al., 1987).

alanine, and 10% aspartate and numerous tandem repeats of AHH and AHHAAD amino acid sequences (Wellems and Howard, 1986). PfHRP2 migrates at lower apparent $M_r$ than PfHRP1, with an $M_r$ of 72 kD for Malayan Camp K$^+$ P. falciparum. Interestingly, the $M_r$ calculated from its genomic sequence is $\sim M_r$ 35,000. Efforts to dissociate the $M_r$ 72-kD Pf-HRP2 molecule (a possible dimer) in the presence of strong protein peturbants and numerous detergents were unsuccessful (L. Panton, unpublished data). The unusual composition of this molecule and constituent repeats may cause a monomeric polypeptide chain to migrate aberrantly on SDS-PAGE. PfHRP2 is distinguished from PfHRP1 not only by its apparent $M_r$ and presence in K$^-$ parasites, but by its secretion from intact infected cells as a water-soluble protein (Howard et al., 1986). PfHRP2 is secreted continuously from intact infected cells, accumulating in culture supernatant from the early ring stage (Howard et al., 1986).

The extracellular form of PfHRP2 appears to be distinct from intracellular PfHRP2, suggesting that processing of the intracellular molecule is required to "label" the molecule for export. The specifically immunoprecipitated PfHRP2 protein from washed, infected cells is apparent as a multiplet of bands extending from $M_r$ 72 kD to faster-moving bands (Howard et al., 1986; Rock et al., 1987). The extracellular PfHRP2, in contrast, is a sharp single band, invariably at the apparent $M_r$ of the slowest-moving PfHRP2 band identified in cells. Nonspecific release of intracellular PfHRP2 into the culture medium (through rupture of a small proportion of cells) would be expected to yield a PfHRP2 multiplet in the culture medium rather than a single band at the slowest mobility. More direct evidence for "labeling" PfHRP2 for export comes from biosynthetic labeling studies with [$^3$H]-monosaccharides as precursors for complex carbohydrates and protein glycoconjugates. Inspection of the PfHRP2 sequence (Wellems and Howard, 1986) revealed one potential amino acid sequence, Asn-Asn-Ser, where the first Asn could serve as recognition site for N-linked glycosylation (i.e., Asn-X-Ser or Asn-X-Thr where X is any of the normal protein amino acids other than Pro). To test whether this Asn (residue 24) or some other site could act as acceptor, we incubated synchronous cultures of asexual P. falciparum parasites with [$^3$H]-mannose, [$^3$H]-galactose, [$^3$H]-glucosamine, or [$^3$H]-fucose; collected the culture supernatants; and washed infected cells and immunoprecipitated with specific anti-PfHRP2 antibodies. Parallel samples were labeled with [$^3$H]-histidine. A [$^3$H]-histidine-labeled PfHRP2 band was specifically precipitated from culture supernatant, and a comigrating band, precipitated by the same antibodies, was labeled by uptake of [$^3$H]-galactose. The [$^3$H]-galactose-labeled band was not evident in the washed cells, even after prolonged fluorographic exposure. We conclude that the extracellular PfHRP2 is labeled by uptake of galactose. This protein was not labeled by biosynthetic labeling in the presence of the other [³H]-monosaccharides, even though [³H]-mannose and [³H]-glucosamine-labeled numerous other malarial proteins. We suggest that the intracellular PfHRP2 is glycosylated just prior to its release from the infected erythrocyte. Glycosylation may even be an obligatory step necessary for trafficking of PfHRP2 along this pathway. Once glycosylated the PfHRP2 is presumably rapidly exported, otherwise we would detect an intracellular pool of [³H]-galactose-labeled PfHRP2. The site of this glycosylation step is unknown, but it is presumably close to the periphery of the infected cell.

Cryo-thin-section immuno-electron microscopy of infected cells using monoclonal antibody specific for PfHRP2 (McAb 87) and protein A-gold (Howard et al., 1986) localized PfHRP2 to several cell compartments. Pf-HRP2 was identified in the parasite cytoplasm, as concentrated "packets" within the host erythrocyte cytoplasm and in association with the cell surface membrane as submembrane packets. These results, and the results of fluorescence microscopy plus quantitation of intracellular versus extracellular Pf-HRP2 (Howard et al., 1986), suggest the following pathway for movement of PfHRP2 within the infected cell: PfHRP2 is synthesized as a parasite protein within the parasite and exported through the topological barriers of the parasite plasmamembrane and parasitophorous vacuole membrane into the host erythrocyte cytoplasm. There, in concentrated "packets" it is exported through the host cytoplasm to the cytoplasmic side of the host cell membrane. How PfHRP2 crosses the cell surface membrane remains a mystery. We have not seen fusion of membrane vesicles enclosing PfHRP2 with the surface membrane (as in the hypothetical scheme for S-antigen release from intact cells shown in Fig. 7). PfHRP2 does appear to be tightly associated with the cell surface membrane. Intact infected cells were labeled by the $Na^{125}I$-$H_2O_2$-lactoperoxidase method under conditions that did not label submembrane proteins. The washed cells were extracted sequentially with Triton X-100 and SDS, and the detergent extracts immunoprecipitated with monoclonal antibodies. A single $^{125}I$-labeled PfHRP2 band was identified in both extracts at the same $M_r$ as the extracellular PfHRP2 (Rock et al., 1987). This surface-exposed PfHRP2 is obviously attached in some way to the lipid bilayer and/or cytoskeleton of the host cell membrane.

These properties of PfHRP2, the fact that its sequence is determined, and the availability of various monoclonal antibodies and rabbit antiserum specific for this molecule, offer an ideal opportunity for detailed analysis of the processing steps involved in subcellular trafficking and export of a malarial protein. The function of PfHRP2 is unknown. Although it may contribute to knob structure, it is clearly not the major determinant of knob formation, since $K^-$ parasites still export PfHRP2 to the culture medium. PfHRP2 may function within the plasma of the infected host in some way advantageous to the malaria parasite: perhaps it modulates the activities of the immune system, or perhaps it performs biochemical functions related to its extraordinary content of charged amino acids (histidine and aspartate). We have found that PfHRP2 binds with unusual tenacity to $Zn^{2+}$-chelate and $Cu^{2+}$-chelate columns (L. Panton and R. Howard, submitted for publication). The concentrations of imidazole, added as competitive ligand, required to elute PfHRP2 from such metal chelate columns are much higher than for elution of other serum proteins known to bind divalent metal ions. The potential importance of this metal ion binding capacity of PfHRP2 for the parasite (or host) remains to be explored.

## PfEMP2

This very large *P. falciparum* protein ($M_r \sim 300,000$) was identified as a constituent of the erythrocyte membrane using specific monoclonal antibodies. PfEMP2 was shown to be different antigenically from PfEMP1 and different in apparent $M_r$ in some parasites (Howard et al., 1987a). Another study identified a similar molecule, called mature-parasite-infected erythrocyte surface antigen (MESA), and obtained a remarkable partial cDNA sequence encoding tandem repeats of the amino acids GESKET (Coppel et al., 1986). We subsequently showed that PfEMP2 and MESA were the same protein (Howard et al., 1987b). Unlike PfEMP1, which is exposed to surface labeling reagents and antibodies at the infected cell's surface, PfEMP2/MESA does not appear to be surface exposed (Coppel et al., 1986; Howard et al., 1987a, 1988). The several antigenic differences between PfEMP1 and PfEMP2 have been discussed at length elsewhere (Howard, 1987). The function of PfEMP2 is unknown. Since immuno-electron microscopy studies localized this antigen to the submembrane electron-dense material at knobs, this protein presumably has a structural and/or functional role at this site. PfEMP2 could conceivably interact with PfHRP1, which is also localized at the electron dense material.

The movement of PfEMP2 from parasite to host cell cytoplasm and cell surface has been studied by indirect fluorescence microscopy and immuno-electron microscopy (Coppel et al., 1986; Howard et al., 1987a). Immunofluorescence reactivity has been studied during the development of parasites from rings to early schizonts (Howard et al., 1987a). Fluorescence appeared first in the parasite cytoplasm at the mature ring stage and thereafter spread to the erythrocyte cytoplasm and erythrocyte membrane. There was no fluorescence reactivity with intact infected cells that had not been permeabilized by fixation.

Immuno-electron microscopy on frozen thin sections using specific monoclonal antibodies and protein A-gold (Fig. 2) reveals a remarkable pattern of antigen localization for PfEMP2. The antigen can be seen associated with electron-dense spheres budding off the parasitophorous vacuole membrane into the erythrocyte cytoplasm (Fig. 2b) and also located distant from the host cell cytoplasm (Fig. 2c,d,g). PfEMP2 appears to be localized throughout these spheres when seen in cross section. In Figure 2c a typical unit membrane vesicle seen in infected cells, or Maurer's cleft, is designated by a pair of arrows. The cleft is devoid of PfEMP2, while an adjacent circle of membrane is densely covered with gold grains. Other membranes decorated with gold grains are seen in Figure 2d. We conclude that at least two types of structure bear PfEMP2 in the host erythrocyte cytoplasm: electron-dense spheres and larger unit membrane structures exemplified by those marked with arrows in Figure 2d. It is also apparent that not all unit membrane structures in the host cell cytoplasm bear PfEMP2. This observation implies that malarial proteins found associated with membranes in the host erythrocyte cytoplasm are not randomly associated with such membranes but are directed to some membranes and not others. Figure 2e–g shows the association of PfEMP2 with the cell surface membrane. Electron-dense structures bearing this protein abut the underside of this membrane (Fig. 2e–g). In some cases the protein is associated with knobs (Fig. 2e,f), although in others it is not possible to unequivocally identify knob protrusion at the site of antigen depositions (Fig. 2g). When a control monoclonal antibody was used, no gold grains were deposited on the infected cell (Fig. 2h).

## ORGANELLES INVOLVED IN TRAFFICKING MALARIAL PROTEINS TO THE HOST CELL MEMBRANE

From the results of cryo-thin-section immuno-electronmicroscopy presented above, it is clear that diverse subcellular structures, or organelles, are involved in the export of malarial proteins from the parasitophorous vacuole membrane to the cell surface membrane. Despite the limitations of the cryo-thin-section technique (thicker sections compared with transmission electron microscopy and only limited staining with heavy metals), we were able to discern whorls of several membranes; single membrane clefts (Maurer's clefts); electron-dense spheres, some of which were delimited by membrane; and circles of unit membrane. With different antibody reagents, each of these

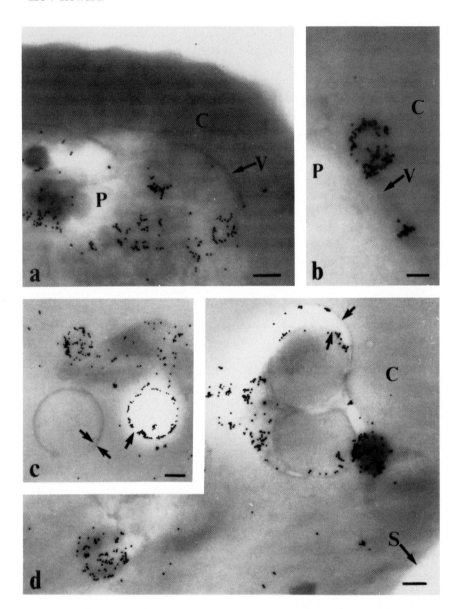

Fig. 2. Localization of PfEMP2 by cryo-thin-section immuno-electron microscopy using monoclonal antibodies 4H9.1 (**a–d**) and 8B7.4 (**e–g**) or a control IgG monoclonal antibody 2D2C5 (**h**). Thin sections of mature trophozoite-infected erythrocytes ($K^+$ Malayan Camp strain) were incubated sequentially with monoclonal antibody 8B7.4 and protein A-gold, or monoclonal antibody 4H9.1, rabbit anti-mouse $\mu$-serum, and protein A-gold. Identical results were obtained with these two antibodies which define different epitopes on PfEMP2 (Howard

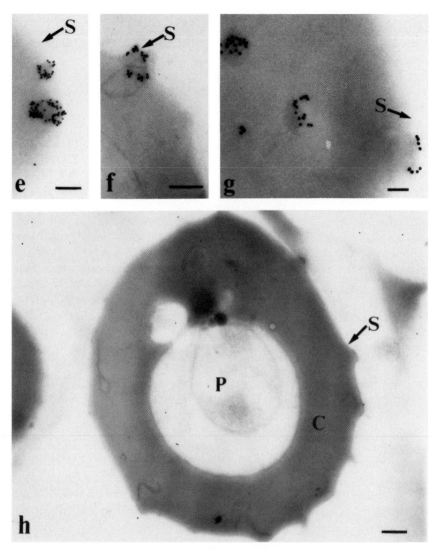

et al., 1988). Panels a through g show in sequence the path of movement of PfEMP2 from parasite cytoplasm (P), through the parasitophorous vacuole membrane (V) and host erythrocyte cytoplasm, (C) to final deposition under knobs at the erythrocyte surface membrane (S). **d:** The membranes bearing PfEMP2 are marked with arrows. **h:** After treatment of thin sections with control monoclonal antibody very few particles of protein A-gold were detected. Bars = a,c–f, 0.5 $\mu$m; b,g, 0.25 $\mu$m; h, 1 $\mu$m (S. Uni, M. Aikawa, and R. Howard, unpublished results).

organelles could be identified as a component involved in antigen trafficking to the cell surface. In this section I describe the results of transmission electron microscopy where we attempt to identify, at high resolution, the detailed structure of these organelles.

Using standard methods of transmission electron microscopy (Aikawa, 1966; Aikawa et al., 1986), we identified the following structures in the erythrocyte cytoplasm.

### Maurer's Cleft

Detailed descriptions of this organelle have been provided by Rudzinska and Trager (1968) and Aikawa (1966, 1971). These structures are exceedingly diverse in shape but all consist of a unit membrane-bounded vesicle with a lumen of very low electron density. The vesicle is often elongated (up to 2 $\mu$m long) (e.g., Fig. 4a, upper right of photograph), but can also appear as a horseshoe (e.g., Fig. 3c) or as a circle, these different shapes presumably representing different cross sections of the same type of organelle. The width of the lumen in Maurer's clefts is generally ~40 nm (Rudzinska and Trager, 1968; also Fig. 4a), but can range from as low as 15 nm (Fig. 4a, membrane whorl) or as high as 75 nm (Fig. 3c). Maurer"s clefts are seen with both $K^+$ and $K^-$ P. falciparum (Aikawa et al., 1986).

### Maurer's Cleft With Electron-Dense Material

Maurer's clefts with electron-dense material on the cytoplasmic face of the unit membrane were first described by Aikawa et al. (1986). This organelle is seen in early- to late-trophozoite–infected cells of $K^+$ P. falciparum but not with $K^-$ parasites. The electron-dense material on these clefts often consists of roughly 100-nm-diameter bodies of a shape, size, and electron density identical with that of the electron-dense material seen under knobs. Maurer's clefts bearing electron-dense material have been identified contiguous with both the parasitophorous vacuole membrane (from where they appear to originate) and the erythrocyte membrane (the point of electron-dense material deposition). Figure 3b shows the typical appearance of this organelle.

### Parasitophorous Vacuole Membrane-Derived Vesicles/Extensions of This Membrane

Figure 3a shows two examples of this structure (labeled G). In cross section one can identify a unit-membrane-bounded, roughly circular structure (0.8–1.5 $\mu$m diameter) with granular electron-dense contents indistinguishable from those of the lumen of the parasitophorous vacuole. The parasitopho-

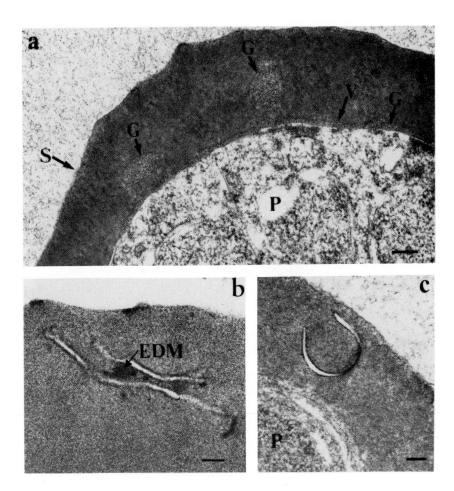

Fig. 3. Transmission electron microscopy of organelles in the host erythrocyte cytoplasm with K+ Malayan Camp strain. **a:** PVM-derived vesicles, or extensions of the PVM in the erythrocyte cytoplasm. The granular contents of these structures are indicated (G), together with the identical appearance of the contents of the PVM (G). The erythrocyte surface membrane (S) and parasite cytoplasm (D) are also labeled. Bar = 0.5 $\mu$m. **b:** Maurer's cleft bearing electron-dense material (EDM). Bar = 0.1 $\mu$m. **c:** Maurer's cleft. Bar = 0.25 $\mu$m (W. Daniel and R. Howard, unpublished results; and Aikawa et al., 1986).

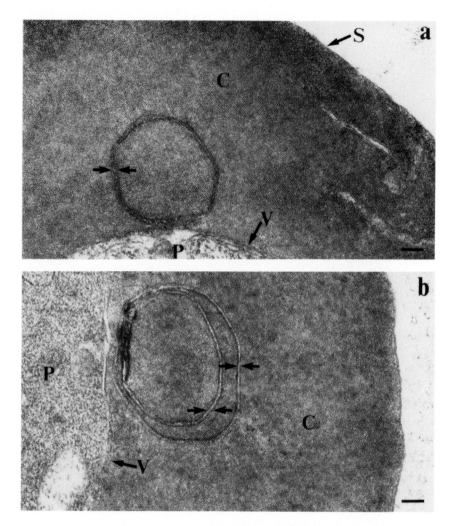

Fig. 4. Transmission electron microscopy of organelles in the host erythrocyte cytoplasm (C) with K$^+$ Malayan Camp strain. **a:** A typical Maurer's cleft is seen on the right, close to the erythrocyte surface membrane (S). A membrane whorl is in direct contact with the parasitophorous vacuole membrane (V). The pair of unit membranes which form the whorl are indicated by arrows. Bar = 0.25 μm. **b:** A complex membrane whorl comprising two pairs of unit membranes (each membrane marked by an arrow). Note electron-dense material and complex pattern of membranes where these whorls abut the vacuole membrane (V). Bar = 0.25 μm. (M. Aikawa, W. Daniel, and R. Howard, unpublished results).

rous vacuole membrane is labeled V in Figure 3a, and granular vacuole contents are also labeled as G. Without more extensive sectioning we cannot distinguish at this stage whether this structure represents fingerlike extensions of the parasitophorous vacuole and vacuole membrane into the host erythrocyte cytoplasm, or sealed membrane vesicles derived from the vacuole.

## Membrane Whorl

In this report I am distinguishing this structure from Maurer's clefts, which it resembles to some extent. We have frequently noted pairs of circular unit membranes that, spaced roughly 15 nm apart, delimit an annular vesicle. The diameter of these membrane circles can range from 1 to 3 $\mu$m. Two features distinguish these whorls from Maurer's clefts: 1) The width of the lumen in whorls is always considerably less than that of most Maurer's clefts (including circular clefts), namely 10–15 nm versus approximately 40 nm. 2) Part of the whorl structure consists of a complex, usually electron dense, of several unit membranes in direct apposition. Such a feature is lacking from Maurer's clefts, which consist of a single unit membrane enclosing a vesicle. The circular membrane structures in Figures 4a,b and 5 show the typical appearance of a whorl. At low magnification the structures in Figures 4a and 5 appear identical with a circular Maurer's cleft (except for the smaller lumen width); however, at the point indicated by paired arrows in Figure 4a and at several points in Figure 5, note the appearance of additional membranes over those required to form a simple vesicle. In Figure 5 the apparent juxtaposition of a third membrane is seen within the lumen, first with one side of the outer membrane and then the other. Most often, these whorls are seen close to the parasitophorous vacuole membrane (e.g., Fig. 4a,b). At the point of apparent contact with the vacuole membrane there is usually a complex pattern of numerous membranes and electron-dense material (Fig. 4a,b). The whorls shown in Figure 4b are even more complex in that two whorls are seen, one within the other. The diameters of these concentric whorls were 1.2 and 1.5 $\mu$m.

## Golgi-Like Membrane Stack

This structure has not been described previously. We have observed stacks of between 5 and 15 elongated vesicles, each demarcated by a single unit membrane, in the erythrocyte cytoplasm of several isolates of $K^+$ *P. falciparum* (Fig. 6). The individual vesicles are approximately 0.5 $\mu$m long and enclose a lumen roughly 15 nm wide. Each membrane vesicle is aligned roughly parallel to adjacent members of the stack on either side. The stacks are generally distant from the parasitophorous vacuole membrane seen within

the same section (Fig. 6). From the higher-power electron micrograph (Fig. 6d), the individual unit membranes composing the stack are seen to be identical in appearance with the erythrocyte surface membrane. The lumen of the vesicles within the stack is of low electron density, i.e., identical in appearance with Maurer's clefts. Figure 6a,b shows both Maurer's clefts and membrane stacks in the same section. In particular, it can be noted in Figure 6b that the appearance of the lumen in Maurer's clefts and the stacks is quite different from that of the lumen of the parasitophorous vacuole. These membrane stacks are identical in appearance with the typical Golgi apparatus seen within the cytoplasm of eukaryotic cells. Experiments are under way to test specifically whether these membrane stacks in the cytoplasm of the host cell are in fact Golgi. The Golgi apparatus has been shown to bear receptors for proteins which are to be modified by Golgi enzymes (generally by glycosylation, deglycosylation) prior to export to the cell surface or extracellular fluid (see Roth, 1987). Perhaps the membrane stacks in the erythrocyte cytoplasm are involved in post-translational processing of malarial proteins en route to the cell surface membrane or destined for secretion into plasma. We emphasize that this structure is absent from uninfected erythrocytes or reticulocytes and must represent membrane synthesis under direction (directly or indirectly) of the malaria parasite. Whether or not these stacks do perform the specific functions of Golgi apparatus described for other cells, the existence of such a complex membrane structure in the host cell cytoplasm illustrates the remarkable extent to which the malaria parasite alters its external environment—the host cell cytoplasm external to the parasite plasma membrane.

### Electron-Dense Circles (Spheres)

By cryo-thin-section immuno-electron microscopy we identified electron-dense circles in the erythrocyte cytoplasm of infected cells; however, the analogous structures were not seen by transmission electron microscopy. A description of these structures is therefore included here for completeness, despite the absence of high-resolution photomicrographs. PfHRP1 was identified as a surface constituent on these circles. In contrast, PfEMP2 was identified both on the surface of and within these electron-dense circles (Fig. 2b–e). The circular organelles bearing PfEMP2 range in size from 0.5 to 1.2 $\mu$m in diameter. Given the different location of PfHRP1 and PfMEP2 with respect to these organelles, it would be of interest to determine whether these proteins are exported to the cell surface membrane on the same or different pathways. In some cryo-thin-sections we could clearly distinguish a unit membrane around electron-dense circular organelles; however, in most cases the resolution was too poor to identify a membrane. If, as suggested from their shape, these organelles are spherical, and if they are limited by a unit membrane, they would be very similar to the spherical, electron-dense secretory vesicles and storage granules described in the cytoplasm of eukaryotic cells (see Lodish et al., 1981; Palade, 1975; Gumbiner and Kelly, 1982). In this case the remarkable distinction would be that such organelles are derived from within an organism and exported into the external environment (the host erythrocyte cytoplasm).

## TOPOGENESIS OF MALARIAL PROTEINS IN THE HOST ERYTHROCYTE SURFACE MEMBRANE

Some malarial proteins, such as the *P. knowlesi* variant antigen (Howard et al., 1983, 1984a) and PfEMP1 of *P. falciparum* (Leech et al., 1984b; Howard et al., 1988), expose antigenic determinants on the surface of intact, nonfixed infected erythrocytes. For other malarial proteins localized to the host cell membrane, such as PfHRP1 and PfEMP2 (reviewed in Howard, 1987), there is no evidence for exposure of these molecules to the cell surface but evidence for their attachment to the cytoplasmic face of the cell membrane. The asymmetric integration of proteins into membranes, a general feature of membrane structure, has been termed "protein topogenesis" (Blobel, 1980). How can we account for the asymmetric integration of malarial proteins into the infected erythrocyte's surface membrane using the much-studied mechanisms for membrane protein topogenesis revealed for other cells? Without discussing the multiple alternative mechanisms of spontaneous or cotranslational protein insertion into membranes (see Blobel, 1980; Engelman and Steitz, 1981; Wickner and Lodish, 1985) I will focus on how a particular membrane orientation of a malarial protein generated within the malaria parasite could produce the membrane topologies we observe at the infected erythrocyte's surface. The special aspect of this discussion that applies to malaria-infected cells but not to membrane protein topogenesis in other cells, is the problem of how the cell-surface-derived malarial proteins traverse the topological barriers of parasite plasmamembrane, parasitophorous vacuole membrane, and cell surface membrane. I have already suggested above that the problem of distance for protein transport between parasite and cell surface has been solved by the genesis of parasite-derived organelles which move from the parasite to their final destination.

The hypothetical scheme shown in Figure 7 accounts for the asymmetric incorporation of the *P. falciparum* schizont glycoprotein into the parasite plasma membrane (Howard et al., 1984b) and the secretion of the *P. falci-*

**TRAFFICKING OF THE MAJOR P. FALCIPARUM
GLYCOPROTEIN AND THE S-ANTIGEN**

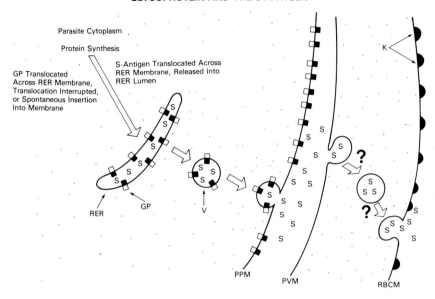

Fig. 7. A schematic diagram showing examples of trafficking of *P. falciparum* proteins to the parasite plasmamembrane (PPM) and parasitophorous vacuole. ▢◼ = the $M_r$ ~ 200,000 *P. falciparum* schizont glycoprotein with an asymmetric membrane disposition. S = the *P. falciparum* S-antigen. RER = rough endoplasmic reticulum; V = unit-membrane-bounded secretory vesicle; PVM = parasitophorous vacuole membrane; RBCM = erythrocyte surface membrane; K = knobs. The $M_r$ ~ 200,000 schizont glycoprotein is expressed on the PPM but is not exported beyond this membrane (see Howard et al., 1984b). The S-antigen accumulates in the parasitophorous vacuole and is released into the external medium when the infected cell ruptures (see Coppel et al., 1983). Contents of the parasitophorous vacuole, such as the S-antigen, may also be exported to the extracellular medium with intact infected cells, if unit membrane vesicles move from PVM to RBCM (indicated with question marks). This scheme is taken from Howard et al. (1987b).

*parum* S-antigen into the parasitophorous vacuole (Coppel et al., 1983). These processes are probably very similar to analogous processes of surface membrane protein topogenesis and protein secretion defined in higher eukaryotic cells. Since other malarial proteins, such as PfHRP1, PfHRP2, PfEMP1, and PfEMP2, are exported *beyond* the parasite plasmamembrane and vacuolar space, the schizont glycoprotein must *lack* special signal sequences which route the other proteins along different subcellular trafficking pathways.

The scheme shown in Figure 8 accounts for the subcellular trafficking and surface membrane topogenesis of PfHRP1, PfEMP1 and PfEMP2. It should

**TRAFFICKING OF P. FALCIPARUM PROTEINS TO THE HOST
ERYTHROCYTE SURFACE MEMBRANE**

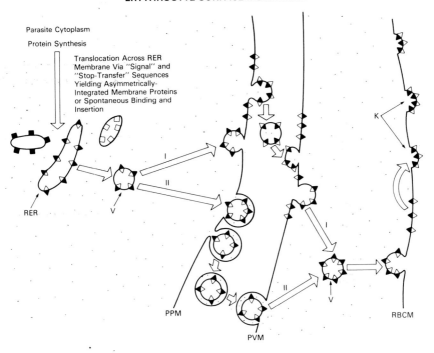

Fig. 8.   Schematic diagram showing trafficking of membrane proteins from the intracellular *P. falciparum* parasite to the RBCM. Three simple asymmetric dispositions of newly synthesized malarial proteins are indicated at the RER: exclusively on the cytoplasmic membrane face, transmembrane, exclusively on the membrane face within the lumen. Only the subsequent trafficking of the transmembrane protein is indicated. Two different schemes for crossing the topological barriers of PPM and PVM are presented (indicated as I and II). Scheme I involves fusion of the vesicle (V) with the PPM, vesicle blebbing from the PPM, fusion with the PVM, and subsequent blebbing from the PVM to produce a vesicle within the erythrocyte cytoplasm. Scheme II involves blebbing of the PPM around the original vesicle without fusion and transport of this bilamellar complex to the PVM, where fusion of the outer membrane would release the vesicle into the erythrocyte cytoplasm. Thereafter the two schemes are indistinguishable. These pathways would account for the trafficking of PfEMP1 to the external face of the RBCM and trafficking of PfEMP2 and PfHRP1 to the EDM under knobs. This figure is taken from Howard et al. (1987b).

be possible to distinguish experimentally the two alternative pathways described in Figure 8. Scheme I predicts that proteins destined for export to the erythrocyte surface membrane are also expressed on the parasite plasma membrane and that these proteins will have the same asymmetry in both membranes. Scheme II predicts that such proteins should not be identified on the surface membrane of intracellular parasites. Scheme I also predicts that this class of malarial proteins are transitory constituents of the parasitophorous vacuole membrane, but, that in this site their asymmetry is the reverse of that at the cell surface. It should be noted, however, that at both locations the same protein domains are located in the "aqueous phase" (external medium and vacuolar space) and host cell cytoplasm. If Scheme II applies, these proteins do not become constituents of the parasitophorous vacuole membrane. Purification of membrane vesicles bearing these proteins should resolve these alternative schemes: in Scheme II all membrane vesicles bear the proteins in the same asymmetrical orientation. In contrast, Scheme I predicts that some vesicles should be identified with the proteins expressed in the opposite asymmetry.

The subcellular trafficking of PfHRP2 has been omitted from the above discussion since several of its properties cannot be accounted for by extrapolation of the typical modes of protein trafficking and secretion seen in higher eukaryotic cells. As discussed above, PfHRP2 is exported from the intracellular malaria parasite, through the host erythrocyte cytoplasm and cell surface membrane to accumulate in the extracellular fluid as a water-soluble protein. PfHRP2 is not identified by immuno-electron microscopy within unit membrane vesicles such as those shown in Figure 7 for the secretion of the S-antigen from intact infected cells. Instead, PfHRP2 is localized to electron-dense spheres in the host cell cytoplasm that do not, with current immuno-electron microscopy technique, appear to be bounded by a unit membrane (Howard et al., 1986). We have no explanation at present for the obvious fact that this protein appears to be able to cross the three topological barriers (parasite plasma membrane, vacuole membrane, and erythrocyte surface membrane) which separate its presumed site of synthesis (the parasite cytoplasm) from the extracellular medium.

## NOVEL IMPLICATIONS OF PROTEIN TRAFFICKING AND TOPOGENESIS IN MALARIA-INFECTED ERYTHROCYTES

In the foregoing sections I have reviewed the evidence for insertion of malarial proteins into the host erythrocyte membrane and, from the results of immuno-electron microscopy, provided direct evidence for their paths of subcellular trafficking. Numerous membrane organelles and novel structures have been identified in the host cell cytoplasm both by cryo-thin-section immuno-electron microscopy and traditional transmission electron microscopy. In some examples particular *P. falciparum* antigens can be associated with particular organelles in the host cell cytoplasm. By extrapolation of the knowledge gained on protein trafficking and membrane topogenesis in the higher eukaryotic cells much studied by cell biologists, I have presented likely schemes for protein movement and topogenesis in the malaria-infected erythrocyte. Several aspects of this scenario blur the conventional distinction made between intracellular malaria parasite and host cell.

### Biogenesis of Extraparasitic Organelles

Mature human erythrocytes lack membrane-bound organelles such as mitochondria, Golgi apparatus, endoplasmic reticulum, and secretory vesicles. The numerous membrane organelles seen in the host cytoplasm of infected erythrocytes, including multilamellar structures of similar appearance to Golgi (Fig. 6), are therefore ultimately of parasite origin. Even if some of these membranes are derived topologically from the surface membrane of malaria-infected erythrocytes, such organelles do not exist, and such an extensive endocytic process does not occur, in the absence of the intracellular parasite. From the numerous electron micrographs presented here and elsewhere it appears that many of these membrane structures bud off the parasitophorous vacuole membrane and/or parasite plasmamembrane and are directly of parasite origin. If the malaria parasite is considered to be delimited by its plasmamembrane, then the biogenesis of membrane organelles in the cytoplasm of the host erythrocyte represents an extraordinary process. An organism is modifying its external environment (the host cell cytoplasm surrounding it) by releasing organelles into that environment! Another alternative, perhaps less disturbing in its novelty, is to propose that the malaria-infected erythrocyte is not a malaria parasite within a host cell, but a new entity, distinct to the component cells involved in its conception (merozoites and uninfected erythrocyte). In this case the parasite plasmamembrane is no longer a boundary between two interacting cells but only one of several membrane complexes within the larger cell.

### Extraparasitic Machinery for Trafficking Organelles and Parasite Proteins

The cytoplasm of eukaryotic cells is no longer thought of as an unstructured solution but has been shown to contain numerous insoluble filaments such as microtubules ($\sim 25$ nm diameter), intermediate-sized filaments ($\sim 10$ nm diameter), and microfilaments ($\sim 7$ nm diameter). Microtubules and microfilaments are dynamic structures, being continuously degraded and resynthesized, while intermediate filaments may have half-lives in excess of cell-cycle times. These filaments, in combination, are thought to provide a mechanically continuous network, often dynamic, that provides cell shape and is involved in organelle movement and cytoplasmic/cell movement (see, for example, Steinert, 1981; Stossel, 1984; Olmsted, 1986).

The presence of numerous membrane-bounded organelles in the erythrocyte cytoplasm of infected cells and the fact that malarial proteins are moved from parasite to cell surface, imply the existence of machinery for organelle and/or protein movement. To date there have been no immunolocalization studies on the occurrence of malarial proteins analogous to actin, tubulin, vimentin, microtubule-associated proteins, etc., in the infected erythrocyte. I predict that such studies will demonstrate an extensive network of such protein filaments in the erythrocyte cytoplasm. This prediction is a corollary of the demonstrated presence of malarial organelles and proteins at this site, and will, if verified, demonstrate again the extent to which the parasite has modified its external environment.

## SUMMARY

Erythrocytes infected with mature asexual stages of *Plasmodium falciparum* express malarial proteins at the cell surface membrane. As these new cell surface antigens are recognized by host immune responses, this phenomenon would appear to act to the disadvantage of the malaria parasite. However, parasite-induced modifications in the structure and functional properties of the host cell outer membrane were probably obligatory for the intracellular growth and replication of a complex eukaryotic parasite within a host cell devoid of biosynthetic organelles and enzyme pathways. Furthermore, one of these malarial cell-surface proteins confers an important immune-evasion property (cytoadherence of infected erythrocytes to venous capillary endothelial cells) and is also antigenically diverse, confounding isolate-specific immune responses to earlier infections of the same host. Insertion of malarial proteins into and under the host erythrocyte membrane implies a subcellular pathway for movement of malarial proteins from their site of biosynthesis within the parasite cytoplasm. Such subcellular trafficking has now been identified using specific monoclonal antibodies and cryo-thin-section immuno-electron microscopy. Even more extensive extraparasitic modifications of the parasite's external environment have been identified in the form of parasite-derived multimembrane organelles in the host erythrocyte cytoplasm. Multiple biochemical steps of protein signaling for different routes of subcellular trafficking via these organelles and stepwise protein modifications in such organelles are postulated. These results blur the classical morphologic distinction between intraerythrocytic malaria parasite and surrounding host cell. The extent and nature of parasite-induced alterations in host cell cytoplasm and cell surface membrane have led to the suggestion that the two interacting cells (host erythrocyte and malaria parasite) have "fused" to create a new entity with special properties lacking in the original component parts. This concept is not unusual when one considers the diverse and intimate interrelationships in nature among symbiotic organelles, viruses, unicellular pro- and eukaryotes, and multicellular organisms.

## ACKNOWLEDGMENTS

The author is grateful in particular to S. Aley, W. Daniel, J. Leech, L. Panton, and E. Rock and my other colleagues at the NIH for their contributions to these studies. I am also indebted to L. Miller, NIH, for his support and encouragement. S. Uni and M. Aikawa at Case Western Reserve University, D. Taylor at Georgetown University, and J. Lyon at the Walter Reed Army Institute of Research were all major contributors to these experiments.

## REFERENCES

Aikawa M (1966): The fine structure of the erythrocytic stages of three avian malarial parasites, *Plasmodium fallax*, *P. lophurae* and *P. cathemerium*. Am J Trop Med Hyg 15:449–471.
Aikawa M (1971): *Plasmodium*: The fine structure of malarial parasites. Exp Parasitol 30:284–320.
Aikawa M, Uni S, Andrutis AT, Howard RJ (1986): Membrane-associated electron-dense material of the asexual stages of *Plasmodium falciparum*: Evidence for movement from the intracellular parasite to the erythrocyte membrane. Am J Trop Med Hyg 35:30–36.
Aley SB, Sherwood JA, Howard RJ (1984): Knob-positive and know-negative *Plasmodium falciparum* differ in expression of a strain-specific malarial antigen on the surface of infected erythrocytes. J Exp Med 160:1585–1590.
Aley SB, Sherwood JA, Marsh K, Eidelman O, Howard RJ (1986): Identification of isolate-specific proteins on sorbitol-enriched *Plasmodium falciparum*-infected erythrocytes from Gambian patients. Parasitology 92:511–525.
Ali SN, Fletcher KA, Maegraith BC (1969): Effect of antimalarials on the carbohydrate metabolism of malaria parasites. Trans R Soc Trop Med Hyg 63:3–4.
Allred DR, Gruenberg JE, Sherman IW (1986): Dynamic rearrangements of erythrocyte membrane internal architecture induced by infection with *Plasmodium falciparum*. J Cell Sci 81:1–16.
Ardeshir F, Flint JE, Matsumoto Y, Aikawa M, Reese RT, Stanley H (1987): cDNA sequence encoding a *Plasmodium falciparum* protein associated with knobs and localization of

the protein to electron-dense regions in membranes of infected erythrocytes. EMBO J 6:1421–1427.

Bignami A, Bastianelli G (1890): Observation of Estivo-Autumnal malaria. Riforma Med 6:1334–1335.

Blobel G (1980): Intracellular protein topogenesis. Proc Natl Acad Sci USA 77:1496–1500.

Bowman IBR, Grant PT, Kermack WO (1960): The metabolism of *Plasmodium berghei*, the malaria parasite of rodents. I. The preparation of the erythrocytic form of *P. berghei* separated from the host cell. Exp Parasitol 9:131–136.

Card RT, Mohandas N, Mollison PL (1983): Relationship of post-transfusion viability to deformability of stored red cells. Br J Haematol 53:237–240.

Chulay JD, Haynes JD, Diggs CL (1985): *Plasmodium falciparum*: Assessment of *in vitro* growth by [$^3$H]hypoxanthine incorporation. Exp Parasitol 55:138–146.

Coppel RL, Culvenor JG, Bianco AE, Crewther PE, Stahl H-D, Brown GV, Anders RF, Kemp DJ (1986): Variable antigen associated with the surface of erythrocytes infected with mature stages of *Plasmodium falciparum*. Mol Biochem Parasitol 20:265–277.

Coppel RL, Cowman AF, Lingelbach KR, Brown GV, Saint RB, Kemp DJ, Anders RF (1983): An isolate-specific S-antigen of *Plasmodium falciparum* contains an exactly repeated sequence of eleven amino acids. Nature 306:751–756.

Cranston HA, Boylan CW, Carroll GL, Sutera SP, Williamson JR, Gluzman IY, Krogstad DJ (1984): *Plasmodium falciparum* maturation abolishes physiologic red cell deformability. Science 223:400–403.

Culvenor JG, Langford CJ, Crewther PE, Saint RB, Coppel RL, Kemp DJ, Anders RF, Brown GV (1987): *Plasmodium falciparum*: Identification and localization of a knob protein antigen expressed by a cDNA clone. Exp Parasitol 63:58–67.

Divo AA, Geary TG, Davis NL, Jensen JB (1985): Nutritional requirements of *Plasmodium falciparum* in culture. I. Exogenously supplied dialyzable components necessary for continuous growth. J Protozool 32:59–64.

Driessen GK, Scheidt-Bleichert H, Sabota A, Inhoffen W, Heidtmann H, Haest CW, Kamp D, Schmid-Schonbein H (1982): Capillary resistance to flow of hardened (diamide treated) red blood cells (RBC). Pflugers Arch 392:261–267.

Elford BC, Haynes JD, Chulay JD, Wilson RJM (1985): Selective stage-specific changes in the permeability to small hydrophilic solutes of human erythrocytes infected with *Plasmodium falciparum*. Mol Biochem Parasitol 16:43–60.

Ellis J, Irving DO, Wellems TE, Howard RJ, Cross GAM (1987): Structure and expression of the knob-associated histidine-rich protein of *Plasmodium falciparum*. Mol Biochem Parasitol 26:203–214.

Engleman DM, Steitz TA (1981): The spontaneous insertion of protein into and across membranes: The helical hairpin hypothesis. Cell 23:411–422.

Geary TG, Divo AA, Bonanni LC, Jensen JB (1985): Nutritional requirements of *Plasmodium falciparum* in culture. III. Further observations on essential nutrients and antimetabolites. J Protozool 32:608–613.

Ginsburg H, Krugliak M, Eidelman O, Cabantchik ZI (1983): New permeability pathways induced in membranes of *Plasmodium falciparum*-infected erythrocytes. Mol Biochem Parasitol 8:177–190.

Ginsburg H, Kutner S, Krugliak M, Cabantchik ZI (1985): Characterization of permeation pathways appearing in the host membrane of *Plasmodium falciparum*-infected red blood cells. Mol Biochem Parasitol 14:313–322.

Gruenberg J, Allred PR, Sherman IW (1983): A scanning electron microscope analysis of the protrusions (knobs) present on the surface of *Plasmodium falciparum*-infected erythrocytes. J Cell Biol 97:795–802.

Gumbiner B, Kelly RB (1982): Two distinct intracellular pathways transport secretory and membrane glycoproteins to the surface of pituitary tumor cells. Cell 28:51–53.

Hadley TJ, Leech JH, Green TJ, Daniel WA, Miller LH, Howard RJ (1983): A comparison of knobby (K+) and knobless (K−) parasites from two strains of *Plasmodium falciparum*. Mol Biochem Parasitol 9:271–278.

Hommel M, David PH, Oligino LD (1983): Surface alteration of erythrocytes in *Plasmodium falciparum* malaria. J Exp Med 157:1137–1148.

Hommel M, David PH, Oligino LD, David JR (1982): Expression of strain-specific surface antigens on *Plasmodium falciparum*-infected erythrocytes. Parasite Immunol 4:409–419.

Howard RJ (1987): Malarial proteins at the membrane of *Plasmodium falciparum*-infected erythrocytes and their involvement in cytoadherence to endothelial cells. Adv in Allergy (in press).

Howard RJ, Barnwell JW (1983): The roles of surface antigens on malaria-infected red blood cells in evasion of immunity. In J.J. Marchalonis (ed): Contemporary Topics in Immunology. Vol. 12. New York: Plenum Publishing Corporation, pp 127–200.

Howard RJ, Barnwell JW, Kao V (1983): Antigenic variation in *Plasmodium knowlesi* malaria: Identification of the variant antigen on infected erythrocytes. Proc Natl Acad Sci USA 80:4129–4133.

Howard RJ, Kao V, Barnwell JW (1984a): Protein antigens of *Plasmodium knowlesi* clones of different variant antigen phenotype. Parasitology 88:221–237.

Howard RJ, Barnwell JW, Rock EP, Neequaye J, Ofori-Adjei D, Maloy WL, Lyon JA, Saul AJ (1988): Two ~300 kDa *Plasmodium falciparum* proteins at the surface membrane of infected erythrocytes. Mol Biochem Parasitol 27:207–224.

Howard RJ, Lyon JA, Diggs CL, Haynes JD, Leech JH, Barnwell JW, Aley SB, Aikawa M, Miller LH (1984b) Localization of the major *Plasmodium falciparum* glycoprotein on the surface of mature intraerythrocytic trophozoites and schizonts. Mol Biochem Parasitol 11:349–362.

Howard RJ, Lyon JA, Uni S, Saul AJ, Aley SB, Klotz F, Panton LJ, Sherwood JA, Marsh H, Aikawa M, Rock EP (1987a) Transport of a Mr ~300,000 *Plasmodium falciparum* protein (PfEMP2) from the intraerythrocytic asexual parasite to the cytoplasmic face of the host cell membrane. J Cell Biol 104:1269–1280.

Howard RJ, Uni S, Aikawa M, Aley SB, Leech JH, Lew AM, Wellems TE, Marsh K, Rener J, Taylor DW (1986): Secretion of a malarial histidine-rich protein (PfHRPII) from *Plasmodium falciparum* infected erythrocytes. J Cell Biol 103:1269–1277.

Howard RJ, Uni S, Lyon JA, Taylor DW, Daniel W, Aikawa M (1987b): Export of *Plasmodium falciparum* proteins to the host erythrocyte membrane: Special problems of protein trafficking and topogenesis. In K.-P. Chang and D. Snary (eds): Host-Parasite Cellular and Molecular Interactions in Protozoal Infections. Nato ASI Series, Vol H11. Berlin: Springer-Verlag, pp 282–296.

Kilejian A (1984): The biosynthesis of the knob protein and a 65,000 Dalton histidine-rich polypeptide of *Plasmodium falciparum*. Mol Biochem Parasitol 12:185–194.

Kreier JP, Green TJ (1980): The vertebrate host's immune response to *Plasmodia*. In J.P. Kreier (ed): Malaria. Immunology and Immunization. Vol. 3. New York: Academic Press, pp 111–162.

Kutner S, Baruch D, Ginsburg H, Cabantchik ZI (1982): Alterations in membrane permeability of malaria infected human erythrocytes are related to growth stage of the parasite. Biochim Biophys Acta 687:82–86.

Kutner S, Ginsburg H, Cabantchik ZI (1983): Permselectivity changes in malaria (*Plasmodium falciparum*) infected human red blood cell membranes. J Cell Physiol 114:245–251.

Langreth SG, Reese RT, Motyl MR, Trager W (1979): *Plasmodium falciparum*: Loss of knobs on the infected erythrocyte surface after long-term cultivation. Exp Parasitol 48:213–219.

Leech JH, Barnwell JW, Aikawa M, Miller LH, Howard RJ (1984a): *Plasmodium falciparum* malaria: Association of knobs on the surface of infected erythrocytes with a histidine-rich protein and the erythrocyte skeleton. J Cell Biol 98:1256–1264.

Leech JH, Barnwell JW, Miller LH, Howard RJ (1984b): Identification of a strain-specific malarial antigen exposed on the surface of *Plasmodium falciparum*-infected erythrocytes. J Exp Med 159:1567–1575.

Lodish HF, Braell WA, Schwartz AL, Strauss GJM, Zilberstein A (1981): Synthesis and assembly of membrane and organelle proteins. In A.L. Muggleton-Harris (ed): Membrane Research, Classic Origins and Current Concepts. Int Rev Cytol Suppl 12. NY: Academic Press, pp 247–309.

Luse SA, Miller LH (1971): *Plasmodium falciparum* malaria. Ultrastructure of parasitized erythrocytes in cardiac vessels. Am J Trop Med Hyg 20:655–660.

MacPherson GG, Warrell MJ, White NJ, Looareesuwan S, Warrell DA (1985): Human cerebral malaria. A quantitative ultrastructural analysis of parasitized erythrocyte sequestration. Am J Pathol 119:385–401.

Marsden PD, Bruce-Chwatt LJ (1975): Cerebral malaria. In R.W. Hornabrook (ed): Topics in Tropical Neurology. Philadelphia: FA Davis Publishing, pp 29–44.

Marsh K, Howard RJ (1986): Antigens induced on erythrocytes by *P. falciparum*: Expression of diverse and conserved determinants. Science 231:150–153.

Olmsted JB (1986): Microtubule-associated proteins. Annu Rev Cell Biol 2:421–457.

Oo MM, Aikawa M, Than T, Aye TM, Myint PT, Igarashi IB, Schoene WC (1987): Human cerebral malaria: A pathological study. J Neuropathol Exp Neurol 46:223–231.

Palade G (1975): Intracellular aspects of the process of protein synthesis. Science 189:347–358.

Pfaller MA, Krogstad DJ, Parquette AR, Nguyen-Dinh P (1982): *Plasmodium falciparum*: Stage-specific lactate production in synchronous cultures. Exp Parasitol 54:391–396.

Pologe LG, Pavlovec A, Shio H, Ravetch JV (1987): Primary structure and subcellular localization of the knob-associated histidine-rich protein of *Plasmodium falciparum*. Proc Natl Acad Sci USA 84:7139–7143.

Rock EP, Marsh K, Saul AJ, Wellems TE, Taylor DW, Maloy WL, Howard RJ (1987): Comparative analysis of the *Plasmodium falciparum* histidine-rich proteins PfHRPI, PfHRPII and PfHRPIII in malaria parasites of diverse origin. Parasitology 95:209–227.

Roth J (1987): Subcellular organization of glycosylation in mammalian cells. Biochim Biophys Acta 906:405–436.

Rudzinska MA, Trager W (1968): The fine structure of trophozoites and gametocytes in *Plasmodium coatneyi*. J Protozool 15:73–88.

Sandza JR, Jr, Clark RE, Weldon S, Sutera P (1974): Subhemolytic trauma of erythrocytes: Recognition and sequestration by the spleen as a function of shear. Trans Am Soc Artif Intern Organs 20:457–462.

Scheibel LW, Ashton SH, Trager W (1979): *Plasmodium falciparum* microaerophilic requirements in human red blood cells. Exp Parasitol 47:410–418.

Schnitzer B, Sodeman T, Mead ML, Contacos PG (1972): Pitting function of the spleen in malaria: Ultrastructural observations. Science 177:175–177.

Sharma YD, Kilejian A (1987): Structure of the knob protein (KP) gene of *Plasmodium falciparum*. Mol Biochem Parasitol 26:11–16.

Sherman IW (1979): Biochemistry of *Plasmodium* (Malarial Parasites). Microbiol Rev 43:453–495.

Sherman IW (1985): Membrane structure and function of malaria parasites and the infected erythrocyte. Parasitology 91:609–645.

Sherman IW, Greenan JRT (1986): *Plasmodium falciparum*: Regional differences in lectin and cationized ferritin binding to the surface of the malaria-infected human erythrocyte. Parasitology 93:17–32.

Steinert PM (1981): Intermediate filaments. In J.R. Harris (ed): Electron Microscopy of Proteins, Vol. 1. London: Academic Press, pp 125–166.

Stossel TP (1984): Contribution of actin to the structure of the cytoplasmic matrix. J Cell Biol 99:15s–21s.

Taylor DW, Parra M, Chapman GB, Stearns ME, Rener J, Aikawa M, Uni S, Panton LJ, Aley SB, Howard RJ (1987): *Plasmodium falciparum* histidine rich protein Pf HRP 1: Its subcellular localization to the host erythrocyte membrane under knobs using a monoclonal antibody. Mol Biochem Parasitol 25:165–174.

Trager W, Rudzinska MA, Bradbury PC (1966): The fine structure of *Plasmodium falciparum* and its host erythrocytes in natural malarial infections in man. Bull WHO 35:883–885.

Vernot-Hernandez JP, Heidrich HG (1985): The relationship to knobs of the 92,000 D protein specific for knobby strains of *Plasmodium falciparum*. Z Parasitenkd 71:41–51.

Wellems TE, Howard RJ (1986): Homologous genes encode two distinct histidine-rich proteins in a cloned isolate of *Plasmodium falciparum*. Proc Natl Acad Sci USA 83:6065–6069.

Wickner WT, Lodish HF (1985): Multiple mechanisms of protein insertion into and across membranes. Science 230:400–407.

Wyler DJ, Oster CN, Quinn TC (1979): The role of the spleen in malaria infections. In The Role of the Spleen in Immunology of Parasitic Diseases. Trop Dis Res, Ser No 1. Basel: Schabe and Company, pp 183–204.

Young JD, Jones SEM, Ellory JC (1980): Amino acid transport in human and sheep erythrocytes. Proc R Soc Lond [Biol] 209:355–375.

The Biology of Parasitism, pages 147–165

# Lyme Disease and Human Babesiosis: Evidence Incriminating Vector and Reservoir Hosts

**Andrew Spielman**

*Department of Tropical Public Health, Harvard School of Public Health, Boston, Massachusetts 02115*

## INTRODUCTION

Even where affluent human populations exploit their surroundings in a seemingly sophisticated manner, new infectious agents may emerge from the landscape. Such a highly prevalent infection has recently come to afflict residents and visitors in many of the most attractive sites in North America and Europe. Each year since the mid 1970's, thousands of cases occur among this well-educated elite, hospitalizing jurists, legislators, medical scientists, prominent entertainers, and captains of industry, often severely debilitating these victims for extended periods of time. Vacationers from remote sites have returned to their homes to present their bewildered physicians with the perplexing signs of this multisystem disorder. The disease is life-threatening and may leave devastating sequellae. Clinical managment of patients may be difficult. Indeed, the public health impact and geographical distribution of Lyme disease continues to increase.

Various kinds of information have accumulated during the past decade detailing the cycle of dependency linking the spirochetal agent of Lyme disease, *Borrelia burgdorferi*, its rodent reservoir hosts, and the *Ixodes* ticks that serve as its vectors. A similarly new, but less prevalent companion infection appeared at the same time and in some of the same locations. This malarialike piroplasm, *Babesia microti*, shares many epizootiological features with the agent of Lyme disease. As with any infectious agent, the basic epidemiological question lies in the mechanism of perpetuation. How are these host relationships linked such that the reproductive rate of the infection exceeds unity? Continuity of the chain of transmission depends upon each tier of infections, producing at least as many secondary infections. Stability of the cycle of transmission derives from the resulting ratio.

The following discussion presents a narrative of the development of the present epizootic of Lyme disease, focusing on the changing numerical relationships between alternative hosts and on the dynamic features of the landscape that support these hosts. The objective is to explain why Lyme disease developed as a human zoonosis during the present decade and in particular sites. This analysis will present the evidence incriminating the vector and reservoir hosts and will concentrate on the developing situation in Eastern North America.

## EMERGENCE OF THE NEW ZOONOSES

### Discovery of Human Babesiosis

The emergence of human babesiosis in 1969 on Nantucket Island, Massachusetts, heralded the advent of Lyme disease. An elderly summer occupant of a home on a moor, surrounded by brush and light forest, experienced a malarialike infection that resisted chloroquine treatment (Western et al., 1970). Giemsa-stained preparations of her blood revealed bicolored piroplasms parasitizing numerous erythrocytes. George C. Healy, of the Center for Disease Control, recognized them as the merozoites of *Ba. microti*, thereby establishing the zoonotic nature of the outbreak and providing crucial epizootiological information that facilitated the early studies.

All *Babesias* are tick-borne, and this newly recognized zoonotic agent parasitizes rodents in various parts of the world. On the nearby island of Martha's Vineyard, *Ba. microti* had been discovered in meadow voles, *Microtus pennsylvanicus* (Tyzzer, 1938). Because the dog tick, *Dermacentor variabilis*, was the only tick thought to attack people in the region, a vole-dog tick cycle, like that of Rocky Mountain spotted fever, was informally assumed. But the incident attracted scientific attention mainly as a clinical oddity because human babesiosis, as it is known in Europe, appears to occur solely in asplenic people and as a fulminating, generally fatal disease. Epidemiological attention was directed toward the new zoonosis only in 1973, when a second human infection was recognized in an acquaintance of the index case summering in a nearby Nantucket village (Dammin et al., 1981). She, too, had an intact spleen and suffered a severe, but self-limiting, febrile episode. Another six such cases were recognized on Nantucket Island in 1975 when similar infections became evident on Shelter Island, near the eastern end of Long Island in New York. Human infection by *Ba. microti* continues to be recorded in various parts of eastern Massachusetts, Long Island, and in

Wisconsin. Some 20 clinical cases come to medical attention each year, generally in elderly, asplenic, or immune-compromised people. Many more asymptomatic than apparent infections occur (Ruebush et al., 1981), including about 5% of some 600 healthy people recently surveyed on Cape Cod, in Massachusetts (unpublished). Geographical clustering of human infections helped establish the venue for epidemiological studies.

## Hosts of Babesiasis in Nature

The unambiguous appearance of the blood stages of the etiological agent of human babesiosis facilitated efforts to describe its reservoir hosts. Although infection was recognized in a variety of mammals, the first Nantucket survey immediately focused attention on the white-footed mouse (*Peromyscus leucopus*), the most abundant mammal present and that most frequently found to be infected (Healy et al., 1976). Prevalence of infection in these mice greatly exceeded that in voles, and cotton-tail rabbits (*Sylvilagus floridanus*) and white-tailed deer (*Odocoileus virginianus*) appeared not to be infected (Piesman and Spielman, 1979). The characteristically depauperate nature of this insular fauna further simplified the task of incriminating reservoir hosts. Chipmunk, squirrel, raccoon, skunk, fox, and opossum, which were abundant in nearby mainland sites, automatically became eliminated as candidate reservoir hosts for *Ba. microti* because they did not exist on Nantucket Island.

These early observations immediately pointed toward an *Ixodes* tick as the likely maintenance vector of human babesiosis. Indeed, immature ticks of this genus abundantly infested Nantucket mice (Spielman, 1976) and were tentatively designated as *I. muris*. Nantucket is the type locality for that species (Bishopp and Smith, 1937), and the tick had been abundant nearby on Martha's Vineyard some 30 years earlier. The late medical entomologist Carrol N. Smith began his illustrious career by completing his doctoral thesis on the bionomics of these common ticks (Smith, 1944). That species alone, however, could not have supported the zoonosis because it seems to parasitize rodent hosts exclusively, mainly focusing on voles. Human infection would not have occurred. Nantucket voles, on the other hand were infested mainly by *D. variabilis* and mice by the *Ixodes* species.

Laboratory cultures of the various local ticks were established in order to help incriminate the vector of this zoonosis; a test of host competence for *Ba. microti* was required. Thus, nymphal *Ixodes* removed from Nantucket mice were reared to the adult stage. Surprisingly, these nymphs appeared to "transmute" into adults of another *Ixodes* species, *I. scapularis*. That tick is common in the southeastern states, but with an isolated northern record on

Naushon Island, a nearby island in the Elizabeth chain located off the Massachusetts coast (Cooley and Kohls, 1945). Systematic comparison of museum specimens of the various stages of these ticks established that three species were at issue (Spielman et al., 1979). The *Ixodes* from the southern part of the United States conformed to criteria originally established for *I. scapularis*. The common species, collected from Naushon Island as well as most other sites located along the New England coast and on eastern Long Island, proved to be a new species, later designated *I. dammini*. The once common mouse tick, *I. muris,* has reciprocally become exceedingly scarce, perhaps reflecting some form of competitive displacement. The species is found mainly on Muskeget Island, an islet located between Nantucket and Martha's Vineyard that is too small to support deer.

Studies on vector competence were hampered, initially, because the developmental cycle of *Ba. microti* in its tick vector was virtually unknown. The design of these early studies relied mainly on field observations, which lead to the hypothesis that larval *Ixodes* may acquire infection by feeding on infected rodents, and that the piroplasm would later be transmitted when the resulting nymph fed on another susceptible rodent. Indeed, these experiments succeeded (Spielman, 1976). Hamsters, inoculated with the blood of *Babesia*-infected patients, served to infect larval *Ixodes* ticks that later transmitted infection to other hamsters. An early report of vector competence in *Dermacentor variabilis* has not since been confirmed. The new tick, then, emerged as the prime candidate for the role as the vector of the new zoonosis.

Because of its small size, conventional light microscopy could not serve to record the pattern of development of the piroplasm. Ultrastructural studies, however, provided the necessary resolution to distinguish these complex developmental events. In vertebrate hosts, *Ba. microti* parasites are evident as intraerythrocytic merozoites that divide by budding, as well as accordian-like gametocytes that do not divide in the vertebrate host (Rudzinska et al., 1979). Proportionately more gametocytes form with the passage of time. After ingestion by vector ticks, the accordian forms develop into gametes as their host red cells become ruptured (Rudzinska et al., 1983). Gametes armed with "arrowhead" organelles then undergo syngamy with other anisogametes that are not so armed. The resulting "arrowhead" kinetes correspond to the "ray bodies" observed by Robert Koch at the turn of the century. Soon thereafter, these primary kinetes appress themselves to the peritrophic membrane of the replete larva, penetrate its gut wall, and eventually invade various remote tissues including those of the salivary glands (Karakashian et al., 1986). These primary ookinetes then dedifferentiate to form kinetoblasts which, coincident with the larval-nymphal molt, redifferentiate into secondary ookinetes. Soon thereafter, these ookinetes invade particular salivary acini, where they develop into a latent sporoblast stage of development characterized by a meshwork form. This diapause condition continues through the larval-nymphal molt and until 2 days after the resulting nymph attaches to a host (Piesman et al., 1986c). Sporogony then resumes, resulting in the formation of thousands of infectious sporozoites (Rudzinska et al., 1983) that are injected into the vertebrate host during the third day of feeding. Sporozoites of *Ba. equi* invade and transform lymphocytes, and this is said to apply to *Ba. microti,* as well (Mehlhorn and Schein, 1984). This exoerythrocytic stage of the piroplasm appears to complete its developmental cycle as it occurs in *Ixodes* ticks.

The major epidemiological features of human babesiosis thereby became established. The ubiquitous white-footed mouse, *P. leucopus* serves as host, and the newly described northern deer tick, *I. dammini,* as vector.

## Discovery of Lyme Disease

Lyme disease first appeared in epidemic form in 1975, when some 51 residents of several communities around the Connecticut village of Old Lyme were found suffering from an atypical arthritic condition generally preceded by an annular rash (Steere et al., 1977a). Another 32 residents came to medical attention during the following year. Since the initial study, the clinical spectrum has become greatly enlarged (Steere et al., 1977b, 1987) to include an initial pathognomonic dermal lesion, and later pathology in dermal, nervous, cardiac, musculoskeletal, and meningeal tissues as well as fetal pathology. The original epidemic site is located some 200 km west of Nantucket Island in coastal New England near the mouth of the Connecticut River. This affluent community experienced clustered foci of human infection, generally occurring near the periphery of the community where forested land lay close to homes. The incidence of Lyme disease in North America has continued to rise (Dammin, 1986 unpublished).

The dearth of diagnoses in non-Connecticut cases, before the 1980s, may have been artifactual because a travel history to this original site of discovery generally was considered essential to clinical differential diagnosis. In retrospect, physicians practicing on the neighboring islands of Nantucket and Martha's Vineyard recall the ringlike rash of Lyme disease on patients suffering from babesiosis. Dual infection is not uncommon (Piesman et al., 1986b). "Spider-bite" was said to be the cause of the rash; but these reports extended back only to the mid 1970s. The travel-history criterion of diagnosis was liberalized by 1980, and cases began to be recognized across much broader regions. Interestingly, the characteristic lesion of Lyme disease was first described in the Americas in a Wisconsin patient in 1968 (Scrimenti, 1970). This early report was remarkably complete, even including a discussion of a possible spirochetal etiology for the patient's condition and proposing antibiotic therapy.

Lyme disease subsequently become recognized as an important clinical entity, affecting people in many temperate parts of the world, including much of the Eurasian, Australian, and North American continents. Diagnoses date back to 1909, when a Swedish physician assigned the name "erythema migrans" to a lesion expressed by one of his patients (Afzelius, 1921). The condition has been diagnosed in numerous Europeans during the next half century. But, the case record greatly expanded there during the 1980s.

Cases in this continent cluster mainly along the Atlantic coast from New Hampshire to Virginia and inland through Pennsylvania. Other regions where cases cluster include Wisconsin and Minnesota and coastal California. Scattered cases have been recorded in virtually all parts of the country (Schmid et al., 1982). Annual incidence, where transmission is intense, may approach 10% (Hanrahan et al., 1984; Steere et al., 1986). These extraordinarily intense zoonotic foci in coastal sites in New York and Massachusetts appear to have developed only in the late 1970's.

A common thread came to connect Lyme disease in Connecticut and human babesiosis on Nantucket when the presence of the new tick, *I. dammini,* was recognized in both locations (Steere et al., 1978; Wallis et al., 1978). Although the spirochetal etiology of Lyme disease had not yet been established, epidemiological considerations provided convincing evidence of a shared mode of transmission.

Willy Burgdorfer's seminal discovery of the etiological agent of Lyme disease was not recorded until 6 years after the disease syndrome was first described (Burgdorfer et al., 1982). Spirochetes were found in the guts of adult *I. dammini* collected in a Lyme disease focus on Shelter Island, located near the eastern end of Long Island in New York. The bacterium was cultured and antisera produced in rabbits. Indeed, sera from patients suffering from the infection in various parts of the United States as well as in Europe reacted with this antigen. This critical demonstration of etiology prepared the way for conclusive epidemiological studies of both infections.

## Hosts of Lyme Disease in Nature

Until its etiological agent became known, direct experiments seeking to explain the mechanism of perpetuation of Lyme disease in nature could not proceed. Instead, the epizootiology of human babesiosis served as the model for both infections, a generalization based on theoretical considerations

(Spielman et al., 1984). This theory assumed that these ticks generally became infected by transstadial transmission and was based on the fact that ixodod ticks feed only once in each of their two subadult instars. Thus, effective transmission required that the vector tick specialize in feeding on a competent reservoir host. A "one vector/one reservoir" rule would thereby be axiomatic. The correlative argument that led to the suggestion that *I. dammini* was the vector of Lyme disease similarly pointed toward white-footed mice as the reservoir hosts.

A description of certain features of the life cycle of *Bo. burgdorferi* in the laboratory was a prerequisite to efforts to demonstrate how the agent is perpetuated in nature. Even now, we have only rudimentary information on the life cycle of the agent in mammalian and acarine hosts. The spirochete is abundantly present in the lumen of the midgut of infected ticks (Burgdorfer et al., 1982), and it disseminates to the hemolymph and salivary glands at 2 days after nymphs or 5 days after adults attach to a host (Ribeiro et al., 1987). Occasionally, the spirochete invades ovarian tissue (Burgdorfer et al., 1983); only about 1% of *I. dammini* inherit infection in nature (Piesman et al., 1986a). Although the agent is said to be transmitted to the mammalian reservoir via regurgitation (Burgdorfer, in Barbour and Hayes, 1986), a salivary route had been proven when topical pilocarpine induced spirochetemic ticks to produce numerous spirochetes in their colorless saliva (Ribeiro et al., 1987). Saliva of infected ticks did not contain spirochetes unless these organisms were present in the hemolymph. In mammalian hosts, the pathogen has been demonstrated in various tissues, including the skin, liver, spleen and kidneys (Anderson et al., 1985). Spirochetemia occasionally can be detected. Much remains to be learned.

The original report on the spirochetal etiology of Lyme disease (Burgdorfer et al., 1982) provided the key for testing ticks for infection, as well as the "xenodiagnostic" method for determining whether mammals become infectious. Spirochetes can dependably be recognized by means of darkfield microscopy in the dissected guts of nonfed ticks. Prevalence of infection by the agent of Lyme disease was measured in white-footed mice by observing the midguts of field caught *I. dammini* that engorged on mice in an earlier stage of development (Levine et al., 1985). Infection in these ticks correlated with the probability that they had fed on these mice, and all such field-derived mice appeared to be highly infectious to ticks. Another study was designed to test reservoir competence of white-footed mice in a laboratory setting (Donahue et al., 1987). Indeed, the bite of one infected *I. dammini* generally rendered these mice infectious to ticks within a week or so, and such mice remained infectious for periods greatly exceeding the length of the natural transmission season. Thus, white-footed mice proved to be highly competent experimental hosts and virtually universally infected where the transmission of the spirochete is intense. These mice appear to be the main reservoir hosts in nature.

## HOST RELATIONSHIPS
### Role of Deer in the Origins of These Zoonoses

The restricted geographical distribution of Lyme disease and especially babesiosis argued for a well-defined ecological determinant for this zoonosis. Indeed, wherever either infection was common, the landscape was parklike, with homes sited near stands of forest. Homes were well-spaced on at least 1-ha parcels of land and were surrounded by conspicuous brush. Such sites favor the abundance of deer.

The timing of the appearance of Lyme disease in North America coincided with the reappearance of deer in the region (Spielman et al., 1985). White-tailed deer had disappeared from much of the northeastern portion of the United States before 1640 (Cronon, 1983). Coincident with the arrival of Europeans, the predatory behavior of the aboriginal inhabitants of the region is said to have exceeded the reproductive resources of many medium and large mammals. Deer hunting, at that time, was remembered as an activity of their fathers. These animals remained scarce for more than three centuries until deliberate conservation efforts combined with favorable changes in the landscape permitted unprecedented proliferation of these aninmals. The Elizabeth Islands, located near Cape Cod in Massachusetts, seem to have constituted the main refugia for deer in the region throughout this period of time (Spielman et al., 1985). The local archives record repeated visits by prominent hunting enthusiasts, including Presidents Ulysses S. Grant and Theodore Roosevelt, who are said to have sought invitations to this uniquely prolific deer habitat. Interestingly, a chalcid parasitoid wasp that parasitizes nymphal *I. dammini* is abundant mainly on this original point of infestation (Mather and Spielman, 1987b). Later, conservation efforts promoted deer abundance in various other island sites located along the New England coast. Shelter Island (New York) and Prudence Island (Rhode Island) became famous hunting grounds during the early 1900s; Nantucket and Marthas Vineyard Islands (Massachusetts) attained that state during the 1950s and 1960s; heavy infestations of deer on Great Island (Massachusetts) and various other mainland sites were recognized first in coastal locations during the 1970s. Deer abundance in inland sites appeared to follow that on the coast, and these animals have been extraordinarily abundant in the northern midwestern part of the United States. The pattern of appearance of epidemic Lyme disease and human babesiosis followed this pattern of distribution. Indeed, these infections cluster solely where deer are abundant and beginning only some time after they became abundant.

As in the northeastern United States, epidemic Lyme disease seems to correlate with the abundance of deer wherever cases cluster in space and time. European and Asian foci of infection similarly include wooded regions and these have become prominent mainly in recent years. The deer herds in many of these sites are recovering after virtual extinction during World War II.

A series of observational studies has established the causality of the observed correlation between the local abundance of deer and risk of human infection. The initial suggestion arose from the observation that the adult stage of U.S. vector tick, *I. dammini*, feeds mainly on deer (Piesman et al., 1979). These large mammals proved to be the most abundant and abundantly parasitized hosts for this tick. Indeed, infestations by the tick were not sustained unless deer were abundant (Anderson et al., 1987). The numerical relationship between the abundance of deer and that of the larval stages of the tick then was precisely established on a series of Massachusetts islands (Wilson et al., 1985). The presence of larvae provides the most sensitive such criterion of a successful infestation because birds disperse these ticks (see below). Together, these observations provided a strong case for the role of deer in supporting foci of the main U.S. vector of Lyme disease.

An experimental proof of the role of deer in determining the distribution of Lyme disease was established in Massachusetts on Great Island (Wilson et al., 1988). All but 1 or 2 of the 35 deer that inhabited this 200-ha island were eradicated during a 2-year span. Soon thereafter, the abundance of larvae infesting mice sharply declined. Nymphs too, declined in abundance. After 3 years, however, the diminishing trend of the infestation appeared to proceed slowly. It may be, however, that the density of the infestation on mice was artificially increased due to the absence of deer! Some of these ticks would have been diverted from mice if deer were present. Another spurious result due directly to removal of deer became evident in a superabundance of questing adult ticks during each of the four winters that followed deer removal. This unexpected finding served as additional proof that these ticks were frustrated in their search for hosts and indicated that the tick population was "buffered" against any temporary absence of deer. The prolonged duration of adult superabundance suggested that these ticks may be capable of surviving for more that one season in the absence of suitable hosts. Any deer remaining after such an intervention would likely harbor extraordinary numbers of ticks. For this reason, temporary removal of deer probably would fail to suppress the transmission of Lyme disease, particularly when practiced in a portion of a continuous land mass. Beginning in the third year after deer were eliminated from Great Island, however, the human population of that site largely became protected against infection. Whereas the 200-odd summer residents of this island previously had suffered four to eight Lyme disease infections annually, none were reported thereafter. This experiment demonstrated conclusively that the zoonosis depends upon the presence of deer.

Continuous presence of deer, paradoxically, may inhibit as well as promote transmission of these zoonoses by serving as hosts for ticks that otherwise might have fed on more competent reservoir hosts. Although deer are parasitized by numerous subadult as well as adult *I. dammini,* larvae that feed upon them fail to become infected by either of these zoonotic agents (Telford et al., 1988). This failure of deer to serve as a source of infection for vector ticks stands in interesting contrast to the reported presence of these spirochetes in their tissues (Bosler et al., 1983). When deer are present in a site during May through July, larvae feeding upon them would fail to become infected. When present during August and September, infected nymphs feeding upon them would thereby fail to infect mice. Of course, these zoonoses depend upon the wintertime presence of deer.

Records of the distribution of deer suggest that the Elizabeth Islands may have served as a principal refuge for *I. dammini* during the centuries in which these mammals were scarce elsewhere in the eastern United States. Lyme disease and human babesiosis then, as now, probably occurred there. Interestingly, *Ba. microti* was found on Marthas Vineyard (Massachusetts) in the 1930s (Tyzzer, 1938) and near Ithaca (New York) in the 1950s (Kirner et al., 1958). The occurrence of *Bo. burgdorferi* would not have been recorded. Transmission in these sites may then have been perpetuated by *I. muris,* the once common mouse tick. But few if any human cases would have occurred. If this speculation is correct, both pathogens would have been enzootic across a broad region in eastern North America. The zoonoses, in the sense of human infections, would have disseminated from the Elizabeth Islands during the past few decades.

### Role of Birds Disseminating These Zoonoses

Passerine birds play an important role in the transmission of Lyme disease because they frequently are parasitized by numerous subadult *I. dammini* (Spielman et al., 1979; Anderson and Magnarelli, 1984; Anderson et al., 1984). Ground-feeding birds, including certain thrush and sparrows are most involved. Occasionally, a dozen or so ticks may engorge on a bird. These relationships are complex; they may enhance as well as inhibit transmission.

During the 1970s, the range of *I. dammini* appears to have spread from its point of origin in coastal Massachusetts about 1,000 km to the south and 100 km to the north (Spielman et al., 1985). This rapid, but asymmetrical change in range is consistent with the movements of birds. Larval *I. dammini* quest for hosts mainly during August and September, at the height of the southward migratory flight of birds. During their 4-day period of feeding, numerous ticks could have been carried across this entire distance. On the other hand, few *I. dammini* quest during the late-winter northward migratory flight of these birds. The season of nymphal questing does not begin until May, by which time many birds are already nesting. This seasonal discordance between the activity of birds and that of the tick explains the asymmetrical pattern of distribution of the main U.S. vector of Lyme disease.

The circumstances through which birds contribute to the establishment of a new infestation of *I. dammini* are complex. The activity of these hosts would serve to disperse engorged larvae or nymphs over vast distances. Ultimately, such "migrant" ticks would become adult males or females and would seek to attach to deer. Although *Ixodes* ticks are capable of mating before contacting a host, proximity on the host would maximize opportunity for mating. Thus, ticks that are dispersed by birds would remain virgin if the density of deer were excessive. The sexes might not meet. A restricted site, such as a fenced enclosure or a small island, would be protected in another way; the engorged tick might not drop from a bird within the "target" area. In this manner, two complementary factors would protect a zoo or a deer park from initial introduction of these ticks. Once *I. dammini* became established, however, the density parameters would then promote the abundance of the tick.

The range of distribution of Lyme disease is much greater than that of the piroplasm. Transmission seems to occur wherever the *Ixodes* vector has become well established, and this suggests that the spirochete is more readily transported by birds than is the piroplasm. Although birds appear to have a certain level of vector competence (Anderson et al., 1984), inherited spirochetal infection may explain the difference. About 1% of larvae sampled in enzootic sites are infected (Piesman et al., 1986a), and these presumably would develop into infected nymphs regardless of the competence of the animal that the larva parasitized. The piroplasm appears not to be inherited by the tick and birds appear not to serve as reservoir hosts. Thus, human babesiosis should invade new regions much more slowly than does Lyme disease, a line of reasoning that conforms to the actual situation in nature.

Although ground-frequenting birds play an important role in establishing new zoonotic foci, the force of transmission of both zoonoses would be reduced due to the activity of these hosts. Even if they were highly competent as reservoir hosts, they would serve to disperse infected vectors away from zoonotic foci. Such ticks would be lost to the transmission cycle in proportion to the frequency of feeding on birds.

Numerous nymphal but no larval *I. dammini,* for example, infest mice inhabiting a site just south of a major enzootic site in Ipswich in northeastern Massachusetts (unpublished). The southern site is devoid of deer, rendering the ticks unable to reproduce. Larval *I. dammini* abundantly parasitized birds in both sites, and these ticks became dispersed into a nonpermissive environment.

## Role of Mice in Perpetuating These Zoonoses

The body of field and laboratory evidence that has now accumulated demonstrates that the agents of human babesiosis and Lyme disease are both perpetuated in white-footed mice, at least in the eastern part of North America. As discussed above, this mouse remains highly infectious for extended periods of time, infection is prevalent in nature, and the larval and nymphal stages of the vector specialize in feeding on these hosts.

The importance of white-footed mice as reservoir hosts for the Lyme disease spirochete was tested experimentally by selectively distributing acaricide in a zoonotic site such that mice alone were exposed to the chemical. The observation that subadult *I. dammini* become replete and detach from their nocturnal hosts during the daylight hours (Mather and Spielman, 1986) indicated that the potentially infected portion of the vector population might be concentrated in the nests of their rodent hosts. To place acaricide near these ticks, cotton-wool was impregnated with permethrin and distributed at 10-m intervals across the study sites (Mather et al., 1987a). The cotton rapidly transferred to the nests of mice, and traces of permethrin came to contaminate their coats. A year later, few nymphal ticks were found questing in the treated site, and even fewer of these were infected. In a recent larger-scale effort, only 2% as many questing infected ticks were present where the treated cotton had been distributed as compared to that in a nontreated comparison site. This finding proves that these mice play the dominant role in perpetuating infection in nature.

The capacity of a vertebrate host to function as a reservoir for a vector-borne infection reflects the parameters considered in estimating the capacity of the invertebrate vector. In the case of tick-borne infection, specificity of the vector-host relationship appears to be the dominant variable and is equally relevant in the case of all infectious agents. White-footed mice are the main hosts for larval and nymphal *I. dammini,* and this is consistent with the finding that spirochetal infection, like that of the piroplasm, is virtually universal in these rodents. Indeed, this reasoning would apply to any infections that were to be transmitted by *I. dammini.* Because an effective vector cannot serve as such for more than one reservoir host, all pathogens horizontally transmitted by a vector would be perpetuated by the same reservoir (Spielman et al., 1984).

Lyme disease seems to be more intensely transmitted than is human babesiosis, even where both are enzootic; point-prevalence of the spirochete exceeds that of the piroplasm in mice, in ticks, and in people (Piesman et al., 1987a). One factor contributing to this differential pattern of transmission may lie in the broad host range of the spirochete; it seems to infect a greater variety of hosts than does the piroplasm. Infection has been recognized in a diverse array of at least eight mammalian species (Anderson et al., 1985) as well as certain birds (Anderson and Magnarelli, 1984). Although each of these animals may have little capacity to replace white-footed mice as the main maintenance host of the infection, ticks that are diverted to them might, occassionally, acquire infection. In contrast, *Ba. microti* seems largely limited to small rodents. A nymph developing from a larva that fed on a sciurid, for example, might transmit the spirochete, but not the piroplasm. In addition, larval *I. dammini* seem to become infected by the spirochete more easily than by the piroplasm. In laboratory experiments, poorly infected lots of ticks are far more frequently encountered in the latter case than in the former (nonpublished). Differential transstaddial vector competence may most powerfully contribute to this observed difference in the force of transmission.

Immune tolerance of white-footed mice for *I. dammini* is another variable crucial in interpreting the capacity of a vertebrate to serve as reservoir for a tick-borne infectious agent. In a landmark study of host-parasite specificity, William Trager (1939) demonstrated that subadult *D. variabilis* effectively fed repeatedly on voles, but not on white-footed mice. *I. dammini* demonstrates the reciprocal adaptation (nonpublished), thereby permitting repeated feeding on individual hosts without rejection. In laboratory experiments, for example, it is difficult to test infectivity of voles that had been infected by the bite of infected ticks. Engorgement of attached ticks is prolonged, and the few ticks that successfully engorge rarely develop to the next stage. Such intolerant hosts might confuse epizootiological studies by accumulating numerous, slow-feeding ticks on their bodies. A simple comparison of the frequency of parasitization would suggest that these hosts contribute more to the force of transmission than is the case. In fact, the presence of "attractive" but immune-intolerant hosts would inhibit transmission.

Although white-footed mice serve as the main reservoir hosts of these zoonotic agents in eastern North America, the role of these rodents in the force of transmission may be complex. If, as appears to be the case, the abundance of *I. dammini* depends more upon the abundance of deer than on that of white-footed mice, transmission of these zoonotic agents would vary inversely with the abundance of the mouse. Fewer ticks would parasitize each mouse where mice are abundant than where they are scarce. Indeed, prevalence of infection in nymphal *I.dammini* seemed greatest in mouse-poor Nantucket Island sites (Spielman et al., 1981).

## Role of Other Arthropods in Transmitting These Zoonoses

Spirochetes, considered to be *Bo. burgdorferi*, have been isolated from a broad variety of arthropods taken from nature. Such evidence seems to implicate the ticks *Amblyomma americanum* (Shultze et al., 1985) and *D. variabilis* (Anderson et al., 198S) as candidate vectors. But neither tick feeds on white-footed mice sufficiently frequently to perpetuate infection. Subadults of the *Amblyomma* species, at least in the northeastern portion of the United States, appear to feed exclusively on deer, as demonstrated in unpublished studies performed on Prudence Island, located in Narraganset Bay in Rhode Island. Those of the *Demacentor* species feed mainly on voles (Piesman and Spielman, 1979). Laboratory demonstrations of vector competence are lacking. Furthermore, this evidence of natural infection is weak because the spirochetes were found in the blood-filled guts of ticks that were taken from spirochete-infected hosts. These spirochetes appear to have been ingested from those hosts and rnay not survive until the tick is capable of refeeding in its next stage of development.

Similarly, the Lyme disease spirochete has been demonstrated in the crushed heads of wild-caught mosquitoes and deer flies of various species (Anderson and Magnarelli, 1984). Although this finding served to excite great lay interest, it seems to have little epidemiological importance. Vector competence has not been demonstrated; the head tissues of mosquitoes are not relevant to transmission; and these insects appear to have little contact with mice. Such spirochetes have even been found in nonhematophagous insects, such as botflies. It seems unlikely that any rapidly feeding hematophagous insect could serve as vector for this agent because the natural vector must remain in contact with the host for at least 2 days (Piesman et al., 1987b).

One series of studies indicates that Lyme disease may be transmitted directly from mouse to mouse, even in the absence of arthropod vectors (Burgess et al., 1986). This finding, which depends entirely on serological evidence of infection, requires confirmation. Using xenodiagnostic criteria for infection, other attempts failed to accomplish such transmission (unpublished). In any event, mice are not infected in the absence of *I. dammini* on islands in an enzootic archipelago in the northeastern United States (Anderson et al., 1987).

*Ixodes* ticks, on the other hand, appear to serve as efficient vectors of the Lyme disease spirochete (Spielman et al., 1985). *I. dammini* is the only proven vector in the northcentral and northeastern section of the United States, and *I. pacificus* in the west (Lane and Burgdorfer, 1986). In Europe, *I. ricinus* transmits the infection (Burgdorfer et al., 1983) and *I. persulcatus* in Asia. The role of mouse-feeding ticks such as *I. trianguliceps* of Europe, *I. muris* of eastern North America, and *I. angustus* of western North America has not been investigated.

The presence of *I. dammini* is more frequently associated with intense transmission of Lyme disease than is that of the other *Ixodes* ticks, and this may derive from a restricted range of available hosts (Spielman et al., 1984). Larval and nymphal *I. dammini* are largely restricted to white-footed mice, in part because no lizards occur in their range, thereby freeing the transmission cycle from any inhibitory effect resulting from diversion of the vector to such potentially noninfected hosts, The reservoir competence of lizards has not been tested. In many parts of the range, however, subadult *I. pacificus* frequently attach to lizards (Arthur and Snow, 1968). Another kind of diversion may depress the ability of *I. ricinus* to transmit any infection horizontally. The larvae of this tick feed mainly on small rodents, the nymphs on lagomorphs and the adults on ungulates. *I. scapularis,* in the southeastern United States, appears not to perpetuate the spirochete, perhaps because it feeds mainly on lizards.

## CONCLUSIONS

It would seem that risk of Lyme disease and of human babesiosis is greatest in sites in which deer are abundant during the winter but absent during the transmission season, where the variety of the indigenous mammal fauna is limited, and where white-footed mice are present but not abundant. The most vulnerable setting would be a summer community situated on an island with houses sited in mowed, parklike clearings surrounded by brushy vegetation. This series of interlocking and qualified requirements reflects the delicate complexity inherent in the perpetuation of vector-borne zoonoses.

Another generality arises from considerations of the importance of these numerically precise ecological relationships. An existing zoonotic cycle represents a transmission opportunity for any infectious agent capable of developing in the local vector-reservoir combination. It follows, therefore, that large segments of North America may be vulnerable to the introduction of certain exotic viral infections. That of European tick-borne encephalitis (TBE), for example, is maintained in hosts that are closely related to those maintaining Lyme disease in eastern North America. The Old World deer tick, *I. ricinus*, which is also the recognized European vector of Lyme disease (Burgdorfer et al., 1983), serves as vector, and various species of *Apodemus* mice as reservoir. The transmission cycle comprising this tick and mouse shares many properties with the corresponding cycles of the *Ixodes* and the *Peromyscus* of North America. The recent proliferation of deer may have rendered much of North America vulnerable, for the first time, to introduced TBE. The European cattle pathogen, *Ba. divergens*, may similarly find fertile ground if introduced into the New World.

In general, ecological islands provide an ideal venue for a vector-borne zoonosis. The restricted variety of the biota permits efficient transmission by vectors that would otherwise feed indiscriminately. Zoonotic leishmaniasis, for example, may emerge from such sites, particularly in desertic environments. But stability of transmission decreases when vector arthropods become diverted from the main reservoir hosts. This is the zoonotic paradox; human infections represent diversions from the maintenance cycle that perpetuate the agent.

## ACKNOWLEDGMENTS

This work was supported in part by grant AI19693 from NIAID and a gift from David Arnold.

## REFERENCES

Afzelius A (1921): Erythema chronicum migrans. Acta Dermatol Venereol (Stockholm) 2: 120–125.

Anderson JF, Magnarelli LA (1984): Avian and mammalian hosts for spirochete-infected ticks and insects in a Lyme disease focus in Connecticut. Yale J Biol Med 57:627–641.

Anderson JF, Johnson RC, Magnarelli LA, Hyde FW (1984): Involvement of birds in the epidemiology of the Lyme disease agent *Borrelia burgdorferi*. Infect Immun 51:394–396.

Anderson JF, Johnson RC, Magnarelli LA, Hyde FW (1985): Identification of endemic foci of Lyme disease: isolation of *Borrelia burgdorferi* from feral rodents and ticks (*Dermacentor variabilis*). J Clin Microbiol 22:36–38.

Anderson JF, Johnson RC, Magnarelli LA, Hyde FW, Myers JE (1987): Prevalence of *Borrelia burgdorferi* and *Babesia microti* in mice on islands inhabited by White-tailed Deer. App Environ Microbiol 53:892–894

Arthur DR, Snow KR (1968): *Ixodes pacificus,* Cooley and Kohls, 1943; its life history and occurrence. Parasitology 58:893–906

Barbour AG, Hayes SF (1986): Biology of *Borrelia* species. Microbiol Rev 50:381–400.

Bishopp FC, Smith CN (1937): A new species of *Ixodes* from Massachusetts. Proc Entomol Soc Wash 39:133–138.

Bosler EM, Coleman JL, Benach JL, Massey DA, Hanrahan JP (1983): Natural distribution of the *Ixodes dammini* spirochete. Science 220:321, 322.

Burgdorfer W, Barbour AG, Hayes SF, Benach JL, Grunwaldt E, Davis J.P. (1982): Lyme disease-a tick-borne spirochetosis? Science 216:1317–1319.

Burgdorfer W, Barbour AG, Hayes SF, Peter O, Aeschlimann A (1983): *Erythema chronicum migrans:* A tick-borne spirochetosis. Acta Trop (Basel) 40:79–83.

Burgess EC, Amundson TE, Davis JP, Kaslow RA, Edelman R (1986): Experimental inoculation of *Peromyscus* spp. with *Borrelia burgdorferi:* Evidence of contact transmission. Am J Trop Med Hyg 35:355–359.

Cooley RA, Kohls GM (1945): The genus *Ixodes* in North America. Natl Inst Health Bull 184:1–246.

Cronon W (1983): Changes in the Land: Indians, Colonists, and the Ecology of New England. New York: Hill and Wang.

Dammin GJ, Spielman A, Benach JL, Piesman J (1981): The rising incidence of clinical *Babesia microti* infection. Hum Pathol 12:398–400.

Donahue JG, Piesman J, Spielman A (1987): Reservoir competence of white-footed mice for Lyme disease spirochetes. Am J Trop Med Hyg 36:92–96.

Hanrahan JP, Benach JL, Coleman JL, Bosler EM, Morse DL, Cameron DJ, Edelman R, Kaslow RA (1984): Incidence and cumulative frequency of endemic Lyme disease in a community. J Infect Dis 150:489–496.

Healy GR, Spielman A, Gleason N (1976): Human babesiosis: reservoir of infection on Nantucket Island. Science 192:479–480.

Karakashian SJ, Rudzinska MA, Spielman A, Lewengrub S, Campbell J (1986): Primary and secondary ookinetes of *Babesia microti* in the larval stags of the tick *Ixodes dammini* Can J Zool 64:328–339.

Kirner SH, Barbehenn KR, Travis BV (1958): A summer survey of the parasites of *Microtus pennsylvanicus pennsytvanicus* (Ord) populations. J Parasitol 44:103–105.

Lane RS, Burgdorfer W (1986): potential role of native and exotic deer and their associated ticks (Acari: Ixodidae) In the ecology of Lyme disease in California, USA. Zentralbl Bakteriol Mikrobiol Hyg [A] 263:55–64.

Levine JF, Wilson ML, Spielman A (1985): Mice as reservoirs of the Lyme disease spirochete. Am J Trop Med Hyg 34:355–360.

Mather TN, Spielman A (1986): Diurnal detachment of immature deer ticks *(Ixodes dammini)* from nocturnal hosts. Am J Trop Med Hyg 35:182–186.

Mather TN, Ribeiro JMC, Spielman A (1987a): Lyme disease & babesiosis: Acaricide focused on potentially infected ticks. Am J Trop Med Hyg 36:609–614.

Mather TN, Piesman J, Spielman A (1987b): Absence of spirochetes *(Borrelia burgdorferi)* and piroplasms *(Babesia microti)* in deer ticks *(Ixodes dammini)* parasitized by chalcid wasps *(Hunterellus hookeri).* Med Vet Entomol 1:3–8.

Mehlhorn H, Schein E (1984): The piroplasms: Life cycle and sexual stages, Adv Parasitol 23:37–103.

Piesman J, Spielman A (1979): Host-associations and seasonal abundance of immature *Ixodes dammini* in southeastern Massachusetts. Ann Entomol Soc Am 72:829–832.

Piesman J, Spielman A, Etkind P, Reubush TK II, Juranek DD (1979): Role of deer in the epizootiology of *Babesia microti* in Massachusetts, USA. J Med Entomol 15:538–540.

Piesman J, Donahue JG, Mather TN, Spielman A (1986a): Transovarially acquired Lyme disease spirochetes (*Borrelia burgdorferi*) in field-collected larval *Ixodes dammini* (Acari: Ixodidae). J Med Entomol 23:219.

Piesman J, Mather TN, Telford SR III, Spielman A (1986b): Concurrent *Borrelia burgdorferi* and *Babesia microti* infection in nymphal *Ixodes dammini*. J Clin Microbiol 24:446–447.

Piesman J, Karakashian SJ, Lewengrub S, Rudzinska MA, Spielman A (1986c): Development of *Babesia microti* sporozoites in adult *Ixodes dammini*. Int J Parasitol 16:381–385.

Piesman J, Mather TN, Dammin GJ, Telford SR III, Lastavica SR, Spielman A (1987a): Seasonal variation of transmission risk of Lyme disease and human babesiosis. Am J Epidemiol 126:1187–1189.

Piesman J, Mather TN, Sinsky RJ, Spielman A (1987b): Duration of tick attachment and *Borrelia burgdorferi* transmission. J Clin Microbiol 25:557,558.

Ribeiro JMC, Mather TN, Piesman J, Spielman A (1987): Dissemination and salivary delivery of Lyme disease spirochetes in vector ticks (Acari: Ixodidae). J Med Entomol 24:201–205.

Rudzinska MA, Spielman A, Lewengrub SJ, Piesman J (1979): Intraerythrocytic "gametocytes" of *Babesia microti* and their maturation in ticks. Can J Zool 57:424–434.

Rudzinska MA, A Spielman, S Lewengrub, W Trager, J Piesman (1983): Sexuality in piroplasms as revealed by electron microscopy in *Babesia microti*. Proc Natl Acad Sci USA 80:2966–2970

Ruebush II TK, Juranek DD, Spielman A, Piesman J, Healy GR (1981): Epidemiology of human babesiosis on Nantucket Island. Am J Trop Med Hyg 30:937–941.

Schmid GP, Horsley R, Steere AC, Hanrahan JP, Davis JP, Bowen GS, Weisfeld MY, Hightower AW, Broome CV (1982): Surveillance of Lyme disease in the United States. J Infect Dis 151:1144–1149.

Schulze TL, Bowen GS, Bosler EM, Lakat MF, Parkin W, Altman R, Ormiston BG, Shisler JK (1985): *Amblyomma americanum:* A potential vector of Lyme disease in New Jersey. Science 224:601–603.

Scrimenti, RJ (1970): Erythema chronicum migrans. Arch Dermatol 102:104–105.

Smith, CN (1944): Biology of *Ixodes muris* Bishopp and Smith. Ann Entomol Soc Am 37:221–234.

Spielman A (1976): Human babesiosis on Nantucket Island: Transmission by nymphal *Ixodes* ticks. Am J Trop Med Hyg 25:781–787.

Spielman A, Clifford CM, Piesman J, Corwin MD (1979): Human babesiosis on Nantucket Island, USA: Description of the vector, *Ixodes (Ixodes) dammini*, n.sp. (Acarina: Ixodidae). J Med Entomol 15:218–234.

Spielman A, Etkind P, Piesman J, Ruebush TK II, Juranek D, Jacobs MS (1981): Reservoir hosts of human babesiosis on Nantucket Island. Am J Trop Med Hyg 30:560–665.

Spielman, A, Levine JF, Wilson ML (1984): Vectorial capacity of North American *Ixodes* ticks. Yale J Biol Med 57:507–513.

Spielman A, Wilson ML, Levine JF, Piesman J (1985): Ecology of *Ixodes dammini*-borne human babesiosis and Lyme disease. Annu Rev Entomol 30:439–460.

Steere AC, Malawista SE, Hardin JA, Ruddy S, Askenase PW, Andiman WA (1977a): Erythema chronicum migrans and Lyme arthritis: The enlarging clinical spectrum. Ann Intern Med 86:685–698.

Steere A, Malawista SE, Snydman DR, Shope RE, Andiman WA, Ross MR, Steele FM (1977b): Lyme arthritis: An epidemic of oligoarticular arthritis in children and adults in three Connecticut communities. Arthritis Rheum 20:7–17.

Steere AC, Broderick TF, Malawista SE (1978): Erythema chronicum migrans and Lyme arthritis: Epidemiological evidence for a tick vector. Am J Epidemiol. 108:312–321.

Steere AC, Taylor E, Wilson ML, Levine JF, Spielman A (1986): Longitudinal assessment of the clinical and epidemiological features of Lyme disease in a defined population. J Infect Dis 154:295–300.

Steere AC, Schoen RT, Taylor E (1987): The clinical evolution of Lyme arthritis. Ann Intern Med 107:725–731.

Telford SR III, Mather TN, Moore SI, Wilson ML, Spielman A (1988): Incompetence of deer as reservoir hosts of the Lyme disease spirochete. Am J Trop Med Hyg in press.

Trager W (1939): Acquired immunity in ticks. J Parasitol 25:57–81.

Tyzzer EE (1938): *Cytoecetes microti* n.g.n.sp.; a parasite developing in granulocytes and infective for small rodents. Parasitology 30:247–257.

Wallis RC, SE Brown, KO Kloter, AJ Main Jr (1978): Erythema chronicum migrans and Lyme arthritis; field study of ticks. Am J Epidemiol 108:322–327.

Western KA, Benson GD, Healy GR, Schulz MG (1970): Babesiosis in a Massachusetts resident. N Engl J Med 283:854–856.

Wilson ML, Adler GH, Spielman A (1985): Correlation between abundance of deer and that of the deer tick, *Ixodes dammini* (Acari:Ixodidae). Ann Entomol Soc Am 78:172–176.

Wilson ML, Telford SR III, Piesman J, Spielman A (1988): Reduced abundance of larval *Ixodes dammini* (Acari: Ixodidae) following elimination of deer. J Med Entomol 25:159–178.

# PARASITE IMMUNOLOGY

The Biology of Parasitism, pages 169–182
© 1988 Alan R. Liss, Inc.

# Vaccination Against Parasites: Special Problems Imposed by the Adaptation of Parasitic Organisms to the Host Immune Response

**Alan Sher**

*Immunology and Cell Biology Section, Laboratory of Parasitic Diseases, National Institute of Allergy and Infectious Diseases, National Institutes of Health, Bethesda, Maryland 20892*

## IMMUNOLOGIC INTERVENTION IN PARASITIC DISEASE

A major goal of current research on parasitic disease is vaccine development. Indeed, few parasitic disease experts would disagree with the premise that mass vaccination represents an ideal means of controlling parasitic infection. However, others (e.g., Marsden, pp. 77–92) argue that from the point of view of both cast and difficulty of implementation, antiparasitic vaccines are unlikely to be practical for use in the developing world and that the intensified deployment of traditional measures (e.g., vector control, sanitation) combined with improved chemotherapy may be a more realistic and ultimately successful approach for dealing with the continued problem of parasitic diseases. A more balanced view is that if cheap, effective vaccines can be developed and administered under the existing mechanism of the World Health Organization's Expanded Programme for Immunization (EPI), they would have a great impact on parasitic infection, particularly when used in conjunction with existing control schemes. Although global eradication as achieved with the smallpox immunization campaign is probably an unrealistic goal for most human parasitic agents, vaccination by reducing morbidity and transmission could itself become a crucially important control measure.

Despite the widespread consensus concerning the need for immunization methods against parasitic infections, it has not yet been possible to develop a successful vaccine for preventing any of the human parasitic diseases. While veterinary vaccines exist for dog hookworm *(A. caninum)* and cattle and sheep lungworms *(D. viviparus, D. filaria)* (Miller, 1971; Urquart et al., 1962), these involve the use of attenuated parasites, a procedure which in most countries never would receive authorization for application in man.

While our difficulty in developing vaccines against parasites is in part attributable to their novelty and complexity as infectious agents, it is the remarkable adaptations which parasites have developed to the vertebrate host immune system which probably best explain their ability to withstand attempts at conventional immunization. In the following chapter I will explore the premise that because of the unique relationship of parasites to the host immune system special approaches are needed for vaccine development. Both the nature of parasite immune evasion mechanisms and the strategies which may be effective in overcoming them will be discussed.

## STAGE-SPECIFIC TARGETS OF VACCINE-INDUCED IMMUNITY

Designs for antiparasitic vaccines fall into several well-defined categories with regard to the life-cycle stages which serve as their targets (Fig. 1). Immunization against *invasive stages* is aimed at destroying parasites immediately upon their entry into the vertebrate host, thereby preventing their subsequent development into pathogenic forms. This type of vaccine is the most logical in design since it blocks infection at its initiation. Moreover, since periods of several days are required for immune responses to develop in naive hosts, there is no immunologic selection pressure acting on invasive stages during natural infection which would favor the development of evasion mechanisms. Thus, theoretically at least, invasive stages should be more susceptible to immune attack than more mature life-cycle forms.

On the negative side, invasive life-cycle stages, since they are usually vector derived, are more difficult to obtain in large quantities. Thus, the initial immunochemical studies on malaria sporozoite and filarial L3/L4 antigens have required the infection of massive numbers of mosquitoes or flies to produce enough parasites for analysis. Moreover, in the case of protozoan parasites which, unlike most helminths, multiply in the vertebrate host, immunity against invasive stages must not only be long-lasting but completely sterilizing. Any parasites which break through the protective immune barrier can initiate infections which ultimately will be as pathogenic as those occurring in the absence of vaccination. In addition, because invasive stages often express their major antigens in a transient fashion, the anamnestic response triggered by infection of hosts previously vaccinated against these antigens in most cases would not occur rapidly enough to contribute to the elimination of the invading organisms. Perhaps the archetypal vaccine

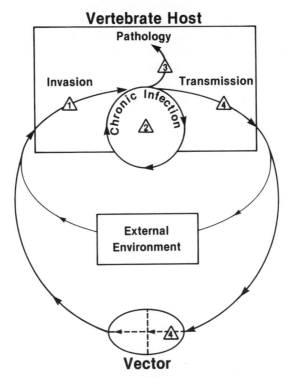

Fig. 1. Stage specificity of antiparasite vaccines. As described in the text, parasite vaccines are directed against 1) invasive stages immediately upon entering the vertebrate host; 2) established stages responsible for maintaining chronic infection; 3) stages responsible for pathology; or 4) the life-cycle forms responsible for maintaining transmission. The immunologic effects on the latter can be expressed in the vertebrate host itself or in the vector by means of antibodies ingested with the blood meal.

against invasive parasite stages is that currently being developed against the circumsporozoite (CS) antigens of malaria sporozoites. The rationale of this vaccine and the strategies currently being employed in its are discussed by Nussenzweig and Nussenzweig, pp. 183–199.

*Established stages* are the second major target against which antiparasite vaccines are directed. These are the life-cycle forms which maintain chronic infection and which in the case of most parasites are required for both the induction of pathology and continued transmission. Since established parasite forms are usually large in size (in the case of worms) or replicative (in the case of protozoa), antigens can usually be obtained in relatively high amounts from these stages. Moreover, with the notable exceptions of African trypanosomes and certain plasmodia, antigen expression by mature parasite forms is usually stable rather than transient (as in the invasive stages). In addition, immunity induced against these stages should be readily boosted by natural challenge infection. Nevertheless, there are several fundamental disadvantages associated with the use of established parasite forms as targets for vaccination. First and foremost is the extent of their adaptation to the host immune response. As the life-cycle stages which maintain chronic infection, mature parasite forms are clearly highly adapted for evasion of immune effector mechanisms and are the stages most closely associated with antigen-specific immunologic suppression. They are therefore usually poor targets of induced immunity, although they can be at least partially controlled, in the case of certain protozoan infections (e.g., malaria, African trypanosomes, *Leishmania*) by immune responses (e.g., anti-SICA responses in malaria) which are induced naturally for the purpose of regulating parasite replication in vivo. Moreover, since established parasite stages are often directly associated with pathology, the immunopathologic consequences of vaccination need to be carefully considered. This is particularly important in *Trypanosoma cruzi* infections, where vaccination against blood-stage trypomastigote forms could conceivably trigger autoimmune pathology comparable to that observed in chronic disease (Nogueira, 1987).

The third major class of stage-specific antiparasite vaccines are those directed against the life-cycle forms which mediate pathology *(antipathology vaccines)*. This type of vaccine, which has been considered almost exclusively for use against diseases caused by helminths (Warren, 1972), is designed to destroy adult worm progeny which induce histopathology or immunologically ameliorate the tissue response to these stages. In infections such as onchocerciasis, where pathology is thought to be produced almost exclusively by these life-cycle forms (i.e., the microfilariae), their elimination as a consequence of vaccination could prevent the disease as well as block transmission (see below). Similarly, if the acute granulomatous response induced by schistosome eggs in naive hosts could be immunologically suppressed, much of the pathology associated with schistosomiasis could be avoided. Since the progeny stages which induce pathology are often less antigenically complex than adult forms and clearly induce good immune responses in the host, they may be easier to immunize against than adult worms which are adapted for chronic infection. Nevertheless, the major problem with antipathology vaccines is the danger of inducing immune responses which will enhance disease rather than diminish it. Thus, in designing such a vaccine not only must the relevant antigens be identified, but also a method of antigen presentation which will consistently elicit responses which kill the target (e.g., microfilariae) without inducing histopathology, or (in the case of schistosomes eggs) which suppress the host reaction against the target without the risk of enhancing it.

The final category of antiparasite vaccine is that designed specifically for *blocking transmission*. This form of immunization is aimed at eliminating the life-cycle stages (e.g., microfilariae, gametocytes, eggs) which by passage to the vector or the external environment maintain transmission. No effect on infective or established forms is necessarily intended, although in the case of antimicrofilarial vaccines prevention of pathology may also occur. This type of immunization would be useful only against parasitic infections which lack significant animal reservoirs (e.g., *Onchocerca P. falciparum, Necator*) but could become an extremely powerful tool, equivalent to vector control or sanitation, in controlling the spread of human disease. Interestingly, in the case of vaccines against sexual stages of malaria (see R. Carter, pp. 225–231) the effector mechanisms which block transmission are directed against target antigens expressed by the parasite in the mosquito rather than in the vertebrate host.

In summary, each of the four vaccine strategies *(anti-invasive stage, anti-established stage, antipathology,* and *transmission blocking)* has its particular advantages and disadvantages. Clearly, the ideal vaccine would be directed against invasive stages. Nevertheless, for protozoa such a vaccine may require the difficult-to-achieve state of a continuously maintained sterilizing immunity. Invasive stage vaccines may thus be more feasible against nonreplicative helminths (e.g., schistosomes) where the primary goal is the reduction of morbidity rather than the induction of complete protection. At the other end of the spectrum, transmission-blocking vaccines, while only useful against certain parasites, may be the easiest to produce. Thus, in the case of malaria and filarial parasites, the evidence suggests that transmission-blocking responses can be more readily stimulated than any of the other forms of antiparasite immunity (Miller et al., 1986; Kazura et al., 1986).

## PARASITE ADAPTATIONS TO THE HOST IMMUNE RESPONSE

An understanding of parasite immune evasion mechanisms is of fundamental concern in the design of effective vaccines. For instance, without knowledge of antigenic variation (see chapters by Turner and Donelson, pp. 349–369 and pp. 371–400) one would suspect a priori that variant surface glycoprotein (VSG), the major surface glycoprotein of African trypanosomes, would induce an excellent protective immunity against infection with the parasite. Similarly, the difficulties which most investigators have experienced vaccinating against schistosomes with tegumental antigens seem surprising without knowledge of the rapid adaptive changes undergone by schistosomula soon after entering the host (Pearce and Sher, 1987).

A list (adapted from Cohen, 1982) of major immune evasion strategies used by parasites is presented in Table 1. While it is probable that many more of these mechanisms will be uncovered as a consequence of future research, a unifying pattern is evident in the available data. Thus, it would appear that a majority of the mechanisms studied function to prevent or counteract the lethal effects of antibody binding to parasite surface epitopes. Indeed, given the knowledge that parasite surface antigens are nearly always immunogenic, it is not surprising that defense against membrane attack would be a major priority for survival.

A correlate of this hypothesis is that the induction of antibody responses against surface epitopes should in most instances (the exceptions are discussed in a later section of this chapter) be an inefficient strategy for vaccination against parasites. What then are the alternative immune responses and immunogens to be utilized in the production of parasite vaccines?

## THE NATURE OF ANTIPARASITIC IMMUNE EFFECTOR MECHANISMS

While considerable effort has been focused during the last two decades on the nature of immune reactions which kill parasites in vitro, it is remarkable how little progress has been made on the definitive identification of effector

TABLE 1. Major Immune Evasion Strategies Employed by Parasites[a]

A. Mechanisms specifically designed to prevent damage to the parasite surface
1. Stage-specific developmental modification in major surface antigen(s)
2. Antigenic variation
3. Rapid turnover and replacement of surface antigens
4. Antigenic disguise
5. Intrinsic resistance of membrane to immune damage[b]

B. Mechanisms which could act at other sites in addition to the parasite surface
1. Antibody cleavage ("fabulation")
2. Inactivation of complement
3. Inactivation of effector cell function
4. Immunosuppression
5. Anatomical sequestration

[a]Adapted from Cohen (1982).
[b]For a description of this type of mechanism see Pearce and Sher (1987).

mechanisms capable of killing the same organisms in vivo. Nevertheless, it is clear from both transfer and depletion experiments that both antibody-dependent and T-lymphocyte-dependent cell-mediated responses can be protective against challenge infection and that these responses can be modulated by certain cytokines. Largely because of the availability of the appropriate technology (e.g., monoclonal antibodies) for their analysis, most research has concentrated on the identification of parasite antigens which are the targets of putative humoral protective responses. Nevertheless, while monoclonal antibodies against many of these antigens are highly effective in passively immunizing against challenge infection, in most instances active immunization with the same molecules fails to induce comparable levels of protection (e.g., Lal et al., 1987; Harn, 1987). While this phenomenon is poorly understood, it is likely that the high antibody levels, antigenic fine specificity, and timing achieved in passive immunization experiments may be difficult to duplicate by direct vaccination. Moreover, recent work on schistosome immunity in the rat model has stressed the critical role of isotypes in both mediating and blocking antibody-dependent parasite killing (Capron and Capron, 1986). Thus, successful vaccination may require the induction of unusually high titers of antibodies as well as a carefully balanced combination of isotypes not stimulated during natural infection.

While not conforming to a fixed set of criteria, the target antigens of protective humoral responses often display similar characteristics. Thus, as surface antigens, many of these molecules are either glycoconjugates in which the carbohydrate portions are most immunogenic (e.g., Snary, 1985; Omer-Ali et al., 1986), or in the case of plasmodia and certain other protozoa, proteins with repeating internal epitopes (see Anders, pp. 201–224). More importantly, the majority of the well-documented antigenic targets of protective humoral immunity appear to be on *invasive* or *transmitted* parasite stages, where the induced immune responses would not directly influence the survival of *established* cycle forms. This situation probably reflects the previously discussed capacity of established stages to evade surface-directed antibody-dependent effector mechanisms, an adaptation needed to maintain chronic infection (see above). Such adaptations should not be required by invasive or transmitted life-cycle forms. The inefficacy of vaccination attempts against established forms is dramatically illustrated by the results (Klotz et al., 1987) of a recent immunization trial in monkeys employing a highly conserved 140-kilodalton (kD) merozoite surface antigen of *Plasmodium knowlesi*. Only half of the vaccinated monkeys showed partial protection when challenged with a malaria clone expressing the antigen. More to the point, parasites eventually appeared in the blood of these protected monkeys which lacked the 140-kD surface antigen and instead expressed new proteins of different molecular size. Similar phenomena have recently been described with giardia parasites after in vitro incubation with monoclonal antibody (Nash et al., 1988). These findings underscore the remarkable adaptability of established parasite stages when confronted with antibody-dependent surface attack.

The importance of cell-mediated as opposed to antibody-dependent effector mechanisms in protective immunity against parasitic infection has only recently begun to be appreciated. Although long assumed to play a major role in the killing of intracellular parasites by activated macrophages, cell-mediated immunity (CMI) is now being considered as a major mechanism in protection helminth infection (see James and Scott, pp. 249–264). Moreover, recently data has emerged from studies of malaria immunity arguing for a new form of antiparasite CMI involving killing of exoerythrocytic forms of the parasite by $CD8^+$ lymphocytes (Weiss et al., 1988; Schofield et al., 1988).

As discussed in detail by James and Scott (pp. 249–264), CMI offers a series of important advantages over antibody-dependent protective immunity in vaccination against parasites. Most relevant to this discussion is the lack of an obligatory dependence on surface antigens as targets of cell-mediated effector mechanisms. That is, in theory any released antigen (membrane associated or not) can serve as an immunogen for CMI. For example, molecules conventionally thought to be internally expressed (e.g., viral nucleoproteins) have been shown to serve as excellent target antigens in the recognition of virus infected host cells by cytotoxic lymphocytes (Townsend et al., 1984). More importantly, soluble/nonmembrane antigens have been shown to be potent stimulators of helper T cells involved in lymphokine production for activation of macrophages for both intracellular and extracellular parasite killing and in that capacity can serve as vaccine immunogens (see James and Scott, pp. 249–264). This feature of CMI-based effector mechanisms is of particular importance to vaccination against parasites where, as emphasized repeatedly above, evasion of antibody-dependent attack against surface antigens is common.

It should be stressed that parasites, in addition to their highly effective adaptations to the humoral response, clearly must have also evolved mechanisms for avoiding the potentially lethal effects of the CMI which they induce during natural infection. Nevertheless, the available evidence suggests that the escape from CMI occurs primarily through immunosuppression rather than from the use of evasion mechanisms which directly interfere with effector function (e.g., *Leishmania*; Howard, 1986). Indeed, suppression of parasite-stimulated T-cell responses is a common occurrence in chronic infection, while the humoral response against surface antigens is only rarely affected in the same parasitized hosts (Sher and Otteson, 1988). However, parasite-induced suppression of cell-mediated responses is probably not an important consideration in the design of CMI-based vaccines since in immunized hosts challenge infections should be curtailed before they have the opportunity to dampen vaccine induced helper T-cell function. The latter hypothesis could be directly tested by assaying the effect of challenge infection on cell-mediated responses in vaccinated experimental hosts.

## SELECTION AND CONSTRUCTION OF VACCINE IMMUNOGENS

In addition to the choice of effector mechanism, successful vaccination against parasites is also dependent, as one would suspect, on the nature of the antigen selected. As already discussed above, internal or excretory/ secretory antigens may offer a distinct advantage for immunization in that parasites may not have evolved mechanisms for evading immune responses directed against them. Moreover, as nonmembrane molecules, these proteins should not be heavily glycosylated and should therefore possess a high density of peptide epitopes available for recognition by helper T cells. A fascinating exception to this generalization is the membrane form of the excretory factor (EF) produced by leishmania promastigotes which induces potent protection against challenge infection apparently based on the stimulation of CMI (Handman and Mitchell, 1985). Interestingly, this molecule is a glycolipid which theoretically at least should not be recognized efficiently by T lymphocytes. The mechanism underlying this novel form of immunization remains to be elucidated.

The choice of antigen also has a direct bearing on whether the protective immune response induced will be genetically restricted. Thus, as would be predicted, the response to the highly repeated epitopes on surface antigens such as the CS molecule on malaria sporozoites is highly major histocompatibility complex (MHC) restricted (Good et al., 1986). In contrast, the response to nonsurface antigens (in this case from malaria sexual stages) shows no major MHC influence (M. Good, personal communication). Since these internal molecules presumably contain multiple, nonidentical T-cell epitopes, the latter observation (which we have confirmed in unpublished studies on the response to schistosome paramyosin) is entirely consistent with current concepts of MHC restriction. Indeed, it is has been hypothesized that parasites may promote their survival in host species by limiting the number of T-cell epitopes contained within their surface-exposed molecules, thereby reducing the capacity of these antigens to induce helper T responses and promoting opportunities for the parasite to encounter genetically unresponsive host individuals (Good et al., 1987).

On the basis of the evidence cited above, the need for multiple/nonidentical T-cell epitopes becomes an important factor in the design of synthetic antiparasitic vaccines. Therefore, in addition to any B epitopes which may be required for antibody-based immunization, the ideal vaccine construct should contain highly immunogenic T epitopes in sufficient number and diversity to avoid any possibilities of MHC restriction and to induce strong CMI, if required. Furthermore, the T epitopes employed should be carefully selected such that they will induce a state of immunologic memory readily recalled upon challenge infection. These elaborate structural requirements would seem difficult to achieve in a subunit vaccine assembled from a series of distinct synthetic peptides. Instead, immunization with recombinant proteins containing the largest portion possible of the native antigen is on theoretical grounds a more simple and practical means of achieving the desired goal of epitope polyvalency.

## "WAKSMAN'S POSTULATE" AND THE SEARCH FOR THE "ACHILLES HEEL"

The optimal choices of immunogen- and vaccine-induced effector mechanism may not be obvious, particularly in the case of parasites (e.g., African trypanosomes, schistosomes, hookworms) which are well-adapted to host defenses. Indeed, the analysis of immune responses occurring during natural parasitic infections may be an inappropriate approach for protective immunogen and effector mechanism identification. I credit Dr. Byron Waksman, the distinguished Yale cellular immunologist (and lifelong summer resident of Woods Hole), as the first modern proponent of this concept. Waksman suggested at a schistosome immunology meeting in the 1970s that rather than examining parasite antigens recognized by infected animals and patients as potential vaccine immunogens, we should focus on those molecules against which little or no response was directed. He based this hypothesis (which at the time seemed heretical) on the very logical concept that any antigen against which parasites allowed the host to mount an immune response is by definition unimportant for the survival of the organism.

Although "Waksman's postulate" in its most strict interpretation represents a radical departure from conventional immunologic thinking, it is a useful concept to bear in mind in the design of vaccines against highly adapted parasites. Thus, as applied to the choice of immunogens, antigens that induce poor or immunosuppressed responses during natural infection should not be ignored since they may be molecules essential for parasite survival. As parasite "Achilles' heels," these apparently poorly immunogenic molecules may in fact represent ideal vaccine antigens. Indeed, a current theme in vaccine development is the search for enzymes (e.g., proteases) whose function is vital for parasite survival (McKerrow, 1987). As emphasized by Waksman, it would be a mistake to expect antibodies from infected laboratory or human hosts to identify these potentially highly effective vaccine immunogens. The same logic applies to immune effector mechanisms. Thus, in situations where infection stimulates strong humoral responses, it may be appropriate to test the role of CMI in vaccine-induced resistance.

## ANTIGEN PRESENTATION AND THE FUTURE OF VACCINE TECHNOLOGY

In his 1987 Presidential Address to the American Association of Immunologists, Dr. William Paul noted that during the past century studies on the immune system have focused on the elucidation of the cellular and molecular foundations of immunologic specificity. During the next century, he argued, immunologists would now utilize this new understanding of the structural basis of immune recognition in reexamining the more complex issues of the physiology of immune responses and their regulation (Paul, 1987). Dr. Paul's proclamation of a new era of physiological studies on the immune system has occurred at a highly appropriate moment for immunoparasitologists involved in the field of vaccine development against parasites. For it is precisely in this area that the future challenges in the field appear to lie. Thus, as summarized above, major progress has been achieved in the identification of parasite antigens which can serve as targets of protective immune responses as well as in the definition of the effector mechanisms involved. Where the field now appears to be stalled is in the induction of high levels of long-lasting immunity with the vaccine immunogens already identified. Our inability to do so stems in large part from our poor understanding of the in vivo process of antigen presentation, i.e., how to administer antigen to induce the exact immune response desired in terms of T-cell subset, antibody isotype, receptor affinity, etc. The latter are major problems in the grey area of immune response physiology referred to by Dr. Paul.

While basic immunology should begin to provide new insights as well as technologies for antigen presentation at the afferent level of the immune response, knowledge of parasite biology and biochemistry will remain essential in understanding the susceptibility of parasite targets to vaccine-induced effector mechanisms as well as the poorly defined events in the elicitation of vaccine primed B- and T-cell responses by parasites during challenge infection. The latter process is of fundamental importance for the successful triggering of cell-mediated effector mechanisms as well as for the boosting by natural infection of previously vaccinated individuals. As demonstrated in experimental vaccination against *Leishmania* (see James and Scott, pp. 249–264), protective molecules can induce T-cell responses indistinguishable from those which fail to induce immunity. The crucial difference may be in the manner in which these antigens are presented by the parasite to primed T cells during challenge infection.

Important beginnings hae been made in the analysis of antigen presentation for vaccination against schistosomes, *Leishmania*, and malaria in murine model systems (e.g., James and Scott, pp. 249–264, Howard, 1986; Playfair and DeSouza, 1987) and in the conceptualization of the relevant issues at the cellular level (Kaye, 1987). It is likely that the next and final stage in parasite vaccine research will be focused around this central problem of the regulation and optimization of parasiticidal responses induced by defined immunogens.

## CONCLUSIONS AND SUMMARY

Largely because of the complexity of the problem of vaccine development against parasitic organisms, a new subdiscipline, immunoparasitology, was created to analyze the special, highly adapted relationship of parasites with the vertebrate immune system. After approximately two decades of intense scientific effort, the field is beginning to show promise of practical payoffs in that it is now possible by immunization with defined antigens partially to protect experimental animals against several major human parasitic infections. The majority of these vaccine models are based on the induction against major parasite surface antigens of antibody responses which are much stronger than those occurring during natural or attenuated infection (see Nussenzweig and Nussenzweig, pp. 183–199). In this chapter, I have hypothesized that more effective and practical vaccines could be designed by utilizing specific immune responses not induced during natural infection and against which parasites have no pre-adaptation. The use of nonmembrane antigens and the induction of cell-mediated immunity are cited as two examples of such a strategy. In addition, the need for further research on the identification of target antigens whose function is critical for parasite survival as well as on the optimization of antigen presentation methodologies is stressed. Because of its strong reliance on knowledge of the structure, function, and host adaptation of protozoa and helminths, the future of vaccine development against these organisms may depend as heavily on advances in the biology of parasitism as on progress in basic immunology and molecular biology.

## ACKNOWLEDGMENTS

I am grateful to my colleagues S. James, P. Scott, M. Good, E. Pearce, L. Miller, T. Nutman, K. Joiner, and D. Sacks for their invaluable discussions and criticism. I also thank the students and faculty of the Biology of Parasitism course for their enthusiasm and inspiration during the period of my association with the MBL summer program.

## REFERENCES

Capron M, Capron A (1986): Rats, mice and men—models for immune effector mechanisms against schistosomiasis. Parasitol Today 2:69–74.
Cohen S (1982): Survival of parasites in the immunocompetent host. In S. Cohen and K.S. Warren (eds): Immunology of parasitic infections, 2nd ed. Oxford: Blackwells, pp 431–474.
Good MF, Berzofsky JA, Maloy WL, Hayashi Y, Fujii N, Hockemeyer WT, Miller LH (1986): Genetic control of the immune response in mice to a *Plasmodium falciparum* sporozoite vaccine: Widespread non-responsiveness to a single malaria T epitope in a highly repetitive vaccine. J Exp Med 164:655–660.
Good MF, Maloy WL, Lunde MN, Margalit H, Cornette JL, Smith GL, Moss B, Miller LH, Berzofsky JA (1987): Construction of synthetic immunogen: Use of new T-helper epitope on malaria circumsporozoite protein. Science 235:1059–1062.
Handman E, Mitchell GF (1985): Immunization with *Leishmania* receptor for macrophages protects mice against cutaneous leishmaniasis. Proc Natl Acad Sci USA 82:5910–5915.
Harn DA (1987): Immunization with schistosome membrane antigens. Acta Trop (Basel) 44(Suppl 12):46–49.
Howard JG (1986): Immunological regulation and control of experimental leishmaniasis. Int Rev Exp Pathol 28:79–116.
Kaye P (1987): Antigen presentation and the response to parasitic infection. Parasitol Today 3:293–299.
Kazura JW, Cicirello H, McCall JW (1986): Induction of protection against *Brugia malayi* infection in jirds by microfilarial antigens. J Immunol 136:1422–1428.
Klotz FW, Hudson DE, Coon HG, Miller LH (1987): Vaccination-induced variation in the 140 kD merozoite surface antigen of *Plasmodium falciparum* malaria. J Exp Med 165:359–367.
Lal AA, De La Cruz VF, Good MF, Weiss WR, Lunde M, Maloy WL, Welsh JA, McCutchan TF (1987): In vivo testing of subunit vaccines against malaria sporozoites using a rodent system. Proc Natl Acad Sci USA 84:8647–8651.
McKerrow JH (1987): The role of proteases in the pathogenesis and immune response to parasitic disease. In N. Agabian, H. Goodman, and N. Nogueira (eds): Molecular Strategies of Parasitic Invasion. New York: Alan R. Liss, Inc., pp 553–557.
Miller LH, Howard RJ, Carter R, Good MF, Nussenzweig V, Nussenzweig RS (1986): Research toward malaria vaccines. Science 234:1349–1356.

Miller TA (1971): Vaccination against canine hookworm diseases. Adv Parasitol 9:153–183.

Nash TE, Aggarwal A, Adam RD (1988): Antigenic variation in *Giardia lamblia*. Immunol, in press.

Nogueira N (1987): American Trypanosomiasis: Antigens and host-parasite interactions. In T.W. Pearson (ed): Parasitic Antigens. New York: Marcel Dekker, Inc., p 105.

Omer-Ali P, Magee Al, Kelly C, Simpson AJG (1986): A major role for carbohydrate epitopes preferentially recognized by chronically infected mice in the determination of *Schistosoma mansoni* schistosomula surface antigenicity. J Immunol 137:3601–3607.

Paul WE (1987): Between two centuries: Specificity and regulation in immunology (Presidential Address, American Association of Immunologists). J Immunol 139:1–6.

Pearce EJ, Sher A (1987): Mechanisms of immune evasion in schistosomiasis. Contrib Microbiol Immunol 8:219–232.

Playfair JHL, DeSouza JB (1987): Recombinant γ IFN is a potent adjuvant for a malaria vaccine in mice. Clin Exp Immunol 67:5–10.

Schofield L, Nussenzweig V, Nussenzweig RS (1988): Antimalarial activity of Lyt2$^+$ (suppressor/cytotoxic) T cells required for immunity to sporozoite challenge. In R. Channock, F. Brown, R. Lerner, and H. Ginsburg (eds): Vaccines 88. Cold Spring Harbor: Cold Spring Harbor Press, in press.

Sher A, Ottesen E (1988): Immunoparasitology. In M. Samter (ed): Immunologic Diseases, 4th edition. Boston: Little, Brown and Co., Chapter 34, in press.

Snary D (1985): Biochemistry of surface antigens of *Trypanosoma cruzi*. Br Med Bull 41:144–148.

Townsend ARM, McMichael AJ, Carter NP, Huddleston JA, Brownlee GG (1984): Cytotoxic T cell recognition of the influenza nucleoprotein and hemagglutinin expressed in transfected mouse L cells. Cell 39:13–25.

Urquhart GM, Jarrett WFH, Mulligan W (1962): Helminth immunity. Adv Vet Sci 4:87–129.

Warren KS (1972): The immunopathogenesis of schistosomiasis: A multidisciplinary approach. Trans R Soc Trop Med Hyg 66:417–434.

Weiss WR, Sedegah M, Beaudowin RL, Miller LH, Good MF (1988): CD8+ T cells (cytotoxic/suppressors) are required for protection in mice immunized with malaria sporozoites. Proc Natl Acad Sci USA 85:573–576.

The Biology of Parasitism, pages 183–199
© 1988 Alan R. Liss, Inc.

# Sporozoite Malaria Vaccines

Victor Nussenzweig and Ruth S. Nussenzweig

Departments of Pathology (V.N.) and Medical and Molecular Parasitology
(R.S.N.), New York University Medical Center, New York, New York 10016

## INTRODUCTION

Malaria continues to be a great killer of children and one of the most incapacitating diseases of adults in the tropical and subtropical areas. Malaria can be acquired outside the endemic areas—for example, by blood transfusion—or even by more peculiar means such as the sharing of syringes among drug addicts, or by the bite of infected mosquitoes carried by airplanes that have landed in endemic areas. Two cases of malaria were reported in a house miles away from Schiphol Airport in Holland, and a single infected mosquito was most likely responsible! Female mosquitoes need blood meals for the development of eggs, and natural selection has made them highly proficient bloodsuckers.

Insecticides and drugs have failed to control the spread of the disease, and in certain areas in Africa, Asia, and Latin America the situation is deteriorating. For example, in the Amazon region drug resistance is widespread, and from 1971 to 1985 infection rates went from 4/1,000 to 24/1,000. This particular epidemic is associated with gold mining, road construction, and agricultural projects, and the affected individuals are mainly male adults who spread the disease when they move to other areas (Marques, 1987).

Vaccines would be a useful addition to chemotherapy and mosquito control in combating malaria. However, they were not considered possible because until recently there was no practical way to obtain sufficiently large amounts of antigen. Moreover, the parasite undergoes complex antigenic changes from one stage of development to the next, and protective immunity is stage specific. Therefore, vaccines might have to contain a mixture of at least two antigens, from the sporozoite and erythrocytic stages of the parasite.

The outlook for vaccine development changed considerably when culture forms of *Plasmodium falciparum* became available to study the molecular basis of the parasite function (Trager and Jensen, 1976) and when antigens involved in protective immunity against sporozoites were identified (Yoshida et al., 1980) and produced on a large scale by applying methods of modern biology (monoclonal antibodies and genetic engineering). In this article we will discuss only progress towards the development of sporozoite vaccines and some of the problems which will have to be overcome to design an ideal immunoprophylactic agent. Studies of other potential malaria vaccines have been reviewed elsewhere (Trager et al., 1986).

## PROTECTIVE IMMUNITY TO SPOROZOITES

When inoculated into the mammalian host, sporozoites develop exclusively inside hepatocytes. Since fewer than 100 sporozoites are sufficient to infect a susceptible host, it seems likely that cell invasion is a specific event mediated by a high-avidity ligand-receptor interaction. In spite of the speed with which infection occurs (sporozoites can be found inside hepatic cells within minutes after inoculation), protective immunity has been achieved by vaccination of rodents, monkeys, and humans with $\gamma$-irradiated parasites (reviewed in Cochrane et al., 1980; Nussenzweig and Nussenzweig, 1986). Mice have also been vaccinated with viable parasites when curative chemoprophylactic regimens have prevented the development of the blood stages.

As expected, immunity was stage specific. Animals fully protected against sporozoite challenge were as susceptible to infection by the erythrocyte stages as naive animals. Immunity was also species specific with a few exceptions; for example, vaccination with attenuated sporozoites of *P. yoelii nigeriensis* also protected mice against challenge with *P. berghei*. This implied that separate vaccines would have to be developed against the various human malaria parasites.

The immunization with irradiated sporozoites did not require adjuvants. Although the protection of rodents against *P. berghei* sporozoites lasted only about 3 months, immunity could be boosted and greatly prolonged by the bite of infected mosquitoes. In endemic areas, the periodic inoculation of sporozoites by mosquito bite could have a similar effect in vaccinated individuals. A few humans have been immunized with *P. falciparum* in a similar way, that is, they were subjected to the bite of hundreds of x-irradiated infected mosquitoes. They were fully protected against challenge with two isolates of *P. falciparum* from distant geographical areas, but were fully susceptible to sporozoites of *P. vivax*. One volunteer was also vaccinated successfully by this unusual procedure with *P. vivax* sporozoites (Clyde et al., 1973, 1975; Rieckmann et al., 1974).

With regard to the mechanism of protective immunity, it appeared to be—at least in part—antibody mediated, since the incubation of sporozoites with the serum of vaccinated and protected animals neutralized their infectivity. In addition this incubation resulted in the formation of a tail-like precipitate at the posterior end of the parasite which increased with time, suggesting that a surface component was being shed by the parasite (circumsporozoite, or CSP, reaction). Also suggestive of a role for antibody in protection was the fact that newborn rodents suckled by sporozoite-immunized foster mothers were totally resistant to sporozoite-induced malaria. Other lines of evidence, however, pointed to a cell-mediated component. For example, in vaccinated mice there was not a good correlation between protection and serum levels of antibodies as measured by the CSP reaction. Also, mice whose antibody production was suppressed at birth with antibodies to the $\mu$-chain of immunoglobulin, and then vaccinated with irradiated sporozoites, were partially protected against challenge (Chen et al., 1977).

These findings were rather puzzling since sporozoites circulate for a very short time, and it would be quite unlikely that they could be directly attacked by specific T cells. Recently, however, it has been shown that the target of the sensitized T cells might instead be the exoerythrocytic forms (EEF) whose development inside hepatocytes is inhibited very efficiently by small doses of $\gamma$-interferon (a T-cell product). In fact, the injection of monoclonal antibodies to $\gamma$-interferon in mice vaccinated with irradiated sporozoites partially reverses the protective immunity (Schofield et al., 1987b). The nature of the sporozoite antigen(s) involved in T-cell sensitization is not known. Considering, however, that vaccination of mice with sporozoites from oocysts of *P. berghei* does not mediate protection, the relevant antigen(s) are stage specific, only expressed in mature salivary gland sporozoites, and may therefore include or consist of the circumsporozoite protein (see below). In short, it appears that vaccination with irradiated sporozoites leads to the production of specific antibodies and T cells. After challenge, the T-cells are triggered to produce $\gamma$-interferon, which acts upon the EEF originating from those sporozoites which escaped the neutralizing activity of the antibodies.

## CIRCUMSPOROZOITE PROTEINS AS TARGETS FOR VACCINE DEVELOPMENT

A monoclonal antibody (3D11) raised against the surface of *P. berghei* sporozoites displayed all the properties of polyclonal antisera obtained from animals vaccinated with irradiated sporozoites. This antibody identified an $M_r$ 44,000 stage- and species-specific protein covering uniformly the surface membrane of sporozoites (circumsporozoite [CS] protein), neutralized their infectivity, and mediated the CSP reaction.

Identical results were obtained when malaria parasites of monkeys (*P. knowlesi*) and humans (*P. vivax* and *P. falciparum*) were studied. Additional studies demonstrated that the various CS proteins had similar physicochemical, biosynthetic, and antigenic properties, and were structurally related. Of greater importance for vaccine development was the evidence that these polypeptides appeared to be involved in the initial interactions between sporozoites and the hepatocyte membrane. Fab fragments of monoclonal antibodies to CS proteins neutralized parasite infectivity in vitro and in vivo and prevented attachment of the parasite to target cells. In addition, there was a close temporal correlation between the acquisition of infectivity and the appearance of CS protein on the membrane of the parasite. Immature sporozoites (found in the midgut of *Anopheles* mosquitoes) bear little or no CS protein on their surface membranes and were not infective.

We know now that CS proteins constitute one of the major biosynthetic products of sporozoites. They are strictly stage specific, are not present in the blood stages of the parasite, and are found in lesser amounts in sporozoites from oocysts. There are two precursor forms of the CS protein, of higher molecular weights and with different isoelectric points. Although the biosynthetic pathways are still obscure, it appears that the membrane form of CS protein is generated by sequential removal of two small basic peptides from the precursors.

CS proteins exhibit unusual immunological properties (Zavala et al., 1983). Not only do they have a strikingly immunodominant epitope, but this epitope is present multiple times in a single molecule. All monoclonal antibodies raised against sporozoites recognize this epitope. Preincubation of crude extracts of sporozoites with any of these monoclonal antibodies inhibits between 70% and 100% of the subsequent binding of antisporozoite antibodies found in human or animal sera. Thus, most or all antisporozoite antibodies in polyclonal antisera also recognize the repetitive epitope. The identification of the CS proteins and the availability of monoclonal antibodies also led to the direct demonstration that the CSP reaction results from the cross-linking and shedding of the CS protein. It is unlikely that this represents a mechanism of escape from the host's immune response as some have proposed, since most or all parasites that have undergone the CSP reaction lose their infectivity. One intriguing possibility is that it mimics an event which normally occurs when the parasite encounters multivalent receptors on the surface of hepatocytes. The CSP reaction could thus represent premature ejection of the surface coat, triggered by a nonphysiological cross-linking signal.

## CLONING OF THE CS GENE

After the identification of the CS protein as a good target for the development of a vaccine, the major practical impediment was that its only source was the mature sporozoite found in the salivary glands of mosquitoes. To overcome this difficulty there were two possible approaches: genetic engineering and/or synthesis of the portion of the molecule that contains the relevant epitopes. As discussed below, both approaches became feasible.

Biosynthetic studies had shown that in mature sporozoites a relatively large proportion of the labeled amino acids incorporated into protein was found in the CS protein and their intracellular precursors, and therefore that the corresponding mRNA should be relatively abundant (Yoshida et al., 1981; Cochrane et al., 1982). mRNA was extracted from the thoraces of mosquitoes infected with sporozoites of *P. knowlesi* to construct a cDNA library, and a clone expressing part of the CS protein of *P. knowlesi* was identified (Ellis et al., 1983; Godson et al., 1983). The cloning of several other CS genes soon followed, including those for the human malaria parasites *P. falciparum* and *P. vivax* (Ozaki et al., 1983; Enea et al., 1984; Dame et al., 1984; Arnot et al., 1985; McCutchan et al., 1985; Galinski et al., 1987; Eichinger et al., 1986; Lal et al., 1987). Only one copy of the CS gene is present per haploid genome and it contains no introns.

There are no sites for N-linked glycosylation or evidence for the presence of O-linked oligosaccharides. In vitro translation from CS mRNA yields polypeptides slightly smaller than the intracellular precursor synthesized by the parasite (51,000 versus 52,000 $M_r$ in the case of *P. knowlesi*). The precise nature of the post-translational modifications which occur in vivo and account for this difference is not known. Biosynthetic studies indicate that radiolabeled palmitic acid and myristic can be incorporated into the CS polypeptides (N. Andrews, unpublished observation).

A schematic representation of a CS protein is shown in Figure 1. On the basis of sequence, charge densities, and secondary structure predictions, the precursor of the CS protein can be divided into several domains between the postulated signal and anchor sequences. The central area consists of the tandem repeats. Two regions have charged residues and may contain $\alpha$-helical structures. One flanks the domain containing the repeats in the direction of the N-terminus. The other is closer to the C-terminus and is flanked on both sides by pairs of cysteine residues. These may form intramolecular disulfide loops, but it should be pointed out that palmitic acid which, as mentioned, is associated with CS proteins, is often linked via a thioester bond to cysteine (Sefton and Buss, 1987). Anchorage of the C5 protein to the membrane by means of a fatty acid may facilitate lateral movement and shedding.

The composition of the postulated anchor sequence (Fig. 2) is similar to that of other intramembranous segments of integral membrane proteins (Mar-

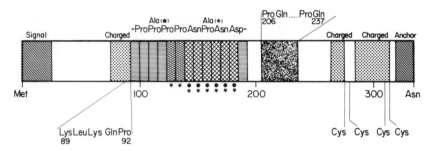

Fig. 1. Structure of the circumsporozoite protein of *Plasmodium berghei* (Eichinger et al., 1986). The aminoterminal amino acids contain a hydrophobic signal sequence, the exact cleavage site of which is unknown. The central part of the protein is composed of two regions of tandemly repeated amino acid units. The largest contains 13 complete 8 amino acid units which vary slightly in sequence (proline → alanine substitutions). Five of these units have the sequence PPPPNPND, six PAPPNAND (●★), and two PPPPNAND (★). Following this series of octamer repeats there is a smaller second, seemingly unrelated, region composed of proline-glutamine pairs. Areas containing large numbers of charged amino acids are found on either side of the repeats, and one contains the highly conserved sequence KLKQP (Dame et al., 1984). Closer to the C-terminal are two pairs of conserved cysteine residues.

| Parasites | disulfide bond ? | | | | | | Surface membrane | Anchor sequence ? | | | | | | | | | | | | | | | | | | | |
|---|---|---|---|---|---|---|---|---|---|---|---|---|---|---|---|---|---|---|---|---|---|---|---|---|---|---|---|
| P. berghei | C | K | M | D | K | C | | S | S | I | F | N | I | V | S | N | S | L | G | F | V | I | L | L | V | L | V F F | N |
| P. yoelii | C | K | M | D | K | C | | S | S | I | F | N | I | V | S | N | S | L | G | F | V | I | L | L | V | L | V F F | N |
| P. falciparum | C | K | M | E | K | C | | S | S | V | F | N | V | V | N | S | S | I | G | L | I | M | V | L | S | F | L F L | N |
| P. cynomolgi | | | | | | | | | | | | | | | | | | | | | | | | | | | | |
| P. knowlesi | C | T | M | D | K | C | | A | G | I | F | N | V | V | S | N | S | L | G | L | I | V | L | L | V | L | A L F. | N |
| P. vivax | ⊖ ⊕ | | | | | | | | | | | | | | | | | | | | | | | | | | | |

Fig. 2. C-terminal sequences of CS proteins. Note that a pair of cysteine residues and a pair of charged amino acids precede the putative anchor sequence. The C-terminal sequences of CS proteins of monkey malarias and *P. vivax* are identical but differ from that of *P. falciparum*.

chesi, 1986), but there is no formal proof that this sequence is present in the mature, membrane-associated CS protein. It contains 23 noncharged amino acids which could span the bilayer and, on the basis of secondary structure prediction programs, they tend to form a helical structure. Most amino acids are hydrophobic and they are interspersed with serines and asparagines. One interesting feature is that the putative anchor is directly flanked on the extracellular side by charged residues located between two conserved cysteine residues which could form a loop if a disulfide bond is established. There is no hydrophilic cytoplasmic domain, suggesting that the CS protein does not transmit directly information across the membrane.

An alternative possibility is that CS proteins are anchored via glycosyl-phosphatidylinositol (GPI), similarly to what has been found for other parasite coat proteins (Ferguson et al., 1988; Low and Saltiel, 1988). In this class of membrane proteins, the GPI-preassembled tail is added soon after protein synthesis, at the same time that the string of predicted genomic COOH-terminal hydrophobic amino acids is removed. One characteristic of the C-terminal of all GPI-anchored proteins is the absence of a cytoplasmic hydrophilic domain and, as pointed out before, this is also the case for CS proteins.

If fatty acids anchor CS proteins to the sporozoite membrane, this would provide an attractive explanation for the striking features of the circumsporozoite reaction. A lipid anchor, spanning only one leaflet of the surface membrane, should facilitate the release of cross-linked CS proteins, and the filamentous and rather rigid structure shed by the parasite might, in fact, consist of antibodies reacting with a CS protein bilayer, whose hydrophobic core is formed by paired CS protein fatty acids.

The gene sequences of the various CS proteins code for a total of about 300–400 amino acids, and one-third to one-half are part of the repeats. The number of repeats found in the CS protein from different isolates of *P. falciparum* and *P. vivax* parasites may vary and this variation is detected as CS protein size polymorphisms (Zavala et al., 1985a). These changes may have arisen by unequal crossover events. The 5'-end of the repeat domain is well defined, but the separation at the 3'-end is less precise and the repeat units at this point appear to degenerate.

As shown in Table 1, the amino acid sequences of the repeats differ markedly among species of *Plasmodium*. However, they share a restricted repertoire of amino acids, that is, Pro, Asn, Gln, Ala, Gly, Asp, Glu, and, more rarely, Arg, Val. In every case the secondary structure prediction programs indicate that the repeats have a strong tendency to form reverse turns. Energy-minimization programs have predicted that six repeats of Asn-

**TABLE 1. Sequence of Repeats of CS Proteins[a]**

| Species | (Strain) | Sequence |
| --- | --- | --- |
| *P. falciparum* | | NA$\overset{\text{VD}}{\text{N}}$P |
| *P. vivax* | | $\overset{\text{D}}{\underset{\text{A}}{}}$GQPAGDRA$\overset{}{\underset{\text{R}}{}}$ |
| *P. knowlesi* | (Nuri) | EQPAAGAGG |
| *P. knowlesi* | (H) | AGQPQAQGDGAN |
| *P. yoelii* | | QGPGAP(QQPP) |
| *P. berghei* | | PPPP$\overset{\text{A}}{\text{N}}$P$\overset{\text{A}}{\text{N}}$D(PQ) |

[a]Amino acids used: P,N,Q,A,G,D,(E),(V),(R).

Ala-Asn-Pro from the *P. falciparum* CS protein would acquire helical conformations, both in polar or nonpolar environments (Gibson and Scheraga, 1986). The unique structural features of the repeats indicate that strong selective pressures severely restrict the amino acid sequences which are compatible with the survival of the parasite. Further evidence for the importance of preserving the repeat sequence stems from the analysis of the *P. falciparum* repeats. In three different isolates (Dame et al., 1984; del Portillo et al., 1987; Lockyer and Schwartz, 1987) the repeats display a large number of silent nucleotide substitutions and a remarkable conserved amino acid sequence.

## FUNCTION AND EVOLUTION OF THE REPEATS

The function of the repeats and the nature of the constraints limiting their variation within a CS protein has been the subject of much speculation. It has been argued that the repeats from neighboring CS molecules may interlock to form a sheath surrounding the parasite. A variation in the structure of the subunit repeat might interfere with the assembly of this quaternary structure and diminish parasite viability.

The repeats could also serve as ligands for the putative sporozoite receptor on hepatocytes. The regular arrangement of the repeats and their high local concentration on the parasite surface membrane would enhance the probability of a fruitful high-avidity encounter with multiple low-affinity hepatocyte receptors. Internal duplications in a gene encoding a ligand for a host cell receptor may be advantageous for parasite survival. The redundant expression of the ligand would enhance the probability of cooperativity in the interaction with a mobile or multimeric receptor. It would also render a key stretch of the CS polypeptide impervious to single amino acid substitutions or to small deletions and insertions, since mutations in one subunit would be of little consequence unless the others were equally affected.

One apparent contradiction to the idea that the CS protein repeats form the ligand for liver cells is the observation that sporozoites from different species (or strains) of malaria parasites which have distinct CS protein repeats can invade hepatocytes of the same host species. It is possible, however, that more than one structure is recognized by different sporozoites on the same liver cells, or that repeats with different amino acid sequences can fold or interact with each other to form similar three-dimensional ligands. Indeed, unexpected cross-reactions have been observed between monoclonal antibodies against CS protein repeats which appear distantly related or unrelated judging from the primary amino acid sequences (Table 2). Because antibodies

TABLE 2. Cross-Reactivities Between Monoclonal Antibodies to Repeats of CS Proteins

| Monoclonal antibody to— | Cross-reactivity with— |
|---|---|
| P. knowlesi (H strain) | P. cynomolgi (NIH strain)[a] |
| P. falciparum | P. berghei[b,c] |
| P. cynomolgi (Berok strain) | P. vivax[d] |
| P. vivax | P. cynomolgi (Berok strain)[d] |
| P. yoelii nigeriensis | P. berghei[e] |
| P. knowlesi (H strain) | Human keratin (57K)[f] |
| P. knowlesi (H strain) | P. falciparum[g] |

[a]Cochrane et al. (1985).
[b]Ballou et al. (1985).
[c]Hockmeyer and Dame (1985).
[d]Cochrane et al. (1986).
[e]Yoshida N, Nussenzweig RS: unpublished observation.
[f]Vergara, U, Bystryn J-C, Sun T, Nussenzweig VN: unpublished observation. There is no significant similarity in sequence between the 57K human keratin and the P. knowlesi repeats, but keratin contains a large number of G, N, Q, E and S residues, and repeated sequences of amino acids.
[g]Cochrane et al. (1982).

TABLE 3. Sequence of the Repeats of the P. cynomolgi Complex (Galinsky et al., 1987)[a]

| Subspecies? Strain? | Sequence |
|---|---|
| Mulligan/NIH | NAGG (D above) |
| Gombak | DGAAAAGGGGN (G above) |
| London | a) ADGARA (E above) |
|  | b) GNQAGGQAGAG (R E above) |
| Ceylon | a) AGNNAAAGE (G A above) |
|  | b) AGNNAAAGEAGAGGAGR |
| Break | a) PDGDGAPA A (G above) |
|  | b) NRAGGQPAAGGNQAGG (A above) |

[a]Amino acids used: P,N,Q,A,G,D,E,R.

are bivalent, cooperativity in binding may reveal subtle similarities between secondary structures of the seemingly unrelated repeats.

Analysis of the repeats of CS genes of the P. cynomolgi complex (Table 3) have shown clearly that this domain is subject to stronger evolutionary pressures than the other domains (Galinski et al., 1987). The sequences flanking the repeats in six strains (or subspecies) of P. cynomolgi are 95%

conserved, while the repeats are markedly diverse. The genetic mechanism leading to the shift from one set of repeat sequences in one strain to a different set in another strain is unknown.

One interesting hypothesis to explain how the CS repeats can evolve rapidly without degenerating is that their maintenance is controlled at the level of the DNA. This process might also be responsible for the spreading of the mutations and the formation of new sets of repeats (Galinski et al., 1987). It should be pointed out that although the repeats are remarkably conserved within a given CS protein, focal changes do occur. One of the most striking was observed in P. falciparum, where NVDP (instead of NANP) appeared a few times in separate regions of the repeat domain of three strains of the parasite originating from widely different geographic locations. This cannot be explained by independent mutational events.

Whatever the genetic mechanisms involved, variation in the repeats is a potential problem for the immunoprophylaxis of malaria. If under selective pressure of the immune system, populations of sporozoites can rapidly change the CS protein repeats, the effectiveness of repeat-containing vaccines would be limited. The available evidence, however, does not justify a pessimistic outlook. The same monoclonal antibodies recognize CS proteins from all isolates of P. falciparum and P. vivax obtained from Latin America, Asia, and Africa (Zavala et al., 1985a). Furthermore, CS proteins in P. falciparum isolates from Thailand and Brazil and other areas contain $(NANP)_3$ repeats (Weber and Hockmeyer, 1985; Yoshida et al., 1987; del Portillo et al., 1987; Lockyer and Schwartz, 1987). In endemic areas there are relatively high levels of antibodies to this epitope in the serum of adults, who after prolonged exposure have developed resistance to malaria infection (Hoffman et al., 1986; Esposito et al., 1988). It appears, therefore, that genetic changes leading to the entire substitution of the repeats of the human malaria CS proteins are either rare or else compromise the viability or infectivity of sporozoites.

## RATIONALE FOR THE DEVELOPMENT OF SYNTHETIC SPOROZOITE VACCINES

In the case of P. falciparum, the immunodominant epitope is defined by only three consecutive repeats $(NANP)_3$. This conclusion is based on studies which showed that $(NANP)_3$ was recognized by polyclonal and monoclonal antibodies to sporozoites, and conversely, that antibodies to (NANP) reacted with sporozoites and prevented their entry into hepatocytes in vitro (Zavala et al., 1985b). In one of these studies, a series of peptides—$(NANP)_2$,

(NANP)$_3$, (NANP)$_4$, and (NANP)$_5$—was synthesized and used to inhibit the binding of antibodies to extracts of *P. falciparum* sporozoites. The results showed that (NANP)$_3$, (NANP)$_4$, and (NANP)$_5$ strongly inhibited the binding of one of the monoclonal antibodies to the antigen with almost equal efficiency on a molar basis. In contrast, (NANP)$_2$ was a poor inhibitor. Similar findings were obtained with several other monoclonal antibodies.

(NANP)$_3$ was then used as the antigen in an immunoradiometric assay (IRMA) to detect antisporozoite antibodies in the sera of humans living in an endemic area, the Gambia, West Africa. In agreement with previous epidemiological studies showing that the immune response of humans to sporozoites is age dependent, it was found that the percentage of positive sera detected by the IRMA increased with age, ranging from 22% in children 1 to 14 years old to 84% in adults older than 34 years. Among many individuals older than 29 years of age, serum titers reached as high as 1/640: the levels of antibodies to (NANP)$_3$ in the serum of a human volunteer (G.Z.) vaccinated with x-irradiated *P. falciparum* sporozoites and protected against malaria infection were also measured. This serum had the highest reactivity and the titer of some samples reached 1/4,096.

Additional observations showed that most or all antibodies to the sporozoites present in the human sera were directed against (NANP)$_3$. Indeed, in these sera, a highly significant correlation ($r_s = 0.87$, $P < 0.001$) was found between the antipeptide and antisporozoite antibody titers. Moreover, the reactivity of the antibodies with sporozoites was abolished or strongly inhibited when the reaction was carried out in the presence of the (NANP)$_3$ peptide.

To study the immunogenicity of (NANP)$_3$, groups of rabbits were immunized with conjugates prepared by coupling the peptide to tetanus toxoid by means of glutaraldehyde. The sera were assayed by an IRMA using (NANP)$_3$ immobilized on the bottom of plastic wells as antigen. All samples from immunized animals were positive, and the positive reactions were totally inhibited by the synthetic peptide. The antibodies to (NANP)$_3$ reacted in a Western blot with *P. falciparum* CS protein and its precursors, and with the surface of glutaraldehyde-fixed sporozoites of *P. falciparum*, as determined by IFA.

Several lines of evidence indicate that the majority of antibodies to this synthetic peptide react with sporozoites. First, *P. falciparum* sporozoite extracts inhibit most of the reactivity of the antibodies with (NANP)$_3$. Second, there is a highly significant correlation between the reactivity (IRMA titers) of antipeptide antibodies with the peptide and their reactivity (IFA titers) with intact sporozoites. Third, among eleven monoclonal antibodies raised against (NANP)$_3$, eight react with the native protein (A. Ruiz and V. Nussenzweig, unpublished observation). It seems, therefore, that the synthetic peptide coupled to the carrier protein adopts a configuration similar to that of native NANP sequences within the repeat domain of the CS protein.

Immunoglobulin was isolated from the serum of one of the rabbits immunized with (NANP)$_3$-tetanus toxoid and tested for its ability to neutralize the infectivity of sporozoites of *P. falciparum* in vitro. The results showed that the immune IgG inhibited parasite invasion of hepatoma cells in a dose-dependent fashion. A strong effect was observed with total IgG concentrations as low as 2 $\mu$g/ml. When the antibodies to (NANP)$_3$ were removed by absorption with peptide bound to Sepharose beads, the inhibitory effect was abolished.

These results strongly suggest that the synthetic peptide (NANP)$_3$ represents the dominant epitope of the domain of the CS protein containing the NANP repeats, and that antibodies to the repeats recognize an uninterrupted sequence of amino acids. The repeat domains most likely contain sequential and not configurational epitopes in all CS proteins, since monoclonal and polyclonal antibodies to the repeats of CS proteins of several species react with the denatured antigen heated at 100°C for 30 min or treated with 6 M urea and 1% $\beta$-mercaptoethanol.

In short, synthetic vaccines containing (NANP)$_3$ are attractive candidates for further development. Most of the antibodies to (NANP)$_3$ raised in experimental animals recognize the native CS protein and neutralize the parasite infectivity. Antibodies in the serum of a volunteer vaccinated with x-irradiated sporozoites and protected against malaria infection were mainly or exclusively directed against (NANP)$_3$. The peptide reacted well with the sera from randomly selected individuals from areas where malaria is endemic. (NANP)$_3$ is repeated many times in each CS molecule, and the CS molecule covers uniformly the surface membrane of sporozoites; therefore, the parasite should be particularly vulnerable to attack by anti-(NANP)$_3$ antibodies. Last, but not least, (NANP)$_3$ is present in strains of *P. falciparum* from all over the world.

A (NANP)$_3$-tetanus toxoid vaccine has recently undergone phase I and limited phase II trials in human volunteers (Herrington et al., 1987). Seroconversions against (NANP)$_3$ occurred in 71% of the recipients of 160 $\mu$g of vaccine. Most positive sera reacted with sporozoites by immunofluorescence and the correlation between antibody titers to (NANP)$_3$ and immunofluorescence titers was highly significant ($r_s = 0.858$, $P < 0.001$). Three vaccinees with the highest antibody titers and four unimmunized controls were challenged with *P. falciparum* sporozoites by the bite of infected mosquitoes.

Blood-stage parasites were detected in all controls after 7–10 days (mean 8.5 days). In contrast, two vaccinated individuals did not show parasites in their blood until day 11, and the third one was not infected. A recombinant vaccine containing multiple NANP repeats was also tested in humans, and in that instance protection also seemed correlated with antibody titers (Ballou et al., 1987).

There was no evidence for parasite-specific T-cell sensitization in any of the peptide-vaccinated individuals, indicating that protection can be achieved with antibodies alone, if the titers are sufficiently high. Similarly, complete protection against *P. berghei* sporozoite challenge has been achieved in a large proportion of mice vaccinated with a repeat peptide conjugated to tetanus toxoid. The serum titers of antibody to the *P. berghei* CS protein were high, but the T-cells did not respond in vitro to the synthetic repeat peptide (Zavala et al., 1987).

These results encourage the development of improved sporozoite vaccine formulations which, if combined with blood stage antigens, may lead to effective control of malaria infection.

A disadvantage of this particular peptide-conjugate vaccine is that it may not prime parasite-specific T-cells, and a boosting of the antibody response and lymphokine production may not occur during subsequent exposure to sporozoites under natural infection. $\gamma$-Interferon, a T-cell product, has a potent inhibitory effect on the exoerythrocytic stages of the parasite, and if produced at the time of challenge, would prevent the development of the sporozoites which escape the neutralizing activity of antibody (Ferreira et al., 1986, 1987; Schofield et al., 1987a).

Perhaps T-cell epitopes from the CS protein (or from other sporozoite molecules) can be synthesized together with the NANP repeats to form a T-B parasite-specific vaccine. In fact, T-cell epitopes from the *P. falciparum* CS protein recognized by certain strains of inbred mice have already been characterized. NANP repeats are recognized by T cells from mice bearing I-A$^b$ in major histocompatibility complex (Del Giudice et al., 1986; Good et al., 1986), while another sequence outside the repeat domain is recognized by mice bearing I-A$^k$ (Good et al., 1987). These observations highlight the fact that T-cell epitopes are genetically restricted, and may not be recognized by all humans. It may take some time until the structure of the most "promiscuous" sporozoite-associated T-cell epitopes is known, and the conditions for assembling these new types of synthetic vaccines are established.

Meanwhile another and perhaps more practical approach to include T and B epitopes in a vaccine would be to use as antigen large segments of genetically engineered CS molecules, in addition to the repeats. An example of the latter type of vaccine for *P. vivax* malaria is now under development (Barr et al., 1987). DNA coding for 234 amino acids of the circumsporozoite (CS) protein of *Plasmodium vivax* was incorporated into yeast expression vectors. The DNA encoded all the repeat domain and codons for sequences 5' and 3' of the repeat domain. If the nucleotides representing the hydrophobic leader and presumed anchor sequences are excluded, it encompasses about 70% of the CS gene. Yeast cells transformed with these autonomously replicating plasmids expressed, upon induction, high levels of the CS polypeptide. The malaria antigen was purified in good yields from yeast extracts and was injected into mice using alum as adjuvant. The antibodies recognized the authentic CS protein, and at high dilutions, they inhibited the invasion of hepatocytes by sporozoites in vitro.

Moreover, the antibody response was boosted by *P. vivax* sporozoites several months following vaccination (Nardin et al., 1987). T-dependent secondary antibody responses in vitro were also obtained by incubation of lymphoid cells from these mice with *P. vivax* sporozoite extracts. It appears, therefore, that the yeast antigen has potential advantages over other types of subunit vaccines consisting of recombinant fusion proteins, or of synthetic peptides coupled to nonrelevant carrier molecules. It remains to be determined whether this type of vaccine will stimulate the production of antibodies of CS protein-specific T cell in a human population.

## REFERENCES

Arnot DE, Barnwell JW, Tam JP, Nussenzweig V, Nussenzweig RS, Enea V (1985): Circumsporozoite protein of *Plasmodium vivax*: Cloning of the immunodominant epitope. Science 230:815–818.

Ballou WR, Rothbard J, Wirtz RA, Gordon DM, Williams JS, Gore RW, Schneider I, Hollingdale MR, Beaudoin RL, Maloy WL, Miller LH, Hockmeyer WT (1985): Immunogenicity of synthetic peptides from circumsporozoite protein of *Plasmodium falciparum*. Science 228:996–999.

Ballou WR, Hoffman SL, Sherwood JA, Hollingdale MR, Neva FA, Hockmeyer WT, Gordon DM, Schneider I, Wirtz RA, Young JF, Wasserman GF, Reeve P, Diggs CL, Chulay JD (1987): Safety and efficacy of a recombinant DNA *Plasmodium falciparum* sporozoite vaccine. Lancet 1:1276–1281.

Barr PJ, Gibson HL, Enea V, Arnot DE, Hollingdale MR, Nussenzweig V (1987): Expression in yeast of a *Plasmodium vivax* antigen of potential use in a human malaria vaccine. J. Exp Med 165:1160–1171.

Chen DH, Tigelaar RE, Weinbaum FI (1977): Immunity to sporozoite-induced malaria infection in mice. I. The effect of immunization of T and B cell-deficient mice. J Immunol 118:1322–1327.

Clyde DF, McCarthy VC, Miller RM, Hornick RB (1973): Specificity of protection of man immunized against sporozoite-induced falciparum malaria. Am J Med Sci 266:398–403.

Clyde DF, McCarthy V, Miller RM, Woodward WE (1975): Immunization of man against falciparum and vivax malaria by use of attenuated sporozoites. Am J Trop Med Hyg 24:397–401.

Cochrane AH, Nussenzweig RS, Nardin EH (1980): Immunization against sporozoites. In J.P. Kreier (ed): Malaria, Vol. 3, Immunology and Immunization. New York: Academic Press, pp 163–202.

Cochrane AH, Santoro F, Nussenzweig V, Gwadz RW, Nussenzweig RS (1982): Monoclonal antibodies identify the protective antigens of sporozoites of Plasmodium knowlesi. Proc Natl Acad Sci USA 79:5651–5655.

Cochrane AH, Gwadz RW, Ojo-Amaize E, Hii J, Nussenzweig VN, Nussenzweig RS (1985): Antigenic diversity of the circumsporozoite proteins in the Plasmodium cynomolgi complex. Mol Biochem Parasitol 14:111–124.

Cochrane AH, Gwadz RW, Barnwell JW, Kamboj KK, Nussenzweig RS (1986): Further studies on the antigenic diversity of the circumsporozoite proteins of the Plasmodium cynomolgi complex. Am J Trop Med Hyg 35:479–487.

Dame JB, Williams JL, McCutchan TF, Weber JL, Wirtz RA, Hockmeyer WT, Maloy WL, Schneider I, Roberts D, Sanders GS, Reddy EP, Diggs CL, Miller LH (1984): Structure of the gene encoding the immunodominant surface antigen on the sporozoite of the human malaria parasite Plasmodium falciparum. Science 225:593–599.

Del Giudice G, Cooper JA, Merino J, Verdini AS, Pessi A, Togna AR, Engers HD, Corradin G, Lambert PH (1986): The antibody response in mice to carrier-free synthetic polymers of Plasmodium falciparum circumsporozoite repetitive epitope is I-A$^b$-restricted: Possible implications for malaria vaccines. J Immunol 137:2952–2955.

del Portillo HA, Nussenzweig RS, Enea V (1987): Circumsporozoite gene of a Plasmodium falciparum strain from Thailand. Mol Biochem Parasitol 24:289–294.

Eichinger DJ, Arnot DE, Tam JP, Nussenzweig V, Enea V (1986): The circumsporozoite protein of Plasmodium berghei: Gene cloning and identification of the immunodominant epitope. J Mol Cell Biol 6:3965–3972.

Ellis J, Ozaki LS, Gwadz RW, Cochrane AH, Nussenzweig V, Nussenzweig RS, Godson GN (1983): Cloning and expression in E. coli of the malarial sporozoite surface antigen gene from Plasmodium knowlesi. Nature 302:536–538.

Enea V, Arnot DE, Schmidt E, Cochrane AH, Gwadz R, Nussenzweig RS (1984): Circumsporozoite gene of Plasmodium cynomolgi (Gombak): cDNA cloning and expression of the repetitive circumsporozoite epitope. Proc Natl Acad Sci USA 81:7520–7524.

Esposito F, Lombardi S, Modiano D, Zavala F, Reeme J, Diallo M, Coluzzi M, Nussenzweig RS (1987): Prevalence and level of antibodies to the circumsporozoite protein of Plasmodium falciparum in an endemic area. Its relationship to resistance against malaria infection. Submitted.

Ferguson MAJ, Homans SW, Dwek RA, Rademacher TW (1988): Glycosyl-phospatidylinositol moiety that anchors Trypanosoma brocei variant surface glycoprotein to the membrane. Science 239:753–759.

Ferreira A, Schofield L, Enea V, Schellenkens H, van der Meide P, Collins WE, Nussenzweig RS, Nussenzweig V (1986): Inhibition of development of exoerythrocytic forms of malaria parasites by gamma interferon. Science 232:881–884.

Ferreira A, Morimoto T, Altszuler R, Nussenzweig V (1987): Use of DNA probe to measure the neutralization of Plasmodium berghei sporozoites by a monoclonal antibody. J Immunol 138:1256–1259.

Galinski MR, Arnot DE, Cochrane AH, Barnwell JW, Nussenzweig RS, Enea V (1987): The circumsporozoite gene of the Plasmodium cynomolgi complex. Cell 48:311–319.

Gibson KD, Scheraga HA (1986): Predicted conformations for the immunodominant region of the circumsporozoite protein of the human malaria parasite *Plasmodium falciparum*. Proc Natl Acad Sci USA 83:5649-5653.

Godson GN, Ellis J, Svec P, Schlesinger DH, Nussenzweig V (1983): Identification and chemical synthesis of a tandemly repeated immunogenic region of *Plasmodium knowlesi* circumsporozoite protein. Nature 306:29-33.

Good MF, Berzofsky JA, Maloy WL, Hayashi Y, Fujii N, Hockmeyer WT, Miller LH (1986): Genetic control of the immune response in mice to a *Plasmodium falciparum* sporozoite vaccine. Widespread nonresponsiveness to single malaria T epitope in highly repetitive vaccine. J Exp Med 164:655-660.

Good MF, Maloy WL, Lunde MN, Margalit H, Cornette JL, Smith GL, Moss B, Miller LH, Berzofsky JA (1987): Construction of a synthetic immunogen: Use of new T-helper epitope on malaria circumsporozoite protein. Science 235:1059-1062.

Herrington DA, Clyde DF, Losonsky G, Cortesia M, Murphy JR, Davis J, Bagar S, Felix AM, Heimer EP, Gillessen D, Nardin EH, Nussenzweig RS, Nussenzweig V, Hollingdale MR, Levine MM (1987): Safety and immunogenicity in man of a synthetic peptide malaria vaccine against *Plasmodium falciparum* sporozoites. Nature 328:257-259.

Hockmeyer WT, Dame JB (1985): Recent efforts in the development of a sporozoite vaccine against human malaria. Adv Exp Med and Biol 185:233-245.

Hoffman SL, Wistar R, Jr, Ballou WR, Hollingdale MR, Wirtz RA, Schneider I, Marwoto HA, Hockmeyer WT (1986): Immunity to malaria and naturally acquired antibodies to the circumsporozoite protein of *Plasmodium falciparum*. Engl J Med 315:601-606.

Lal AA, de la Cruz FV, Welsh JA, Charoenvit Y, Maloy WL, McCutchan TF (1987): Structure of the gene encoding the circumsporozite protein of *Plasmodium yoelii*. J Biol Chem 262:2937-2940.

Lockyer MJ, Schwartz RT (1987): Strain variation in the circumsporozoite protein gene of *Plasmodium falciparum*. Mol Biochem Parasitol 22:101-108.

Low MG, Saltiel AR (1988) Structural and functional roles of glycosyl phosphatidylinositol in membranes. Science 239:268-275.

Marchesi VT (1986): Functional adaptations of transbilayer proteins. In D.S. Dhindsa and O.P. Bahl (eds): Molecular and Cellular Aspects of Reproduction. New York: Plenum Publishing Corp., pp 107-120.

Marques AC (1987): Human migration and the spread of malaria in Brazil. Parasitol Today 3:166-170.

McCutchan TF, Altaf AL, de la Cruz VF, Miller LH, Maloy WL, Charoenvit Y, Beaudoin RL, Guerry P, Wistar R, Jr, Hoffman SL, Hockmeyer WT, Collins WE, Wirth D (1985): Sequence of the immunodominant epitope of the surface protein on sporozoites of *Plasmodium vivax*. Science 230:1381-1383.

Nardin EH, Barr PJ, Gibson HL, Collins WE, Nussenzweig RS, Nussenzweig V (1987) Induction of sporozoite-specific memory cells in mice immunized with a recombinant *Plasmodium vivax* circumsporozoite protein. Eur J Immunol 12:1763-1767.

Nussenzweig V, Nussenzweig RS (1986): Development of a sporozoite malaria vaccine. Am J Trop Med Hyg 35:678-688.

Ozaki LS, Svec P, Nussenzweig RS, Nussenzweig V, Godson GN (1983): Structure of the *Plasmodium knowlesi* gene coding for the circumsporozoite protein. Cell 34:815-822.

Rieckmann KH, Carson PE, Beaudoin RL, Cassells J, Sell KW (1974): Sporozoite induced immunity in man against an Ethiopian strain of *Plasmodium falciparum*. Trans R Soc Trop Med Hyg 68:258-259.

Schofield L, Ferreira A, Altszuler R, Nussenzweig V, Nussenzweig RS (1987a): Gamma-interferon inhibits the intrahepatocytic development of malaria exoerythrocytic forms *in vitro*. J Immunol 139:2020–2025.

Schofield L, Villaquiran J, Schellekens H, Nussenzweig RS, Nussenzweig V (1987b): γ-interferon, CD8⁺ T-cells and antibodies are required for immunity to malaria sporozoites. Nature 330:664–666.

Sefton BM, Bass JE (1987): The covalent modification of eukaryotic protein with lipid. J Cell Biol 104:1449–1453.

Trager W, Jensen JB (1976): Human malaria parasites in continuous culture. Science 193:673–675.

Trager W, Perkins ME, Lanners HN (1986): Malaria vaccine. Prog Clin Biochem Med 4:57–70.

Weber JL, Hockmeyer WT (1985): Structure of the circumsporozoite gene in 18 strains of *Plasmodium falciparum*. Mol Biochem Parasitol 15:305–316.

Yoshida N, Nussenzweig RS, Potocnjak P, Nussenzweig V, Aikawa M (1980): Hybridoma produces protective antibodies directed against the sporozoite stage of malaria parasite. Science 207:71–73.

Yoshida N, Potocnjak P, Nussenzweig V, Nussenzweig RS (1981): Biosynthesis of Pb44, the protective antigen of sporozoites of *Plasmodium berghei*. J Exp Med 154:1225–1236.

Yoshida N, del Portillo HA, di Santi SM, Nussenzweig RS, Enea V (1987): *Plasmodium falciparum*: Epidemiological studies on the CS gene. Exp Parasitol 64:510–513.

Zavala F, Cochrane AH, Nardin EH, Nussenzweig RS, Nussenzweig V (1983): Circumsporozoite proteins of malaria parasites contain a single immunodominant region with two or more identical epitopes. J Exp Med 157:1947–1957.

Zavala F, Masuda A, Enea V, Graves PM, Nussenzweig V, Nussenzweig RS (1985a): Ubiquity of the repetitive epitope of the CS proteins. J Immunol 135:2790–2793.

Zavala F, Tam JP, Hollingdale MR, Cochrane AH, Quakyi I, Nussenzweig RS, Nussenzweig V (1985b): Rationale for development of a synthetic vaccine against *Plasmodium falciparum* malaria. Science 228:1436–1440.

Zavala F, Tam JP, Barr PJ, Romero PJ, Ley V, Nussenweig RS, Nussenzweig V (1987): Synthetic peptide vaccine confers protection against murine malaria. J Exp Med 166:1591–1596.

The Biology of Parasitism, pages 201–224
© 1988 Alan R. Liss, Inc.

# Antigens of *Plasmodium falciparum* and Their Potential as Components of a Malaria Vaccine

## R.F. Anders

*The Walter and Eliza Hall Institute of Medical Research, Melbourne, Victoria 3050, Australia*

## INTRODUCTION

The discovery by Ross almost a century ago that malaria was transmitted by mosquitoes led to the belief that malaria could be eradicated by the elimination of mosquitoes from areas of human habitation. Twice more in the last century new tools—the residual insecticide dichloro-diphenyl-tri-chloro-ethane (DDT) and efficient chemoprophylactic drugs—led to renewed optimism that malaria could be eliminated or at least controlled. Each of these battles against malaria saw impressive victories in some areas, but today malaria remains one of the world's greatest health problems with several hundred million people infected and approximately 40% of the world's population living in areas where they are at risk of infection.

The control of mosquito populations by removing or treating breeding areas and the use of residual insecticides together with chemophrophylaxis remain central to malaria control programs in many parts of the tropical world. However, because of vector resistance to insecticides, changes in vector behaviour, the development of drug resistance by malaria parasites, financial and manpower constraints, and political turmoil, conventional control programs are failing in many parts of the world. For this reason new methods of controlling malaria are being sought. The most cost-effective way of controlling infectious diseases is through immunization. Under special circumstances, immunization may eradicate a disease from the face of the earth, as evidenced by the campaign against smallpox. A number of other viral vaccines are also spectacularly effective, but there are no vaccines for any of the parasitic infections of humans. Nevertheless, great hopes are now held that malaria vaccines, which are being developed in various parts of the world, will provide the new tool that is urgently required to stem the advance of malaria in many parts of the world.

Individuals infected with malaria parasites develop specific immunity that decreases the parasites' ability to survive in the human host and also decreases the clinical consequences of infection. This fact underlies the belief that eventually an effective vaccine against malaria will be developed. However, in contrast to the rapidly developing and long-lived immunity induced by most viral infections, immunity to malaria only develops after years of repeated infections and may wane rapidly during periods of nonexposure. Therefore, what is required in a malaria vaccine is something that will be more effective at inducing immunity than the natural infection. Some people consider this a difficult or perhaps impossible task. However, I believe that the progress in the decade since the development of methods for the culture of malaria parasites in vitro (Trager and Jensen, 1986) provides reasons to be optimistic that an effective vaccine against malaria is on the horizon. In this chapter I will discuss some of this progress.

The malaria parasite has a complex life cycle developing through several different life-cycle stages both in the definitive mosquito host and in the intermediated vertebrate host. Within the human host the parasite invades and multiplies asexually within two different types of cell. Sporozoites transmitted to humans by the feeding female mosquito enter a liver cell within minutes and there initiate the exoerythrocytic cycle wherein a single *Plasmodium falciparum* sporozoite can develop into as many as 40,000 merozoites over the course of 6–9 days. Merozoites released from the rupturing hepatocyte then invade erythrocytes, and in *P. falciparum*, the erythrocytic cycle of asexual multiplication which takes 48 hours usually gives rise to 10–20 new merozoites, each capable of invading a new erythrocyte. The synchronous rupturing of many schizont-infected erythrocytes causes the characteristic signs and symptoms of clinical malaria. For reasons that are not clear, occasional merozoites, after invading an erythrocyte, do not undergo a further cycle of asexual multiplication but instead develop into male or female gametocytes, the sexual forms that perpetuate the cycle through a mosquito when taken up in a blood meal.

The complexity of the life cycle is being exploited in attempts to develop a malaria vaccine because it provides several different potential targets for immune attack. Currently the antigens of five different life-cycle stages— sporozoites, exoerythrocytic (liver) stages, asexual blood stages, gametes, and zygotes—are under investigation to determine their potential as vaccine components. There is much evidence that naturally acquired immunity to malaria is largely directed against antigens of the asexual blood stages, however, during infections antibodies develop to antigens of all the life-cycle stages listed above (McGregor et al., 1965; Nardin et al., 1979; Mendis et al., 1987). Thus, eventually it may be possible to incorporate in the one vaccine antigens of several different life-cycle stages. Such a novel approach to vaccine development is one possible way in which vaccination may achieve more than short-term exposure to the natural infection. In this review I will only deal with the developments concerning the sporozoite and asexual blood-stage vaccines as this work is the most advanced.

## THE CIRCUMSPOROZOITE PROTEIN

Antisporozoite immune responses induced by natural exposure or by immunization with irradiated sporozoites are essentially directed against a single sporozoite molecule, the circumsporozoite (CS) protein. This protein, which forms the surface coat of sporozoites, has now been characterized in great detail as a result of the cloning of the CS protein gene for many different species of *Plasmodium* (Ellis et al., 1983; Dame et al., 1984; Enea et al., 1984; Arnot et al., 1985; McCutchan et al., 1985; Eichinger et al., 1986; Galinski et al., 1987; Lal et al., 1987). The primary structure of the CS protein, deduced from the nucleotide sequence of the gene, is unusual in that much of the polypeptide chain is composed of an array of short sequence repeats which in *P. falciparum* is the tetrameric sequence NANP (Dame et al., 1984; Enea et al., 1984). The sequence repeats in the CS proteins of different species vary in sequence and length of the repeat unit; however, the number of repeats also varies so that the overall length of the block of repeats (and the full length polypeptide chain) is relatively conserved. Although antibodies have been elicited which react with epitopes within the flanking nonrepeat sequences of some CS proteins (Ballou et al., 1985; Vergara et al., 1985; Sharma et al., 1986) the repeats are very much the immunodominant region of the molecule (Ballou et al., 1985; Zavala et al., 1985).

A considerable body of experimental data has accumulated indicating that the repetitive epitopes in CS proteins, including that of *P. falciparum*, are the targets of protective antibodies (reviewed in Nussenzweig and Nussenzweig, 1984, 1985). Because of this data the repeat region became the focus of research aimed at developing an antisporozoite vaccine. During the last year this research has culminated in the first clinical trials of a subunit vaccine against malaria. Two different experimental vaccines have been tested in separate trials. In one trial the test material was a recombinant antigen (R32tet32) which consisted of the sequence MDP(NANP)$_{15}$NVDP (NANP)$_{15}$NVDP fused to a 32-amino-acid sequence derived from the tetracycline resistance gene read out of frame (Ballou et al., 1987). In the other trial the antigen tested was a synthetic peptide $(NANP)_3$ linked to tetanus toxoid (Herrington et al., 1987). In each trial the antigen was administered together with alum as the adjuvant. Overall the antibody responses were poor, although in some individuals the antibody levels were comparable to those found in individuals who have had long-term natural exposure to infected mosquitos. Because of the relatively low antibody responses only a limited number of individuals in each trial were subsequently challenged with *P. falciparum* to determine the efficacy of the vaccines. Upon challenge there was one individual in each trial who was protected and a small number of others who had a prolonged prepatent period. The protected individuals were among the highest antibody responders, and it seems likely that a much better level of protection would be achieved if a much higher antibody response to the CS protein repeat could be induced (Ballou et al., 1987; Herrington et al., 1987).

Although titers of antibodies to the CS protein increase with age (Nardin et al., 1979; Tapchaisri et al., 1983; Del Guidice et al., 1987), the higher titers of these antibodies do not appear to correlate with resistance to reinfection (Hoffman et al., 1987). Thus, although antibodies have been shown to have an effect against sporozoites in vitro and in animal models (Potocnjak et al., 1980; Yoshida et al., 1980; Hollingdale et al., 1982; Gysin et al., 1984; Mazier et al., 1986; Barr et al., 1987), they may not have an important role in naturally acquired immunity or, indeed, in immunity induced by immunizing with irradiated sporozoites. Recent studies in murine models have indicated that immunity induced by immunization with sporozoites is mediated by T cells of the cytotoxic $(CD8^+)$ phenotype (Schofield et al., 1987; Weiss et al., 1987). These cells may act against the parasite in the hepatocyte by releasing $\gamma$-interferon, which is known to act against exo-erythrocytic stages of malaria parasites (Ferreira et al., 1986; Maheshwari et al., 1986). Clarification of these immune mechanisms and the development of immunization strategies which effectively induce these cell-mediated effects offers a new and exciting direction to sporozoite vaccine development.

## ASEXUAL BLOOD-STAGE ANTIGENS

In contrast to the relatively monospecific antibody response induced by exposure to sporozoites, antibody responses to the asexual blood stages of *P. falciparum* and other malaria parasites are extremely polyspecific. The first definition of the complexity of this antibody response came from the application of immunoprecipitation and two-dimensional gel electrophoresis techniques to the analysis of biosynthetically labelled antigens recognized by antibodies in the serum of individuals naturally exposed to *P. falciparum* infections (Brown et al., 1981). With the use of these techniques, in excess of 100 antigenic polypeptides have been visualized in nonionic detergent extracts of asynchronously growing cultures of the asexual blood stages of *P. falciparum*. The current realization that a number of asexual blood-stage antigens are processed into several antigenic fragments (Holder and Freeman, 1984; Braun-Breton et al., 1986; Delplace et al., 1987; Bushell et al., 1988) indicates that the number of spots seen on two-dimensional gel analysis of immunoprecipitates is an overestimate of the number of different abundant antigens, but nevertheless there are a very large number of antigens, and this is a major complication for the development of a vaccine against the asexual blood stages of malaria. Because antibodies to many of these antigens are abundant in the serum of nonimmune individuals (Brown et al., 1983) and also because the titre of antibodies to the asexual blood stages does not correlate with immune status, it has been assumed that the majority of these antigens induce responses which either play no role in protecting the human host or perhaps even favour survival of the parasite. The challenge has been to identify which of these many antigens are capable of inducing host-protective immune responses.

## Identification of Candidate Vaccine Antigens in the Asexual Blood Stages

Several different approaches have been used to identify antigens that are the targets of protective immune responses in the asexual blood stages of *P. falciparum*. One approach used has been differential immunoprecipitation, in which the antigens recognized by human sera that inhibit the growth of *P. falciparum* in vitro are compared with those recognized by noninhibitory sera (Brown et al., 1982). Alternatively, comparisons have been made of the specificities in acute and convalescent sera (Brown et al., 1983) or monkey sera effective and ineffective in passively immunizing against infection (Jendoubi et al., 1985). This approach has been less rewarding than was originally anticipated possibly because nonprotective antibodies to the immunodominant regions of candidate vaccine molecules are prevalent in individuals exposed to malaria irrespective of their immune status.

A second approach has been to use immunochemical procedures to identify molecules involved in critical events in the parasite life cycle that may be disrupted by specific immune responses. Of particular interest are the molecules on the merozoite surface that have affinity for receptors on the erythrocyte membrane. A number of different putative glycophorin-binding proteins have been identified in *P. falciparum* (Jungery et al., 1983; Perkins,

1984; Ravetch et al., 1985), although it has proven difficult to establish that any of these molecules play a critical role in the invasion event in vivo (van Schravendijk et al., 1987). Also of interest are parasite molecules that have a function in modifying the membrane of the parasitized erythrocyte. As the asexual blood stages of *P. falciparum* mature, the parasitized erythrocyte develops knobs on its surface (Trager et al., 1966). These knobs provide sites of interaction with the vascular endothelium and thereby enable the parasitized erythrocyte to cytoadhere to endothelial cells (Luse and Miller, 1971). This leads to the sequestration of the mature stages of *P. falciparum* out of the peripheral circulation (Bignami and Bastianelli, 1889; David et al., 1983) a process that is central to the life-threatening complications, such as cerebral malaria, that are seen in nonimmune individuals (Spitz, 1946; Macpherson et al., 1985). The parasite molecules that mediate cytoadherence have not yet been identified unequivocally but are of obvious interest as potential vaccine components. The knob-associated histidine-rich protein (KAHRP or HRP1), one of several histidine-rich proteins (HRPs) identified in *P. falciparum*, has been shown to be a component of the knob (Kilejian, 1979; Leech et al., 1984; Culvenor et al., 1987), but KAHRP is located on the cytoplasmic face of the erythrocyte membrane, and although anti-KAHRP antibodies are generated in man by the natural infection (Culvenor et al., 1987), this molecule is not thought to be a target of protective immune responses.

Another approach used by several laboratories to identify antigens of importance in the asexual blood-stage parasites has been to produce panels of monoclonal antibodies, screen them for antiparasitic activity, and then use inhibitory monoclonal antibodies to identify the corresponding parasite antigens. The value of this approach was first demonstrated with antigens of the murine malaria *P. yoelii* (Freeman et al., 1980), and it has led to the identification of several *P. falciparum* antigens that are currently being assessed for efficacy in experimental vaccines (Perrin et al., 1981; Schmidt-Ullrich et al., 1986; Schofield et al., 1986). Another approach that we at The Walter and Eliza Hall Institute have exploited involved making as complete a library as possible of recombinant cDNA clones expressing various *P. falciparum* asexual blood-stage antigens (Kemp et al., 1983). The original selection for recombinant clones was immunoreactivity with antibodies in the serum of adult Papua New Guineans presumed to be immune to *P. falciparum*. By preparing monospecific antibodies to these cloned antigens it has been possible to identify the corresponding parasite molecules (Anders et al., 1984; Crewther et al., 1986). This approach has provided a library of clones expressing fragments of many different (currently approximately 30) asexual blood-stage antigens (Kemp et al., 1987). Not all of these antigens are potential vaccine components, but the availability of the individual antigens and monospecific antibodies to them has enabled us to identify several antigens that because of their location on the merozoite surface or presence in secretory organelles in merozoites are being assessed as vaccine components. Others have used a similar immunological approach to generate panels of clones expressing various malaria antigens (McGarvey et al., 1984; Ardeshir et al., 1985; Langsley et al., 1985; Ozaki et al., 1986).

Preclinical testing of an asexual blood-stage vaccine for *P. falciparum* can be carried out in *Aotus* or *Saimiri* monkeys; however, it is important to use the minimum number of monkeys for this purpose. Furthermore, trials in monkeys are expensive and difficult to carry out, and *P. falciparum* infections in these monkeys differ in some aspects from the human disease. For these reasons it is not feasible to move directly to monkey trials with every asexual blood-stage antigen that becomes available. Instead, indirect methods have had to be used to assess antigens as potential vaccine candidates. One important criteria that has been used to identify antigens of potential interest is the ultrastructural location. Antigens on the surfaces of the merozoite or the infected erythrocyte are of obvious interest because such antigens are directly accessible to immune attack, and there is considerable evidence that host-protective immune responses can be directed against antigens in these locations (Cohen et al., 1969; Miller et al., 1975). Another group of antigens that have attracted interest as potential vaccine components are those located in the rhoptries and micronemes, secretory organelles that form part of the apical complex in merozoites. A rhoptry location is indicated by a fine "double dot" punctate appearance when examined by immunofluorescence microscopy and has been confirmed in several cases by immuno-electron microscopy (Oka et al., 1984; Coppel et al., 1987; Roger et al., 1988; Bushell et al., 1988). Listed in Table 1 are some of the antigens that have been located to one or other of these subcellular sites and that are actively under investigation by various laboratories as potential vaccine components. The characteristics of these antigens will now be discussed.

### Precursor of the Major Merozoite Surface Antigens (PMMSA)

PMMSA—also referred to as the polymorphic schizont antigen (McBride et al., 1985), gp 185 (Howard et al., 1985), and gp 195 (Lyon et al., 1986)—was first identified in the murine parasite *Plasmodium yoelii* with a monoclonal antibody. Although this antibody did not passively protect mice, immunization with the antigen isolated by affinity chromatography did protect mice against a challenge infection (Holder and Freeman, 1981). The *P.*

**TABLE 1. Asexual Blood-Stage Candidate Vaccine Antigens**

| Antigen | Characteristics |
|---|---|
| Precursor of the major merozoite surface antigen (PMMSA) | $M_r$ 190–210,000; considerable antigenic diversity amongst isolates of P. falciparum; very limited repeat structure; presumed to have a glycosyl-phosphatidylinositol anchor |
| The ring-infected erythrocyte surface antigen (RESA) | $M_r$ 155,000; no antigenic diversity described; contains two blocks of sequence repeats; involved in an extensive network of cross-reactivities |
| $M_r$ 41,000 and $M_r$ 83,000 rhoptry antigens | Form a complex which also contains proteolytic fragments of the $M_r$ 83,000 antigen; no published sequences; extent of antigenic diversity not established |
| $M_r$ 105,000, $M_r$ 135,000, and $M_r$ 155,000 rhoptry antigens | Form a complex; inhibitory monoclonal antibody described; extent of antigenic diversity not established |
| $M_r$ 45,000 merozoite surface antigen | Established as integral membrane protein by sequence and solubility in Triton X-114; has a glycosyl-phosphatidylinositol anchor |

*falciparum* equivalent of the *P. yoelii* antigen has been the subject of detailed study by several laboratories. The antigen isolated from parasites and a synthetic peptide corresponding to part of the known sequence have been used to protect monkeys partially against infection with *P. falciparum* (Hall et al., 1984; Perrin et al., 1984; Cheung et al., 1986). The PMMSA is synthesized during schizogony (Holder and Freeman, 1982) as a polypeptide with an approximate M*r* of 200,000 but varying slightly in size among different isolates (McBride et al., 1985; Howard et al., 1986a; Schwartz et al., 1986). Close to the time of merozoite release the polypeptide is processed into a series of fragments (Lyon et al., 1986, 1987; Holder et al., 1987), three of which are major surface antigens identified by radio-iodination of free merozoites (Freeman and Holder, 1983; Heidrich et al., 1983; McBride and Heidrich, 1987). The PMMSA is acylated (Haldar et al., 1985; Schwarz et al., 1986), probably as a result of the attachment of a glycosyl phosphati-dylinositol moiety to the C-terminus of the molecule (Ferguson and Williams, 1988). It appears that a C-terminal fragment remains associated with the merozoite and is carried over into the ring stage, whereas the other fragments of PMMSA are lost, presumably at the time of merozoite invasion (McBride and Heidrich, 1987).

In addition to being polymorphic in size, PMMSA is antigenically diverse. McBride and her colleagues have used a panel of monoclonal antibodies to identify many different serotypes of *P. falciparum* based on antigenic differences in the PMMSA of different isolates (McBride et al., 1982, 1985). Recent sequencing studies have revealed that much of the variation in this antigen among different isolates of *P. falciparum* has been generated by intragenic recombination between different alleles (Tanabe et al., 1987; Peterson et al., 1988). The extent to which the antigenic diversity of this antigen will frustrate attempts to use PMMSA as a component of a vaccine is unknown.

### The Ring-Infected Erythrocyte Surface Antigen (RESA)

This antigen has been located to the micronemes of merozoites and after the merozoite invades to form the ring-stage parasite, RESA is found associated with the membrane of the infected erythrocyte (Perlmann et al., 1984; Brown et al., 1985). Presumably RESA is released via the merozoite apical pore at the time of invasion. This molecule is ranked highly as a potential vaccine component for several reasons in addition to its location. First, antibodies to RESA (also called Pf155) have been shown to be potent inhibitors of merozoite invasion in vitro (Wåhlin et al., 1984). Second, fragments of RESA produced in *Escherichia coli* as $\beta$-galactosidase fusion proteins have been used to protect *Aotus* monkeys partially against overwhelming infection with *P. falciparum* (Collins et al., 1986). Third, a polypeptide believed to be the homolog of RESA has been isolated by affinity chromatography from *P. chabaudi* and used protectively to immunize mice against this murine malaria (Wanidworanum et al., 1987). In contrast to PMMSA, RESA has not been observed to exhibit antigenic heterogeneity among different isolates of *P. falciparum*. This adds further to the potential of this antigen for vaccination studies. The sequence of the RESA gene revealed that the polypeptide contains three sets of related sequence repeats, two of which are at the C-terminal end of the molecule (3' repeats) and the other in the middle of the polypeptide (5' repeats) (Coppel et al., 1984; Cowman et al., 1984). These repeats encode naturally immunogenic antigenic epitopes that are immunodominant. Antibody responses to the major tetrameric repeat are most prominent in infected individuals (Anders et al., 1986); however, experiments in *Aotus* monkeys indicate that antibodies of this specificity are not important to protection in contrast to antibodies with specificity for the two minor repeats (Collins et al., 1986). The epitopes encoded by the repetitive sequences of RESA are involved in a network of intra- and intermolecular cross-reactions (reviewed in Anders et al., 1987). We believe that this network of cross-reactivities may impair the development of high-affinity antibody responses and thereby favour survival of the parasite in the infected host (Anders, 1986).

## Rhoptry Antigens

The first monoclonal antibody reported to have an antiplasmodial effect was directed against an $M_r$ 235,000 antigen of *P. yoelii* (Holder and Freeman, 1981). Subsequently, this antigen was identified as a component of the rhoptry contents by immuno-electron microscopy (Oka et al., 1984). The analogous antigen in *P. falciparum* has not been identified, but a number of other antigens identified in the rhoptries in *P. falciparum* merozoites are being studied as potential vaccine antigens. At least five different antigens have been identified that appear to exist in two soluble complexes. A low molecular weight complex includes an antigen of approximate $M_r$ 40,000 and an antigen of approximate $M_r$ 83,000, together with large proteolytic fragments of the $M_r$ 83,000 antigen (Campbell et al., 1984; Howard et al., 1984; Braun-Breton et al., 1986; Schofield et al., 1986; Bushell et al., 1988). An $M_r$ 41,000 antigen, presumed to be the lower molecular weight component of this complex, has been isolated from cultured parasites by affinity purification using a monoclonal antibody together with electro-elution from acrylamide gels and tested in monkey vaccination trials (Perrin et al. 1985). This antigen was very effective at protecting monkeys against challenge infections, particularly if the antigen was not first solubilized in SDS. The larger component of this complex has not been tested in vivo for vaccine efficacy but monoclonal antibodies to this polypeptide that block merozoite invasion in vitro have been described (Schofield et al., 1986).

Three other antigens identified in the rhoptries of *P. falciparum* appear to be associated in a high molecular weight complex (Campbell et al., 1984; Holder et al., 1985; Cooper et al., 1988). One of these antigens ($M_r$ 155,000) has been isolated from cultured parasites (Holder et al., 1985), and another ($M_r$ 105,000) has been cloned in *E. coli* (Coppel et al., 1987). Apart from their location the only evidence supporting the candidacy of these high molecular weight complex antigens as vaccine components is the finding that a monoclonal antibody precipitating the complex weakly inhibited merozoite invasion in vitro (Cooper et al., 1988). As this monoclonal antibody failed to immunoblot, the component of the complex with which it reacts has not been established. Two other antigens have been located to rhoptries that are not associated with these high and low molecular weight complexes. One of these antigens is synthesized as a $M_r$ 240,000 polypeptide that is processed into a $M_r$ 225,000 polypeptide (Roger et al., 1988). The other, an antigen of $M_r$ 55,000, has been shown to have the solubility properties of an integral membrane protein (Smythe et al., 1988).

## Antigens Associated With the Erythrocyte Membrane

Parasite antigens may become associated with the membrane of the erythrocyte in several different ways. First, they possibly may be acquired by passive adherence of soluble proteins released by the rupturing schizont or transferred from the surface of the merozoite. Such antigens would then be accessible to immune effector mechanisms; however, there exists no well-documented example of such an antigen. Second, antigens released from the secretory organelles of merozoites at the time of invasion may interact with components of the erythrocyte membrane (and thereby facilitate the invasion process). This appears to be the case for RESA, but this antigen interacts with the membrane skeleton and does not appear to be exposed on the external face of the erythrocyte plasma membrane (Brown et al., 1985). Third, the parasite developing within the erythrocyte synthesizes proteins which are transported out to the erythrocyte membrane (Howard et al., 1987). One reason for this in *P. falciparum* is to generate the knob structure in the erythrocyte membrane which is involved in cytoadherence. KAHRP and the mature erythrocyte surface antigen (MESA—also called Pf EMP 2) are associated with these knobs but are located on the cytoplasmic side of the erythrocyte membrane and therefore are not accessible as targets of immune effector mechanisms (Leech et al., 1984; Howard et al., 1987). Cytoadherence is presumed to be mediated by another parasite protein that spans the erythrocyte membrane to provide an external domain with affinity for vascular endothelium and a cytoplasmic domain anchoring the molecule to components of the knob on the cytoplasmic face of the membrane (reviewed in Howard et al., 1986b). Cytoadherence can be blocked by antibodies in a strain-specific manner (Udeinya et al., 1983). Using such antibodies, Howard and his colleagues have identified an antigen they have called erythrocyte membrane protein 1 (Pf EMP 1), which may mediate cytoadherence (Leech et al., 1984; Aley et al., 1986). Pf EMP 1 is polymorphic in size, bears strain-specific epitopes, and has very similar characteristics to the SICA (schizont-infected cell agglutination) antigen of *P. knowlesi* (Leech et al., 1984; Aley et al., 1986; Howard et al., 1988). Clones of *P. knowlesi* undergoing antigenic variation express a different SICA phenotype (Howard et al., 1983); thus it seems possible that antigenic variation in *P. falciparum* will be associated with the expression of different forms of Pf EMP 1.

An antigen exposed on the external face of the infected erythrocyte membrane that varies among different *P. falciparum* isolates has also been identified by immunofluorescence microscopy (Hommel et al., 1983) and agglutination (Sherwood et al., 1985; Marsh and Howard, 1986). It seems likely that Pf EMP 1 is the antigen being recognized in these studies; however, this remains to be established, as does the role of Pf EMP 1 in the process of cytoadherence. Antibodies that block cytoadherence might be very effective in preventing much of the morbidity associated with *P. falciparum* infections, although variation in Pf EMP 1 or other antigens involved in this process may frustrate attempts to use it in a vaccine.

## DEVELOPMENT OF AN ASEXUAL BLOOD-STAGE VACCINE

### Assessment of Antigens

One reason for optimism concerning the future development of a malaria vaccine is that several different asexual blood-stage antigens have now been identified that are capable of inducing immune responses with an antiparasitic effect, as briefly reviewed above. Immunization with a vaccine containing several of these antigens may not only induce a more protective immune response but also should make it less likely that the effects of immunization will be frustrated by the emergence of parasites bearing variant forms of the antigens. An even more effective vaccine may eventually be developed by combining several asexual blood-stage antigens with the CS protein or antigens of other life-cycle stages. Because it is difficult to obtain reproducible infections in nonhuman primates with sporozoites of *P. falciparum*, candidate sporozoite vaccine antigens have been tested in human volunteers without evidence of efficacy in monkeys. However, monkeys of the *Aotus* and *Saimiri* species can be reproducibly infected with *P. falciparum* by blood challenge, and therefore asexual blood-stage antigens that are candidate vaccine molecules are being tested for efficacy in preclinical trials using monkeys of these two species.

A number of trials have been carried out in which some protection of immunized monkeys against challenge infection with *P. falciparum* has been achieved with well-defined antigens. The antigens used in these trials have either been isolated from cultured parasites by affinity chromatography on monoclonal antibodies (Hall et al., 1984; Perrin et al., 1984, 1985; Siddiqui et al., 1986, 1987), fragments of antigens expressed from genes cloned in *E. coli* (Collins et al., 1986), or synthetic peptides corresponding to known epitopes or N-terminal sequences of blood-stage antigens (Cheung et al., 1986; Patarroyo et al., 1987). In one of these trials $\beta$-galactosidase fusion proteins containing fragments of RESA were used to immunize *Aotus* monkeys (Collins et al., 1986). The parasitemias showed only a partial protection, but protection correlated with antibody responses to repeat sequences in two different regions of the RESA polypeptide. A particularly impressive protective effect was achieved in another trial in which synthetic peptides conjugated to tetanus toxoid were used as the immunogen (Patarroyo et al., 1987). Peptides corresponding to the N-terminal sequences of PMMSA and two other uncharacterized antigens were more effective when used together than when the peptides were used alone.

These trials in monkeys (Table 2) have provided important information about antigens that subsequently will be tested in clinical trials. The first clinical trials will be carried out in human volunteers from nonendemic countries, and subsequently vaccines will be tested in individuals living in countries where malaria is endemic. There are several considerations that make this seemingly straightforward sequence of events problematic. First, the monkey models are less than ideal in that only a relatively limited number of strains of *P. falciparum* will give reproducible infections in the available species and the number of monkeys that can be used for vaccine trials is limited (reviewed in Collins and Pappaionou, 1985). Thus it is difficult to carry out trials in monkeys on all the forms of the many different antigens that have now been identified as of potential interest, e.g., fusion proteins, near-native proteins, or synthetic peptides. Neither *Aotus* or *Saimiri* monkeys are natural hosts for *P. falciparum*. Thus it may be inappropriate to discard an antigen as a potential vaccine component because of lack of efficacy in these monkey systems, particularly if the homologous antigen has been shown to immunize effectively in a rodent model system or if other data, such as location on the merozoite surface, suggested that it was a likely target of protective immune responses.

A second consideration concerns the endpoint of a vaccine trial testing an asexual blood stage antigen in human volunteers. With the sporozoite vaccine trials that have been carried out in the last year (Ballou et al., 1987; Herrington et al., 1987) the target of the induced immunity was the sporozoite or possibly the exoerythrocytic stage. Thus, the appearance of a single blood-stage parasite indicated that the immunized individual was still susceptible to infection and the volunteer could be treated at that early stage without risk of severe morbidity as a result of the experimentally induced infection. It is

**TABLE 2. Recent Vaccine Trials in Monkeys Using Defined Asexual Blood Stage Antigens**

| Antigen | Monkey | Reference |
|---|---|---|
| Purified with monoclonal antibody | | |
| PMMSA | *Saimiri* | Hall et al. (1984) |
| PMMSA | *Saimiri* | Perrin et al. (1984) |
| PMMSA and rhoptry antigens | *Aotus* | Siddiqui et al. (1987) |
| Synthetic peptides | | |
| PMMSA | *Saimiri* | Cheung et al. (1986) |
| PMMSA plus peptides from undefined 35K and 55K antigens | *Aotus* | Patarroyo et al. (1987) |
| Recombinant proteins | | |
| RESA | *Aotus* | Collins et al. (1986) |

possible that an effective asexual blood-stage vaccine may increase the period of latency for individuals challenged with sporozoites, in which case efficacy could be determined with the infections being terminated at the earliest stage of patency, as in the sporozoite vaccine trials. However, it is more probable that a clear indication of efficacy with a blood-stage vaccine will only be obtained if the trial protocol allows parasitemias in nonprotected individuals to progress considerably beyond the threshold of detection. Inevitably this increases the risk of morbidity in the volunteer group and adds an ethical dilemma to the many other problems confronting the development of such a vaccine.

As discussed above, it could be argued that failure to protect in a monkey model does not necessarily eliminate an antigen from being tested in clinical trials. On the other hand, some efficacy in nonexposed human volunteers subjected to experimental challenge will be required before a vaccine is tested in populations living in areas where malaria is endemic. However, lack of efficacy in semi-immune adults living in an endemic area should not necessarily exclude a vaccine from further testing. The prime target population for a malaria vaccine in endemic countries is the young children, and the major aim is to reduce the mortality that is attributable to malaria in these children. One could envisage a vaccine that failed to have any impact on the largely asymptomatic parasitemias suffered by the adult population but that still had a significant impact on the level of mortality and morbidity in the very susceptible, nonimmune, young children. It will be important to see that the procedures used for evaluating the experimental vaccines do not overlook this possibility.

### Form of Antigen To Be Used in a Vaccine

The unusual repeat structures in many malaria antigens enabled the immunodominant epitopes in some of the molecules being assessed as vaccine components to be easily identified. It is believed that the numerous immunodominant repeats, and the cross-reacting epitopes they encode, are part of a parasite evasion mechanism (Anders, 1986), although repetitive sequences in the CS protein (Ballou et al., 1987; Herrington et al., 1987) and in RESA (Collins et al., 1986) appear to be targets of protective antibody responses. Thus, synthetic peptides corresponding to critical epitopes within the polypeptide chain of candidate molecules are being investigated for their efficacy as vaccines. It is too early to know whether this approach will be successful, although on theoretical grounds the approach has both advantages and disadvantages. A major advantage is being able to work with a small, defined chemical entity the sequence and length of which can be chosen to optimize the type of immune response that is required for protection. This may be particularly relevant in the case of malaria if, as suspected, some parts of the polypeptide chains, by virtue of their immunodominance, are counterproductive to optimal immune responses to the critical "protective" epitopes (Kemp et al., 1987). The immunogenicity of a synthetic peptide can be further enhanced by coupling the peptide to a highly immunogenic carrier.

Despite these advantages the synthetic peptide approach to a malaria vaccine may not provide a very effective vaccine for use in populations living in areas where malaria is endemic. Because the immunity induced by a malaria vaccine is unlikely to be very long-lived, and certainly not lifelong, it may be important that immunity induced by the vaccine is boosted by exposure to the infection. In order for an infection to boost an antibody response induced by vaccination with a synthetic peptide, the peptide will need to include T-cell as well as B-cell epitopes (Miller and Mitchell, 1969; Mitchison, 1971). To be effective the T-cell epitope should be derived from the same malaria protein as the B-cell epitope or from another protein which is simultaneously processed by an antigen-presenting cell. It is feasible to identify T-cell epitopes and construct such hybrid peptides (Good et al., 1987); however, identifying a T-cell epitope which provides help for an antibody response of the appropriate specificity may be difficult. In a recent study using "hybrid" peptides to immunize mice against foot-and-mouth-disease virus (FMDV) Francis et al. (1987) identified T-cell epitopes in unrelated proteins that provided help for antibody responses to FMDV. However, the antibody responses were not equivalent, as only two of three hybrid peptides tested produced high titers of neutralizing antibodies. Even if a functional T-cell epitope can be incorporated into a hybrid peptide, immune response gene effects make it very unlikely that a subunit vaccine of this type would be equally effective in all members of an outbred population.

The alternative to a synthetic peptide vaccine is to produce a suitably immunogenic polypeptide in *E. coli,* yeast, or from the gene cloned in some other host cell. Two of the prime vaccine candidates, PMMSA and RESA, are large polypeptide chains which will make the high level expression of the full-length polypeptide difficult. However, the generation of large fragments of these antigens containing appropriate epitopes hopefully will effectively immunize individuals of diverse genetic makeup and also provide for effective boosting by subsequent infections with *P. falciparum.*

## CONCLUDING COMMENTS

It is very probable that one or more of the asexual blood-stage antigens discussed above will induce in humans some level of protection against malaria when included in a vaccine. Although we have complete knowledge of the primary structure for several of these polypeptides, and know their ultrastructural location, we know little about their function and nothing of their three-dimensional structure. In the next few years I anticipate this situation will change dramatically, and the new knowledge gained will provide opportunities for the rational development of vaccines with increased efficacy. Most of the asexual blood-stage antigens under study as candidate vaccine molecules interact, or are presumed to interact, with the membrane of the host erythrocyte. The specific nature of these interactions and their precise functions remain to be defined. Although some information is being gained using currently available techniques, the future development of a transfection system for *Plasmodium* and the application of site-directed mutagenesis will enable rapid dissection of the functional role of different regions of these antigens.

## ACKNOWLEDGMENTS

I thank my many colleagues in the Malaria Group at The Walter and Eliza Hall Institute for stimulating discussions. This work was supported by the Australian National Health and Medical Research Council, the John D. and Catherine T. MacArthur Foundation, the Australian National Biotechnology Research Grants Scheme, and the Australian Malaria Joint Venture.

## REFERENCES

Aley SB, Sherwood JA, Marsh K, Eidelman O, Howard RJ (1986): Identification of isolate-specific proteins on sorbitol-enriched *Plasmodium falciparum* infected erythrocytes from Gambian patients. Parasitology 92:511–525.

Anders RF (1986): Multiple cross-reactivities amongst antigens of *Plasmodium falciparum* impair the development of protective immunity against malaria. Parasite Immunol 8:529–539.

Anders RF, Coppel RL, Brown GV, Kemp DJ (1987): Antigens with repeated amino acid sequences from the asexual blood stages of *Plasmodium falciparum*. Prog Allergy 41:1–26.

Anders RF, Coppel RL, Brown GV, Saint RB, Cowman AF, Lingelbach KR, Mitchell GF, Kemp DJ (1984): *Plasmodium falciparum* complementary DNA clones expressed in *Escherichia coli* encode many distinct antigens. Mol Biol Med 2:177–191.

Anders RF, Shi P-T, Scanlon DB, Leach SJ, Coppel RL, Brown GV, Stahl H-D, Kemp DJ (1986): Antigenic repeat structures in proteins of *Plasmodium falciparum*. Ciba Found Symp 119:164–175.

Ardeshir F, Flint JE, Reese RT (1985): Expression of *Plasmodium falciparum* surface antigens in *Escherichia coli*. Proc Natl Acad Sci USA 82:2518–2522.

Arnot DE, Barnwell JW, Tam JP, Nussenzweig V, Nussenzweig RS, Enea V (1985): Circumsporozoite protein of *Plasmodium vivax:* Gene cloning and characterization of the immunodominant epitope. Science 230:815–818.

Ballou WR, Hoffman SL, Sherwood JA, Hollingdale MR, Neva FA, Hockmeyer WT, Gordon DM, Schneider I, Wirtz RA, Young JF, Wasserman GF, Reeve P, Diggs CL, Chulay JD (1987): Safety and efficacy of a recombinant DNA *Plasmodium falciparum* sporozoite vaccine. Lancet i:1277–1281.

Ballou WR, Rothbard J, Wirtz RA, Gordon DM, Williams JS, Gore RW, Schneider I, Hollingdale MR, Beaudoin RL, Maloy WL, Miller LH, Hockmeyer WT (1985): Immunogenicity of synthetic peptides from circumsporozoite protein of *Plasmodium falciparum*. Science 228:996–999.

Barr PJ, Gibson HL, Enea V, Arnot DE, Hollingdale MR, Nussenzweig V (1987): Expression in yeast of a *Plasmodium vivax* antigen of potential use in a human malaria vaccine. J Exp Med 165:1160–1171.

Bignami A, Baastianelli A (1889): Observations of Estivo-Autumnal malaria. Riforma Med 6:1334–1335.

Braun Breton C, Jendoubi M, Brunet E, Perrin L, Scaife J, Pereira Da Silva L (1986): In vivo time course of synthesis and processing of major schizont membrane polypeptides in *Plasmodium falciparum*. Mol Biochem Parasitol 20:33–43.

Brown GV, Anders RF, Mitchell GF, Heywood PF ( 1982): Target antigens of purified human immunoglobulins which inhibit growth of *Plasmodium falciparum* in vitro. Nature 297:591–593.

Brown GV, Anders RF, Stace JD, Alpers MP, Mitchell GF (1981): Immunoprecipitation of biosynthetically-labelled proteins from different Papua New Guinea *Plasmodium falciparum* isolates by sera from individuals in the endemic area. Parasite Immunol 3:283–298.

Brown GV, Culvenor JG, Crewther PE, Bianco AE, Coppel RL, Salnt RB, Stahl H-D, Kemp DJ, Anders RF (1985): Localization of the Ring-infected Erythrocyte Surface Antigen (RESA) of *Plasmodium falciparum* in merozoites and ring-infected erythrocytes. J Exp Med 162:774–779.

Brown GV, Stace JD, Anders RF (1983): Specificities of antibodies boosted by acute *Plasmodium falciparum* infection in man. Am J Trop Med Hyg 32:1221–1228.

Bushell GR, Ingram LT, Fardoulys CA, Cooper JA (1988) An antigenic complex in the rhoptries of *Plasmodium falciparum*. Mol Biochem Parasitol 28:105–112.

Campbell GH, Miller LH, Hudson D, Franco EL, Andrysiak PM (1984): Monoclonal antibody characterization of *Plasmodium falciparum* antigens. Am J Trop Med Hyg 33:1051–1054.

Cheung A, Leban J, Shaw AR, Merkli B, Stocker J, Chizzolini C, Sander C, Perrin LH (1986): Immunization with synthetic peptides of a *Plasmodium falciparum* surface antigen induces antimerozoite antibodies. Proc Natl Acad Sci USA 83:8328–8332.

Cohen S, Butcher GA, Crandall RB (1969): Action of malarial antibody in vitro. Nature 223:368–371.

Collins WE, Anders RF, Pappaioanou M, Campbell GH, Brown GV, Kemp DJ, Coppel RL, Skinner JC, Andrysiak PM, Favaloro JM, Corcoran LM, Broderson JR, Mitchell GF, Campbell CC (1986): Immunization of *Aotus* monkeys with recombinant proteins of an erythrocyte surface antigen of *Plasmodium falciparum*. Nature 323:259–262.

Collins WE, Pappaioanou M (1985): Non-human primate models for testing malarial vaccines. In W.A. Siddiqui (ed): Proceedings of the Asia and Pacific Conference on Malaria:

Practical Considerations on Malaria Vaccines and Clinical Trials. Honolulu: University of Hawaii, pp 254–265.

Cooper JA, Ingram LT, Bushell GR, Fardoulys CA, Stenzel D, Schofield L, Saul AJ (1988): The 140/130/105 kilodalton soluble complex in the rhoptries of *Plasmodium falciparum* consists of discrete polypeptides. Mol Biochem Parasitol (in press).

Coppel RL, Bianco AE, Culvenor JG, Crewther PE, Brown GV, Anders RF, Kemp DJ (1987): A cDNA clone expressing a rhoptry protein of *Plasmodium falciparum*. Mol Biochem Parasitol 25:73–81.

Coppel RL, Cowman AF, Anders RF, Bianco AE, Saint RB, Lingelbach KR, Kemp DJ, Brown GV (1984): Immune sera recognize on erythrocytes a *Plasmodium falciparum* antigen composed of repeated amino acid sequences. Nature 310:789–791.

Cowman AF, Coppel RL, Saint RB, Favaloro J, Crewther PE, Stahl HD, Bianco AE, Brown GV, Anders RF, Kemp DJ (1984): The Ring-infected Erythrocyte Surface Antigen (RESA) polypeptide of *Plasmodium falciparum* contains two separate blocks of tandem repeats encoding antigenic epitopes that are naturally immunogenic in man. Mol Biol Med 2:207–221.

Crewther PE, Bianco AE, Brown GV, Coppel RL, Stahl H-D, Kemp DJ, Anders RF (1986): Affinity purification of human antibodies directed against cloned antigens of *Plasmodium falciparum*. J Immunol Methods 86:257–264.

Culvenor JG, Langford CJ, Crewther PE, Saint RB, Coppel RL, Kemp DJ, Anders RF, Brown GV (1987): *Plasmodium falciparum*: Identification and localization of a knob protein antigen expressed by a cDNA clone. Exp Parasitol 63:58–67.

Dame JB, Williams JL, McCutchan TF, Weber JL, Wirtz RA, Hockmeyer WT, Maloy WL, Haynes JD, Schneider I, Roberts D, Sanders GS, Reddy EP, Diggs CL, Miller LH (1984): Structure of the gene encoding the immunodominant surface antigen on the sporozoite of the human malaria parasite *Plasmodium falciparum*. Science 225:593–599.

David PH, Hommel M, Miller LH, Udeinya IJ, Oligino L (1983): Parasite sequestration in *Plasmodium falciparum* malaria: Spleen and antibody modulation of cytoadherence of infected erythrocytes. Proc Natl Acad Sci USA 80:5075–5079.

Del Guidice G, Engers HD, Tougne C, Biro SS, Weiss N, Verdini AS, Pessi A, Degremont AA, Freyvogel TA, Lambert P-H, Tanner M (1987): Antibodies to the repetitive epitope of *Plasmodium falciparum* circumsporozoite protein in a rural Tanzanian community: A longitudinal study of 132 children. Am J Trop Med Hyg 36:203–212.

Delplace P, Fortier B, Tronchin G, Dubremetz J-F, Vernes A (1987): Localization, biosynthesis, processing and isolation of a major 126 kDa antigen of the parasitophorous vacuole of *Plasmodium falciparum*. Mol Biochem Parasitol 23:193–201.

Ellis J, Ozaki LS, Gwadz RW, Cochrane AH, Nussenzweig V, Nussenzweig RS, Godson GN (1983): Cloning and expression in E. coli of the malarial sporozoite surface antigen gene from *Plasmodium knowlesi*. Nature 302:536–538.

Eichinger DJ, Arnot DE, Tam JP, Nussenzweig V, Enea V (1986): Circumsporozoite protein of *Plasmodium berghei*: Gene cloning and identification of the immunodominant epitopes. Mol Cell Biol 6:3965–3972.

Enea V, Ellis J, Zavala F, Arnot DE, Asavanich A, Masuda A, Quakyi I, Nussenzweig RS (1984): DNA cloning of *Plasmodium falciparum* circumsporozoite gene: Amino acid sequence of repetitive epitope. Science 225:628–630.

Ferguson MAJ, Williams AF (1988): Cell surface anchoring of proteins via glycosylphosphatidylinositol structures. Annu Rev Biochem 57:285–320.

Ferreira A, Schofield L, Enea V, Schellekens H, Van Der Meide P, Collins WE, Nussenzweig RS, Nussenzweig V (1986): Inhibition of development of exoerythrocytic forms of malaria parasites by γ-interferon. Science 232:881–884.

Francis MJ, Hastings GZ, Syred AD, McGinn B, Brown F, Rowlands DJ (1987): Non-responsiveness to a foot-and-mouth disease virus peptide overcome by addition of foreign helper T-cell determinants. Nature 300:168- 170.

Freeman RR, Holder AA (1983): Surface antigens of malaria merozoites. A high molecular weight precursor to an 83000 m.w. form expressed on the surface of *Plasmodium falciparum* merozoites. J Exp Med 158:1647–1653.

Freeman RR, Trejdosiewicz AJ, Cross GAM ( 1980): Protective monoclonal antibodies recognising stage-specific merozoite antigens of a rodent malaria parasite. Nature 284:366–368.

Galinski MR, Arnot DE, Cochrane AH, Barnwell JW, Nussenzweig RS, Enea V (1987): The circumsporozoite gene of the *Plasmodium cynomolgi* complex. Cell 48:311–319.

Good MF, Maloy WL, Lunde MN, Margalit H, Cornette JL, Smith GL, Moss B, Miller LH, Berzofsky JA (1987): Construction of a synthetic immunogen: Use of a new T-helper epitope on malaria circumsporozoite protein. Science 235:1059–1062.

Gysin J, Barnwell J, Schlesinger DH, Nussenzweig V, Nussenzweig RS (1984): Neutralization of the infectivity of sporozoites of *Plasmodium knowlesi* by antibodies to a synthetic peptide. J Exp Med 160:935–940.

Haldar K, Ferguson MAJ, Cross GAM (1985): Acylation of a *Plasmodium falciparum* merozoite surface antigen via sn- 1,2-diacyl glycerol. J Biol Chem 260:4969–4974.

Hall R, Hyde JE, Goman M, Simmons DL, Hope IA, Mackay M, Scaife J ( 1984): Major surface antigen gene of a human malaria parasite cloned and expressed in bacteria. Nature 311:379–382.

Heidrich HG, Strych W, Mrema JEK (1983): Identification of surface and internal antigens from spontaneously released *Plasmodium falciparum* merozoites by radioiodination and metabolic labelling. Z Parasitenkd 69:715–725.

Herrington DA, Clyde DF, Losonsky G, Cortesia M, Murphy JR, Davis J, Baqar S, Felix AM, Heimer EP, Gillessen D, Nardin E, Nussenzweig RS, Nussenzweig V, Hollingdale MR, Levine MM (1987): Safety and immunogenicity in man of a synthetic peptide malaria vaccine against *Plasmodium falciparum* sporozoites. Nature 328:257–259.

Hoffman SL, Oster CN, Plowe CV, Woolett GR, Beier JC, Chulay JD, Wirtz RA, Hollingdale MR, Mugambi M (1987): Naturally acquired antibodies to sporozoites do not prevent malaria: Vaccine development implications. Science 237:639–642.

Holder AA, Freeman RR (1981): Immunization against blood-stage rodent malaria using purified parasite antigens. Nature 294:361–364.

Holder AA, Freeman RR (1982): Biosynthesis and processing of a *Plasmodium falciparum* schizont antigen recognized by immune serum and a monoclonal antibody. J Exp Med 156:1528–1538.

Holder AA, Freeman RR (1984): The three major antigens on the surface of *Plasmodium falciparum* merozoites are derived from a single high molecular weight precursor. J Exp Med 160:624–629.

Holder RR, Freeman RR, Uni S, Aikawa M ( 1985): Isolation of a *Plasmodium falciparum* rhoptry protein. Mol Biochem Parasitol 14:293–303.

Holder AA, Sandhu JS, Hillman Y, Davey SL, Nicholls SC, Cooper H, Lockyer MJ (1987): Processing of the precursor to the major merozoite surface antigens of *Plasmodium falciparum*. Parasitology 94:199–208.

Hollingdale MR, Zavala F, Nussenzweig RS, Nussenzweig V (1982): Antibodies to the protective antigen of *Plasmodium berghei* sporozoites prevent entry into cultured cells. J Immunol 128:1929-1930.

Hommel M, David PH, Oligino LD (1983): Surface alterations of erythrocytes in *Plasmodium falciparum* malaria. Antigenic variation, antigenic diversity, and the role of the spleen. J Exp Med 157:1137-1148.

Howard RF, Stanley HA, Campbell GH, Reese RT (1984): Proteins responsible for a punctate fluorescence pattern in *Plasmodium falciparum* merozoites. Am J Trop Med Hyg 33:1055-1059.

Howard RF, Stanley HA, Campbell GH, Langreth SG, Reese RT (1985): Two *Plasmodium falciparum* merozoite surface polypetides share epitopes with a singe $M_r$ 185,000 parasite glycoprotein. Mol Biochem Parasitol 17:61-77.

Howard RJ, Barnwell JW, Kao V (1983): Antigenic variation in *Plasmodium knowlesi* malaria: Identification of the variant antigen on infected erythrocytes. Proc Natl Acad Sci USA 80:4129-4133.

Howard RJ, Barnwell JW, Rock EP, Neequaye J, Ofori-Adjei D, Maloy WL, Lyon JA, Saul A (1988): Two approximately 300 kilodalton *Plasmodium falciparum* proteins at the surface membrane of infected erythrocytes. Mol Biochem Parasitol 27:207-224.

Howard RJ, Lyon JA, Uni S, Saul AJ, Aley SB, Klotz F, Panton LJ, Sherwood JA, Marsh K, Aikawa M, Rock EP (1987): Transport of an $M_r$ ~ 300,000 *Plasmodium falciparum* protein (Pf EMP 2) from the intra-erythrocytic asexual parasite to the cytoplasmic face of the host cell membrane. J Cell Biol 104:1269-1280.

Howard RJ, McBride JS, Aley SB, Marsh K (1986a): Antigenic diversity and size diversity of *P. falciparum* antigens in isolates from Gambian patients. II. The schizont surface glycoprotein of molecular weight approximately 200,000. Parasite Immunol 8:57-68.

Howard RJ, Uni S, Lyon JA, Taylor DW, Daniel W, Aikawa M (1987): Export of *Plasmodium falciparum* proteins to the host erythrocyte membrane: Special problems of protein trafficking and topogenesis. In K.-P. Chang, D. Stary (eds): Host-parasite cellular and molecular interactions in protozoal infections. Berlin, Springer 281-296.

Jendoubi M, Dubois P, Pereira da Silva L (1985): Characterization of one polypeptide antigen potentially related to protective immunity against the blood infection by *Plasmodium falciparum* in the squirrel monkey. J Immunol 134:1941-1945.

Jungery M, Boyle D, Patel T, Pasvol G, Weatherall DJ (1983): Lectin-like polypeptides of *P. falciparum* bind to red cell sialoglycoproteins. Nature 301:704-705.

Kemp DJ, Coppel RL, Anders RF (1987): Repetitive proteins and genes of malaria. Annu Rev Microbiol 41:181-208.

Kemp DJ, Coppel RL, Cowman AF, Saint RB, Brown GV, Anders RF (1983): Expression of *Plasmodium falciparum* blood-stage antigens in Escherichia coli: Detection with antibodies from immune humans. Proc Natl Acad Sci USA 80:3787-3791.

Kilejian A (1979): Characterization of a protein correlated with the production of knob-like protrusions on membranes of erythrocytes infected with *Plasmodium falciparum*. Proc Natl Acad Sci USA 76:4650-4653.

Lal AA, de la Cruz VF, Welsh JA, Charoenvit Y, Maloy WL, McCutchan TF (1987): Structure of the gene encoding the circumsporozoite protein of *Plasmodium yoelii*. A rodent model for examining antimalarial sporozoite vaccines. J Biol Chem 262:2937-2940.

Langsley G, Scherf A, Mercereau-Puijalon O, Koenen M, Kahane B, Mattei D, Guillotte M, Sibilli L, Garner I, Müller-Hill B, Pereira da Silva L (1985): Characterization of *P.*

*falciparum* antigenic determinants isolated from a genomic expression library by differential antibody screening. Nucleic Acids Res 13:4191–4201.

Leech JH, Barnwell JW, Aikawa M, Miller LH, Howard RJ (1984): *Plasmodium falciparum* malaria: Association of knobs on the surface of infected erythrocytes with a histidine-rich protein and the erythrocyte skeleton. J Cell Biol 98:1256- 1264.

Luse S, Miller LH (1971): *Plasmodium falciparum* malaria: Ultrastructure of parasitized erythrocytes in cardiac vessels. Am J Trop Med Hyg 20:660–665.

Lyon JA, Geller RH, Haynes JD, Chulay JD, Weber JL (1986): Epitope map and processing scheme for the 195,000-dalton surface glycoprotein of *Plasmodium falciparum* merozoites deduced from cloned overlapping segments of the gene. Proc Natl Acad Sci USA 83:2989–2993.

Lyon JA, Haynes JD, Diggs CL, Chulay JD, Haidaris CG, Pratt-Rossiter J (1987): Monoclonal antibody characterization of the 195-kilodalton major surface glycoprotein of *Plasmodium falciparum* malaria schizonts and merozoites: Identification of additional processed products and a serotype-restricted repetitive epitope. J Immunol 138:895–901.

Macpherson AA, Warrell MJ, White NJ, Looareesuwan S, Warrell DA (1985): Human cerebral malaria. A quantitative ultrastructural analysis of parasitized erythrocyte sequestration. Am J Pathol 119:385–401.

Maheshwari RK, Czarniecki CW, Dutta GP, Puri SK, Dhawan BN, Friedman RM (1986): Recombinant human gamma interferon inhibits simian malaria. Infect Immun 53:628–630.

Marsh K, Howard RJ (1986): Antigens induced on erythrocytes by *P. falciparum:* Expression of diverse and conserved determinants. Science 231:150–153.

Mazier D, Mellouk S, Beaudoin RL, Texier B, Druilhe P, Hockmeyer W, Trosper J, Paul C, Charoenvit Y, Young J, Miltgen F, Chedid L, Chigot JP, Galley B, Brandicourt O, Gentilini M (1986): Effect of antibodies to recombinant and synthetic peptides on *P. falciparum* sporozoites in vitro. Science 231:156–159.

McBride JS, Heidrich H-G (1987): Fragments of the polymorphic $M_r$ 185000 glycoprotein from the surface of isolated *Plasmodium falciparum* merozoites form an antigenic complex. Mol Biochem Parasitol 23:71–84.

McBride JS, Newbold CI, Anand R (1985): Polymorphism of a high molecular weight schizont antigen of the human malaria parasite *Plasmodium falciparum*. J Exp Med 161:160–180.

McBride JS, Walliker D, Morgan G (1982): Antigenic diversity in the human malaria parasite *Plasmodium falciparum*. Science 217:254–257.

McCutchan TF, Lal AA, de la Cruz VF, Miller LH, Maloy WL, Charoenvit Y, Beaudoin RL, Guerry P, Wistar R, Hoffman SL, Hockmeyer WT, Collins WE, Wirth D (1985): Sequence of the immunodominant epitope for the surface protein on sporozoites of *Plasmodium vivax*. Science 230:1381–1383.

McGarvey MJ, Sheybani E, Loche MP, Perrin L, Mach B (1984): Identification and expression in *Escherichia coli* of merozoite stage-specific genes of the human malaria parasite *Plasmodium falciparum*. Proc Natl Acad Sci USA 81:3690–3694.

McGregor I, Williams K, Voller A, Billewicz W (1965): Immunofluorescence and the measurement of immune response to hyperendemic malaria. Trans R Soc Trop Med Hyg 59:395–414.

Mendis KN, Munesinghe YD, de Silva YNY, Keragalla I, Carter R (1987): Malaria transmission blocking immunity induced by natural infections of *Plasmodium vivax* in humans. Infect Immun 55:369–372

Miller LH, Aikawa M, Dvorak JA (1975): Malaria *(Plasmodium knowlesi)* merozoites: Immunity and the surface coat. J Immunol 114:1237–1242.

Miller JFAP, Mitchell GF (1969): The thymus and the precursors of antigen reactive cells. Nature 219:659–663.

Mitchison NA (1971): The carrier effect in the secondary response to hapten-protein conjugates. II. Cellular cooperation. Eur J Immunol 1:18–27.

Nardin EH, Nussenzweig R, McGregor I, Bryan J (1979): Antibodies to sporozoites: Their frequent occurrence in individuals living in an area of hyperendemic malaria. Science 206:597–599.

Nussenzweig RS, Nussenzweig V (1984): Development of sporozoite vaccines. Philos Trans R Soc Lond [Biol] 307:117–128.

Nussenzweig V, Nussenzweig RS (1985): Circumsporozoite proteins of malaria parasites. Cell 42:401–403.

Oka M, Aikawa M, Freeman RR, Holder AA, Fine E (1984): Ultrastructural localization of protective antigens of *Plasmodium yoelii* merozoites by the use of monoclonal antibodies and ultrathin cryomicrotomy. Am J Trop Med Hyg 33:342–346.

Ozaki LS, Mattei D, Jendoubi M, Druihle P, Blisnick T, Guillotte M, Puijalon O, Pereira da Silva L (1986): Plaque antibody selection: Rapid immunological analysis of a large number of recombinant phage clones positive to sera raised against *Plasmodium falciparum* antigens. J Immunol Methods 89:213–219.

Patarroyo ME, Romero P, Torres ML, Clavijo P, Moreno A, Martinez A, Rodriguez R, Guzman F, Cabezas E (1987): Induction of protective immunity against experimental infection with malaria using synthetic peptides. Nature 328:629–632.

Perkins ME (1984): Surface proteins of *Plasmodium falciparum* merozoites binding to the erythrocyte receptor, glycophorin. J Exp Med 160:788–798.

Perlmann H, Berzins K, Wahlgren M, Carlsson J, Björkman A, Patarroyo ME, Perlmann P (1984): Antibodies in malarial sera to parasite antigens in the membrane of erythrocytes infected with early asexual stages of *Plasmodium falciparum* . J Exp Med 159:1686–1704.

Perrin LH, Merkli B, Gabra MS, Stocker JW, Chizzolini C, Richle R (1985): Immunization with a *Plasmodium falciparum* merozoite surface antigen induces a partial immunity in monkeys. J Clin Invest 75:1718–1721.

Perrin LH, Merkli B, Loche M, Chizzolini C, Smart J, Richle R (1984): Antimalarial immunity in *Saimiri* monkeys. Immunization with surface components of asexual blood stages. J Exp Med 160:441–451.

Perrin LH, Ramirez E, Lambert PH, Miescher PA (1981): Inhibition of *P.falciparum* growth in human erythrocytes by monoclonal antibodies. Nature 289:301–303.

Peterson MG, Coppel RL, McIntyre P, Langford CJ, Woodrow G, Brown GV, Anders RF, Kemp DJ (1988): Variation in the precursor to the major merozoite surface antigens of *Plasmodium falciparum*. Mol Biochem Parasitol 27:291–302.

Potocnjak P, Yoshida N, Nussenzweig RS, Nussenzweig V (1980): Monovalent fragments (Fab) of monoclonal antibodies to a sporozoite surface antigen (Pb44) protect mice against malarial infection. J Exp Med 151:1504–1513.

Ravetch JV, Kochan J, Perkins M (1985): Isolation of the gene for a glycophorin-binding protein implicated in erythrocyte invasion by a malaria parasite. Science 227:1593–1597.

Roger N, Dubremetz J-F, Delplace P, Fortier B, Tronchin G, Vernes A (1988): Characterization of a 225 kilodalton rhoptry protein of *Plasmodium falciparum*. Mol Biochem Parasitol 27:135–141.

Schofield L, Bushell GR, Cooper JA, Saul AJ, Upcroft JA, Kidson C (1986): A rhoptry antigen of *Plasmodium falciparum* contains conserved and variable epitopes recognized by inhibitory monoclonal antibodies. Mol Biochem Parasitol 18:183–195.

Schofield L, Nussenzweig V, Nussenzweig RS (1987): Antimalarial activity of Lyt2$^+$ (suppressor/cytotoxic) T cells required for immunity to sporozoite challenge. In: R. Chanock, R. Lerner, F. Brown, and H. Ginsberg (eds). Vaccines 88. Cold Spring Harbor, NY: Cold Spring Harbor Laboratory, in press.

Schmidt-Ullrich R, Brown J, Whittle H, Lin P-S (1986): Human-human hybridomas secreting monoclonal antibodies to the $M_r$ 195,000 *Plasmodium falciparum* blood stage antigen. J Exp Med 163:179–188.

Schwarz RT, Riveros-Moreno V, Lockyer MJ, Nicholls SC, Davey LS, Hillman Y, Sandhu JS, Freeman RR, Holder AA (1986): Structural diversity of the major surface antigen of the *Plasmodium falciparum* merozoites. Mol Cell Biol 6:964–968.

Sharma S, Gwadz RW, Schlesinger DH, Godson GN (1986): Immunogenicity of the repetitive and nonrepetitive peptide regions of the divergent CS protein of *Plasmodium knowlesi*. J Immunol 137:357–361.

Sherwood JA, Marsh K, Howard RJ, Barnwell JW (1985): Antibody mediated strain-specific agglutination of *Plasmodium falciparum*-parasitized erythrocytes visualized by ethidium bromide staining. Parasite Immunol 7:659–663.

Siddiqui WA, Tam LQ, Kan S-C, Kramer KJ, Case SE, Palmer KL, Yamaga KM, Hui GSN (1986): Induction of protective immunity to monoclonal-antibody-defined *Plasmodium falciparum* antigens requires strong adjuvant in *Aotus* monkeys. Infect Immun 52:314–318.

Siddiqui WA, Tam LQ, Kramer KJ, Hui GSN, Case SE, Yamaga KM, Chang SP, Chan EBT, Kan S-C (1987): Merozoite surface coat precursor protects *Aotus* monkeys against *Plasmodium falciparum* malaria. Proc Natl Acad Sci USA 84:3014–3018.

Smythe J, Coppel RL, Brown GV, Ramasamy R, Kemp DJ, Anders RF (1988): Identification of two integral membrane proteins of *Plasmodium falciparum*. Proc Natl Acad Sci USA (in press).

Spitz S (1946): The pathology of acute falciparum malaria. Milit Surg 99:555–572.

Tanabe K, Mackay M, Goman M, Scaife JG (1987): Allelic dimorphism in a surface antigen gene of the malaria parasite *Plasmodium falciparum*. J Mol Biol 195:273–287.

Tapchaisri P, Chomcharn Y, Poonthong C, Asavanich A, Limsuwan S, Maleevan O, Tharavanij S, Harinasuta T (1983): Anti-sporozoite antibodies induced by natural infection. Am J Trop Med Hyg 32: 1203–1208.

Trager W, Jensen JB (1976): Human malaria parasites in continuous culture. Science 193:673–675.

Trager W, Rudzinska MA, Bradbury PC (1966): The fine structure of *Plasmodium falciparum* and its host erythrocytes in natural malaria infections in man. Bull WHO 35:883–885.

Udeinya IJ Miller LH, McGregor IA, Jensen JB (1983): *Plasmodium falciparum* strain-specific antibody blocks binding of infected erythrocytes to amelanotic melanoma cells. Nature 303:429–431.

van Schravendijk MR, Wilson RJM, Newbold CI (1987): Possible pitfalls in the identification of glycophorin-binding proteins of *Plasmodium falciparum* . J Exp Med 166:376–390.

Vergara U, Ruiz A, Ferreira A, Nussenzweig RS, Nussenzweig V (1985): Conserved group-specific epitopes of the circumsporozoite proteins revealed by antibodies to synthetic peptides. J Immunol 134:3445–3448.

Wåhlin B, Wahlgren M, Perlmann H, Berzins K, Björkman A, Patarroyo ME, Perlmann P (1984): Human antibodies to a $M_r$ 155,000 *Plasmodium falciparum* antigen efficiently inhibit merozoite invasion. Proc Natl Acad Sci USA 81:7912–7916.

Wanidworanun C, Barnwell JW, Shear HL (1987): Protective antigen in the membranes of mouse erythrocytes infected with *Plasmodium chabaudi*. Mol Biochem Parasitol 25:195–201.

Weiss WR, Sedegah M, Beaudoin RL, Miller LH, Good MF (1988): CD8+ T cells (cytotoxic/suppressors) are required for protection in mice immunized with malaria sporozoites. Proc Natl Acad Sci USA 85:573–576.

Yoshida N, Nussenzweig RS, Potocnjak P, Nussenzweig V, Aikawa M (1980): Hybridoma produces protective antibodies directed against the sporozoite stage of malaria parasite. Science 207:71–73.

Zavala F, Tam JP, Hollingdale MR, Cochrane AH, Quakyi I, Nussenzweig RS, Nussenzweig V (1985): Rationale for development of a synthetic vaccine against *Plasmodium falciparum* malaria. Science 228:1436–1440.

The Biology of Parasitism, pages 225–231
© 1988 Alan R. Liss, Inc.

# Immunity to Sexual Stages of Malaria Parasites in Relation to Malaria Transmission

### Richard Carter

Medical Research Council External Scientific Staff, Institute of Animal Genetics, University of Edinburgh, Edinburgh, Scotland

Immunity against the sexual stages of malaria parasites has, until recently, received relatively little attention. This is because the sexual stages are nonpathogenic and do not, therefore, directly contribute to the disease in an infected individual. By contrast, the sporozoites, liver stages, and asexual blood-stage parasites are, respectively, the indirect and the direct agents of the pathology of malaria; from a clinical point of view, these are the obvious stages for immunological investigation. Epidemiologically, however, the prevalence of malarial infections depends upon the availability and infectivity to mosquitoes of gametocytes in the human population. Recent evidence from studies of human populations exposed to malarial infection have indicated not only that immunity against the sexual stages of malaria parasites is readily induced during infections in man (Mendis et al., 1987) but that such immunity may play an important role in curtailing the severity of malarial epidemics. Thus, the medical relevance of immunity to these stages can no longer be ignored. In this essay, I shall discuss the nature of immunity to the sexual stages of malaria parasites and the significance of this immunity during malaria transmission in human populations.

Immunity against any disease involves two distinct, but not always conceptually distinguished, sets of questions: i) what are the stages or products of the aetologic agent and the immune effector mechanisms which are involved in immunity against the disease? and ii) what are the immunogens and the circumstances under which such immunity is induced? The relevance of this distinction is very evident in the discussion of immunity to the sexual stages of malaria parasites. For the purposes of such discussion, I will use the term *sexual* to apply to all stages involved in the sexual development of the parasites beginning with the gametocytes in the bloodstream of the vertebrate host, to the gametes and fertilized zygote in the mosquito midgut, and up to the transformation of the zygote into a mature ookinete (Fig. 1). Gametocytes, like the asexual blood stages of malaria parasites, develop following invasion of an RBC by a merozoite; the basis for differentiation of a merozoite into a sexual or an asexual parasite is unknown. In the course of a normal infection, gametocytes grow to maturity, circulate in the blood, and are eventually cleared and destroyed, presumably by scavenging cells of the immune system. Occasionally, however, mature and functional gametocytes may be ingested by an appropriate species of mosquito and stimulated to transform into the stages which establish the parasites in their vector.

Under the influence of changes in temperature and chemical environment, including a potent factor present in the mosquito midgut, the gametocytes in

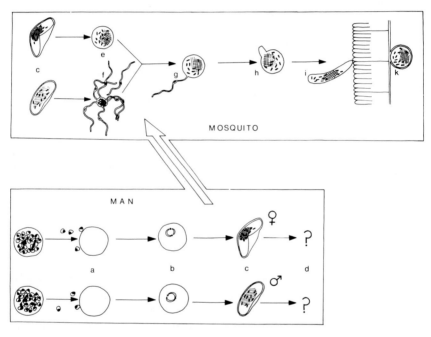

Fig. 1. The development of sexual stages of malaria parasites as represented by the human parasite *Plasmodium falciparum*. In man, (**a**) merozoites invade RBC's (**b**) young "ring"-stage parasites growing inside RBCs; maturation of ring stages into mature male or female gametocytes (**c**) takes place over about 8 days; (**d**) senescent gametocytes disposed of in the host by unknown route; in the mosquito (c) mature gametocytes ingested in a blood meal emerge from host RBC and form female (**e**) or male (**f**) gametes; fertilization (**g**) takes place within minutes and the zygote undergoes transformation (**h**) to form a mature ookinete (**i**) about 16 to 20 h after the blood meal; mature ookinete penetrates the mosquito midgut to form a young oocyst (**k**) on the serosal surface adjacent to the haemocoel.

a blood meal become extracellular within minutes of ingestion. Although the "trigger" is set off by environmental factors, the processes of gametocyte emergence and transformation into gametes is mediated by the gametocytes themselves. Probably under the action of substances released by the gametocytes, the host RBC membranes surrounding the parasites are disrupted. Minutes after emergence from the RBCs, the male gametocytes transform into eight highly motile male gametes; fertilization of female gametes, derived from female gametocytes, ensues and is complete within 10–30 minutes after ingestion of blood by the mosquito.

From emergence onwards, the parasites are completely extracellular and are, therefore, directly exposed to the contents of the mosquito blood meal, including immunologically active components such as phagocytes and other immune cells, antibodies, and complement. During the 16–20 hours following fertilization, each zygote transforms into an ookinete, an elongate and motile stage which crosses the mosquito midgut and continues its development on the outer side of the midgut adjacent to the haemocoel.

The biology of the sexual stages of malaria parasites thus suggests two general arenas within which immunity against these stages could be effective: i) the intracellular gametocytes circulating in the blood of the vertebrate host and ii) the extracellular gametes, zygotes, and ookinetes in the midgut of a mosquito. Induction of immunity to these stages during malarial infection, on the other hand, could only be due to parasite immunogens present in the vertebrate host. The importance of this fact will become evident during discussion of immunity against the sexual stages of the parasites in the mosquito midgut.

As already implied, immunity against the sexual stages of malaria parasites, commonly called "malaria transmission blocking immunity" for obvious reasons, may operate against gametocytes in the vertebrate host as well as against the extracellular forms in a mosquito midgut. The results of studies with various laboratory systems involving malaria parasites of birds, rodents, or monkeys, have shown that transmission blocking immunity can be induced by artificial immunization with preparations containing any of the sexual stages of the parasites, as defined here. Depending upon the immunogens used, immunity has been demonstrated against i) intracellular gametocytes during circulation in the vertebrate host (Harte et al., 1985), ii) extracellular gametes in the mosquito midgut (Carter and Chen, 1976; Gwadz, 1976; Mendis and Targett, 1979), iii) newly fertilized zygotes in the mosquito midgut (Kaushal et al., 1983), and iv) transforming zygotes or ookinetes in the mosquito midgut (Grotendorst et al., 1984; Vermeulen et al., 1985). Immunity in each of these situations involves either different target antigens and/or different immune effector agents.

Little is known about the mechanisms or target antigens of immunity against the gametocytes in the bloodstream of the vertebrate host except that they cannot be entirely the same as those involving the mosquito midgut stages; transmission blocking immunity against the circulating gametocytes appears to be T-cell dependent and largely, if not entirely, antibody independent.

More is known about the mechanisms and targets of immunity against the mosquito midgut stages of malaria parasites. Our information comes mainly from studies on the chicken malaria parasite *Plasmodium gallinaceum* and the human parasite *P. falciparum*; the situations for these two parasites are closely analogous for most considerations. Transmission blocking immunity against the gametes of the parasites is mediated by antigamete antibodies which function in either i) complement-independent reactions with gamete surface antigens to prevent fertilization or ii) in complement-fixing reactions which destroy by lysis both male and female gametes and, should any form, the newly fertilized zygotes also (Grotendorst et al., 1984; Rener et al., 1983). Since gametogenesis and fertilization occur within a few minutes, and generally less than half an hour after ingestion of a blood meal, such antibodies are effective within a correspondingly short space of time after ingestion by a mosquito.

The antigenic targets of these antigamete antibodies are two sets of proteins of analogous properties and similar sizes in both *P. falciparum* and *P. gallinaceum* (Grotendorst et al., 1984; Rener et al., 1983). These comprise in each species a protein of about $M_r$ 230,000 and a pair of apparently closely related proteins of $M_r$ 48,000 and 45,000. Both the $M_r$ 230,000 and the $M_r$ 48/45,000 proteins have been shown to be targets of antigamete transmission blocking antibodies.

A third type of antigen is the target of antibodies which are effective against the fertilized zygotes. These antigens are proteins of relatively low $M_r$ being about 31,000 in *P. gallinaceum* (Grotendorst et al., 1984) and 25,000 in *P. falciparum* (Vermeulen et al., 1985). Notwithstanding the difference in $M_r$ between the two species, the proteins appear to be biologically analogous in these species of parasite. Other than the fact that complement is not involved, the mechanism of action of transmission blocking antibodies against these proteins is not known. The proteins themselves do not appear on the zygote surface until after gametogenesis and fertilization and continue to be expressed on the mature ookinete. It is possible that antibodies against these proteins operate, at least in part, to inhibit the passage of ookinetes through the mosquito midgut wall.

These, therefore, are the known effector mechanisms and target antigens of malaria transmission blocking immunity. The extent to which these effects may be induced during natural malarial infections depends upon the presence of corresponding immunogens in the parasite stages in the vertebrate host. In this regard evidence is available from two types of investigation: i) the study of antibodies and transmission blocking immunity induced during malarial infections in animals and in man and ii) immunobiochemical studies on the expression of the target antigens (presumably also potential immunogens) of transmission blocking antibodies in the sexual stage parasites found in the vertebrate host, namely, the gametocytes.

The immunobiochemical evidence from *P. falciparum* is clear concerning the $M_r$ 230,000 and $M_r$ 48/45,000 gamete surface proteins. Both of these proteins are synthesized and are present in the gametocytes of this species from an early stage in their development (Vermeulen et al., 1985; Kumar and Carter, 1984). Moreover, antibodies which immunoprecipitate the $M_r$ 230,000 and/or the $M_r$ 48/45,000 gamete surface proteins of *P. falciparum* are present in a high proportion of sera from individuals exposed to intense *P. falciparum* transmission, as in Papua New Guinea (P.G. Graves and R. Carter, unpublished results). Furthermore, many of these sera mediate strong suppression of infectivity of gametocytes of *P. falciparum* to mosquitoes. The target antigens of antigamete immunity in *P. falciparum* appear, therefore, to be capable of functioning as effective immunogens of transmission blocking immunity in human infections due to their presence in the gametocytes in the blood infections.

The presence of the $M_r$ 25,000 ookinete surface protein in gametocytes of *P. falciparum* is, on the other hand, an open question. Readily detectable and present as a relatively major protein in gametes and zygotes of *P. falciparum* (Vermeulen et al., 1985), the $M_r$ 25,000 protein is barely detectable in gametocytes by immunofluorescence or by immuno (Western) blot. It is not clear whether the traces or $M_r$ 25,000 found in gametocytes are due to artifactual activation of the gametocytes and induction of gamete protein synthesis during preparation or whether they represent the genuine presence of small amounts of the $M_r$ 25,000 protein in *P. falciparum* gametocytes. If the latter, the antigen could be immunogenic during malarial infection. However, the Papua New Guinea sera which carried antibodies to the $M_r$ 230,000 and $M_r$ 48/45,000 gamete surface proteins showed little evidence of reactivity with the *P. falciparum* $M_r$ 25,000 zygote protein.

It appears, therefore, that infections of *P. falciparum* are naturally immunogenic for transmission blocking immunity against the target antigens of antigamete fertilization blocking antibodies; on the other hand, immunogens for the target antigen of postfertilization transmission blocking antibodies may not occur during natural *P. falciparum* infections.

The effectiveness with which malaria transmission blocking immunity may be induced during malarial infection in man has been most clearly shown in studies on *P. vivax* in Sri Lanka (Mendis et al., 1987). Primary infections of *P. vivax* were frequently accompanied by transmission blocking antibodies which effectively eliminated or greatly reduced the infectivity of the gametocytes of *P. vivax* infections to mosquitoes. In secondary attacks, following within a few weeks or months of a first *P. vivax* attack, the level of transmission blocking antibodies was sufficiently enhanced that the majority of such cases were noninfectious to mosquitoes.

Other than being antibody mediated and frequently, but not always, complement dependent, it is not clear in the case of naturally induced transmission blocking immunity during *P. vivax* infections in man whether the immunity is against antigens of the gametes or against postfertilization developmental antigens of the zygotes and ookinetes. The target antigens of transmission blocking immunity against *P. vivax* remain to be clearly defined and so, therefore, do the specificities of the antibodies in the transmission blocking sera from natural *P. vivax* infections.

The lack of this information does not, however, diminish the significance of transmission blocking immunity induced during natural *P. vivax* infections. The degree of transmission blocking immunity, as shown from observations on about 50 individuals with *P. vivax* infections, is such that the epidemic or seasonal outbreaks of *P. vivax* malaria characteristic of many parts of the tropical and subtropical world would probably involve multifold increases in the numbers of cases transmitted in the absence of immunity to the sexual stages of the parasites.

In conclusion, it now appears that immunity to the sexual stages of malaria parasites commonly accompanies malarial infections in man. The nature of this immunity is such that the infectivity of gametocyte carriers to mosquitoes is often greatly reduced. Naturally induced transmission blocking immunity against the sexual stages of malaria parasites may, therefore, represent a significant aspect of antimalarial immunity in protecting human populations against the spread of malarial infections under epidemic conditions.

## REFERENCES

Carter R, Chen DH (1976): Malaria transmission blocked by immunization with gametes of the malaria parasite. Nature 263:57.

Grotendorst CA, Kumar N, Carter R, Kaushal DC (1984): A surface protein expressed during the transformation of zygotes of *Plasmodium gallinaceum* is a target of transmission-blocking antibodies. Infect Immun 45:775.

Gwadz RW (1976): Malaria: Successful immunization against the sexual stages of *Plasmodium gallinaceum*. Science 193: 1150.

Harte PG, Rogers NC, Targett GAT (1985): Role of T cells in preventing transmission of rodent malaria. Immunology 56:1.

Kaushal DC, Carter R, Rener J, Grotendorst CA, Miller LH, Howard RJ (1983): Monoclonal antibodies against surface determinants on gametes of *Plasmodium gallinaceum* block transmission of malaria parasites to mosquitoes. J Immunol 131:2557.

Kumar N, Carter R (1984): Biosynthesis of the target antigens of antibodies blocking transmission of *Plasmodium falciparum*. Mol Biochem Parasitol 13:333.

Mendis KN, Targett GAT (1979): Immunization against gametes and asexual erythrocytic stages in rodent malaria parasites. Nature 277:389.

Mendis KN, Munasinghe YD, de Silva YNY, Keragalla I, Carter R (1987): Malaria transmission blocking immunity induced by natural infections of *Plasmodium vivax* in humans. Infect Immun 55:369.

Rener J, Graves PM, Carter R Williams JL, Burkot TR (1983): Target antigens of transmission-blocking immunity in gametes of *Plasmodium falciparum*. J Exp Med 158:976.

Vermeulen AN, Ponnudurai T, Beckers PJA, Verhave J-P, Smits MA, Meuwissen JHETh (1985): Sequential expression of antigens in sexual stages of *Plasmodium falciparum* accessible to transmission-blocking antibodies in the mosquito. J Exp Med 162:1460.

The Biology of Parasitism, pages 233–248
© 1988 Alan R. Liss, Inc.

# Immunity to Murine Malarial Parasites

Carole A. Long

*Malaria Research Group, Department of Microbiology and Immunology,
Hahnemann University, Philadelphia, Pennsylvania 19102*

## CHARACTERISTICS OF MURINE MALARIAL PARASITES

The discovery of *Plasmodium berghei* by Vincke in 1943 was an example of a protozoan infection which was seen first in the invertebrate vector, *Anopheles dureni*, and only later in its mammalian host (reviewed by Killick-Kendrick, 1978). Five years later Vincke and Lips found malaria parasites in the blood of an African rodent, *Thamnomys surdaster*; and subsequently other rodents were shown to harbor this parasite. Later, other plasmodial species—*P. yoelii* (closely related to the *P. berghei* group), *P. vinckei*, and *P. chabaudi*—were isolated in African rodents and adapted to laboratory animals. These isolates have generally been classified on the basis of morphological and biological characteristics. Table 1 summarizes host cell preference and outcome of infection for the four different species of rodent plasmodia. As can be seen from the table, the erythrocytic stages of some species prefer reticulocytes, while others preferentially infect mature erythrocytes. The basis for this host cell preference is not yet understood. The outcome of infection varies with the species, with some resolving spontaneously and others resulting in a fulminating lethal parasitemia. In addition, two variants of *P. yoelii* have been isolated, one of which is nonlethal (17X); the other, lethal (YM or XL) (Yoeli et al., 1975).

More recently, other techniques have been used to try to establish evolutionary relatedness of species of plasmodia. For example, McCutchan and colleagues (1984) have examined the base composition of DNA from various plasmodial species. Three different rodent parasites displayed a $dG \cdot dC$ content of approximately 18%. In contrast, *P. vivax*, a parasite of man, and a number of the plasmodia infecting monkeys were reported to have a $dG \cdot dC$ content of about 30%. This classification also correlated with the extent of cross-hybridization observed on Southern blots using probes for several conserved genes such as actin. Interestingly, *P. falciparum*, the most significant malarial parasite infecting man, was found to have a base composition plasmodial species and for the utilization of repeated epitopes of the *P. falciparum* CS protein in phase I vaccine trials in man (Dame et al., 1984; Enea et al., 1984).

Despite the advances made in characterizing the CS proteins of the plasmodia infecting primates, the murine model systems still have much to offer with regard to our understanding of host responses to sporozoite antigens. Although immunization studies with the immunodominant repeat of the *P. falciparum* CS protein had elicited strong humoral responses in CS7BL/6 mice (Young et al., 1985), immunization of human volunteers with the same construct did not result in comparable levels of antibodies (Ballou et al., 1987; Herrington et al., 1987). Subsequent studies in mouse strains with a variety of haplotypes showed that the response to the repeated NANP epitope is genetically restricted so that only mice with the I-A$^b$ genotype gave good antibody responses to the CS repeat (Good et al., 1986,; Del Giudice et al., 1986). Additional studies suggested that this restriction was due to a lack of T-cell sensitization. The relatively poor responses of humans to the same antigen might thus be explained on the basis of poor antigen presentation by the class II molecules of the population or by the repertoire of T-cell receptors available. Such considerations must be addressed in the immunogens selected for human trials by the inclusion of epitopes which can be recognized by and presented to T lymphocytes. An additional complicating factor is the possibility that the T cell epitopes on the CS protein may lie within regions of this molecule which vary between strains (Good et al., 1988). Such polymorphism in the T cell epitopes of the CS protein from different isolates of *P. falciparum* might preclude boosting by natural infection due to the lack of memory T cells with the appropriate specificity.

In addition to assisting in the definition of antigenic epitopes in CS molecules relevant to immunization, the murine model systems contribute to our understanding of the complexities of the host immune responses to these antigens. Although low concentrations of monoclonal antibodies to the CS repeat antigen of *P. berghei* were passively protective in mice, the observation that some B-cell-deficient mice could be protected by immunization with irradiated sporozoites suggested that additional cell-mediated mechanisms of immunity might be operative as well (Chen et al., 1977). More recently, Egan and colleagues (1987) have compared the efficacy of immunization of mice using recombinant CS antigen from *P. berghei* or irradiated sporozoites from the same strain. While all mice developed high titers of antibodies to the CS protein and the repeat, those animals which received the irradiated sporozoites were more resistant to challenge infection. This result also suggested that additional mechanisms of immunity, perhaps cell-mediated, could contribute to the enhanced resistance displayed by mice given the irradiated sporozoites. Passive administration of an IgG fraction from the immunized animals did result in the transfer of protective activity. However, T-cell-enriched fraction of spleen cells obtained from mice which had received the sporozoite vaccine protected naive recipients, while B-cell-enriched fractions did not. Thus, the response to sporozoite antigens appears to be complex, and successful immunization may require the elicitation of both cell-mediated and humoral mechanisms of immunity.

Further support for a role of T cells beyond their helper functions has recently been provided by the studies of Schofield et al., 1987 and Weiss and colleagues (1988), who immunized mice with sporozoites and showed that in vivo depletion of the CD8+ subset of T cells using a specific monoclonal antibody significantly reduced host resistance to sporozoite challenge. These cells may be functioning as cytotoxic T cells or as producers of lymphokines such as γ-interferon. Development of systems to study differentiation of murine plasmodial species in hepatic cells will facilitate the elucidation of such cellular effector mechanisms. Other studies have implicated γ-interferon as an inhibitor of plasmodial development within hepatocytes both in vivo and in vitro (Schofield et al., 1987).

## Immunity to Asexual Erythrocytic Stages

**P. yoelii, an antibody-dependent model.** The erythrocytic stages of *P. yoelii* represent a model of plasmodial infection which requires a T-cell-dependent humoral immune response for resolution. Evidence that this is the case has been extensively reviewed elsewhere (Jayawardena, 1981; Deans and Cohen, 1983,; Weidanz and Long, 1987) and includes the following findings:

1. Athymic nude mice develop a fulminating lethal infection with this parasite in contrast to the self-limiting parasitemias seen in normal littermates. Grafting of the nude recipients with thymus cells prior to parasite challenge reduced the mortality and increased levels of antimalarial antibodies. However, these studies did not reveal whether the T cells were required for their helper functions or whether they played an alternate role.

2. Mice made B-cell deficient by anti-μ treatment succumbed to this normally avirulent infection, indicating the importance of the humoral immune system.

3. CBA/N mice lacking certain B-cell subsets developed severe, long-lasting infections when challenged with *P. yoelii.*

4. Hyperimmune sera obtained from animals repeatedly infected with this parasite passively transferred protection to normal or T-cell-deprived recipients.

The construction of a monoclonal antibody which provided passive protective activity against challenge infection with *P. yoelii* provided additional evidence that humoral immunity could play a major role in the host's armamentarium. Majarian and colleagues (1984) reported a hybridoma producing antibodies which could protect naive recipients against challenge with either the lethal or nonlethal variants of this parasite. Moreover, the antibody was still effective when administered after the parasitemia had reached 1%. Interestingly, mice which were rechallenged long after the passively administered antibody had disappeared were still protected, suggesting that the limited and controlled exposure to plasmodial antigen allowed them to develop an active immune response. This potent antibody was of the IgG3 isotype and immunoprecipitated a parasite antigen of 230 kilodaltons (kD) (PY230) as well as a series of smaller polypeptides. Other investigators showed that the purified PY230 antigen could actively immunize mice against challenge infection, pointing to this large merozoite surface antigen as potentially protective (Holder and Freeman, 1981).

Identification of the target plasmodial antigen for a protective monoclonal antibody established a model system for examination of the role of this polypeptide in the host's immune response. This antigen is cross-reactive with high molecular weight, merozoite surface proteins described in other plasmodial species which are synthesized late during the erythrocytic life cycle and processed to a series of smaller polypeptides (Holder et al., 1983). The analogous 195-kD protein of *P. falciparum* (PF195) has been cloned from a number of parasite strains and sequenced by several laboratories (Holder et al., 1985; Mackay et al., 1985; Lyon et al., 1986). While it has been evident for some time from studies with monoclonal antibodies that this antigen is polymorphic in different variants of this species (McBride et al., 1985), recent molecular analysis has begun to define the nature and extent of the variability as well as its distribution within the molecule. From the predicted amino acid sequence data, it is apparent that one significant site of variability in this protein is the repeated sequence in the N-terminal region. Recently, Tanabe and colleagues (1987) have carried out a detailed analysis of all of the sequence data available for PF195 and have concluded that the gene consists of variable blocks separated by conserved or semiconserved sequences. According to their analysis, the variable sequences can be classified into two types, and they therefore suggest that the protein is encoded by dimorphic alleles which give rise to variants by recombination. This hypothesis has recently been supported by investigators at the Walter and Eliza Hall Institute in Australia who have prepared synthetic oligonucleotides to the various regions and tested them on different *P. falciparum* isolates (R. Coppel, personal communication). Thus the potential variability among the population of PF195 molecules appears to be limited, a finding which is encouraging to those who view this antigen as a significant erythrocytic-stage vaccine candidate. That immunity to this molecule can be protective is supported by a recent report in which *Aotus* monkeys immunized with this antigen in complete Freund's adjuvant were totally protected from lethal challenge infection with the homologous parasite (Siddiqui et al., 1987).

It is unclear, however, which areas of this large molecule may contain important epitopes for recognition by the host's immune system and how well such epitopes may he conserved between strains. One item of evidence is that a monoclonal antibody to the PF195 antigen has been reported which inhibits invasion in vitro and which recognizes a determinant in the C-terminal portion of this molecule (Pirson and Perkins, 1985). However, extrapolation of this observation to protection in vivo is not readily possible, and the epitope recognized has not been defined. In contrast, the murine system offers an opportunity to map epitopes which are recognized by a protective antibody or by T lymphocytes. Moreover, immune responses to such epitopes can be correlated with protective immunity in vivo.

To approach this issue Burns and colleagues (1988) have constructed an expresssion library from *P. yoelii* DNA and have isolated a clone encoding the epitope recognized by monoclonal antibody (McAb) 302. The open reading frame from the insert of this clone was sequenced and identified as the 3′ portion of the gene encoding PY230. The deduced amino acid sequence revealed that this antigen contains a tandemly repeated tetrapeptide of Gly-Ala-Val-Pro, a series of ten cysteine residues located within the terminal 110 amino acids, and a potential membrane anchor of 18 hydrophobic residues. Comparison of this C-terminal sequence with that of PF195 revealed significant nucleic acid and amino acid homology. Of particular note is the observation that nine of the terminal cysteine residues are conserved in analogous positions in both antigens, suggesting that this region of the molecule may provide a structural domain necessary for function. More recently, Mr. Steve Nicholls and Dr. Anthony Holder of Wellcome Biotech have collaborated with Burns and colleagues in comparing the structures of PY230 and PF195. Examination of hydropathy plots of the C-terminal sequences from both proteins has revealed an overall similarity of pattern between the two molecules. This case has been strengthened by additional computer analysis predicting secondary structure from a primary amino acid sequence. Using nine different algorithms, the predicted structural pattern of PY230 is very similar to that of PF195, particularly in the most C-terminal portion of the sequence. In general, this analysis is compatible with the classification of Tanabe et al. (1987), and even in areas where the amino acid sequences diverge, the predicted protein structures are quite homologous. Taken together, this evidence supports the contention that the two proteins, PY230 and PF195, are very similar in structure.

The *P. yoelii* system thus provides a unique model to study the role of the major merozoite surface antigen, to localize significant epitopes on the molecule, to try to elicit protective host responses, and to define the nature of the protective immunity. Based on the data obtained with PY230 and the protective monoclonal antibody, it can be hypothesized that recognition of epitopes in the C-terminal region of this molecule is required for protective immunity. This agrees with other data indicating that the processed fragments from the C-terminal region are retained on the merozoite surface following schizont rupture (Howard et al., 1985). It is also supported by a recent report by McBride and Heidrich (1987) suggesting that the C-terminal PF195 fragment noncovalently associates with other parasite surface antigens, forming an antigenic complex which might have a role in invasion of erythrocytes. This hypothesis can be tested directly by immunization of mice with defined portions of the PY230 antigen, since such polypeptides can be obtained from bacteria expressing recombinant proteins. The critical epitopes can be further localized by employing the products of smaller constructs as immunogens. These empirical immunization studies can be correlated with experiments localizing the epitope recognized by the protective monoclonal antibody. Moreover, relevant T cell epitopes can also be identified. Once the critical regions of the antigen have been identified, it will then be possible to address the question of strain variability of such epitopes. Given the known polymorphism of PF195, this issue can be approached using the available variants of *P. yoelii*. Moreover, if protection can be achieved in vivo with defined portions of this antigen, then the mechanisms involved in this immunity—whether humoral or cell-mediated—can be dissected.

Another significant plasmodial antigen of the erythrocytic stage is one which has been given the acronym RESA (ring-infected erythrocyte surface antigen) or alternately referred to as PF155 (Coppel et al., 1984; Perlmann et al., 1984). This molecule of 155 kD was initially identified by immunofluorescence on the surface of glutaraldehyde-fixed and air-dried erythrocytes infected with *P. falciparum*. Additional evidence suggested that this molecule is localized within the micronemes of developing merozoites and is subsequently transferred to the surface of the newly invaded erythrocyte. Antibodies to this polypeptide have been reported to inhibit merozoite invasion in vitro (Wahlin et al., 1984) and immunization with recombinant peptides containing the repeated epitopes of this protein provided some protection to Aotus monkeys against challenge infection (Collins et al., 1986).

A similar molecule has been described in the membranes of *P. chabaudi*-parasitized erythrocytes by Gabriel et al. (1986). The same antigen has been reported by Wanidworanun et al. (1987), who immunized mice with membranes from parasitized erythrocytes and generated a series of monoclonal antibodies to a 96-kD (PC96) antigen. One of these antibodies cross-reacted with the RESA antigen of *P. falciparum*. Immunization with affinity-purified PC96 antigen provided some protective immunity to mice challenged with the homologous parasite. The nature of this immunity has not yet been defined. Recently, Goldring and colleagues have utilized the monoclonal antibodies of Wanidworanun et al. to isolate a clone from a recombinant expression library of *P. chabaudi* genomic DNA. The β-galactosidase fusion protein isolated carried epitopes recognized by several of the monoclonal antibodies and further characterization of this clone is in progress. This may provide a model to investigate the immunology of a RESA-like antigen in a murine system. Moreover, since portions of the *P. falciparum* RESA antigen have been reported to bear structural similarity to erythrocyte band 3, the major anion transporter, the murine system may aid in clarifying the biological and functional significance of this relationship.

*P. chabaudi adami*, **an antibody-independent model.** It is becoming increasingly apparent that antibody-independent mechanisms of immunity also make a significant contribution to host defenses against the erythrocytic stages of the parasite. Evidence supporting this has been reviewed elsewhere (Weidanz and Long, 1987), and includes the observation by Grun and Weidanz (1981) that mice made B-cell deficient by lifelong treatment with anti-$\mu$ sera were able to resolve acute *P. chabaudi adami* infections with the same kinetics as normal animals. This experiment demonstrated that while the immunodeficient animals could never totally sterilize their infections, humoral immunity was not required to suppress acute infection with this species of parasite. Weidanz and colleagues subsequently showed that T cells were necessary to reduce the parasite burden. This system was then used to define further the role of cellular immunity in response to malarial infection. Since athymic nude mice are deficient in T lymphocytes, these animals are very susceptible to *P. chabaudi adami* and infections with this parasite are lethal. Therefore, nude mice can be used as recipients for adoptive transfer experiments. Cavacini et al. (1986) showed that spleen cells from normal or immune mice could transfer immunity to histocompatible nude animals. Fractionation of the transferred cells demonstrated that T cells, in particular the L3T4+ subset of these cells, could allow nude mice to survive subsequent challenge infection. Moreover, in all the reconstitution studies, the recipients developed patent parasitemias, indicating that the transferred cells were probably activating additional effector systems within the host and were not likely to be directly cytotoxic for parasitized erythrocytes.

Brake et al. (1986) have carried such experiments further by maintaining T cells from immune mice in culture using interleukin 2 (IL-2) and cyclic stimulation with plasmodial antigen and syngeneic antigen-presenting cells. The cell lines thus established were predominantly of the CD4+, helper-inducer phenotype and only responded to *P. chabaudi adami* antigen processed by syngeneic antigen-presenting cells. When these cultured cells were adoptively transferred to nude mice, the recipients were able to survive homologous parasite challenge. Since the cell lines possessed protective activity, they were subsequently used to develop T cell clones, one of which was shown to transfer protection similarly (Brake et al., 1988). As in the case of the immune spleen cells or the T-cell lines, the nude recipients always developed patent parasitemias, again suggesting the activation of host effector systems. Such activation is presumably the result of lymphokine secretion by the grafted cells. In this case, the protective CD4+ clone was shown to secrete γ-interferon and IL-2 in response to homologous malarial antigen.

To examine the specificity of immunologic reconstitution of these animals, they were challenged with *P. chabaudi adami*, and later rechallenged with the homologous parasite or with *P. yoelii*. Mice were immune to the homologous challenge infection but fully susceptible to the heterologous parasite, indicating that the cells possessed immunologic memory. The grafted animals could not mount a delayed-type hypersensitivity response to a strong contact allergen, dinitrofluorobenzene, nor could they produce antibodies to keyhole limpet hemacyanin. These observations are consistent with the limited heterogeneity of the T cells used for reconstitution. However, mice reconstituted with the clone and infected with *P. chabaudi adami* did develop antibodies capable of recognizing a large number of malarial polypeptides as judged by radioimmunoprecipitation assays. It may be that the grafted cells have been exposed to a specific plasmodial epitope as a result of infection and have consequently elaborated lymphokines which have supported the polyclonal differentiation of B-cells. The basis for this help is currently under investigation.

The *P. chabaudi adami* antigens recognized by these protective T-cell lines and the one clone isolated to date have not yet been identified. Once additional protective clones are established, it will be possible to determine whether a variety of plasmodial antigens and epitopes can elicit a protective T cell response or whether the number of protective antigens is very limited. Moreover, one can then correlate lymphokine production with protective capacity. For example, do all protective clones secrete $\gamma$-interferon or other mediators and thus belong to the subtype of T cells recently designated TH 1 cells (Mosmann et al., 1986)? Another relevant issue when dealing with cultured cells is their capacity to migrate to appropriate sites. Many cloned T cells display aberrant trafficking patterns, a finding which may in part be due to the loss of the MEL-14 antigen described by Gallatin et al. (1986). Much evidence exists pointing to the role of the spleen in resolution of malarial infections. Therefore, it may be important for the grafted cells to reach the spleen and perhaps even microenvironments within the spleen in order to fulfill their protective capacity. Such questions can be addressed using *P. chabaudi adami* and cultured T cells as a model system.

The question also arises as to the mechanism by which these cells promote resolution of infection. Since all the reconstituted animals develop significant parasitemias, it is likely that the grafted cells are not directly cytotoxic for parasitized erythrocytes but rather respond to specific plasmodial epitopes resulting in the synthesis of nonspecific lymphokines. It has been suggested that these biological response modifiers in turn mediate activation of cells of the mononuclear phagocytic system leading to the production of other substances inhibitory to the parasites (Clark et al., 1981; Dockrell and Playfair, 1983; Allison and Eugui, 1983; Taverne et al., 1987; Ockenhouse and Shear, 1984). These factors include toxic metabolites of oxygen, hydrogen peroxide, tumor necrosis factor/cachectin, and other toxic products of macrophages. The elucidation of these mechanisms should be facilitated by the use of T-cell clones and lines in the murine model system, and such experiments may provide insight into ways of eliciting similar responses which would suppress parasite infections in man.

**Repertoire of responses to murine plasmodia.** Responses to infections by the two rodent plasmodial species discussed—*P. yoelii* 17X and *P. chabaudi adami*—appear to represent extremes in the repertoire of host immune responses, one predominantly cell mediated and one predominantly humoral. Why should resolution of infections by these two rodent parasites require different types of responses by the host? While the answer to this question is unclear, it may relate to the phylogenetic distance between the two plasmodia and the fact that they preferentially parasitize different types of erythrocytes. However, it is also possible that both antibody-independent and humoral immune mechanisms participate in resolution of all murine malarial parasite infections but that one or the other may predominate in particular cases. Antibody-dependent mechanisms are demonstrably important in asexual *P. yoelii* infections, and these B-cell responses have been shown to require T-cell help (Taylor et al., 1982). In addition to collaborating in the synthesis of antibodies, other cell-mediated mechanisms may also be activated in this infection. For example, some B-cell deficient mice which were cured of their *P. yoelii* infections by drugs were able to resist a secondary challenge with the homologous parasite (Roberts and Weidanz, 1979; Grun and Weidanz, 1983). Additional immune mechanisms, presumably cell mediated in nature, were elicited, even though that immunity was not sufficient to allow the mice to survive a primary infection. In other studies, Freeman and Holder (1983) showed that mice immunized with purified PY230 were resistant to challenge infection, but this immunity could not be transferred with serum.

In the case of *P. chabaudi adami*, resolution of acute infection is primarily independent of antibody, although humoral immunity appears to be required to sterilize the infections completely. Weidanz and colleagues (personal communication) have recently shown that these cell-mediated mechanisms may be effective against a number of other hemosporidia. Such cell-mediated mechanisms may be important effectors against the erythrocytic stages of a variety of plasmodial species. Thus, it may be necessary to induce appropriate cellular and humoral responses in order to achieve optimum protection. The plasmodial antigens and epitopes required to elicit these diverse mechanisms in the repertoire of immune responses have not yet been characterized. The murine models provide systems to investigate these questions and to explore the nature and range of protective host responses.

## SUMMARY

The plasmodial species discussed here possess both advantages and disadvantages as experimental models of malarial infection. On the one hand, laboratory animals are not the natural hosts for these parasites, and the infection profiles are not the same as would be seen in the Thamnomys species. On the other hand, the extensive immunobiology and genetics available for inbred strains of mice and rats allows experimental manipulation of the host immune system and a more detailed understanding of the range of host defense mechanisms. Experiments to date using the sporozoite and the erythrocytic stages of infection have suggested that both cell-mediated and humoral mechanisms of immunity can play a role in suppressing or eliminating parasites. T cells serve both as helper cells for the production of antibody and have additional roles either as effector cells or as inducer cells. Thus it would seem that strategies for human malaria vaccines must incorporate relevant epitopes which can be presented to and recognized by both T cells and B cells. The murine malarial parasites should continue to provide insights into host response mechanisms and to assist in the definition of important plasmodial antigens and epitopes. In a broader sense, these models can also be used to explore fundamental immunological questions using defined antigens of a complex pathogen. Our knowledge of and our capacity to manipulate host recognition of foreign antigens may be furthered by studies with the immune malarial parasites.

## ACKNOWLEDGMENTS

I am happy to acknowledge the continuing collaboration of Dr. William P. Weidanz and Dr. Akhil Vaidya in studies on murine malarial parasites, along with students and associates in the laboratory including Mr. James Burns, Jr., Mr. Thomas Daly, and Drs. David A. Brake, J.P. Dean Goldring, and William R. Majarian. This work was supported by NIH grant AI21089 and by the UNDP/World Bank/WHO Special Programme for Research and Training in Tropical Diseases.

## REFERENCES

Allison AC, Eugui EM (1983): The role of cell-mediated immune responses in resistance to malaria, with special reference to oxidant stress. Annu Rev Immunol 1:361–392.

Ballou WR, Hoffman SL, Sherwood JA, Hollingdale MR, Neva FA, Hockmeyer WT, Gordon DM, Schneider I, Wirtz RA, Young JF, Wasserman GF, Reeve P, Diggs CL, Chulay JD (1987): Safety and efficacy of a recombinant DNA Plasmodium falciparum sporozoite vaccine. Lancet i:1277–1281.

Brake DA, Weidanz WP, Long CA (1986): Antigen-specific, interleukin 2-propagated T lymphocytes confer resistance to a murine malarial parasite, Plasmodium chabaudi adami. J Immunol 137:347–352.

Brake DA, Long CA, Weidanz WP (1988): Adoptive protection against Plasmodium chabaudi adami malaria in athymic nude mice by a cloned T-cell line. J Immunol 140:1989–1993.

Brown GV, Nossal GJV (1986): Malaria—Yesterday, today, and tomorrow. Perspect Biol Med 30:65–76.

Burns JM, Daly TM, Vaidya AB, Long CA (1988): The 3' portion of the gene for a Plasmodium yoelii merozoite surface antigen encodes the epitope recognized by a protective monoclonal antibody. Proc Natl Acad Sci USA 85:602–606.

Cavacini LA, Long CA, Weidanz WP (1986): T-cell immunity in murine malaria: Adoptive transfer of resistance to Plasmodium chabaudi adami in nude mice with splenic T cells. Infect Immun 52:637–643.

Chen DH, Tigelaar RE, Weinbaum FI (1977): Immunity to sporozoite-induced malaria infection in mice. I. The effect of immunization of T and B cell-deficient mice. J Immunol 118:1322–1327.

Clark IA, Virelizier JL, Carswell EA, Wood, PR (1981): Possible importance of macrophage-derived mediators in acute malaria. Infect Immun 32:1058-1065.

Collins WE, Anders RF, Papaioanou M, Campbell GH, Brown GV, Kemp DJ, Coppel RL, Skinner JC, Andrysiak PM, Faraloro JM, Corcoran LM, Broderson JR, Mitchell GF, Campbell CC (1986): Immunization of *Aotus* monkeys with recombinant proteins of an erythrocyte surface antigen of *Plasmodium falciparum*. Nature 323:259-262.

Coppel RL, Cowman AF, Anders RF, Bianco AE, Saint RB, Lingelbach KR, Kemp DJ, Brown GV (1984): Immune sera recognize on erythrocytes a *Plasmodium falciparum* antigen composed of repeated amino acid sequences. Nature 310:789-792.

Dame JB, Williams JL, McCutchan TF, Weber JL, Wirtz RA, Hockmeyer WT, Maloy WL, Haynes JD, Schneider I, Roberts D, Sanders GS, Reddy EP, Diggs CL, Miller LH (1984): Structure of the gene encoding the immunodominant surface antigen on the sporozoite of the human malaria parasite *Plasmodium falciparum*. Science 225:593-599.

Deans JA, Cohen S (1983): Immunology of malaria. Annu Rev Microbiol 37:25-49.

Del Giudice G, Cooper JA, Merino J, Verdini AS, Pessi A, Togna AR, Engers HD, Corradin G, Lambert PH (1986): The antibody response in mice to carrier-free synthetic polymers of *Plasmodium falciparum* circumsporozoite repetitive epitope is I-Ab-restricted: Possible implications for malaria vaccines. J Immunol 137:2952-2955.

Dockrell H, PLayfair JHL (1983): Killing of blood stage murine malaria parasites by hydrogen peroxide. Infect Immun 39:456-459.

Egan JE, Weber JL, Ballou WR, Majarian W, Gordon DM, Hoffman SL, Wirtz PA, Schneider I, Woollett G, Hollingdale MR, Young JF, Hockmeyer WT (1987): Efficacy of murine malaria sporozoite vaccines: Implications for human vaccine development. Science 236:453-456.

Enea V, Ellis J, Zavala F, Arnot DE, Asanavich A, Masuda A, Quakyi I, Nussenzweig RS (1984): DNA cloning of *Plasmodium falciparum* circumsporozoite gene: Amino acid sequence of repetitive epitope. Science 225:628-630.

Freeman RR, Holder AA (1983): Characteristics of the protective response of BALB/c mice immunized with a purified *Plasmodium yoelii* schizont antigen. Clin Exp Immunol 54:609-616.

Gabriel JA, Holmquist G, Perlmann H, Berzins K, Wigzell H, Perlmann P (1986): Identification of a *Plasmodium chabaudi* antigen present in the membrane of ring stage infected erythrocytes. Mol Biochem Parasitol 20:67-75.

Gallatin M, St. John TP, Siegelman M, Reichert R, Butcher EC, Weissman IL (1986): Lymphocyte homing receptors. Cell 44:673-680.

Good MF, Berzofsky JA, Maloy WL, Hayashi Y, Fujii N, Hockmeyer WT, Miller LH (1986): Genetic control of the immune response in mice to a *Plasmodium falciparum* sporozoite vaccine. J Exp Med 164:655-660.

Good MF, Pombo D, Quakyi IA, Riley FM, Houghten RA, Menon A, Alling DW, Berzofsky JA, Miller LH (1988): Human T cell recognition of the circumsporozoite protein of *Plasmodium falciparum*. Immunodominant T cell domains map to the polymorphic regions of the molecule. Proc Natl Acad Sci USA 85:1199-1203.

Grun JL, Weidanz WP (1981): Immunity to *Plasmodium chabaudi* in the B-cell-deficient mouse. Nature 290:143-145.

Grun JL, Weidanz WP (1983): Antibody-independent immunity to reinfection malaria in B-cell-deficient mice. Infect Immun 41:1197-1204.

Herrington DA, Clyde DF, Losonsky G, Cortesia M, Murphy FR, Davis J, Baqar A, Felix AM, Heimer DP, Gillessen D, Nardin E, Nussenzweig RS, Nussenzweig V, Hollingdale MR, Levine MM (1987): Safety and immunogenicity in man of a synthetic peptide malaria vaccine against *Plasmodium falciparum* sporozoites. Nature 328:257–259.

Holder AA, Freeman RR (1981): Immunization against blood-stage rodent malaria using purified parasite antigens. Nature 294:361–364.

Holder AA, Freeman RR, Newbold CI (1983): Serological cross-reaction between high molecular weight proteins synthesized in blood schizonts of *Plasmodium yoelii*, *Plasmodium chabaudi*, and *Plasmodium falciparum*. Mol Biochem Parasitol 9:191–196.

Holder AA, Lockyer MJ, Odink KG, Sandhu JS, Riveros-Moreno V, Nicholls SC, Hillman Y, Davey LS, Tizard MLV, Schwarz RT, Freeman RR (1985): Primary structure of the precursor to the three major surface antigens of *Plasmodium chabaudi* merozoites. Nature 317:270–273.

Howard RF, Stanley HA, Campbell GH, Langreth SG, Reese RT (1985): Two *Plasmodium falciparum* merozoite surface polypeptides share epitopes with a single Mr185,000 parasite glycoprotein. Mol Biochem Parasitol 17:61–77.

Jayawardena AN (1981): Immune responses in malaria. In J. Mansfield (ed): Parasitic Diseases, Vol I: Immunology. New York: Marcel-Dekker, pp 85–136.

Killick-Kendrick R (1978): Taxonomy, zoogeography, and evolution. In R. Killick-Kendrick and W. Peters (eds): Rodent Malaria. New York: Academic Press, pp 1–52.

Lyon JA, Geller RH, Haynes JD, Chulay JD, Weber JL (1986): Epitope map and processing scheme for the 195,000 dalton surface glycoprotein of *Plasmodium falciparum* merozoites deduced from cloned overlapping segments of the gene. Proc Natl Acad Sci USA 83:2989–2993.

Mackay M, Goman M, Bone N, Hyde JE, Scaife J, Certa U, Stunnenberg H, Bujard H (1985): Polymorphism of the precursor for the major surface antigens of *Plasmodium falciparum* merozoites: Studies at the genetic level. EMBO J 4:3823–3829.

Majarian WR, Daly TM, Weidanz WP, Long CA (1984): Passive immunization against murine malaria with an IgG3 monoclonal antibody. J Immunol 132:3131–3137.

McBride JS, Heidrich HG (1987): Fragments of the polymorphic Mr185,000 glycoprotein from the surface of isolated *Plasmodium falciparum* merozoites form an antigen complex. Mol Biochem Parasitol 23:71–84.

McBride JS, Newbold CI, Anand R (1985): Polymorphism of a high molecular weight schizont antigen of the human malarial parasite *Plasmodium falciparum*. J Exp Med 161:160–180.

McCutchan TF, Dame JB, Miller LH, Barnwell J (1984): Evolutionary relatedness of *Plasmodium* species as determined by the structure of DNA. Science 225:808–811.

Miller LH, Howard RJ, Carter R, Good MF, Nussenzweig V, Nussenzweig R (1986): Research toward malaria vaccines. Science 234:1349–1356.

Mosmann TR, Cherwinski H, Bond MW, Giedlin MA, Coffman RL (1986): Two types of murine helper T-cell clone. I. Definition according to profiles of lymphokine activities and secreted proteins. J Immunol 136:2348–2357.

Ockenhouse CF, Shear HL (1984): Oxidative killing of the intraerythrocytic malaria parasite *Plasmodium yoelii* by activated macrophages. J Immunol 132:424–431.

Perlmann H, Berzins K, Wahlgren M, Carlsson J, Bjorkman DA, Patarroyo ME, Perlmann P (1984): Antibodies in malarial sera to parasite antigens in the membrane of erythrocytes infected with early asexual stages of *Plasmodium falciparum*. J Exp Med 159:1686–1704.

Pirson PJ, Perkins ME (1985): Characterization with monoclonal antibodies of a surface antigen of *Plasmodium falciparum* merozoites. J Immunol 134:1946–1951.

Potocnjak P, Yoshida N, Nussenzweig RS, Nussenzweig V (1980): Monovalent fragments (Fab) of monoclonal antibodies to a sporozoite antigen (Pb44) protect mice against malarial infection. J Exp Med 151:1504–1513.

Roberts DW, Weidanz WP (1979): T-cell immunity to malaria in the B-cell deficient mouse. Am J Trop Med Hyg 28:1–3.

Schofield L, Villaquiran J, Ferreira A, Schellekens H, Nussenzweig R, Nussenzweig V (1987): γ-interferon, CD8+ T cells, and antibodies required for immunity to malaria sporozoites. Nature 330:664–666.

Siddiqui WA, Tam LQ, Kramer KJ, Hui GSN, Case SE, Yamaga KM, Chang EBT, Kan SC (1987): Merozoite surface coat precursor protein completely protects *Aotus* monkeys against *Plasmodium falciparum* malaria. Proc Natl Acad Sci USA 84:3014–3018.

Tanabe K, Mackay M, Goman M, Scaife JG (1987): Allelic dimorphism in a surface antigen gene of the malarial parasite *Plasmodium falciparum*. J Mol Biol 195:273–287.

Taverne J, Tavernier J, Fiers W, Playfair JHL (1987): Recombinant tumor necrosis factor inhibits malaria parasites in vivo but not in vitro. Clin Exp Immunol 67:1–14.

Taylor DW, Bever CT, Rollwagen FM, Evans CB, Asofsky R (1982): The rodent malaria parasite *Plasmodium yoelii* lacks both types 1 and 2 T-independent antigens. J Immunol 128:1854–1859.

Wahlin B, Wahlgren M, Perlmann H, Berzins K, Björkman A, Patarroyo ME, Perlmann P (1984): Human antibodies to a Mr155,000 *Plasmodium falciparum* antigen efficiently inhibit merozoite invasion. Proc Natl Acad Sci USA 81:7912–7916.

Wanidworanun C, Barnwell JW, Shear HL (1987): Protective antigen in the membranes of mouse erythrocytes infected with *Plasmodium chabaudi*. Mol Biochem Parasit 25:195–201.

Weidanz WP, Long CA (1988): The role of T cells in immunity to malaria. In P. Perlmann and H. Wigzell (eds): Malaria Immunology. Prog Allergy. New York: Marcel-Dekker.

Weiss WR, Sedegah M, Beaudoin RL, Miller LH, Good MF (1988): CD8+ T cells (cytotoxic/suppressors) are required for protection in mice immunized with malaria sporozoites. Proc Natl Acad Sci USA 85:573–576.

Yoeli M, Hargreaves B, Carter R, Walliker D (1975): Sudden increase in virulence in a strain of *Plasmodium berghei yoelii*. Ann Trop Med Parasitol 69:173.

Yoshida N, Nussenzweig RS, Potocnjak P, Nussenzweig V, Aikawa M (1980): Hybridoma produces protective antibodies directed against the sporozoite stage of malaria parasite. Science 207:71–73.

Young JF, Hockmeyer WT, Gross M, Ballou WR, Wirtz RA, Trosper JH, Beaudoin RL, Hollingdale MR, Miller LH, Diggs CL, Rosenberg M (1985): Expression of *Plasmodium falciparum* circumsporozoite proteins in *Escherichia coli* for potential use in a human malaria vaccine. Science 228:958–962.

The Biology of Parasitism, pages 249–264
© 1988 Alan R. Liss, Inc.

# Induction of Cell-Mediated Immunity as a Strategy for Vaccination Against Parasites

Stephanie L. James and Phillip Scott

*Departments of Medicine and Microbiology, The George Washington University Medical Center, Washington, D.C. 20037 (S.L.J.); Laboratory of Parasitic Diseases, National Institute of Allergy and Infectious Diseases, National Institutes of Health, Bethesda, Maryland 20892 (P.S.)*

## INTRODUCTION

### When Antibody Isn't Enough

Most of the successful vaccines currently in use are understood to be based on induction of neutralizing, complement-fixing, or opsonizing antibodies against viral or bacterial pathogens (Zanetti et al., 1987). Moreover, stimulation of antibody production is generally the criterion by which the usefulness of new candidate vaccine antigens is initially judged. However, many of the world's most common pathogens of man are simply not vulnerable in vivo to humoral effector mechanisms alone. Protection in these cases requires the participation of cell-mediated immunity for recruitment and stimulation of effector cells that function either alone or in combination with antibody to destroy the pathogen.

### Cell-Mediated Immune Effector Mechanisms

Thymus-dependent (T) lymphocytes play the central role in cell-mediated immunity (Fig. 1). T cells recognize foreign antigen in the context of products of the major histocompatibility complex (I-A or I-E) on the surface of antigen-presenting cells (macrophage-like cells or B lymphocytes). Antigen-responsive T lymphocytes will proliferate to expand the reaction and produce multiple mediator substances (lymphokines) that affect the function of various cells. Certain lymphokines will influence antibody production by B lymphocytes. Other lymphokines are able to recruit accessory (e.g., phagocytic) cells to the area of antigen presentation and stimulate the cytotoxic capabili-

INDUCTION PHASE    ANTIGEN SPECIFIC EFFECTOR PHASE    NON-SPECIFIC
EFFECTOR PHASE

Immunization    T-cell Activation    Macrophage Activation    Macrophage Killing

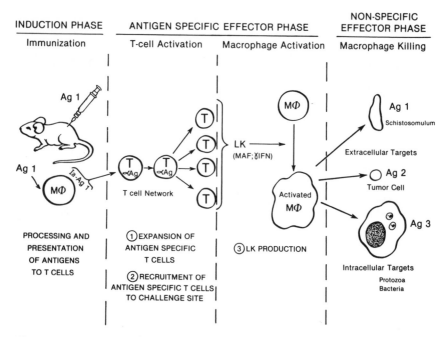

PROCESSING AND PRESENTATION OF ANTIGENS TO T CELLS

① EXPANSION OF ANTIGEN SPECIFIC T CELLS

② RECRUITMENT OF ANTIGEN SPECIFIC T CELLS TO CHALLENGE SITE

③ LK PRODUCTION

Intracellular Targets
Protozoa
Bacteria

Fig. 1. Induction and effector phases of macrophage microbicidal activity. Protection is induced by immunization with parasite antigen (e.g., Ag 1 = schistosome antigen), which is processed by antigen-presenting cells (e.g., macrophage = MΦ) and expressed on their surfaces as a complex with products of the MHC (Ia antigens). T cells with surface receptors specific for this complex interact with antigen-presenting cells, and respond by proliferation to expand the clone of antigen-responsive lymphocytes. At the site of specific Ag presentation (i.e., challenge infection) stimulated T cells also produce lymphokine (macrophage-activating factors = MAFs, including IFN-γ) that recruit effector cells to the area and activate them for enhanced microbicidal function. Such lymphokine-activated MΦ's can kill targets bearing the original antigen (schistosomula) or unrelated targets (Ag 2 = tumor-specific Ag, or Ag 3 = leishmanial Ag) nonspecificially.

ties of these cells. Cells that have developed enhanced killing function as a result of exposure to lymphokine are referred to as "activated." In the case of microbial infection, activated effector cells may eliminate the original pathogen and thereby control the infection. These activated cells can also destroy other microorganisms or foreign cells that happen to be present in the vicinity of the immune response. Thus there are both antigen-specific and nonspecific aspects of cell-mediated immunity: T cells are stimulated to produce lymphokines as a result of interaction with the particular antigen that they are genetically programmed to recognize, while effector cells that have been activated by exposure to these lymphokines are capable of killing nonspecifically any susceptible target in the area. Perhaps the classic experiments in this field were performed by Mackaness, who showed the influence of immunologically committed T cells on macrophage-mediated resistance to *Listeria monocytogenes* (Mackaness, 1969).

### Cell-Mediated Immunity in Vaccination Against Parasites

Notable among those pathogens which generally require the participation of effector cells for protective immunity are parasites. In the following chapter, we will discuss two examples of parasites that are susceptible to killing by activated immune effector cells and how this observation has been exploited to design experimental vaccines based on induction of cell-mediated immunity. The first is the schistosome, a multicellular helminth parasite that transits through the skin to reside in the vascular system of the mammalian host, and thus represents an extracellular target for immune effector mechanisms. The second is *Leishmania*, a protozoal parasite that grows only within macrophages, so that resistance must be manifested through intracellular killing mechanisms. For both parasites, excellent animal models exist for the study of protective immunization. Both cases illustrate a major complication in development of vaccines based on cell-mediated immunity — identification of antigens responsible for stimulating T cells. All of the currently used immunochemical methodologies are geared toward identification of antibody specificities. Thus, it is easy to identify the major parasite antigens recognized by antibodies during infection by immunoprecipitation or immunoblotting techniques, and subsequently to screen expression DNA libraries with antibodies to isolate potentially protective protein antigens. However, since T cells recognize processed (not native) antigens only in association with host histocompatibility antigens on the surface of an antigen-presenting cell, alternative techniques to identify T lymphocyte specificities are required. Approaches for the identification of protective T cell antigens will be illustrated below in relation to the development of vaccines against schistosomiasis and leishmaniasis. Both models employ a similar strategy beginning with the establishment of a successful vaccination protocol using crude antigenic preparations followed by characterization of the protective moieties involved.

## SCHISTOSOMIASIS
### Immunization Against a Multicellular Helminth Parasite

The ability of the mammalian host to develop protective immunity against challenge schistosome infection as a result of ongoing primary infection (concomitant immunity) was noted in rhesus monkeys over two decades ago (Smithers and Terry, 1965), and has subsequently been established in other laboratory hosts including mice and rats (Smithers and Doenhoff, 1982). Epidemiologic data indicate that man also acquires resistance to subsequent infection as a result of prior exposure to the parasite (Butterworth and Hagan, 1987). Moreover, it has been shown that infection with radiation-attenuated schistosome larvae, which die within days after penetrating the skin, will also stimulate strong and long-lasting protective immunity in experimental and veterinary hosts (Smithers and Doenhoff, 1982). These studies provide good indication that a vaccine against schistosomiasis employing defined antigens could decrease the incidence of disease as well as parasite transmission in populations living in endemic areas.

## Mechanisms of Resistance to Schistosomiasis

The mechanism of acquired resistance to schistosomiasis has been studied in detail. While schistosome larvae can be killed in vitro by various combinations of complement, antibody, and immune cells (Smithers and Doenhoff, 1982), these mechanisms have not all proven effective in vivo. The inability to achieve the same high levels of protective immunity present in mice exposed to infective or attenuated larvae by serum transfer (Mangold and Dean, 1986) or by immunization resulting in production of "lethal" antibodies capable of killing schistosomula in vitro (Sher et al., 1974) suggests a requirement for effector cells in resistance in vivo. Two major cell types have been implicated in this role—the eosinophil, from studies in human and rat schistosomiasis (Butterworth and Hagan, 1987; Capron and Capron, 1986), and the macrophage, from studies in mice and rats (James, 1986b; Capron and Capron, 1986). Eosinophils can kill skin-stage schistosomula in vitro in the presence of antibody (Butterworth and Hagan, 1987; Capron and Capron, 1986). Mouse macrophages can kill both skin-stage and older, post-lung-stage, larvae in the absence of antibody when activated by prior exposure to macrophage-activating lymphokine ($\gamma$-interferon) (Pearce and James, 1986). T lymphocytes from infected mice or mice vaccinated with irradiated cercariae produce gamma interferon in vitro when incubated with schistosome antigens (James, 1986b), indicating that this effector mechanism could be operative in vivo. Moreover, an in vivo role for activated macrophages has been demonstrated in genetics studies involving inbred strains of mice that are either high or low responders to the irradiated cercariae vaccine against schistosomiasis. In F2 or backcrosses between these strains, a significant association was observed in individual animals between the level of resistance to subsequent infection and the ability to produce activated, larvacidal macrophages at the site of antigen challenge (James, 1986b).

To a large extent, both eosinophil and macrophage responsiveness are dependent upon T-cell-mediated immune reactivity. Early studies established a link between T-lymphocyte function and eosinophilia, in that T-cell but not B-cell depletion abolished eosinophil response and eosinophilia correlated with delayed hypersensitivity to immunizing antigen (Colley and James, 1979). More recent studies indicate that eosinophilopoiesis is normally under strict T-cell control, which can be ablated by cyclophosphamide treatment (Vadas, 1981). Finally, the cytotoxic potential of peripheral blood or tissue eosinophils can be enhanced by exposure to T-cell products which remain largely undefined (Colley and James, 1979; Spry, 1985). As mentioned above, macrophages are the effector cell most commonly associated with cell-mediated immunity and form a large proportion of the typical delayed hypersensitivity inflammatory response. The major lymphokine responsible for activating macrophages to kill both intracellular and extracellular targets appears to be γ-interferon, although other macrophage-activating factors may also participate in activation for intracellular killing (Nacy et al., 1985).

Thus, it appears that production of an effective defined vaccine against schistosomiasis will require manipulation of cellular as well as humoral immune response. Immune regulation of eosinophilia is not currently understood; and, moreover, it is clear that an overabundance of eosinophils (e.g., hypereosinophilic syndrome) is detrimental due to their toxicity for normal cells as well as pathogens (Gleich et al., 1980; Spry, 1985). Therefore, any attempt immunologically to manipulate eosinophil response as part of a vaccine strategy could present great risk at this time. While people living in areas where schistosomiasis is endemic often demonstrate elevated eosinophil levels as a result of concurrent infection with various helminths, there is evidence that these cells may be rendered less functional due to the presence of cytophilic antibodies against irrelevant antigens or of nonprotective isotypes (Capron and Capron, 1986; Butterworth and Hagan, 1987), and therefore may not serve as a reliable source of immune effector cells. In contrast, the mechanism of macrophage recruitment and activation has been extensively studied, and a major macrophage activating lymphokine (γ-interferon) has been sequenced and is available in recombinant form. Furthermore, good evidence indicates that activated macrophages are only toxic to foreign or "abnormal" targets while failing to harm normal cells in culture (Adams and Marino, 1981). Finally, macrophages persist in an activated state for a limited period of time before they decay to nonkiller, resident cell status (Adams and Marino, 1981). Therefore, induction of activated macrophages as immune effector cells is self-limiting and will cease rapidly following elimination of the target pathogen. For these reasons, production of an antischistosome vaccine based on induction of activated macrophage effector cells would seem to be a wholly reasonable strategy.

## A Vaccine Against Schistosomiasis Based on Cell-Mediated Immunity

We have followed this strategy in developing an experimental vaccine against *Schistosoma mansoni*. Initial studies involved a crude nonliving antigenic preparation made by freezing and thawing 3-h old schistosomula. This antigen was chosen because it contains both soluble and membrane components of the parasite, and because this stage of the parasite is recognized as antigenic in vitro by T cells from mice immunized by prior infection or exposure to irradiated cercariae (James, 1986b). Therefore it is likely to contain T-cell immunogens. Moreover, there is evidence that early schistosomula are a target of protective immunity in vivo (Smithers and Doenhoff, 1982). In this vaccine model, response was predisposed toward development of cell-mediated immunity by immunizing via the intradermal route and including a bacterial adjuvant (bacillus Calmette-Guérin, BCG) known to promote T-cell reactivity. As predicted, this protocol sensitized for cell-mediated immunity, as measured by antigen-specific delayed hypersensitivity response and macrophage activation in vivo, as well as lymphocyte blastogenesis and production of $\gamma$-interferon in vitro (James, 1986a). Furthermore, intradermal vaccination induced up to 85% (mean of 50%) resistance to challenge *S. mansoni* infection in outbred and several inbred strains of mice (James, 1985; 1986a). Production of antibodies toward parasite surface antigens, which might be involved in an antibody-dependent mechanism of resistance was minimal (James 1985, 1986a) and sera from immunized mice did not transfer resistance to naive recipients. Further substantiation that resistance in this model was based upon induction of cell-mediated immunity was obtained by observation that under several conditions those responses were associated (Table 1). For example, when alternative adjuvants were substituted for BCG, only those that stimulated T-cell responsiveness (*Bordetella pertussis* vaccine, complete Freund's adjuvant, saponin) were capable of promoting resistance. Moreover, inoculation of the same antigens by other routes (intravenous, intramuscular) failed to produce protection or cell-mediated immunity, although high levels of antibody production were stimulated by these protocols (James, 1985). Finally, two mouse strains (P and BALB/c) were identified as low responders to the intradermal vaccine (developing negligible resistance to challenge infection), and both of these strains revealed defects in lymphokine production and macrophage activation in this model.

As previously mentioned, further characterization of the protective immunogens involved in induction of cell-mediated immunity is not so straight-

**TABLE 1. Correlation Between Induction of Resistance and Cell-Mediated Immunity (Interferon Production) in Mice Vaccinated Against *S. mansoni* With Nonliving Antigens**

| Immunization protocol[a] | % resistance[b] | Interferon titer[c] |
|---|---|---|
| C57BL/6; FT larvae + BCG i.d. | 61 | 1:50 |
| C57BL/6; FT larvae + *B. pertussis* i.d. | 72 | 1:50 |
| C57BL/6; FT larvae + saponin i.d. | 70 | 1:50 |
| C57BL/6; FT larvae + alum i.d. | 10 | <1:5 |
| C57BL/6; FT larvae + glucan i.d. | 0 | <1:5 |
| C57BL/6; FT larvae + BCG i.m. | 4 | <1:5 |
| C57BL/6; FT larvae i.v. | 6 | <1:5 |
| BALB/c; FT larvae + BCG i.d. | 2 | <1:5 |
| P; FT larvae + BCG i.d. | 5 | <1:5 |

[a]Mouse strain, antigen and adjuvant used for immunization and route of inoculation (i.d., intradermal; i.m., intramuscular; i.v., intravenous).
[b]Percent resistance, determined by comparison of adult worm recovery in immunized mice versus mice receiving adjuvant only following percutaneous infection with 120 cercariae.
[c]Interferon (macrophage-activating factor) titer of supernatant fluids prepared by culturing spleen cells from immunized mice with soluble adult worm antigens in vitro.

forward as is the case with antibody-dependent resistance. Two major possibilities present themselves: a) serial fractionation of crude antigen followed by testing of each fraction for T-cell reactivity and induction of protective immunity, or b) identification of antigens recognized by those antibodies produced in the cell-mediated-immunity–based model, on the assumption that since antibody production requires the participation of antigen-specific helper T cells, these antigens are likely to be immunogenic for T cells producing lymphokines involved in macrophage activation as well. In our model of immunization against schistosomiasis, we have proceeded along both of these lines. Initially, it was determined that the protective antigens involved in sensitization for cell-mediated immunity are not stage specific, being present in cercariae, 3-h schistosomula, and adult worms (James, 1986a). When crude preparations of homogenized schistosomula or adult parasites were separated into soluble or membrane-containing portions by ultracentrifugation, it was observed that protective antigens were present in both fractions (James, 1986a). Soluble antigens were subsequently used for characterization of protective immunogens on the basis that these would be easier to identify and purify than membrane-associated molecules. When soluble adult worm antigens were further fractionated by gel filtration, it was shown that protective reactivity separated in a high molecular weight fraction which also contained immunogens for T-cell response (production of macrophage activating factor) as well as the only antigen recognized by sera from immunized mice (James, 1986a).

One peculiarity of this model is that intradermally immunized mice produce a weak humoral response, which is largely if not entirely directed against a single internal parasite antigen of $M_r$ 97,000 identified by gene cloning experiments (Lanar et al., 1986) as paramyosin, a muscle cell component of invertebrates (James, 1986a). Since paramyosin was present in the protective fraction, and since the antibody response was T-cell dependent, as shown by its absence in immunized nude mice, the immunogenicity of this humorally recognized antigen for cell-mediated immunity and resistance was further investigated. Schistosome paramyosin was isolated by affinity chromatography (James, 1986a) and used to immunize mice by intradermal injection in combination with BCG. Recent experiments have shown that purified paramyosin is indeed protective in quantities more than 50-fold lower than the amount of crude soluble protein antigen usually used for vaccination. Moreover, paramyosin was immunogenic for sensitization of T cells to respond either to itself or to crude antigen in vitro in lymphocyte proliferation and lymphokine production assays (Pearce et al., 1987). Thus, one purified antigen has been identified as protective in this vaccination model. However, paramyosin-depleted soluble antigens have also been found to immunize when injected intradermally with BCG. The identity of other schistosome antigens with protective potential in this model must subsequently be determined by fractionation, or other procedures as discussed below.

## LEISHMANIASIS

Cutaneous leishmaniasis is the only human parasitic infection in which immunization has long been used as a means of controlling the disease (Greenblatt, 1980). Protective immunity is achieved by injection of low doses of viable parasites at selected sites, which leads to the induction of immunity to challenge infections with higher doses of parasites. Although there are several limitations to this vaccine, including the development of complications and the inability to vaccinate against many of the more severe forms of the disease, these studies have clearly demonstrated the feasibility of vaccination as a means of control. The challenge now is to develop defined vaccines that can be used to protect people not only against simple cutaneous leishmaniasis, but also against severely mutilating forms of the disease, such as mucocutaneous leishmaniasis, and fatal visceral leishmaniasis.

## Mechanisms of Resistance to Leishmaniasis

Immunity to leishmaniasis is mediated primarily through cellular mechanisms, although antibodies are produced during infection and following vaccination. During infection, T cells are stimulated by leishmanial antigens, proliferate, and produce macrophage-activating factors (MAFs), such as $\gamma$-interferon. These factors are responsible for activating macrophages to kill their intracellular parasites. Thus, the development of a vaccine would appear to depend upon identification of the antigens capable of stimulating T cells to produce MAF, and the administration of these immunogens in such a manner as to stimulate the appropriate T cells. However, at each step of the induction and effector phases of macrophage microbicidal activity (Fig. 1), there are host or parasite factors that are capable of circumventing the successful outcome of immunization. These include: a) decreased expression of Ia antigens required for appropriate presentation of antigens to T cells (Reiner et al., 1987; Handman et al., 1979); b) induction of suppressor cells that inhibit expansion and/or recruitment of antigen specific effector cells to challenge sites (Howard et al., 1980; Scott and Farrell, 1981); c) recruitment of macrophages that are unresponsive to activation signals (Hoover and Nacy, 1984); d) decreased ability of activated macrophages to kill *Leishmania* at cutaneous temperatures (Scott, 1985); and e) resistance of the parasite to activated macrophage killing (Scott et al., 1983; Scott and Sher, 1986). Thus, to be successful a vaccine must overcome the factors that normally contribute to the chronicity of leishmanial infections.

## Immunization of BALB/c Mice Against *Leishmania major*

Excellent murine models of human leishmaniasis are available that can be used to develop experimental vaccines. For example, intravenous or intraperitoneal immunization with attenuated parasites can completely protect BALB/c mice against a normally fatal infection with *Leishmania major* (Mitchell et al., 1981; Howard et al., 1982; Scott et al., 1987a). Intraperitoneal immunization with a soluble, membrane-free, parasite extract was found to induce protection against *L. major* challenge equal to that obtained with whole organisms (Scott et al., 1987a,b). Induction of immunity was most effective with 100 $\mu$g of the soluble leishmanial antigen (SLA) and required concomitant injection of the bacterial adjuvant, *Corynebacterium parvum*, followed by an intraperitoneal boost of SLA alone 1 week later. Vaccinated animals exhibited *Leishmania*-specific cell-mediated immunity and the production of antibodies against metabolically labeled proteins, but failed to stimulate a detectable humoral immune response against parasite surface antigens. As expected, these experiments demonstrated that vaccine-induced immunity against cutaneous leishmaniasis is strongly associated with the induction of cellular responses, but does not require the development of an antibody response to parasite surface antigens. In addition, the studies established the feasibility of employing soluble, non-membrane-derived parasite material as a source of protective immunogens.

## Identification of Protective Immunogens by Antigen Fractionation

SLA was separated into nine distinct fractions by anion-exchange chromatography, and the fractions were analyzed for their ability to stimulate T cells obtained from immunized mice, to be recognized by antibodies, and to induce resistance (Scott et al., 1987b). While all but one of the fractions were recognized by antibodies from vaccinated mice, only two fractions (fractions 1 and 9) stimulated T cells. When mice were immunized with the fractions, only fraction 9 stimulated significant immunity (Table 2). Isolation of this fraction represented a significant purification step in defining the protective immunogens(s) within SLA, since fraction 9 represented less than 2% of the total protein in SLA and was as protective as SLA at one-tenth the concentration. Proteins were found to be responsible for protection in this fraction, since protease treatment destroyed its immunogenicity. Thus, a partially purified protective protein antigen fraction was obtained, and protection with this fraction correlated with cell-mediated immunity. However, these results also demonstrate that the ability of leishmanial antigens to be recognized by

TABLE 2. Immune Responses to FPLC Fractions and the Protection Induced by These Antigens

| Antigen | Immunological responses | | | Protection[d] (% mice healed or controlled) |
| | MAF[a] | DTH[b] | AB[c] | |
|---|---|---|---|---|
| SLA | + | + | 1:5,120 | 96 |
| 1 | + | + | < 1:40 | 0 |
| 2–8 | − | − | 1:2,560 | 0 |
| 9 | + | + | 1:1,280 | 76 |

[a]MAF activity was determined by assessing the ability of supernates of antigen (25 mg/ml)-stimulated T cells from SLA-immunized mice to activate macrophages to kill *L. major*; $\geq 50\%$ leishmanicidal activity was considered positive.
[b]SLA-immunized and challenged mice were injected in the footpad with 20 mg of antigen 15 weeks following infection. Footpad thickness was measured at 24 h, and $\geq 0.2$ mm swelling was considered positive.
[c]ELISA antibody titers to different antigens of sera from SLA immunized mice.
[d]Results of seven experiments with 35–40 mice per group. Mice were immunized with either SLA (100 mg protein) or fractions 1 through 9 ($OD_{208}$ equivalent of 100 $\mu$g SLA). The adjuvant CP was included in all immunizations. Mice were subsequently challenged with *L. major*.

T cells, and produce MAF, does not in itself predict whether such molecules will induce immunity, since fraction 1 was recognized by T cells from immunized mice but failed to induce any protection. This observation clearly indicates that empirical immunization studies in vaccine models are crucial in defining a leishmanial vaccine, and suggests that there are, as yet, undefined factors which contribute to the protective nature of leishmanial vaccine immunogens. Nevertheless, it is clear that identification of the antigens within fraction 9 recognized by T cells would be an important step in further defining a protective leishmanial immunogen. Recently, Lamb and Young (1987) have shown that T-cell-reactive antigens can be identified by separation of antigens by SDS-PAGE, immunoblotting to nitrocellulose, and stimulation of T cells with small sections of the nitrocellulose corresponding to defined molecular weights. This approach overcomes some of the obstacles of identifying individual T-cell specificities. For example, when T cells from SLA immunized mice were assayed for their stimulation by SDS-PAGE-separated SLA proteins, reactivity was found with multiple antigens of less than 40,000 molecular weight. Subsequently, when a T-cell line from fraction 9 immunized mice was assayed, only three major peaks of reactivity were obtained, corresponding to the molecular weight regions of 8,000–14,000; 20,000–30,000 and 40,000–45,000. Finally, a T-cell clone derived from this cell line was found to respond selectively to antigens which appear to be unique to fraction 9, and are of molecular weight 8,000 to 10,000. Further studies to define protective antigens within fraction 9 will require isolation of the antigens recognized by this T-cell clone, as well as others, and an assessment of their protective capacity in vivo.

## CONCLUSIONS

In this chapter, we have reviewed two examples of experimental antiparasite vaccines based on induction of protective cell-mediated immune responses. Several similarities are apparent in the genesis of these two models. In both cases, there was reason to believe that cell-mediated immunity was involved in resistance due to previous in vivo evidence in models immunizing with attenuated parasites; however, no prior assumptions were made as to the nature of the protective response as the model was developed; i.e., humoral as well as cellular responses were analyzed. Moreover, the models were not biased toward the use of any particular type of antigen; membrane-associated surface antigens as well as soluble internal antigens were present in the original immunizing inocula. Thus, in both cases, there was a progression from the use of living attenuated parasites to a crude antigenic preparation to a more purified, or in the case of schistosomiasis, a defined antigen within the context of an in vivo animal model. This is in contrast to the alternative approach of arbitrarily identifying antibody-reactive surface antigens on the basis of their involvement in in vitro parasiticidal assays and then testing for protection in vivo, a method which has led investigators astray in the past. Thus, the benefit of initially establishing an animal model system in which the appropriate method of antigen presentation for elicitation of protective immunity has been determined and proceeding within that system to empirically identify protective antigens becomes obvious.

It should be pointed out that it is not always simple to test theories developed in such animal models in man. This is particularly true in the case of cell-mediated immunity, where one is limited both in terms of the cell types available for testing as well as in the sample population. For example, T cells and antigen-presenting cells retrieved from peripheral blood may not show the same immune characteristics as those present in lymph nodes or spleen or at the site of secondary response, and peripheral blood monocytes are certainly biochemically distinct from tissue-stage macrophages. Moreover, one is limited to testing the reactivity of cells from individuals that were exposed to parasite antigens as a result of natural injection, where antigen presentation may proceed in an entirely different manner than would be the case in an experimental vaccination situation, and which may therefore show no correlation whatsoever with protective immunity. Finally, as has been shown countless times, no in vitro assay of immune response can establish the role of a given antigen or effector mechanism in resistance as conclusively as can in vivo protection trials. Thus, generalized assays of T-cell responsiveness in patients, such as lymphocyte proliferation, and in vitro parasiticidal assays should be interpreted with care in attempting to evaluate the potential role of cell-mediated mechanisms in human immunity.

For parasite targets that are susceptible to cell-mediated immune responses, basing a vaccine on induction of cell-mediated immunity offers certain theoretical advantages. First, it is known that T-cell reactivity is likely to be directed against protein rather than carbohydrate antigens, and that T cells tend to recognize primary and secondary rather than tertiary, conformational, epitopes (DeLisi and Berzofsky, 1985). Therefore, protective antigens may be easily mass-synthesized by recombinant DNA technology or protein chemistry. Second, for activated macrophage effector cells, cytotoxicity need not be directed toward specific surface epitopes on the target. This is an advantage in vaccinating against parasites, such as schistosomes, that have evolved complex immune evasion schemes for avoiding surface recognition. Moreover, this means that nonsurface antigens, such as excretory/

secretory products, may also be protective in provoking cell-mediated immunity. Finally, the protective reaction is compartmentalized in the area of secondary antigen recognition by sensitized T cells, which produce macrophage chemotactic and activating lymphokines locally. Thus, the response is unlikely to produce significant systemic pathology (e.g., immune complex disease, asthma), such as might be associated with maintenance of high antibody titers.

As discussed previously, technical difficulties exist in defining T-cell epitopes due to the requirement that these are recognized in the context of an antigen presenting cell that processes complex antigen and presents the peptide immunogen in association with MHC products. It is obviously more straightforward to identify and purify serologically reactive antigens that combine directly with antibody receptors. Recent technology offers several advances toward defining T-cell antigens. These include cloning of T cells performing a critical immune response (i.e., lymphokine production or transfer of resistance) (Infante et al., 1982) and using these to screen antigen fractions (Lamb and Young, 1987) or recombinant libraries (Mustafa et al., 1986) to identify crucial antigens. An example of how the antigen specificities of T cells can be determined using immunoblots was described in the development of a vaccine against leishmaniasis. Alternatively, one can follow the approach discussed in the identification of paramyosin as a protective T-cell antigen in the schistosomiasis vaccine, whereby molecules purified on the basis of antibody recognition are characterized for T-cell reactivity on the assumption that antibody production requires helper T-cell function and therefore these antigens must contain T-cell epitopes. It should be cautioned, however, that this approach may lead to identification of antigens stimulatory for Th 2 cells, as opposed to Th 1 cells involved in DTH.

The examples discussed here involve representative helminth and protozoal parasites. Other tropical diseases in which a vaccine based on induction of cell-mediated immunity might be beneficial include those where at some stage in their life cycles the causative agents are interiorized by monocyte/macrophages (e.g., *Toxoplasma gondii*, *Trypanosoma cruzi*, *Mycobacterium leprae*) and thus are potential targets for intracellular killing, or invade other cells that may become transformed into recognizable targets for extracellular killing by effector cells as a result of infection (e.g., malaria, HIV). In particular, there is a renewed interest in the role of T cells in protective immunity against malaria in light of the possible lack of correlation between antibody production and resistance to reinfection following immunization with recombinant circumsporozoite antigen (Egan et al., 1987). The susceptibility of helminths other than schistosomes to macrophage-mediated killing has not been well studied, although it is known that chronic nematode infection (*Trichinella spiralis, Nippostrongylis brasiliensis*) does result in macrophage activation (Wing et al., 1979). These examples illustrate the need for continued study at the basic level of the role of cell-mediated immunity in resistance to parasites. The induction of cell-mediated immunity, as opposed to or in addition to antibody production, should be a major direction in the development of vaccines in the future.

## REFERENCES

Adams D, Marino P (1981): Evidence for a multistep mechanism of cytolysis by BCG-activated macrophages: The interrelationship between the capacity for cytolysis, target binding, and secretion of cytolytic factor. J Immunol 126:981–987.

Butterworth AE, Hagan P (1987): Immunity in human schistosomiasis. Parasitol Today 3:11–16.

Capron M, Capron A (1986): Rats, mice and men-models for immune effector mechanisms against schistosomes. Parasitol Today 1:69–75.

Colley DG, James SL (1979): Participation of eosinophils in immunological systems. In S. Gupta and R.A. Good (eds): Cellular, Molecular and Clinical Aspects of Allergic Disorders. New York: Plenum Publishing Corp., pp 55–86.

DeLisi C, Berzofsky J (1985): T cell antigenic sites tend to be amphipathic structures. Proc Natl Acad Sci USA 82:7048–7052.

Egan JE, Weber JL, Ballover WR, Holligdale MR, Majarian WR, Gordon DM, Maloy WL, Hoffman SL, Wirtz RA, Schneider I, Woollett GR, Young JF, Hockmeyer WT (1987): Efficacy of murine malaria sporozoite vaccines: Implications for human vaccine development. Science 236:453–456.

Gleich G, Loegering D, Frigaf I, Wassom D, Solley G, Mann K (1980): The major basic protein of the eosinophil granule: Physicochemical properties, localization and function. In A. Mahmoud and K. Austen (eds): The Eosinophil in Health and Disease. New York: Grune and Stratton, pp 79–97.

Greenblatt CL (1980): In A. Mizrahi, I. Hertman, M.A. Klingberg, and A. Kohn (eds): New Developments with Human and Veterinary Vaccines. New York: Alan R. Liss, Inc., pp 259–285.

Handman E, Ceredig R, Mitchell GF (1979): Murine cutaneous leishmaniasis: Disease patterns in intact and nude mice of various genotypes and examination of some differences between normal and infected macrophages. Aust J Exp Biol Med Sci 57:9–29.

Hoover DL, Nacy CA (1984): Macrophage activation to kill *Leishmania tropica*: Defective intracellular killing of amastigotes by macrophages elicited with sterile inflammatory agents. J Immunol 132:1487–1493.

Howard JG, Hale C, Liew FY (1980): Immunological regulation of experimental cutaneous leishmaniasis III. Nature and significance of specific suppression of cell mediated immunity. J Exp Med 152:594–607.

Howard JG, Nicklin S, Hale C, Liew FY (1982): Prophylactic immunization against experimental Leishmaniasis: I. Protection induced in mice genetically vulnerable to fatal *Leishmania tropica* infection. J Immunol 129:2206–2212.

Infante AJ, Kimoto M, Fathman CG (1982): An analysis of T cell antigen recognition utilizing T cell clones. In C.G. Fathman and F.W. Fitch (eds): Isolation, Characterization and Utilization of T Lymphocyte Clones. New York: Academic Press, pp 368–374.

James SL (1985): Induction of protective immunity against Schistosoma mansoni by a nonliving antigen is dependent on the method of antigen presentation. J Immunol 134:1956–1960.

James SL (1986a): A review: Schistosoma spp progress toward a defined vaccine. Exp Parasitol 63:247–252.

James SL (1986b): Activated macrophages as effector cells of protective immunity to schistosomiasis. Immunol Res 5:139–148.

Lamb JR, Young DB (1987): A novel approach to the identification of T-cell epitopes in Mycobacterium tuberculosis using human T-lymphocyte clones. Immunology 60:1–5.

Lanar D, Pearce EJ, James SL, Sher A (1986): Identification of paramyosin as the schistosome antigen recognized by intradermally vaccinated mice. Science 234:593–596.

Mackaness G (1969): The influence of immunologically committed lymphoid cells on macrophage activity in vivo. J Exp Med 129:973–992.

Mangold B, Dean D (1986): Passive transfer with serum and IgG antibodies of irradiated cercariae-induced resistance against Schistosoma mansoni in mice. J Immunol 136:2644–2648.

Mitchell GF, Curtis JM, Handman E (1981): Resistance to cutaneous leishmaniasis in genetically susceptible BALB/c mice. Aust J Exp Biol Med Sci 59:555–565.

Mustafa AS, Jill HK, Nerland A, Britton WJ, Mehra V, Bloom BR, Young RA, Godal T (1986): Human T-cell clones recognize a major M. lepre protein antigen expressed in E. coli. Nature 319:63–66.

Nacy CA, Fortier AH, Meltzer MS, Buchmeier NA, Schreiber RD (1985): Macrophage activation to kill Leishmania major: Activation of macrophages for intracellular destruction of amastigotes can be induced by both recombinant interferon-gamma and non-interferon lymphokines. J Immunol 135:3505–3511.

Pearce EJ, James SL (1986): Post lung stage schistosomula of Schistosoma mansoni exhibit transient susceptibility to macrophage-mediated cytotoxicity in vitro that may relate to late phase killing in vivo. Parasite Immunol 8:513–527.

Pearce EJ, James SL, Hieny S, Lanar D, Sher A (1988): Induction of protective immunity against Schistosoma mansoni by vaccination with schistosome paramyosin (Sm 97), a non-surface parasite antigen. Proc Natl Acad Sci (in press).

Reiner NE, Ng W, McMaster WR (1987): Parasite-accessory cell interactions in murine leishmaniasis II. Leishmania donovani suppresses macrophage expression of class I and class II major histocompatibility complex gene products. J Immunol 138:1926–1932.

Scott P (1985): Impaired macrophage leishmanicidal activity at cutaneous temperature. Parasite Immunol 7:277–288.

Scott PA, Farrell JP (1981) Experimental cutaneous leishmaniasis I. Non-specific immunodepression in BALB/c mice infected with Leishmania tropica. J Immunol 127:2395–2400.

Scott P, Sher A (1986): A spectrum in the susceptibility of leishmanial strains to intracellular killing by murine macrophages. J Immunol 136:1461–1466.

Scott P, Pearce E, Natovitz P, Sher A (1987a): Vaccination against cutaneous leishmaniasis in a murine model I. Induction of protective immunity with a soluble extract of promastigotes. J Immunol 139:221–227.

Scott P, Pearce E, Natovitz P, Sher A (1987b): Soluble leishmania antigens that elicit T-cell reactivity and protective immunity against *Leishmania major* in BALB/c mice. In Vaccines 87. Cold Spring Harbor, NY: Cold Spring Harbor Laboratories, pp 107–111.

Scott PA, Sacks D, Sher A (1983): Resistance to macrophage-mediated killing as a factor influencing the pathogenesis of chronic cutaneous leishmaniasis. J Immunol 131:966–971.

Sher A, Kusel JR, Perez H, Clegg JA (1974): Partial isolation of a membrane antigen which induces the formulation of antibodies lethal to schistosomes. Clin Exp Immunol 18:357–369.

Smithers SR, Doenhoff MJ (1982): Schistosomiasis. In S. Cohen and K.S. Warren (eds): Immunology of Parasitic Infections. Oxford: Blackwell Scientific Publications, pp 527–607.

Smithers SR, Terry RJ (1965); Naturally acquired resistance to experimental infections of Schistosoma mansoni in the rhesus monkey. Parasitology 55:701–708.

Spry CJF (1985): Synthesis and secretion of eosinophil granule substances. Immunol Today 6:332–335.

Vadas MA (1981): Cyclophosphamide pretreatment induces eosinophilia to nonparasitic antigens. J Immunol 127:2083–2085.

Wing EJ, Krahenbuhl J, Remington JS (1979): Studies of macrophage function during *Trichinella spiralis* infection in mice. Immunology 36:479–485.

Zanetti M, Sercarz E, Salk J (1987): The immunology of new generation vaccines. Immunol Today 8:18–25.

The Biology of Parasitism, pages 265–284
© 1988 Alan R. Liss, Inc.

# Leprosy

**Barry R. Bloom**

*Department of Microbiology and Immunology, Albert Einstein College of Medicine, Bronx, New York 10461*

## CHALLENGES

The leprosy bacillus (*Mycobacterium leprae*) was the first identified bacterial pathogen of man, yet it remains one of the very few human pathogens that cannot be grown in culture. Like its relative, the tubercle bacillus, discovered 10 years later in 1882, *M. leprae* is an acid-fast bacillus with general characteristics in common with gram-positive bacteria, its characteristic staining being related to the fact that all mycobacteria have unusual waxy lipid coats. There is generally a long latency, perhaps 5 years between putative infection and the manifestations of clinical disease, as a consequence of which the mode of transmission remains largely unknown. Although between 10 and 13 million people in the world are currently estimated to suffer from leprosy, the disease has a relatively low prevalence, seldom exceeding 1–5 per thousand in endemic areas. In contrast, *M. tuberculosis*, which can grow in culture, albeit slowly, causes 10 million cases of tuberculosis a year, leading to the death of 3 million people per year worldwide.

There is a unique fear and stigma associated with leprosy, partially deriving from the deformities that occur in approximately 30% of the people afflicted with it. There is no other disease whose victims were burned at the stake or buried alive in medieval Europe, or given the last rites of the dead and turned outside the city walls. Because of the importance of mycobacterial diseases, and the unique horror and fascination associated with leprosy, it has become essential to apply the tools of modern immunology and molecular biology towards understanding them.

## THE DISEASE SPECTRUM

There is an additional interest as well, in that leprosy is not a single clinical entity, but rather forms part of a spectrum that presents a diversity of clinical manifestations (Bloom and Godal, 1983); (Hastings, 1985; Bloom, 1986). The major parameter that varies across this spectrum is the state of the patients' immunologic responsiveness; hence, leprosy represents an extraordinary model for understanding immunoregulation in man (Fig. 1). At one pole of the spectrum, tuberculoid leprosy, patients develop high levels of cell-mediated immunity that ultimately kills and clears the bacilli tissues, although often with concomitant immunological damage to the nerves. Lesions of tuberculoid leprosy are characteristically granuloma containing many macrophages and lymphocytes, with very few discernible acid-fast bacilli. At the lepromatous pole, patients exhibit a selective immunological unresponsive to antigens of *M. leprae* and the organisms inexorably multiply in the skin, often to extraordinarily high numbers, e.g., $10^{10}$/g. tissue. Lesions of lepromatous leprosy are characterized by large numbers of macrophages containing prodigious numbers of bacilli, with very few lymphocytes evident. The vast majority of patients fall in the spectrum somewhere in between. While antibodies reactive with *M. leprae* are found throughout the spectrum, the highest levels occurring in the lepromatous end of disease, there is a striking inverse correlation between the level of cell-mediated immunity to antigens of *M. leprae* and the growth of the organism in the tissues. These results strongly suggest that protection against leprosy is mediated exclu-

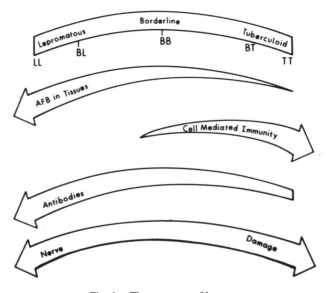

Fig. 1. The spectrum of leprosy.

sively by cell-mediated immunity. In endemic areas or within families of lepromatous patients, it is likely that a very large number of people are infected with the leprosy bacillus, but due to appropriate immune responsiveness they are protected from any of the obvious manifestations of the disease. It remains unclear why the vast majority of exposed individuals fail to develop clinical disease, and why, of the minority who do develop clinical leprosy, a proportion that varies in different parts of the world become multibacillary lepromatous patients and immunologically unresponsive to the antigens of the organisms.

## MECHANISMS OF NERVE DAMAGE

Since the most striking aspect of leprosy is the tragic deformity resulting from untreated infection, it was long believed that damage to nerves was caused by the invasion of the axons by bacilli. *M. leprae* grows primarily in two cells of the body, mononuclear phagocytes, which presumably are designed to kill intracellular pathogens, and Schwann cells surrounding peripheral nerve. The paucity of bacilli in nerves in either pole of the spectrum suggests that mechanisms alternative to direct nerve damage by the organisms must be considered.

In order to understand how nerves could be damaged in tuberculoid leprosy under conditions where there are a few discernible bacilli, we established a model in which guinea pigs or rabbits were sensitized to foreign antigens in complete Freund's adjuvant using killed mycobacteria (Wisniewski and Bloom, 1975; Cammer et al., 1978). When such animals were tested with PPD (purified protein derivative of tuberculin) on top of the sciatic nerve or in the ventricles, a delayed-type hypersensitivity reaction was produced in the tissues. Although this represented a specific cell-mediated immune response to a non-neural antigen, there was invariably primary focal demyelination observed to the local nerves, which were damaged as innocent bystanders. This has been termed "bystander" demyelination. We established that one mechanism responsible for the damage to myelin was the secretion of plasminogen activator, a neutral protease, from activated, but not resting macrophages (Mehra et al., 1979). We believe that this may represent a model for tissue damage in a variety of autoimmune diseases in which a cell-mediated immune response occurs to a foreign antigen leading to damage to host tissues, as in multiple sclerosis or rheumatoid arthritis, or in infectious diseases such as mucocutaneous leishmaniasis. In the animal model, when large amounts of antigens were used to challenge, damage to the nerves was irreversible, which suggests that at least some of the nerve damage in reactions of tuberculoid leprosy may be mediated by bystander demyelina-tion. Recent findings suggest that cytotoxic T lymphocytes from tuberculoid leprosy patients can lyse *M. leprae* infected Schwann cells, suggesting a second mechanism of immunological damage to nerves (Kaufmann, 1988). The mechanism of nerve damage in lepromatous leprosy may well be im-munological as well. In this circumstance, damage occurs to small peripheral sensory nerves, in the absence of white cells' infiltration. A major clue as to the possible mechanism underlying this form of nerve damage was provided by the finding of highly elevated levels of C3d, a breakdown product of the major complement component, in the blood of lepromatous patients undergo-ing reactions involving nerves (Bjorvatn et al., 1976). Since lepromatous patients have elevated levels of antibodies to *M. leprae* antigens, the model for tissue damage suggested would be the release of mycobacterial antigens in the tissues, for example, spontaneously or perhaps following chemother-apy, leading to local antibody-antigen reactions in the tissues in the vicinity of the small sensory nerve fibers. The fixation of complement and release of pharmacologically active complement components and mediators would lead to permeability change and edema. This sequence of events is likely to produce compression neuropathy that would irreversibly damage the small sensory nerve fibers.

There is at present no evidence for an autoimmune component, and progression of nerve damage seems invariably to cease when appropriate chemotherapy for the infection is instituted. While the experiments described by no means constitute formal experimental proof for mechanisms of nerve damage, they are suggestive, at least, that much of the nerve damage in leprosy, and resulting anesthesia, mutilation, and deformity, may be immu-nologically mediated, and could be prevented by appropriate treatment.

## CONTROL OF SUSCEPTIBILITY TO LEPROSY AND ITS FORMS

Probably a more accurate representation of the spectrum of leprosy is that of 1,000 people infected with *M. leprae* in a highly endemic area, 5 develop clinical leprosy and only 1–2 of those develop lepromatous leprosy. What are the factors that determine susceptibility to the disease or control the type of disease? Because of the long latent period, it is difficult to study transmission accurately and consequently difficult to address some of the more important epidemiological issues (Fine, 1982). Because of our inability to grow the organism, it has not yet been possible to ascertain whether genetic differences in *M. leprae* exist between organisms isolated from patients with different forms of disease or from different geographical areas of the world. There is, however, intriguing information on the role of the genes of the major histocompatibility complex (MHC) that bears on the question of genetic susceptibility. What emerges is that there is no convincing evidence that MHC genes have any effect on the susceptibility of individuals to acquiring leprosy. However, there is evidence from studies of segregants in multicase families that a linkage disequilibrium exists between certain MHC class II antigens and tuberculoid leprosy, and a weaker association between another MHC class II haplotype and susceptibility to lepromatous leprosy (De Vries et al., 1985; Van Eden and De Vries, 1984). There is thus formal evidence that the form of disease is controlled to some degree by human immune response (Ir) genes.

In the mouse, the principal genetic determinants of susceptibility to a variety of infectious agents are, surprisingly, genes that map outside the MHC complex. In the mouse, there is a gene on chromosome 1, variously termed *ity*, *leish*, or *bcg*, that controls resistance to *Salmonella*, *Leishmania donovani*, and to the Bacillus Calmette–Guérin (BCG) vaccine strain of *M. bovis* (Skamene, 1986). While the mechanism of action of that gene is unclear, its effect is radiation resistant and thought to be mediated by macrophages. That segment of chromosome 1 has been restriction mapped and by use of appropriate DNA probes it has recently been found that a precisely homologous segment of 0.5 cm has been identified on human chromosome 2. Thus there is the possibility that a homologous gene will be found in man that could play a role in controlling infection to these intracellular parasites. Studies have been initiated by the IMMTUB committee at the World Health Organization (WHO) to obtain B-cell lines from multicase families with tuberculosis and leprosy to provide an infinite source of DNA for restriction fragment length polymorphism (RFLP) analysis of human genes that may influence susceptibility or resistance to these diseases.

## THE SPECIFIC IMMUNOLOGIC UNRESPONSIVENESS IN LEPROMATOUS LEPROSY

The most puzzling feature of patients with lepromatous leprosy is their selective and specific cell-mediated unresponsiveness to antigens of *M. leprae*. *M. leprae* and *M. tuberculosis* are related, and antibodies to the proteins and glycoproteins of *M. tuberculosis* cross-react with those of *M. leprae*. Yet, while patients with lepromatous leprosy fail to respond to skin test or lymphocyte transformation tests to antigens of *M. leprae*, the vast majority are perfectly capable of responding to antigens of *M. tuberculosis* or tuberculin PPD. Since leprosy is an acquired disease, and the development of specific unresponsiveness in patients with lepromatous leprosy appears to be acquired as well, understanding lepromatous leprosy could provide important insights into fundamental mechanisms governing the acquisition of tolerance and unresponsiveness in man. To explain the immunological paradox underlying the ability of lepromatous patients to respond to cross-reactive antigens but be unable to respond to specific antigens, we suggested the hypothesis that there might be one or a small number of unique antigens or epitopes associated with *M. leprae* capable of inducing T suppressor cells that had the ability to suppress responses of potentially cross-reactive helper T cell clones (Bloom and Mehra, 1984; Mehra et al., 1979). Dr. V. Mehra and I developed a simple in vitro model for examining the ability of *M. leprae* antigens (lepromin) to induce suppressor activity in vitro. Lacking human leukocyte antigen (HLA)–matched individuals with leprosy, we tested the ability of *M. leprae* antigens to induce suppression of a mitogen response induced by concanavalin A (Con A). In over 200 patients studied, suppression was found in 84% of lepromatous and borderline patients, but not in tuberculoid patients, lepromin-positive contacts, or normal donors. Separation of the peripheral mononuclear cell populations into adherent and nonadherent subsets indicated that suppression in vitro could be produced by both the populations separately. Our recent experience suggests that the in vitro suppression mediated by the adherent mononuclear phagocytes may be related to the extent of disease and bacillary load, and is found less frequently in patients whose disease is detected at an early stage. Because this type of suppression is not antigen-specific it fails to explain the selective immunological unresponsiveness seen in lepromatous patients.

Consequently, we sought to characterize the nonadherent suppressor cell population. By measuring suppression of mitogen responses, it was possible to mix cells from individuals who were not fully histocompatible and detect suppression at 72 h without the confounding effects of mixed lymphocyte culture reactions. Our results indicated that the suppressor activity was contained in the 30% CD8 subset of peripheral T cells. When obtained directly from the blood of lepromatous leprosy patients, 50% of the CD8+ cells were Fc receptor positive and expressed HLA-DR (Ia) monomorphic antigens. We suggested that the expression of Fc receptors and DR reflected activation of these cells in vivo, and could be useful markers to monitor the degree of suppression in vivo.

The very existence, nature, and function of suppressor cells remains a controversial issue in immunology. One critical test as to whether our hypothesis might be consistent with the disease was examination of T cell subsets in leprosy lesions. A variety of investigators have established by immunoperoxidase staining that the preponderant cell type in lesions of tuberculoid leprosy are CD4 cells, with a small number of CD8 cells forming a mantle around the tuberculoid granulomata (Van Voorhis et al., 1982; Modlin et al., 1983). There are very many fewer lymphocytes in lesions of lepromatous leprosy, and the predominant lymphocyte type there is clearly a CD8 cell. A phenotype does not imply a function; consequently, the function of these cells and lesions was explored in more detail. When such biopsies were stained for the presence of lymphokines, specifically interleukin 2 (IL-2), T cells in tuberculoid lesions but not in lepromatous lesions stained positively (Modlin et al., 1984). Finally, techniques have been developed for isolating lymphocyte subsets directly from biopsies of lesions, and CD4 and CD8 T-cell lines and clones have been established in culture. When CD8 lines from lepromatous and tuberculoid lesions were expanded solely with IL-2, and then tested for in vitro lepromin-induced suppressor activity, about half the lines from lepromatous lesions have suppressor activity, whereas none from tuberculoid lesions did, formally providing evidence that cells with suppressor function existed within the lesions (Modlin et al., 1986). Finally, when individual clones were established from the CD8 lines derived from lesions, the ability to suppress the responses of *M. leprae*-specific CD4 clones to specific *M. leprae* antigens was restricted by the MHC class II antigens, providing the first information on possible genetic restrictions on suppressor cells in man (Modlin et al., 1986). Finally, it was gratifying to note that when patients were treated for long times with chemotherapy or with immunotherapy (vide infra), the suppressor activity of their lymphocytes disappeared, along with exposure of activation markers or CD8 T cells. Overall, these results support the view that the spectrum of leprosy represents a subtle regulatory balance of T-helper (and possibly cytotoxic T cells) and T-suppressor cells within lesions that determines the ultimate course of the disease.

Nevertheless, these findings raise more questions than are answered. The question becomes what factors determine the predominant development of suppressor cells in a very small percentage of people infected with *M. leprae*. Is it a matter of genetic or environmental predisposition of the individual; a variation in the genetics of the organism; route of infection; escape from cytocidal mechanisms of host macrophages; or subversion of antigen-presenting macrophages, dendritic cells, or Langerhans cells leading to suppressor cell dominance? What is the mechanism by which the suppressor cells block potentially protective responses of T helper cells? Are there other relevant mechanisms for inducing unresponsiveness, e.g., functional clonal deletion (Lamb et al., 1983; Jenkins and Schwartz, 1987).

## VACCINES AGAINST LEPROSY—THE FIRST GENERATION

The key finding that initially permitted the possibility of producing a vaccine against leprosy to be contemplated was that *M. leprae* can be grown in very large amounts in the armadillo. Initially, the armadillo was chosen because of its low body temperature (comparable to that of human skin), but it is likely that the undeveloped immune system in this host contributes as well. The first vaccine strategy was developed by one of the truly great experts in leprosy, Dr. Jacinto Convit in Caracas, Venezuela, on the basis of observations he made earlier that killed *M. leprae* inoculated into the skin of lepromatous patients persisted as acid-fast bacilli for long periods of time. However, when live BCG was inoculated together with killed *M. leprae*, a granulomatous response was produced to the BCG and all acid-fast bacilli were rapidly degraded, including the *M. leprae*. This suggested the possibility that immunization with a mixture of live BCG plus killed *M. leprae* might bring about a state of both cross-reactive and specific reactivity to *M. leprae* and thus serve as a therapeutic vaccine.

Convit and his colleagues have now vaccinated several hundred lepromatous leprosy patients with a mixture of live BCG and killed purified *M. leprae*. Their results have shown immunologic conversion to positive skin test reactivity, clearance of bacilli from the skin, histopathological upgrading towards the tuberculoid end of the spectrum, and clinical improvement in approximately 75% of borderline lepromatous leprosy patients, and in 57% of the polar lepromatous patients (Convit et al., 1982). These are rather dramatic results in patients who have been immunologically unresponsive, often for long periods of time. We have had the opportunity, in collaboration with Dr. Convit, to examine the in vitro suppressor activity and activation markers on peripheral T cells from ten such patients in a completely blind fashion, prior and subsequent to immunotherapy (Mehra et al., 1982b). The results indicate that in vitro suppressor activity in 10/10 successfully vaccinated patients decreased to normal levels following immunotherapy. These results provide objective evidence for immunological changes brought about by this vaccine, and establish a correlation between the in vitro T cell suppressor activity and the degree of immunologic unresponsiveness in vivo. Thus, leprosy remains the only example of specific immunological unresponsiveness in man which can, with reasonable success, be overcome by immunologic intervention. That the results are generalizable to other situations of unresponsiveness is suggested by recent work on American cutaneous leishmaniasis in which immunotherapy with killed promastigotes plus live BCG effected a cure rate of 95% (Convit et al., 1987).

Analysis of data on the protection afforded by BCG vaccine against leprosy from four trials indicates that BCG confers significant, but variable degrees of protection (24–85%) in different parts of the world (Fine, 1985). Although it would have been inconceivable even 5 years ago, there are currently two major controlled preventive vaccine trials underway against leprosy sponsored by WHO involving 200,000 people in Venezuela and Malawi. On the assumption that the combined vaccine or a cross-reactive cultivable mycobacteria will be as effective as *M. leprae* in providing immunity to noninfected individuals and likely to be therapeutic in those in the population who have subclinical infection, all three trials are testing the efficacy of killed *M. leprae* + BCG. However, even if the armadillo-derived vaccine should be effective, there is real question whether its cost and the inevitable shortage of the bacilli would permit it to be widely used. For this reason, it is important to search for potentially protective antigens in *M. leprae*.

## THE SEARCH FOR A UNIQUE ANTIGEN

Brennan and his colleagues established that *M. leprae* produces an abundance of a unique antigen, termed phenolic glycolipid I (Hunter et al., 1982). This antigen, the structure of which was established in elegant experiments, consists of a 29-carbon backbone with 2 hydrophobic 30–34-carbon mycolic ester groups, phenolicly linked to a trisaccharide (Fig. 2). The terminal trisaccharide is unique to *M. leprae*, and distinct from the lipids of other known bacteria.

The demonstration of the first unique epitope in *M. leprae* opened a new vista in seroepidemiology of the disease. Because of the low prevalence and incidence of leprosy, it would be enormously useful if seroepidemiology could address the following crucial problems: i) identifying patients with very early infection, especially those who are unaware that they have leprosy and are harboring the bacillus and capable of transmitting the infection; ii)

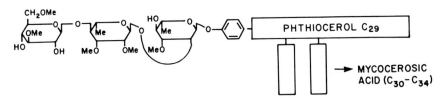

Fig. 2. *M. leprae* phenolic glycolipid I. PGL-I consists of the trisaccharide 3,6,-di-0-methylglucose, 2.3-di-0-methylrhamnose, 3-0-methylrhamnose linked to phenol dimycocerosyl phthiocerol. The specific trisaccharide is unique to *M. leprae*.

native parasite molecules. One clone bound to some extent to all ES antigens, indicating the presence of a common oligosaccharide side chain on structurally distinct proteins. A particularly interesting comparison was made between monoclonal binding to *T. canis* ES and a homologous set of antigens from the cat ascarid *T. cati* (Kennedy et al., 1987b). A pair of anticarbohydrate antibodies, Tcn-2 and Tcn-8, bind related epitopes on *T. canis* with Tcn-2 showing a much greater affinity. However, Tcn-2 fails bind to *T. cati* antigen, although Tcn-8 readily does so. Thus Tcn-2 defines a species-specific determinant which may be envisaged as an evolutionary change in composition, branching, or modification of a dominant carbohydrate structure.

## PHOSPHORYLCHOLINE

The presence of phosphorylcholine (PC) in nematode parasites was revealed by the reactivity of PC-binding myeloma proteins to extracts of *Nippostrongylus brasiliensis* (Pery et al., 1974) and by the production of abundant anti-PC antibody in infected hosts (Brown and Crandall, 1976). Subsequent work localised much of the PC reactivity to the gut and the female reproductive tract (Gutman and Mitchell, 1977; Pery et al., 1979; Forsyth et al., 1984; Gualzata et al., 1986), although monoclonal antibodies to PC also bind avidly to a circulating antigen released by adult filariae of *Onchocerca* (Forsyth et al., 1985), *Brugia* (Maizels et al., 1987c), and *Wuchereria* (Dissanayake et al., 1984; Morgan et al., 1986).

More recent studies on PC in our laboratory have examined the distribution of PC in other helminth species, and characterised in more detail the PC-bearing molecule(s) from *Brugia* and *Onchocerca* species. All nematodes so far tested show high levels of somatic PC (Table 2), although only some species secrete PC-positive molecules in vitro. The contrast is best exemplified by the infective larvae of *Toxocara canis*, which are rich in internalized PC, but which are negative in assays for in vitro-released PC (Sugane and Oshima, 1983) (our unpublished observations).

These comparative studies have shown that a wide variety of molecules in different nematode species may be substituted with the phosphorylcholine determinant. The filarial secreted PC-bearing molecules deserve further study because of their potential significance as diagnostic targets, and their measurable presence in the circulation of infected animals (Forsyth et al., 1984; Maizels et al., 1985) and human patients (Dissanayake et al., 1984; Forsyth et al., 1985; Selkirk et al., 1986). Only a subset of the whole worm PC-coupled molecules are detected in the serum of infected hosts, and while

**TABLE 2. Distribution of Phosphorylcholine in Nematodes[a]**

| Species | Natural host | Stage | Location of PC | Technique | Reference |
|---|---|---|---|---|---|
| *Acanthocheilonema viteae*[b] | Jird | L3,Ad | Somatic | IFAT | Gualzata et al. (1986) |
| *Ascaris suum* | Pig | L2 | Somatic | IFAT | Gutman and Mitchell (1977) |
| | | L2 | Somatic | Antibody | Brown and Crandall (1976) |
| *Brugia malayi* | Man | Adult | Somatic, ES | IRMA | Maizels et al. (1987c) |
| *Brugia pahangi* | Cat, dog | Adult | Somatic, ES | IRMA | Maizels et al. (1987c) |
| *Heligmosomoides polygyrus*[c] | Mouse | Adult | Somatic, ES | IRMA | Maizels and Burke, unpublished |
| *Nippostrongylus brasiliensis* | Rat | Adult | Somatic | ID | Pery et al. (1974, 1979) |
| | | Adult | Somatic, ES | IRMA | Maizels and Burke, unpublished |
| *Onchocerca gibsoni* | Cow | Adult | Uterine, egg | IRMA, IFAT | Forsyth et al. (1985) |
| *Trichinella spiralis* | Carnivore | L1 | Somatic | Antibody | Ubeira et al. (1987) |
| *Toxocara canis* | Dog | L2 | Somatic, *not* ES | | Sugane and Oshima (1983) |
| | | | Somatic, *not* ES | IRMA | Maizels and Burke, unpublished |
| *Wuchereria bancrofti* | | Adult | Somatic, ES | IRMA | Morgan et al. (1986) |

[a]Techniques referred to: IFAT = immunofluorescent antibody staining of sections; IRMA = immunoradiometric assay of soluble preparations with anti-PC monoclonal antibody; ID = immunodiffusion in gel (Ouchterlony or immunoelectrophoresis); Antibody = assays for anti-PC antibody in infected hosts.
[b]*Dipetalonema viteae*.
[c]*Nematospiroides dubius*.

## ACKNOWLEDGMENTS

The studies described in this chapter have been supported by the Wellcome Trust, the Medical Research Council, the UNDP-World Bank-World Health Organization Special Programme for Research and Training in Tropical Diseases, and by the Directorate General for Science Research and Development of the Commission of the European Communities. The authors would also like to thank Dr. D. J. McLaren of the National Institute of Medical Research, Mill Hill, London, for her kind permission to present Figure 2.

## REFERENCES

Anderson RM, May RM (1982): Population dynamics of human helminth infections: Control by chemotherapy. Nature 297:557–563.

Badley JE, Grieve RB, Rockey JH, Glickman LT (1987): Immune-mediated adherence of eosinophils to *Toxocara canis* infective larvae: The role of excretory-secretory antigens. Parasite Immunol 9:133–143.

Betschart B, Jenkins JM (1987): Distribution of iodinated proteins in *Dipetalonema viteae* after surface labelling. Mol Biochem Parasitol 22:1–8.

Betschart B, Glaser M, Keifer R, Rudin W, Weiss N (1987): Immunochemical analysis of the epicuticle of parasitic nematodes. J Cell Biochem [Suppl] 11A:164.

Bird AF (1980): The nematode cuticle and its surface. In B.M. Zuckerman (ed): "Nematodes as Biological Models." New York: Academic Press, Vol 2, pp 213–236.

Brown AR, Crandall CA (1976): A phosphorylcholine idiotype related to TEPC 15 in mice infected with *Ascaris suum*. J Immunol 116:1105–1109.

Cabrera Z, Parkhouse RME (1987): Isolation of an antigenic fraction for diagnosis of Onchocerciasis. Parasite Immunol 9:39–48.

Campbell DH (1936): An antigenic polysaccharide fraction of *Ascaris lumbricoides* (from hog). J Infect Dis 59:266–280.

Campbell CH (1955): The antigenic role of the excretions and secretions of *Trichinella spiralis* in the production of immunity in mice. J Parasitol 41:483–491.

Carlow CKS, Franke ED, Lowrie RC, Partono F, Philipp M (1987): Monoclonal antibody to a unique surface epitope of the human filaria *Brugia malayi* identifies infective larvae in mosquito vectors. Proc Natl Acad Sci USA, 84:6914–6918.

Carroll SM, Howse DJ, Grove DI (1984): The anticoagulant effects of the hookworm, *Ancylostoma ceylanicum:* observations on human and dog blood *in vitro* and infected dogs *in vivo*. Thromb Haemost 51:222–227.

Croll NA, Matthews BE (1977): Biology of Nematodes. Glasgow: Blackie & Son.

Cox GN, Hirsh D (1985): Stage-specific patterns of collagen gene expression during development of *Caenorhabditis elegans*. Mol Cell Biol 5:363–372.

Cox GN, Kusch M, Denevi K, Edgar RS (1981): Temporal regulation of cuticle synthesis during development of *Caenorhabditis elegans*. Dev Biol 84:277–285.

Dawkins HJS, Grove DI (1982): Immunisation of mice against *Strongyloides ratti*. Z Parasitenkd 66:327–333.

Day KP, Howard RJ, Prowse SJ, Chapman CB, Mitchell GF (1979): Studies on chronic versus transient intestinal nematode infections in mice. I. A comparison of responses to excretory/secretory (ES) products of *Nippostrongylus brasiliensis* and *Nematospiroides dubius* worms. Parasite Immunol 1:217-239.

Denham DA, Chusattayanond W (1986): Attempted vaccination of jirds (*Meriones unguiculatus*) against *Brugia pahangi* with radiation attenuated infective larvae. J Helminthol 60:149-155.

Despommier DD, Muller M (1976): The stichosome and its secretion granules in the mature muscle larvae of *Trichinella spiralis*. J Parasitol 62:775-785.

Devaney E (1985): Lectin-binding characteristics of *Brugia pahangi* microfilariae. Trop Med Parasitol 36:25-28.

Dissanayake S, Forsyth KP, Ismail MM, Mitchell GF (1984): Detection of circulating antigen in Bancroftian filariasis by using a monoclonal antibody. Am J Trop Med Hyg 33 (6):1130-1140.

Dow C, Jarrett WFH, Jennings FW, Mclntyre WIM, Mulligan W (1961): Studies on immunity to *Uncinaria stenocephala* infection in the dog—double infection with irradiated larvae. Am J Vet Res 22:352-354.

Durham CP, Murrell KD, Lee CM (1984): *Trichinella spiralis:* Immunization of rats with an antigen fraction enriched for allergenicity. Exp Parasitol 57:297-306.

Edwards AJ, Burt JS, Ogilvie BM (1971): The effect of immunity upon some enzymes of the parasitic nematode, *Nippostrongylus brasiliensis*. Parasitology 65: 547-550.

Fattah DL, Maizels RM, McLaren DJ, Spry CJF (1986): Interaction of human blood eosinophils with *Toxocara canis* larvae *in vitro*. Exp Parasitol 61:421-433.

Faubert GM (1976): Depression of the plaque-forming cells to sheep red blood cells by the new-born larvae of *Trichinella spiralis*. Immunology 30:485-489.

Foo DY, Nowak M, Copeman B, McCabe M (1983): A low molecular weight immunosuppressive factor produced by *Onchocerca gibsoni*. Vet Immunol Immunopathol 4:445-451.

Forsyth KP, Copeman DB, Mitchell GF (1984): *Onchocerca gibsoni:* Increase of circulating egg antigen with chemotherapy in bovines. Exp Parasitol 58:41-55.

Forsyth KP, Spark R, Kazura J, Brown GV, Peters P, Hetwood P, Dissanayake S, Mitchell GF (1985): A monoclonal antibody-based immunoradiometric assay for detection of circulating antigen in *Bancroftian filariasis*. J Immunol 134:1172-1177.

Fujimoto D, Kanaya S (1973): Cuticlin: A noncollagen structural protein from *Ascaris* cuticle. Arch Biochem Biophys 157:1-6.

Fujimoto D, Horiuchi K, Hirama M (1981): Isotrityrosine, a new cross-linking amino acid isolated from Ascaris cuticle collagen. Biochem Biophys Res Commun 99:637-643.

Furhman J, Piessens WF (1985): Chitin synthesis and sheath morphogenesis in *Brugia malayi* microfilariae. Mol Biochem Parasitol 17:93-104.

Furhman JA, Urioste SS, Hamill B, Spielman A, Piessens WF (1987): Functional and antigenic maturation of *Brugia malayi* microfilariae. Am J Trop Med Hyg 36:70-74.

Gamble HR (1985): *Trichinella spiralis:* Immunization of mice using monoclonal antibody affinity-isolated antigens. Exp Parasitol 59:398-404.

Gasbarre LC, Romanowski RD, Douvres FW (1985): Suppression of antigen- and mitogen-induced proliferation of bovine lymphocytes by excretory-secretory products of *Oesophagostomum radiatum*. Infect Immun 48:540-545.

Grencis RK, Crawford C, Pritchard DI, Behnke JM, Wakelin D (1986): Immunization of mice with surface antigens from the muscle larvae of *Trichinella spiralis*. Parasite Immunol 8:587-596.

## 304 / Maizels and Selkirk

Gualzata M, Weiss N, Heuser CH (1986): *Dipetalonema viteae:* Phosphorylcholine and non-phosphorylcholine antigenic determinants in infective larvae and adult worms. Exp Parasitol 61:95–102.

Gutman GA, Mitchell GF (1977): *Ascaris suum:* location of phosphorylcholine antigenic determinants in infective larvae and adult worms. Exp Parasitol 61:95–102.

Hagan P, Behnke JM, Parish HA (1981): Stimulation of immunity to *Nematospiroides dubius* in mice using larvae attenuated by cobalt 60 irradiation. Parasite Immunol 3:149–156.

Hayashi Y, Noda K, Shiresaka A, Nogami S, Nakamura M (1984): Vaccination of BALB/c mice against *Brugia malayi* and *B. pahangi* with larvae attenuated by gamma irradiation. Jpn J Exp Med 54:177–181.

Himmelhoch S, Zuckerman BM (1978): *Caenorhabditis briggsae:* Aging and the structural turnover of the outer cuticle surface and the intestine. Exp Parasitol 45:208–214.

Hinck LW, Ivey MH (1976): Proteinase activity in *Ascaris suum* eggs, hatching fluid and excretions secretions. J Parasitol 62:771–774.

Hotez PJ, Cerami A (1983): Secretion of a proteolytic anticoagulant by *Ancylostoma* hookworms. J Exp Med 157:1594–1603.

Hotez PJ, Le Trang N, McKerrow JH, Cerami A (1985): Isolation and characterization of a proteolytic enzyme from the adult hookworm *Ancylostoma caninum.* J Biol Chem 260:7343–7348.

Howells RE, Chen SN (1981): *Brugia pahangi:* Feeding and nutrient uptake *in vitro* and *in vivo.* Exp Parasitol 51:42–58.

Inglis WG (1983): The design of the nematode body wall: The ontogeny of the cuticle. Aust J Zool 31:705–716.

Jarrett WFH, Jennings FW, McIntyre WIM, Mulligan W, Sharp NCC, Urquhart GM (1959): Immunological studies on *Dictyocaulus viviparus* infection in calves—double infection with irradiated larvae. Am J Vet Res 20:522–526.

Jungery M, Clark NWT, Parkhouse RME (1983): A major change in surface antigens during the maturation of newborn larvae of *Trichinella spiralis.* Mol Biochem Parasitol 7:101–109.

Keith RK, Bremner KC (1973): Immunization of calves against the nodular worm, *Oesophagostomum radiatum.* Res Vet Sci 15:123–124.

Kennedy MW, Foley M, Kuo Y-M, Kusel JR, Garland PB (1987a): Biophysical properties of the surface lipid of parasitic nematodes. Mol Biochem Parasitol 22:233–240.

Kennedy MW, Maizels RM, Meghji M, Young L, Qureshi F, Smith HV (1987b): Species-specific and common epitopes on the secreted and surface antigens of *Toxocara canis* and *Toxocara cati* infective larvae. Parasite Immunol 9:407–420.

Klesius PH, Haynes TB, Cross DA, Ciordia H (1986): *Ostertagia ostertagi:* Excretory secretory chemotactic substances from infective larvae as cause of eosinophil locomotion. Exp Parasitol 61:120–125.

Lal RB, Paranjape RS, Briles DE, Nutman TB, Ottesen EA (1987): Circulating parasite antigen(s) in lymphatic filariasis: Use of monoclonal antibodies to phosphorylcholine for immunodiagnosis. J Immunol 138:3454–3460.

Lee DL, Wright KA, Shivers RR (1984): A freeze-fracture study of the surface of the infective-stage larva of the nematode *Trichinella.* Tissue Cell 16:819–828.

Lewart RM, Lee C-L. (1956): Quantitative studies of the collagenase-like enzymes of cercariae of *Schistosoma mansoni* and the larvae of *Strongyloides ratti.* J Infect Dis 99:1–14.

Mackenzie CD, Preston PM, Ogilvie BM (1978): Immunological properties of the surface of nematodes. Nature 276:826–828.

Mackenzie CD, Jungery M, Taylor PM, Ogilvie BM (1980):Activation of complement, the induction of antibodies to the surface of nematodes and the effect of these factors and cells on worm survival in vitro. Eur J Immunol 10:594–601.

Maizels RM, Philipp M, Ogilvie BM (1982): Molecules on the surface of parasitic nematodes as probes of the immune response in infection. Immunol Rev 61:109–136.

Maizels RM, Meghji M, Ogilvie BM (1983a): Restricted sets of parasite antigens from the surface of different stages and sexes of the nematode Nippostrongylus brasiliensis. Immunology 48:107–121.

Maizels RM, Partono F, Oemijati S, Denham DA, Ogilvie BM (1983b): Cross-reactive surface antigens on three stages of Brugia malayi, B. pahangi and B. timori. Parasitology 87:249–263.

Maizels RM, De Savigny D, Ogilvie BM (1984): Characterisation of surface and excretory-secretory antigens of Toxocara canis infective larvae. Parasite Immunol 6:23–37.

Maizels RM, Denham DA, Sutanto I (1985): Secreted and circulating antigens in the filarial parasite Brugia pahangi: Analysis of in vitro released components and detection of parasite products in vivo. Mol Biochem Parasit 17:277–288.

Maizels RM, Kennedy MW, Meghji M, Robertson BD, Smith HV (1987a): Shared carbohydrate epitopes on distinct surface and secreted antigens of the parasitic nematode Toxocara canis. J Immunol 139:207–214.

Maizels RM, Bianco AE, Burke J, Flint JE, Gregory WF, Kennedy MW, Lim GE, Robertson BD, Selkirk ME (1987b): Glycoconjugate antigens from parasitic nematodes. UCLA Symposia on Molecular Biology 59:267–279.

Maizels RM, Burke J, Denham DA (1987c): Phosphorylcholine-bearing antigens in filarial nematode parasites: analysis of somatic extracts and in vitro secretions of Brugia malayi and B. pahangi. Parasite Immunol 9: 49–66.

Marshall E, Howells RE (1985): An evaluation of different methods for labelling the surface of the filarial nematode Brugia pahangi with [125]iodine. Mol Biochem Parasitol 15:295–304.

Martinez-Palomo A (1978): Ultrastructural characterization of the cuticle of Onchocerca volvulus microfilariae. J Parasitol 64:127–136.

Martzen MR, Peanasky RJ (1985): Ascaris suum: Biosynthesis and isoinhibitor profile of chymotrypsin/elastase isoinhibitors. Exp Parasitol 59:313–320.

Matsumura K, Kazyta Y, Endo R, Tanaka K (1984): Detection of circulating toxocaral antigen in dogs by sandwich enzyme-immunoassay. Immunology 51:609–613.

Matthews BE (1982): Skin penetration by Necator americanus larvae. Z Parasitenkd 68:81–86.

Matthews BE (1984): The source, release and specificity of proteolytic enzyme activity produced by Anisakis simplex larvae (Nematoda: Ascaridida) in vitro. J Helminthol 58:175–185.

McGreevy PB, Ratiwayanto S, Tuti S, McGreevy MM, Dennis DT (1980): Brugia malayi: Relationship between anti-sheath antibodies and amicrofilariae in natives living in an endemic area of South Kalimantan, Borneo. Am J Trop Med Hyg 29:553–562.

McLaren DJ (1974): The anterior glands of Necator americanus (Nematoda: Strongyloidea) 1. Ultrastructural studies. Int J Parasitol 4:25–37.

Meghji M, Maizels RM (1986): Biochemical properties of larval excretory-secretory (ES) glycoproteins of the parasitic nematode Toxocara canis. Mol Biochem Parasitol 18:155–170.

Miller TA (1965): Effect of route of immunization of vaccine and challenge of the immunogenic efficiency of double vaccination with irradiated *Ancylostoma caninum* larvae. J Parasitol 51:200–206.

Miller TA (1971): Vaccination against the canine hookworm diseases. Adv Parasitol 9:153–183.

Morgan TM, Sutanto I, Purnomo, Sukartono, Partono F, Maizels RM (1986): Antigenic characterization of adult *Wuchereria bancrofti* filarial nematodes. Parasitology 93:559–569.

Nicholas WL, Stewart AC, Mitchell GF (1984): Antibody responses to *Toxocara canis* using sera from parasite infected mice and protection from toxocariasis by immunization with ES antigens. Aust J Exp Biol Med Sci 62:619–626.

Ogilvie BM, Rothwell TLW, Bremner KC, Schnitzerling HJ, Nolan J, Keith RK (1973): Acetylcholinesterase secretion by parasitic nematodes. I. Evidence for secretion of the enzyme by a number of species. Int J Parasitol 3:589–597.

Oothuman P, Denham DA, McGreevy PB, Nelson GS, Rogers R (1979): Successful vaccination of cats against *Brugia pahangi* with larvae attenuated by irradiation with 10 krad cobalt 60. Parasite Immunol 1:209–216.

Ortega-Pierres G, Chayen A, Clark NWT, Parkhouse RME (1984): The occurence of antibodies to hidden and exposed determinants of surface antigens of *Trichinella spiralis*. Parasitology 88:359–369.

Ozerol NH, Silverman PH (1970): Further characterization of active metabolites from histotropic larvae of *Haemonchus contortus* cultured in vitro. J Parasitol 56:1199–1205.

Parkhouse RME, Philipp M, Ogilvie BM (1981): Characterization of surface antigens of *Trichinella spiralis* infective larvae. Parasite Immunol 3:339–352.

Pasternack MS, Verret CR, Liu MA, Eisen HN (1986): Serine esterase in cytolytic T lymphocytes. Nature 322:740–743.

Peacock R, Poynter D (1980): Field experience with a bovine lungworm vaccine. In A.E.R. Taylor and R. Muller (eds): Symp Brit Soc Parasitol, vol 18:141–148.

Pery P, Petit A, Poulain J, Luffau G (1974): Phosphorylcholine-bearing components in homogenates of nematodes. Eur J Immunol 4:637–639.

Pery P, Luffau G, Charley J, Petit A, Rouze P, Bernard S (1979): Phosphorylcholine antigens from *Nippostrongylus brasiliensis*. II. Isolation and partial characterization of phosphorylcholine from adult worm. Ann Immunol Inst Pasteur 130C:889–900.

Petralanda I, Yarzabal L, Piesssens WF (1986): Studies on a filarial antigen with collagenase activity. Mol Biochem Parasitol 19:51–59.

Philipp M, Rumjaneck FD (1984): Antigenic and dynamic properties of helminth surface structures. Mol Biochem Parasitol 10:245–268.

Philipp M, Parkhouse RME, Ogilvie BM (1980): Changing proteins on the surface of a parasitic nematode. Nature 287:538–540.

Philipp M, Taylor PM, Parkhouse RME, Ogilvie BM (1981): Immune response to stage-specific surface antigens of the parasitic nematode *Trichinella spiralis*. J Exp Med 154:210–215.

Porter PR (1980): The complex proteases of the complement sytem. Proc R Soc Lond [Biol] 210:477–498.

Pritchard Dl (1986): Antigens of gastrointestinal nematodes. Trans R Soc Trop Med Hyg 80:728–734.

Pritchard Dl, Maizels RM, Behnke JM, Appelby P (1984): Stage-specific antigens of *Nematospiroides dubius*. Immunology 53:325–355.

Pritchard Dl, Crawford CR, Duce IR, Behnke JM (1985): Antigen stripping from the nematode epicuticle using the cationic detergent cetyltrimethylammonium bromide (CTAB). Parasite Immunol 7:575-585.

Rathaur S, Robertson BD, Selkirk ME, Maizels RM (1987): Secretory acetylcholinesterases from Brugia malayi adult and microfilarial parasites. Mol Biochem Parasitol 26:257-265.

Raybourne R, Desowitz RS, Kilks MM, Deardoff TL (1983): Anisakis simplex and Terranova sp.: Inhibition by larval excretory-secretory products of mitogen-induced rodent lymphoblast proliferation. Exp Parasitol 55:289-298.

Rhoads ML (1983): Trichinella spiralis: identification and purification of superoxide dismutase. Exp Parasitol 56:41-54.

Robertson BD, Rathaur S, Maizels RM (1987): Antigenic and biochemical analyses of the excretory-secretory molecules of Toxocara canis infective larvae. In: S. Geerts, J. Kumar, and J. Brandt (eds): Helminth Zoonoses. Dordrecht: Martinus Nijhoff Publishers, pp 167-174.

Rothwell TLW, Love RJ (1974): Vaccination against the nematode Trichostrongylus colubriformis - I. Vaccination of guinea pigs with worm homogenates and soluble products released during in vitro maintenance. Int J Parasitol 4:293-299.

Sadun EH (1963): Fluorescent antibody technique for helminth infections. Exp Parasitol 13:72-82.

Schwartz B (1921): Effects of secretions of certain parasitic nematodes on coagulation of blood. J Parasitol 7:144-150.

Selkirk ME, Denham DA, Partono F, Sutanto I, Maizels RM (1986): Molecular characterization of antigens of lymphatic filariae. Parasitology 91:S15-S38.

Silberstein DS, Despommier DD (1984): Antigens from Trichinella spiralis that induce a protective response in the mouse. J Immunol 132:898-890.

Smith HV, Quinn R, Kusel JR, Girdwood RWA (1981): The effect of temperature and antimetabolites on antibody binding to the outer surface of second stage Toxocara canis larvae. Mol Biochem Parasitol 4:183-193.

Sokolic A, Jovanovic M, Cuperlovic K, Movesesijan M (1965): Vaccination against Dictyocaulus filaria with irradiated larvae. Br Vet J 121:212-222.

Storey DM, Al-Mukhtar AS (1982): Vaccination of jirds, Meriones unguiculatus, against Litomosoides carinii and Brugia pahangi using irradiated larvae of L. carinii. Tropen Med Parasitol 33:23-24.

Stromberg BE, Soulsby EJL (1977): Ascaris suum: Immunization with soluble antigens in the guinea pig. Int J Parasitol 7:287-291.

Sugane K, Oshima T (1983): Activation of complement in C-reactive protein positive sera by phosphorylcholine-bearing component isolated from parasite extract. Parasite Immunol 5:383-395.

Sulston JE, Schierenberg E, White JG, Thomson JN (1983): The embryonic cell lineage of the nematode Caenorhabditis elegans. Dev Biol 100:64-119.

Sutanto I, Maizels RM, Denham DA (1985): Surface antigens of a filarial nematode: Analysis of adult Brugia pahangi surface components and their use in monoclonal antibody production. Mol Biochem Parasitol 15:203-214.

Tanner M, Weiss N (1981): Dipetalonema viteae (Filaroidea): Evidence for serum-dependent cytotoxicity against developing third and fourth stage larvae in vitro. Acta Trop (Basel) 38:325-328.

308 / Maizels and Selkirk

Ubeira FM, Leiro J, Santamarina MT, Villa TG, Sanmartin-Duran ML (1987): Immune response to *Trichinella spiralis* epitopes: the antiphosphorylcholine plaque-forming cell response during the biological cycle. Parasitology 94:543–553.

Urban JF, Tromba FG (1984): An ultraviolet-attenuated egg vaccine for swine ascariasis: Parameters affecting the development of protective immunity. Am J Vet Res 45:2104–2108.

Urban JF, Romanowski RD (1985): *Ascaris suum:* Protective immunity in pigs immunized with products from eggs and larvae. Exp Parasitol 60:245–254.

Urquhart GM, Jarrett WFH, Jennings FW, McIntyre WIM, Mulligan W (1966): Immunity of *Haemonchus contortus* infection: Relationship between age and successful vaccination with irradiated larvae. Am J Vet Res 27:1645–1648.

Vetter JCM, Klaver-Wesseling JCM (1978): IgG antibody binding to the outer surface of infective larvae of *Ancylostoma caninum.* Z Parasitenkd 58:91–96.

Wakelin D, Selby GR (1973): Functional antigens of *Trichuris muris.* The stimulation of immunity by vaccination of mice with somatic antigen preparations. Int J Parasitol 3:711–715.

Weil GJ, Liftis F (1987): Identification and partial characterization of a parasite antigen in sera from humans infected with *Wuchereria bancrofti.* J Immunol 138:3035–3041.

Wenger JD, Forsyth KP, Kazura JW (1988): Identification of phosphorylcholine epitope-bearing antigens in *Brugia malayi* and relation of serum epitope levels to infection status of jirds with Brugian filariasis. Am J Trop Med Hyg 38:133–141.

Wong MM, Suter PF (1979): Indirect fluorescent antibody test in occult dirofilariasis. Am J Vet Res 40:414–420.

Wong MM, Guest MF, Lavoipierre MMJ (1974): *Dirofilaria immitis:* Fate and immunogenicity of irradiated infective stage larvae in beagles. Exp Parasitol 35:465–474.

Yates JA, Higashi GI (1985): *Brugia malayi:* Vaccination of jirds with [60]Cobalt-attenuated infective stage larvae protects against homologous challenge. Am J Trop Med Hyg 34:1132–1137.

Yeates RA, Ogilvie BM (1976): Nematode acetylcholinesterases. In H. Van de Bossche (ed): Biochemistry of Parasites and Host-Parasite Relationships. Amsterdam: Elsevier/North-Holland Biomedical Press, pp 307–310.

The Biology of Parasitism, pages 309–328
© 1988 Alan R. Liss, Inc.

# Molecular Basis for Interactions Between Parasites and the Complement Cascade

Keith A. Joiner

*Laboratory of Clinical Investigation, National Institute of Allergy and Infectious Diseases, National Institutes of Health, Bethesda, Maryland 20892*

## INTRODUCTION

The topic of complement-parasite interaction has been one of recent and growing interest (Santoro, 1982). Not only have novel strategies for parasite evasion of the complement system been identified, but the critical importance of the complement cascade in establishing intracellular infection with a number of parasites has been elucidated. It is instructive to consider the complement-parasite interaction from the perspective of the parasite, which must devise strategies to evade direct destruction by complement and may or may not utilize the complement system and complement receptors to gain access to intracellular compartments. In this paper, I will first review the complement cascade, concentrating on the known or anticipated features of particular relevance to the immunobiology of parasite interactions. Selected aspects will be emphasized, including C1 binding and activation, C1 chemistry, C3 receptors, regulatory molecules, and C5b-9 formation. These topics will be summarized only in brief fashion but are the subject of recent reviews (Fearon and Wong, 1983; Fries and Frank, 1987; Müller-Eberhard and Schreiber, 1980; Ross and Medof, 1985). I will then discuss in some detail the ongoing studies in our laboratory examining the interaction of complement with *Trypanosoma cruzi* and *Leishmania major*.

In its simplest state, the complement system can be thought of as two distinct but nonetheless related pathways, both of which lead to the generation of opsonic, chemotactic, anaphylotoxic and lytic molecules which participate in host defense against invading pathogens. In its most complex state, the complement system can be thought of as 25 separate activation and regulatory proteins and at least 5 receptors involved in a complex cascade of proteolytic cleavage events and conformational alterations leading to biologically active molecules (Fig.1). Any fundamental understanding of the role of complement in host defense requires that one be cognizant of both ends of this spectrum. The intricate biochemical details of complement activation and control must be dissected for each microbial surface, yet this must be done with careful attention to the biological relevance of functionally active complement molecules in the system under study. For example, the capacity of an insect-stage parasite or an intracellular stage parasite to evade or succumb to complement-mediated lysis should be studied at the biochemical level only with the knowledge that these stages may never come in contact with the host complement system. Nonetheless, comparison of complement interactions between these forms with stages which will be exposed to the host complement cascade provides valuable lessons in developmental regulation and in parasite adaptation to changing environmental stresses.

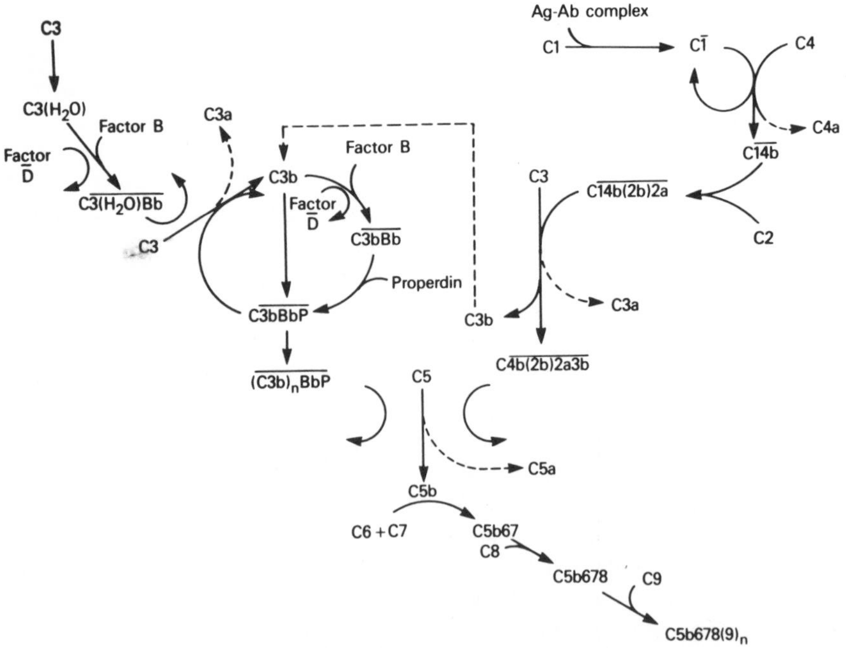

Fig. 1. Diagram of the human complement cascade: A scheme with extensive detail is shown, in order to illustrate those facets of the cascade which are not discussed in the text. Nonetheless, the reader is referred to several recent reviews (Fries and Frank, 1987; Müller-Eberhard and Schrieber, 1980) for a description of the individual steps involved in activation of the classical and alternative pathways.

## ALTERNATIVE AND CLASSICAL PATHWAY

As just mentioned, the complement system is composed of two distinct pathways—the classical (CP) and alternative (ACP)—which converge at the step of C3b deposition and C5 cleavage (Fig. 1). The alternative pathway is the older of the two phylogenetically—proteins related in function to human alternative complement pathway proteins have been identified in elasmobranchs. In addition, because the alternative pathway operates primarily as an antibody-independent cascade, the alternative pathway is often portrayed as an early-warning system which allows the naive host to respond to an invading pathogen before specific antibody has developed. Most studies on the interaction of complement and parasites have presumed that complement activation in nonimmune normal human serum occurs via the alternative pathway, since natural antibody is presumed to be absent. This presumption is clearly incorrect in some circumstances—for example, promastigotes of *Leishmania donovani* were shown to activate the classical pathway in nonimmune serum in an antibody-dependent fashion, presumably due to natural antibodies (Pearson and Steigbigel, 1980). Furthermore as indicated below, some parasites may bind and activate C1 in the absence of antibody. Finally, in some cirumstances, such as with metacyclic or bloodstream forms of *T. cruzi*, specific antibody facilitates alternative pathway activation. Hence, generalizations about the complement pathway being activated in nonimmune serum require direct experimental testing in each case.

## C1 ACTIVATION AND CONTROL

A growing number of parasites/helminths have been shown to interact directly and in an antibody-independent fashion with C1, the first component of the classical pathway, or with its subcomponent C1q. These include *Trypanosoma brucei*, *Trypanosoma cruzi*, and *Schistosoma mansoni* (Rimoldi et al., 1987; reviewed in Cooper, 1985). Although direct binding of C1 or C1q does not necessarily translate into activation of the C1 molecule in serum, or to activation beyond the C1 step, C1q on a particle may enhance the interaction of that particle with cells bearing C1q receptors or with basement membrane proteins (Bohnsack et al., 1985). Therefore, direct C1 binding and activation may play a role in the immunobiology of infections even in the absence of further activation of the complement cascade beyond the C1 step.

## C3 CHEMISTRY

Regardless of whether the classical or alternative pathway is activated, the pathways converge at the step of C3 deposition. Understanding C3 chemistry is key to a fundamental understanding of complement activation and control. Cleavage of C3 by either the classical (C14b2a) or alternative pathway (PC3bBb) C3 convertase results in release of the small anaphylotoxic C3a molecule, and exposure/activation of an internal thioester group within the resultant C3b fragment. This labile thioester group undergoes one of three fates—hydrolysis by $H_2O$, covalent ester bond formation with hydroxyl groups on acceptor molecules, or covalent amide bond formation with amino groups on acceptor molecules (Fig. 2). Ester linkages are preferred at physiologic pH, which in part explains the propensity of C3 to react with carbohydrate residues. Nonetheless, it is of interest that C3 attached to glycoproteins on *L. mexicana* (Russell, 1987) and *L. donovani* (Puentes, et al., manuscript in preparation) is amide linked, whereas C3 attached to

BINDING OF C3b TO ACCEPTOR MOLECULES

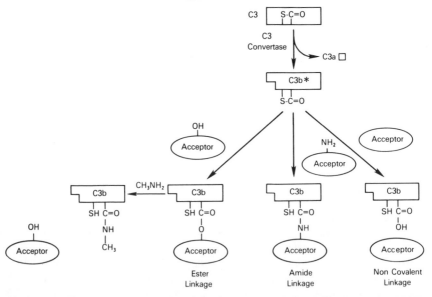

Fig. 2. Mechanisms of C3 interaction with acceptor surfaces: Native C3 contains an internal thioester group which is exposed when the molecule is cleaved to C3b by either the CP or ACP C3 convertase. The internal thioester then undergoes one of three fates: 1) the thioester may be hydrolyzed by water, thereby precluding covalent attachment of C3b to the surface, but allowing a noncovalent interaction to occur **(right)**, 2) a covalent amide bond forms between C3b and exposed amino groups on acceptor molecules **(center)**, and 3) a covalent ester bond forms between the C3b and exposed hydroxyl groups on acceptor molecules; ester linkages are susceptible to cleavage by nucleophiles such as methylamine, liberating the acceptor from the covalently bound C3b fragment **(left)**.

glycoproteins in *T. cruzi* (Joiner et al., 1985a), *Toxoplasma gondii* (Fuhrman and Joiner, 1987), and a glycolipid in *L. major* (Puentes et al., 1987) is ester-linked. Covalently bound C3 may or may not undergo further processing. Ester linkages are inherently less stable than other covalent bonds, and the C3 fragment will spontaneously release from surfaces in a time-, tempera-ture-, and pH-dependent fashion. C3 may also be proteolytically cleaved by the concerted action of the active complement enzyme factor I and either of the necessary co-factors, factor H or CR1 (Figs. 1, 3). Cleavage of C3b to iC3b and further modification/cleavage to C3d/C3dg block the capacity of the C3 fragment to participate in further C3 convertase formation (ACP) or

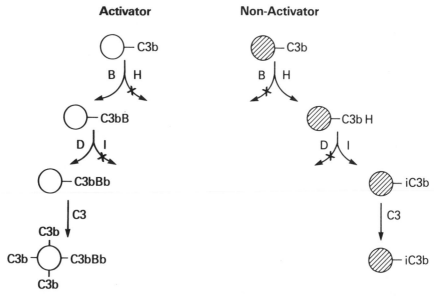

Fig. 3. Control of alternative pathway activation: Control of alternative pathway activation is exerted on C3b or "C3b-like" molecules which have bound randomly to the surface of both activating and nonactivating particles. On activating surfaces, factor B binds to C3b in preference to factor H (**left**) thereby permitting cleavage of factor B to Bb by the active enzyme factor D. The complex of C3bBb is in itself a C3 convertase and is capable of cleaving additional C3 molecules, leading to amplification of the initial C3b deposition. In contrast, on nonactivating surfaces, factor H binds to C3b in preference to factor B (**right**), leading to cleavage of C3b to the hemolytically inactive iC3b fragment by the active enzyme, factor I. The lytic cascade stops at this point, although iC3b and other C3 degradation fragments are capable of interacting with cellular receptors.

C5 convertase formation (ACP and CP). Therefore, the direct cidal activity of the cascade is delimited by cleavage of C3b to iC3b. However, the resultant fragments are not inert with respect to receptor interactions, since distinct receptors exist for the C3b, iC3b, C3dg, and C3d fragments of complement.

## CONTROL OF CLASSICAL PATHWAY AND ALTERNATIVE PATHWAY ACTIVATION

The mechanisms of classical and alternative pathway activation are fundamentally different. Activation of the classical pathway requires the direct interaction of the first component of complement, C1, with specific antibody bound to antigen, or in some cases, the direct antibody-independent activation of the C1 molecule, as discussed above. Activation has a precise biochemical definition—the proteolytic cleavage of the C1r and C1s subcomponents within the trimolecular $C1qC1r_2C1s_2$ complex comprising C1 (reviewed in Cooper, 1985). Once activation has been initiated, two control proteins specific for the classical pathway—C1 esterase inhibitor and C4 binding protein—limit further activation of the cascade (Fig. 1). The extent of limitation is dictated by the nature of the activating surface and by the presence of antibody. Nonetheless, control of opsonic, chemotactic, and lytic component generation is exerted after the process of activation has been initiated *de novo* by a specific surface.

The mechanism of alternative pathway activation is fundamentally different. Activation occurs when the normal control processes limiting the continual low-level ACP activation occurring in the fluid phase system or on a surface are overridden. Low-level deposition of C3b, or of water-hydrolysed C3, which is "C3b-like," occurs continually on both activating and nonactivating surfaces (Fig. 3). On a nonactivating surface, the C3b fragment is cleaved to iC3b by the action of factors H and I (Fig. 3). In contrast, on an activating surface, the affinity of B for C3b exceeds the affinity of H, and activation ensues. On erythrocytes the affinity of H for C3b is low on activating surfaces and high on nonactivating surfaces, while B affinity for C3b does not change. In microbial systems, this scenario does not always apply.

## C3 RECEPTORS

The known receptors for C3 fragments, their ligand specificity, chain structure, and cellular distribution are listed in Table 1. The best studied

TABLE 1. C3 Receptors

| Receptor | Cells[a] | MW (kD) | Specificity |
|----------|----------|---------|-------------|
| CR1 | RBC, PMN, Mono/Mac all B lymphs, mast cells, some T cells | 160–250 | C3b > iC3b |
| CR2 | B lymphs | 140 | C3dg > C3d > iC3b |
| CR3 | PMN, Mono/Mac, LGL | 165 $\alpha$ 95 $\beta$ | iC3b |

[a]RBC = red blood cells; PMN = polymorphonuclear cells; Mono = monocytes; Mac = macrophages; LGL = large granular lymphocytes.

receptors are the CR1 or C3b receptor and the CR3 or iC3b receptor. CR1 performs a dual role functionally—it serves not only as a receptor for attachment/adherence of complement-coated ligands but also as a cofactor for further proteolytic cleavage of C3b to iC3b. Because of its prominence as the only complement receptor on erythrocytes, which outnumber all other circulating cells by a factor of 1,000-fold, CR1 has been implicated in degradation and clearance of immune complexes from the bloodstream. Furthermore, erythrocyte levels of CR1 are genetically controlled, and appear to be lower in patients with the autoimmune disease, systemic lupus erythematosis.

CR3 is a member of a family of receptors, the LFA-1, p150, 95, Mo-1 (CD 18) family, which share a common $\beta$-chain of 95 kilodaltons (kD) and have variable $\alpha$-chain (CD 11) size. Several functions of these receptors have been clearly defined by the identification of patients lacking the entire family of molecules. Neutrophils from these patients display markedly deficient adherence and aggregation, and an abnormal oxidative burst and phagocytic capacity for iC3b-coated targets, although not for soluble stimuli.

There is a growing list of microorganisms whose attachment/entry into phagocytic cells is mediated by the CR3 receptor. Studies from the laboratory of J. Blackwell first implicated CR3 in uptake of *Leishmania donovani* promastigotes by murine peritoneal macrophages (Blackwell et al., 1985). Blackwell and co-workers have argued that attachment of promastigotes to cells in this serum free system is complement dependent and a consequence of activation of macrophage-derived complement proteins by the parasite surface. Several other microorganisms are known to attach and enter phagocytic cells through the CR3 receptor; the possible contribution of other members of the CD 11/CD 18 family (LFA-1, p150,95) in this process has also been suggested, since antibodies to the common $\beta$-chain or a combination of $\alpha$-chain monoclonals are more inhibitory than a single $\alpha$-chain monoclonal directed against CR3 (Bullock and Wright, 1987). Furthermore, the contribution of lectin-like binding activity within CR3 or other members of the CD 11/CD 18 family has been suggested and may contribute synergistically or separately to parasite attachment and entry.

Complement receptors mediate attachment of complement coated particles without necessarily triggering internalization or phagocytosis. However, a number of different second signals may trigger ingestion of particles attached to cells via complement receptors. Antibody is the most specific second signal. A lymphokine generated from mixed lymphocyte reaction supernatants can activate monocytes to ingest via complement receptors. Fibronectin, laminin, and perhaps other basement membrane proteins can also activate monocytes and macrophages to ingest via complement receptors (Pommier et al., 1983; reviewed in Brown, 1986). This activation process requires binding of the fibronectin to the phagocytic cell; nonetheless, the role for parasite attached fibronectin in the internalization process, as may occur for example with *T. cruzi* trypomastigotes (Ouaissi et al., 1986), is an area of recent interest.

Complement receptors for C3 fragments have been identified on a number of parasites/helminths. The suggestion of C3 receptors in schistosomes (Ouaissi et al. 1980), *Babesia* (Jack and Ward, 1980), and *T. cruzi* (Krettli and Pontes de Carvalho, 1985) is based on the functional capacity of parasites to bind C3b-coated particles. In contrast, in *Leishmania major* (Puentes et al., 1987), there is evidence that whole parasites directly bind fluid phase C3 or C3 fragments. For *Babesia* and *Leishmania*, the presence of parasite C3 receptors may be related conceptually to cell invasion if bound C3 fragments serve as a bridging molecule between parasite and host cell complement receptors.

## C5b-9 FORMATION/POLY C9

Formation of the lytic C5b-9 complex involves the sequential high-affinity association of C5b, C6, C7, C8, and C9. Recent interest in the terminal complement cascade has centered around the formation of poly C9. Podack and co-workers (Podack and Tschopp, 1982) first demonstrated that the classical "doughnut" lesion of complement visualized by electron microscopy on complement-lysed erythrocytes, and thought to represent the entire C5b-9 complex, could be reproduced in vitro by heat, heavy metal, or chaotrope-induced polymerization of purified C9 alone. It is now clearly recognized that anywhere from 1 to 16 molecules of C9 can bind to the C5b-8 complex, leading both to progressive enlargement of the channel created by the complex as well as to detergent and protease-resistance of the ring-polymerized molecule. This latter attribute of poly C9, protease resistance, may be important in cytolysis of protease-rich targets, and both attributes may be of critical importance for effective direct complement killing of microorganisms with complex cell walls, such as gram negative bacteria (Joiner et al., 1985b).

Resistance to direct complement-mediated killing may result because a completely formed C5b-9 complex fails to insert into or damage critical membrane structures. This has been most clearly demonstrated for enteric gram negative bacteria, in which polysaccharide side chains within bacterial lipopolysaccharide sterically block access of C5b-9 to hydrophobic domains of the bacterial outer membrane. As a consequence, the C5b-9 complex is released from the bacterial surface (Joiner et al., 1982, 1986b; Grossman et al., 1987). A similar process may be operative in *Leishmania major* (Puentes et al., 1988). In virulent, complement-resistant trophozoites of *Entameba histolytica*, a formed C5b-9 complex apparently binds stably to the parasite membrane, but is not lytic (Reed et al., 1986). The mechanism for serum resistance in this circumstance is still unclear, but conceivably could result because the trophozoites produce a molecule analogous to homologous restriction factor (HRF) (see below).

## CONTROL PROTEINS

There are a variety of control mechanisms exerted on the process of complement activation. Soluble control proteins of the complement cascade include factor H, C4 binding protein, and C1 esterase inhibitor, all of which have been discussed above and which can limit complement in the fluid phase or on surfaces. Two more recently described molecules are of particular interest: decay-accelerating factor (DAF) and homologous restriction factor (HRF). These molecules are unique because they limit complement activation and lysis on cells by homologous but not by heterologous complement components. Decay-accelerating factor (DAF) is a 70,000-kD glycoprotein present in nearly all mammalian cells in which it has been sought. DAF serves multiple functions, including limiting formation and accelerating the normal intrinsic decay of CP and ACP C3 convertases. The presence of DAF thus limits complement activation by limiting the extent of C3 cleavage and deposition (Medof et al., 1984). Although the exact mechanism of action of DAF is yet to be determined, it is presumed that the molecule binds to C3b and C4b (Kinoshita et al., 1986) and inhibits their interaction with factor B, C2 or other C3b/C4b binding molecules. Support for this hypothesis comes from the observation that the DAF sequence contains four repeats of the 60–63 amino acid partial homology sequence shared by all of the evolutionarily related C3b/C4b binding molecules (H, B, C2, C4 binding protein, CR1, DAF) (Caras et al., 1987; Reid et al., 1986).

The second and more recently described molecule mediating control of homologous complement activation is homologous restriction factor (HRF). This glycoprotein, which has also been most carefully studied in the human system, is an integral membrane protein which limits C9 polymerization, apparently by binding to both C8 and C9 from homologous but not heterologous species (Shin et al., 1986; Zalman et al., 1986). HRF thus limits the extent of C9 binding and membrane insertion. In those circumstances in which polymerized C9 is necessary for effective cytoloysis or in which the extent of C9 polymerization affects the efficiency of C5b-9 endocytosis or shedding, HRF would exert an important regulatory role. We and others have searched in preliminary fashion for molecules which cross-react antigenically with human HRF in serum-resistant bacteria and parasites, and no such molecules have been identified. It would not be surprising, however, if parasites such as *E. histolytica* expressed molecules with activity analogous to human HRF, as one mechanism of subverting effective attack by the terminal complement cascade

## COMPLEMENT RESISTANCE IN *T. CRUZI* AND *L. MAJOR*—EXAMPLES OF PREADAPTATION

In the second half of this paper, I will compare in detail two strategies of parasite preadaptation to complement resistance which we have been studying in our laboratory. Preadaptation is a term applied by Dr. David Sacks to indicate that during the process of differentiation within the insect vector from a noninfective to an infective form, the parasite undergoes changes which preadapt it to survive and establish infection within the vertebrate host. One important strategy of preadaptation is development of serum resistance in infective forms. This phenomenon operates broadly for trypanosomids since infective but not noninfective vector stages of *Trypanosoma cruzi* (Nogueira et al., 1975), *Leishmania* species (Franke et al. 1985), and *T. brucei* (Ferranti and Allison, 1983) are complement resistant. The term also has applicability when applied broadly to helminths such as schistosomes. An interesting reverse analog of this situation applies for malaria parasites, in which acquisition of the complement-susceptible state within the insect vector must be preceded by destruction of lytic complement activity within the blood meal if productive infection is to ensue (Grotendorst and

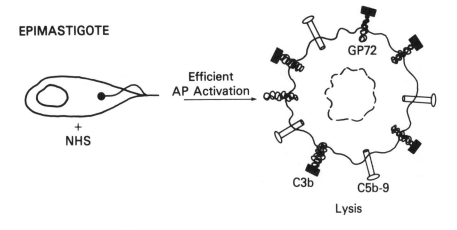

EPIMASTIGOTE

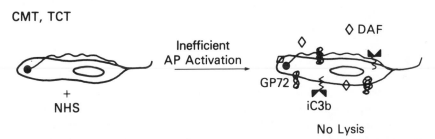

CMT, TCT

Fig. 4.  Model of the interaction of complement with epimastigotes, metacyclic trypomastigotes, and tissue culture trypomastigotes of *T. cruzi*: Epi are lysed by the alternative complement pathway in nonimmune serum, whereas both CMT and TCT resist lysis. On Epi, the major C3 fragment on the parasite surface is C3b, and this molecule is attached exclusively to the developmentally regulated glycoprotein, GP72. The presence of C3b leads to C5 convertase formation, C5b-9 generation, and lysis of the parasite. On CMT and TCT, 6–8-fold less C3 is deposited on the parasite surface in nonimmune serum, and the vast majority of bound C3 is present as the lytically inactive iC3b fragment, which cannot participate in C5 convertase formation or lead to C5b-9 generation. We do not currently know the molecules to which iC3b is attached in CMT and TCT. Inefficient C3 deposition on CMT and TCT results from the production of a molecule which has functional similarity to human decay-accelerating factor, as discussed in detail in the text.

# Potential Mechanisms of Parasite Resistance to Complement — Mediated Killing

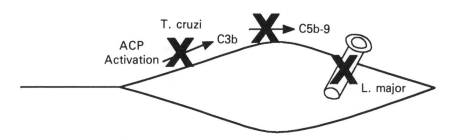

Fig. 5. Strategies for complement evasion by *T. cruzi* and *L. major*: There are three conceptual mechanisms for serum resistance in parasites: 1) the parasites may fail to activate the complement cascade, as is the case with *T. cruzi*; 2) the parasites may activate the cascade through the C3 step but fail to generate C5b-9, as has been suggested with *T. brucei*; and 3) the parasites may activate complement through to the end of the cascade, with formation of a complete C5b-9 complex, but the complex fails to damage vital membrane structures and cause parasite lysis, as with *L. major*.

Carter). We have been studying in detail the process and control of complement activation in *T. cruzi* and *L. major*, and the results of these studies will be presented in detail here.

### Trypanosoma Cruzi

It was recognized nearly 40 years ago that the epimastigote, the noninfective vector form of *Trypanosoma cruzi*, was susceptible to lysis by complement in fresh normal human serum (Muniz and Borrello, 1945). Studies by Nogueria and co-workers (1975) demonstrated, using a mixed population of epimastigotes and axenically derived metacyclic trypomastigotes, that this latter stage, which represents the infective form for the vertebrate host, resists complement-mediated lysis. Thus, the infective stage of the trypomastigote found within the insect vector is preadapted to eventual residence in a complement-containing environment—the vertebrate host. It is also of note that bloodstream trypomastigotes, and their in vitro analog, the tissue culture trypomastigote (TCT), are complement resistant, as would be expected of stages which must reside within the hostile environment of host serum.

Several years ago, Alan Sher and I initiated studies comparing complement activation on serum sensitive epimastigotes and culture-derived serum resistant metacyclic trypomastigotes (CMT) of the M88 clone of *T. cruzi*, kindly provided by Dr. Jim Dvorak. In initial studies (Joiner et al., 1985a) we found that 5–6-fold more C3 was deposited on epimastigotes than on CMT during incubation in either normal human serum, in which epimastigotes (Epi) lysed, or in C8-deficient serum, in which no parasite lysis could occur. Of note was the observation that a single 72-kD glycoprotein (GP72) acted as the C3 acceptor in epimastigotes (Fig. 4). This molecule was of interest because of the previous demonstration by Sher and Snary (1982) that GP72 was an important molecule in developmental regulation: monoclonals to GP72 blocked transformation from epimastigotes to CMT in an in vitro/in vivo transformation system (Sher et al., 1983). We presume that the predominance of GP72 as an acceptor molecule reflects a high content of carbohydrate and a prominent surface exposure.

These initial studies clearly suggested that serum resistance in CMT was a consequence of poor activation of complement on the parasite surface, rather than a block later in the cascade. We thus sought to identify the mechanisms of control of complement activation on the parasite surface. Analysis of the bound C3 fragments indicated that the lytically active C3b fragment predominated on Epi, whereas the lytically inactive iC3b fragment predominated on CMT (Joiner et al., 1986b). In addition, the inefficient C3 convertase formation and C3 deposition on CMT was a consequence of poor binding of factor B to C3b on the parasite surface, and not a consequence of augmented factor H binding. This represents an unusual mechanism of ACP C3 convertase control, since in most systems in which ACP activators and nonactivators are compared, nonactivation is a consequence of augmented H binding rather than diminished B binding. One possibility to explain these results was the presence of a developmentally regulated molecule which was differentially expressed in CMT but not Epi and which influenced B binding to C3b.

Pronase or endoglycosidase treatment of CMT rendered this form susceptible to serum lysis, and incubation of pronase-treated parasites in media containing tunicamycin to inhibit N-linked glycosylation blocked regeneration of serum resistance (Sher et al., 1986). When the interaction of factor B was tested on pronase-treated CMT bearing C3b, the affinity and extent of the interaction resembled that on Epi. Taken as a whole, these results suggested that a surface glycoprotein present on CMT but not Epi was restricting the efficiency of factor B binding to C3b on CMT, either by altering the site of C3b deposition or by influencing the interaction between B and C3b.

Subsequent experiments showed that supernatants from CMT and TCT but not Epi inhibited formation and accelerated decay of an alternative pathway C3 convertase on zymosan (Rimoldi et al., 1987b). Further analysis indicated that these same supernatants accelerated classical pathway C3 convertase decay but did not have factor H-like co-factor activity for factor I-mediated cleavage of C3b to iC3b. Thus in all respects, the supernatants from CMT and TCT contained functional activity more analogous to human decay-accelerating factor (DAF), than to other human complement regulatory molecules such as factor H, C4 binding protein, and CR1, which have both C3b/C4b binding capability and factor I cofactor activity.

We therefore sought to identify molecules with decay-accelerating activity in CMT and TCT supernatants. Using [$^{35}$S]-methionine-labeled parasites, we found that a limited and selective population of bands between 86 and 150 kD were shed by CMT and TCT. Labeled molecules shed by Epi were generally of lower molecular weight than those shed by trypomastigotes, varying between 20 and 86 kD. Separation of CMT and TCT supernatants by chromatofocusing using fast performance liquid chromatography demonstrated that a single $^{35}$S band of 87–93 kD, pI 5.6–5.7, co-eluted with decay-accelerating activity (Joiner et al., 1987). This band, which was recognized by lytic Chagasic sera, was totally removed on Con A-Sepharose, and was susceptible to proteolysis with papain, both of which removed functional activity. We therefore presume that the 87–93 kD $^{35}$S band contains functional decay-accelerating activity. Our current goal is to prepare monoclonal and polyclonal antibodies to the molecule to facilitate molecular cloning.

One of the most interesting issues to explore is the relationship between the C3 convertase inhibitor produced by *T. cruzi* and human complement regulatory molecules with related functions, such as DAF, factor H, C4 binding protein, and CR1. Regardless of the ultimate structural relationship between the *T. cruzi* C3 convertase inhibitor and human regulatory complement molecules, our results raise the intriguing possibility that trypomastigotes of *T. cruzi* resist direct complement-mediated lysis by mimicking a host strategy involved in restriction of complement-mediated lysis by homologous complement proteins. The extent to which other parasites may employ similar strategies to evade complement-mediated lysis has not been investigated, but it is clearly an area of interest for further study.

### Leishmania major

The second model for parasite preadaptation to a complement resistant state currently under study in our laboratory is with promastigotes of *Leishmania major*. Working closely with Dr. David Sacks, we have compared the interaction of the ACP with log phase, noninfective promastigotes and with infective parasites harvested from stationary-phase cultures (Sacks and da

Silva, 1987). As described in much greater detail in another chapter (see Sacks, this volume) log phase parasites which are agglutinated by peanut agglutinin (PNA+) are complement susceptible, whereas a subset of stationary-phase promastigates which are not agglutinated by peanut agglutinin (PNA−) are relatively complement resistant. The PNA− population of infective *L. major* promastigotes, which have been termed metacyclic promastigotes, are preadapted to a complement-resistant form in a fashion highly analogous to that just described for *T. cruzi* metacyclic trypomastigotes. David Sacks and Rosangela da Silva (1987) have demonstrated that the change in parasite agglutination by PNA reflects a developmental modification of an excreted lipophosphoglycan. They have also shown that the developmentally regulated expression of a new epitope on the *L. major* glycolipid is both a specific marker for metacyclogenesis and a necessary marker for parasite infectivity. Handman and co-workers have provided evidence that the *L. major* excreted factor is a protective immunogen in mice, and is necessary for infection (Handman and Mitchell, 1985; Handman et al., 1986).

We were therefore interested in analyzing in detail the relationship of lipophosphoglycan modification to complement deposition and to the serum-resistant phenotype. Dr. Steven Puentes, working in my laboratory, has compared the interaction of complement with log and PNA-metacyclic promastigotes (Puentes et al., 1988). In initial studies, he demonstrated that binding of C3 to log and metacyclic promastigotes during incubation in serum followed different kinetics but that the total extent of C3 deposition did not differ substantially between the two forms of the parasite. Surprisingly, C3 deposition on PNA− was mediated totally through the classical pathway, whereas C3 binding on log parasites occurred via alternative pathway activation. Of note was the observation that bound C3 was released from the log form during incubation in buffer, apparently due to proteolytic cleavage by a parasite-derived protease. The majority of C3 on both forms was the hemolytically active C3b fragment, which was ester linked to the lipophosphoglycan, raising the possibility that the cell attachment function of the surface glycolipid is mediated through C3. Finally, direct binding experiments indicated that whereas nearly $4 \times 10^5$ molecules of C9 bound to log phase promastigotes, minimal binding of C9 occurred to metacyclic forms, despite complete consumption of C9, and the small amount of initially bound C9 was released with continued incubation (Fig. 5). This strongly implies, and experiments are under way to examine this possibility, that C5b-9 is formed on the parasite surface but is rapidly released into the fluid phase as an SC5b-9 complex, exactly as we had reported earlier with serum resistant strains of *Salmonella* and *Escherichia coli* (Joiner et al. 1982, 1986b).

Regardless of the precise mechanism for evading lysis by the C5b-9 complex, metacyclic promastigotes of L. *major* have evolved a mechanism permitting deposition of large amounts of the hemolytically active C3b fragment. This fragment appears to be of fundamental importance for L. *major* attachment and entry into macrophages, which is the obligatory host cell for the *Leishmania* parasite. As described by Dr. David Sacks (see Sacks, this volume) in work done by Dr. R. da Silva and Dr. Lee Hall, monoclonal antibodies to CR1, the C3b receptor, but not to CR3, the iC3b receptor, effectively block PNA− attachment to and entry into both human monocytes and macrophages as well as the murine macrophage J774 cell line. This contrasts with the results reported by Blackwell and co-workers (1985) for L. *donovani* and by Mosser and Edelson for L. *major* (1984) in which anti-CR3 monoclonals effectively inhibited parasite attachment. Although we do not have a ready explanation for the difference in results with L. *major*, our findings with L. *donovani* infection of macrophages parallel those of Blackwell (1985). Furthermore, in studies not reported here, we have determined that iC3b is the main C3 fragment on L. *donovani* promastigotes (Puentes et al., manuscript in preparation).

### Comparison of Complement Resistance in *T. cruzi* and *L. major*

As outlined in the introductory statements in this chapter, parasites have evolved strategies for circumventing the harmful effects of the complement cascade. Clearly, preadaptation to a complement resistant form within the insect vector is a common theme. Two mechanisms for the developmental regulation of complement resistance have been described in detail here: a) limited initiation of the complement cascade in *T. cruzi* trypomastigotes due to production of molecule with functional activity analogous to decay accelerating factor, and b) efficient complement activation by metacyclic promastigotes of L. *major* leading to deposition of large amounts of C3b, but presumably resulting in generation of nonlytic C5b-9. This has led us to speculate that the mechanism of complement resistance for these two parasites is dictated, at least in part, by the host cell range of intracellular forms. *T. cruzi* trypomastigotes, which invade and survive within a wide variety of nonphagocytic cells, do not require surface C3 fragments to establish infection. L. *major* promastigotes, which reside in obligatory fashion within phagocytic cells bearing complement receptors, require surface-bound complement fragments for optimal cell attachment and invasion. Thus it is advantageous for *Leishmania* promastigotes to evade complement lysis at a step distal to C3 deposition, since this strategy facilitates cell attachment and invasion via complement receptors. Whether this theme can be more generally applied to distinguish between parasites which have an obligatory intracellular location within phagocytic cells or nonphagocytic cells remains to be determined.

## CONCLUDING REMARKS

I have attempted in this chapter to outline general concepts related to interaction of parasites with the complement cascade. The discussion of relevant aspects of the complement system has focused around those biologically important systems or molecules which are most clearly related to microbial evasion or invasion strategies. The extent to which parasites mimic host strategies for evading complement damage or employ the complement cascade to establish successful infection will be productive areas for future research.

## REFERENCES

Blackwell JM, Ezekowitz RAB, Roberts MB, Channon JY, Sim RB, Gordon S (1985): Macrophage complement and lectin-like receptors bind *Leishmania* in the absence of serum. J Exp Med 162:324–331.

Bohnsack, JF, Tenner AJ, Laurie G, Kleinman HK, Martin GR, Brown EI (1985): C1q binds to laminin. A mechanism for the deposition and retention of immune complexes in basement membrane. PNAS 82:3824–3828.

Brown EJ (1986): The role of extracellular matrix proteins in the control of phagocytosis. J Leukocyte Biol 39(5):579–591.

Bullock WE, Wright SD (1987): Role of the adherence-promoting receptors, CR3, LFA-1, and p150,95, in binding of *Histoplasma capsulatum* by human macrophages. J Exp Med 165:195–210.

Caras IW, Davitz MA, Rhee L, Weddell G, Martin DW Jr, Nussenzweig V (1987): Cloning of decay-accelerating factor suggests novel use of splicing to generate two proteins. Nature 325:545–549.

Cooper NR (1985): The classical complement pathway: "Activation and regulation of the first complement component. In Advances in Immunology, F. Dixon ed. New York: Academic Press, Inc., pp 151–216.

Fearon DT, Wong WW (1983): Complement ligand-receptor interactions that mediate biological responses. Annu Rev Immunol 1:243–271.

Ferrante A, Allison AC (1983): Alternative pathway activation of complement by African trypanosomes lacking a glycoprotein coat. Parasite Immunol 5:491–498.

Franke ED, McGreevy PB, Katz P, Sacks DL (1985): Growth cycle-dependent generation of complement-resistant *Leishmania* promastigotes. J Immunol 134:2713–2717.

Fries LF, Frank MM (1987): Molecular mechanisms of complement action. In Stamatoyannopoulos, Nienhuis, Leder, Majerus, (eds): The Molecular Basis of Blood Diseases. Philadelphia: W.B. Saunders Co., pp 450–498.

Fuhrman SA, Joiner KA (1987): Interaction of complement with tachyzoites of toxoplasma gondii. Complement 4:156.

Grossman N, Schmetz MA, Foulds J, Klima EN, Jiminez V, Leive LL, and Joiner KA (1987): Lipopolysaccharide size and distribution determine serum resistance in *Salmonella montevideo*. J Bacteriol 169(2):856–863.

Grotendorst CA, Carter R (1987): Complement effects on the infectivity of *Plasmodium gallinaceum* to *Aedes aegypti mosquitoes*. II. Changes in sensitivity to complement-like factors during zygote development. J Parasitol. 73:980–984.

Handman E, Mitchell GF (1985): Immunization with Leishmania receptor for macrophages protects mice against cutaneous leishmaniasis. Proc Natl Acad Sci USA 82(17):5910–5914.

Handman E, Schnur LF, Spithill TW, Mitchell GF (1986): Passive transfer of Leishmania lipopolysaccharide confers parasite survival in macrophages. J Immunol 137:3608–3613.

Jack RM, Ward PA (1980): The role in vivo of C3 and the C3b receptor in babesial infection in the rat. J Immunol 124:1574–1578.

Joiner KA, Hammer CH, Brown EJ, Cole RJ, Frank MM (1982): Studies on the mechanism of bacterial resistance to complement-mediated killing. I. Terminal complement components are deposited and released from *Salmonella minnesota* S218 without causing bacterial death. J Exp Med 155:797–804.

Joiner K, Hieny S, Kirchhoff LV, Sher A (1985a): gp72, the 72 kilodalton glycoprotein, is the membrane acceptor site for C3 on *Trypanosoma cruzi*. J Exp Med 161:1196–1212.

Joiner KA, Schmetz MA, Sanders ME, Murray TG, Hammer CH, Dourmashkin R, Frank MM (1985b): Multimeric complement component C9 is necessary for killing of *Escherichia coli* J5 by terminal attack complex C5b-9. Proc Natl Acad Sci USA 82:4808–4812.

Joiner KA, Grossman N, Schmetz M, Leive L (1986a): C3 binds preferentially to long-chain lipopolysaccharide during alternative pathway activation by *Salmonella montevideo*. J Immunol 136(2):710.

Joiner K, Sher A, Gaither T, Hammer C (1986b): Evasion of alternative complement pathway by *Trypanosoma cruzi* results from inefficient binding of factor B. Proc Natl Acad Sci USA 83:6593–6597.

Joiner KA, Dias da Silva W, Rimoldi MT, Hammer CH, Sher A, Kipnis T, (1988): Identification and characterization of a molecule produced by trypomastigotes of *Trypanosoma cruzi* which accelerates the decay of complement C3 convertases.

Kinoshita T, Medof ME, Nussenzweig V (1986): Endogenous association of decay-accelerating factor (DAF) with C4b and C3b on cell membranes. J Immunol 136:3390–3395.

Krettli AU, Pontes de Carvalho LC (1985): Binding of C3 fragments to the *Trypanosoma cruzi* surface in the absence of specific antibodies and without activation of the complement cascade. Clin Exp Immunol 62:270–277.

Medof ME, Kinoshita T, Nussenzweig V (1984): Inhibition of complement activation on the surface of cells after incorporation of decay-accelerating factor (DAF) into their membranes. J Exp Med 160:1558–1578.

Mosser DM, Edelson PJ (1984): Activation of the alternative complement pathway by *Leishmania* promastigotes: Parasite lysis and attachment to macrophages. J Immunol 132:1501–1505.

Müller-Eberhard HJ, Schreiber RD (1980): Molecular Biology and Chemistry of the alternative pathway of complement. Adv Immunol 29:1–53.

Muniz J, Borriello A (1945): Estudo sobre a acão litica de diferentes soros sobre as formas de cultura e sanguicolas do *S. cruzi*. Rev Bras Biol 5:563–569.

Nogueira N, Bianco C, Cohn Z (1975): Studies on the selective lysis and purification of *Trypanosoma cruzi*. J Exp Med 142:224–229.

Ouaissi MA, Santoro F, Capron A (1980): Interaction between *Schistosoma mansoni* and the complement system receptors for C3b on Cercariae and Schistosomula. Immunol Lett 1:197–210.

Ouaissi MA, Cornette J, Capron, A (1986): Identification and isolation of *Trypanosoma cruzi* trypomastigote cell surface protein with properties expected of a fibronectin receptor. Mol Biochem Parasitol 19:201–211, 1986.

Pearson RD, Steigbigel RT (1980): Mechanism of lethal effect of human serum upon *Leishmania donovani*. J Immunol 125:2195–2201.

Podack ER, Tschopp J (1982): Polymerization of the ninth component of complement. Formation of poly(C9) with a tubular ultrastructure resembling the membrane attack complex of complement. Proc Natl Acad Sci USA 79:574–578.

Pommier C, Inada S, Fries LF, Takahashi T, Frank MM, Brown EJ (1983): Plasma fibronectin enhances phagocytosis of opsonized particles by human peripheral blood monocytes. J Exp Med 157:1844–1854.

Puentes SM, Sacks D, da Silva R, Joiner KA (1988): Binding of complement by two developmental stages of *Leishmania major* varying in expression of a cell surface glycolipid. J Exp Med 167:887–902.

Reed SL, Curd JG, Gigli I, Gillin FD, Braude AI (1986): Activation of complement by pathogenic and nonpathogenic *Entamoeba histolytica*. J Immunol 136:2265–2270.

Reid KBM, Bentley DR, Campbell RD, Chung LP, Sim RB, Kristensen T, Tack BF (1986): Complement system proteins which interact with C3b or C4b: A superfamily of structurally related proteins. Immunol Today 7:230–234.

Rimoldi MT, Tenner A, Bobak D, Joiner KA (1987a): Role of C1 in *Trypanosoma cruzi*—mononuclear cell interaction. Submitted for publication.

Rimoldi MT, Sher A, Heiny S, Lituchy A, Hammer CH, Joiner K (1988): Developmentally regulated expression by *Trypanosoma cruzi* of molecules which accelerate the decay of complement C3 convertases. Proc. Natl. Acad Sci USA. 85:193–197.

Ross GD, Medof ME (1985): Membrane complement receptors specific for bound fragments of C3. In Advances in Immunology, F. Dixon ed. New York: Academic Press, Inc., 37:217–267.

Russell DG (1987): The macrophage-attachment glycoprotein gp63 is the predominant C3-acceptor site on *Leishmania mexicana* promastigotes. Eur J Biochem 164:213–221.

Sacks DL, da Silva R (1987): Developmentally regulated expression of a surface glycolipid on *L. major* promastigotes. J Immunol 139:3099–3106.

Santoro F (1982): Interaction of complement with parasite surfaces. Clin Immunol Allergy 2(3):639–653.

Sher A, Snary D (1982): Specific inhibition of the morphogenesis of *Trypanosoma cruzi* by a monoclonal antibody. Nature 300:639–641.

Sher A, Crane M St J, Kirchhoff LV (1983): Incubation in mice provides a signal for the differentiation of *Trypanosoma cruzi* epimastigotes to trypomastigotes. J Protozool 30:278–288.

Sher A, Hieny S, Joiner K (1986): Evasion of the alternative complement pathway by metacyclic trypomastigotes of *Trypanosoma cruzi*: Dependence on the developmentally regulated synthesis of surface protein and N-linked carbohydrate. J Immunol 137:2961–2967.

Shin ML, Hänsch G, Hu VW, Nicholson-Weller A (1986): Membrane factors responsible for homologous species restriction of complement-mediated lysis: Evidence for a factor other than DAF operating at the stage of C8 and C9. J Immunol 136:1777–1782.

Zalman LS, Wood LM, Müller-Eberhard HJ (1986): Isolation of a human erythrocyte membrane protein capable of inhibiting expression of homologous complement transmembrane channels. Proc Natl Acad Sci USA 83:6975–6979.

The Biology of Parasitism, pages 329–346
© 1988 Alan R. Liss, Inc.

# Genetic Control of Immunity to Parasite Infection: Studies of *Trichinella*-Infected Mice

## Donald L. Wassom

*Department of Pathobiological Sciences, School of Veterinary Medicine,
University of Wisconsin, Madison, Wisconsin 53706*

## INTRODUCTION

The immune response to parasites may be very complex. Parasites express many different antigens, and this array of antigens may change continually as the parasite develops through successive stages in its life history. The host's immune system may recognize and respond to each of these different antigens, even those giving rise to responses which have no effect on the parasite's ability to live, develop, or reproduce. In fact, parasites are often bathed in antibodies and/or cells reactive with their antigens yet suffer no ill effects from the experience. Consequently, it can be difficult to distinguish important immune responses from those which are of little functional significance. An approach to unraveling the complexity of these antiparasite responses is to identify and characterize the functions of specific genes which influence relevant immunological events. If such genes can be identified, comparative studies among hosts with allelic variants of these genes may provide insight into how functional antiparasite responses are controlled and effected. This chapter on genetic control of immunity to parasite infections is not a survey of the literature. Rather, examples are given from studies of *Trichinella spiralis*-infected mice to illustrate the factors which must be considered when conducting studies to identify host genes which influence the outcome of host-parasite interactions.

It is known from studies of many host-parasite systems that individuals may vary in their ability to resist infection. This is true for outbred populations of wild or domestic animals as well as for humans. It is known that genetic factors may influence these individual differences (Wakelin, 1978), and classical genetic analysis of outbred populations has demonstrated, in at least one case, that a single gene may have a dominant influence on the outcome of such host-parasite interactions (Wassom et al., 1974). However, in most cases multiple genes influence the host's response to the parasite, and it is this collection of genes which determines the host's ability to resist infection. Of course, when multiple genes segregate independently in a genetic cross, a range of phenotypes will be produced, and formal genetic analysis becomes difficult. For this reason inbred strains of hosts, particularly mice, have been used to study the genetics of host resistance to parasitic infections. A carefully selected panel of inbred mouse strains may approximate the range of genetic differences expressed in the outbred population, and yet allows for genetically identical individuals to be tested in different ways. In addition, a great deal is known about the mouse genome, and breeding schemes have been devised for the production of congenic mouse strains which are genetically identical, or nearly so, except for the alleles expressed at one or more selected gene loci. When congenic mouse strains are compared, it is often possible to evaluate the influence of a single gene on the response of interest.

Genetic studies of host-parasite interactions require a clear understanding of what is meant by the terms "susceptible" and "resistant." This distinction is not so simple as it first appears. In studies of inbred mice, for example, it may be difficult to identify a strain that is absolutely refractile to infection. Absolutely permissive strains of mice may be equally rare, particularly if the absence of a host response allows for uncontrolled growth of the parasite. Consequently, experiments to assess genetically controlled differences in the host's response to infection must rely on assays which measure relative, not absolute, differences in levels of susceptibility. For this reason, care must be taken to select assays which distinguish "susceptible" from "resistant" hosts on the basis of biologically meaningful parameters. As discussed below, the selection of relevant assays requires extensive knowledge of the parasite's life history.

When selecting criteria to differentiate susceptible from resistant hosts, it is helpful to distinguish between assays that measure resistance to infection and those which measure resistance to disease! In the former case, assays will focus on measurable antiparasite effects, while in the latter case, assays will focus the condition of the host. It must be considered that a strong immune response need not be correlated with resistance. In fact, the opposite may be true if the host response results in immune-mediated damage to host tissue. In addition, a strong immune response against irrelevant parasite antigens may also be misleading if it is not realized that this response does not mediate measurable antiparasite effects.

Assays to measure levels of host resistance to parasite infection should distinguish between innate resistance and immunity to infection. Innate resistance is not specific to the invading organism and no memory is induced. For example, in *Leishmania donovani* infections, parasites invade and multiply rapidly in macrophages from some mouse strains and slowly in macrophages from others (Bradley, 1980; Blackwell, 1985). The ability of inbred mice to limit reproduction of *L. donovani* within macrophages is not specific to the invading organism as strains which best limit the reproduction of *L. donovani* also limit the reproduction of other intracellular pathogens such as *Salmonella typhimurium* or *Mycobacterium bovis* (Skamene et al., 1982). Thus, the innate response is triggered by infection but is not specific to a single pathogen. In addition, later challenge of a previously infected host does not result in enhanced responsiveness. Nonetheless, these innate responses are genetically controlled; the ability of mice to resist the early stages of infection with *L. donovani*, *S. typhimurium*, or *M. bovis* is regulated by what appears to be a single gene on chromosome 1 of the mouse (Skamene et al., 1982). Innate resistance may provide an important first line of defense against parasite invasion. However, the host's ability to mount an effective immune response once the parasite(s) has established is a critical factor in determining the outcome of the host-parasite interaction.

In contrast to innate resistance, immune responses are characterized by their specificity and by the induction of immunological memory. Immune responses against most parasites, at least those with demonstrable antiparasite effects, are dependent on the ability of host T cells to recognize parasite antigen, and to provide appropriate activation signals ("help") to B cells, macrophages, or other T cells which effect the antiparasite response. Congenitally athymic nude mice, for example, are T-cell deficient and extremely susceptible to most parasitic infections. Yet nude mice possess populations of cells which can effect antiparasite responses when exogenous T cells are supplied.

Unlike B cells, which recognize antigen via cell bound immunoglobulin, T cells are not activated by soluble parasite antigens or by antigen on the parasite's surface. Rather, the receptors on T cells recognize antigen only when it is associated with an MHC-encoded molecule on the surface of antigen-presenting cells. In the mouse, there are two kinds of MHC molecules involved in antigen presentation. The class I molecules are the products of the K, D, and L genes expressed on somatic cell surfaces in association with $\beta_2$-microglobulin. The class II molecules are the products of the $A_\beta A_\alpha$ (I-A) and $E_\beta E_\alpha$ (I-E) genes and are expressed as heterodimers on the surface of antigen-presenting cells such as macrophages, dendritic cells, and B cells.

It is now recognized that the $A_\beta$, $A_\alpha$, $E_\beta$, and $E_\alpha$ genes are the "immune response genes" and that the expressed I-A and I-E molecules are the "Ia antigens." The ability of host cells to process and properly present antigen to T cells is a critical step in the generation of functional antiparasite responses.

A large number of H-2 congenic mouse strains are available for studies of how various H-2 genes regulate immune responses. Since these mouse strains share common genetic backgrounds and differ only in the expression of specific H-2 alleles, a difference in the ability of different H-2 congenic mouse strains to resist parasite infection must indicate that H-2 genes influence the response observed. Conversely, when mice which share common H-2 genes differ in their response to infection, non-H-2 genes are implicated. Many different inbred and H-2 congenic strains of mice have been used to study the role of H-2 and non-H-2 genes in the immune response to parasitic infections. Genetic studies of *Trichinella*-infected mice are detailed below to provide an example of how such genetic analyses may be conducted.

## THE BIOLOGY OF *T. SPIRALIS* INFECTIONS

All strains of mice are susceptible to a primary infection with *T. spiralis*. The infection is initiated when a suitable host ingests $L_1$ larvae in the muscle of a previously infected host. In the laboratory, an acid-pepsin solution is used to digest worms from the muscle, and a measured number of the $L_1$ larvae obtained is fed to experimental mice through a curved blunt needle inserted into the esophagus. The $L_1$ worms, after passing through the stomach, are activated by enzymes in the small intestine and enter columnar epithelial cells where they mature to adult worms.

The development of *T. spiralis* within the small intestine is a remarkable process. The worms molt four times within the first 30 h; mate shortly thereafter; and beginning day 5 or 6 postinfection, female worms begin to produce living young called newborn larvae. The newborn larvae enter blood and lymphatic vessels and are distributed throughout the tissues by the general circulation. When the newborn larvae reach striated muscle, they leave the circulation and penetrate individual muscle cells, where they develop over a period of several weeks to the infective stage.

The life history of *T. spiralis* is unique in several respects. First, it has an extremely broad host range; virtually all species of mammals can be infected. This broad host range is unusual for a helminth parasite. Second, the worms live most of their lives within host cells. The adult worms live as intramulticellular parasites inside columnar epithelial cells of the small intestine, and the $L_1$ larvae live within striated muscle cells. Only the migrating newborn larvae occupy an extracellular niche, and these larvae are in the circulation for a very short time. It is unusual for a helminth parasite to be found within host cells. Third, all stages in the parasite's life history are found in a single host individual; there are no intermediate hosts and no stages to be found in the environment. This is unusual for a eukaryotic parasite of any kind.

## THE IMMUNE RESPONSE TO *T. SPIRALIS*

The immature and adult worms of *T. spiralis* in the small intestine and the migrating newborn larvae induce a series of strong immune responses in the host. Each stage in the *Trichinella* life history expresses unique antigens (Bell et al., 1979; Philipp et al., 1980, 1981), and the response to these antigens is specific, sensitive, and leads to the induction of memory. For example, separate immune responses expel adult worms from the gut, limit the ability of female worms to reproduce, and limit the ability of newborn larvae to migrate to the muscle (Wassom et al., 1984a). When previously infected hosts are challenged at a later time, each of these responses occurs at an accelerated rate and with greater strength. Resistance to *Trichinella* infection can be transferred to naive mice using T cells from previously infected syngeneic hosts (Wakelin and Donachie, 1980), and antibodies reactive with *Trichinella* antigens are present in the serum of infected mice (Jungery and Ogilvie, 1982). As would be expected for a specific immune response, T cells from infected mice will proliferate in vitro when cultured with *Trichinella* antigens (Krco et al., 1982). The latter response is so specific that antigens from different *Trichinella* isolates may induce responses of different magnitude, and antigens from other parasites fail to stimulate the response at all (Wassom et al., 1987a).

Although all strains of mice mount a functional immune response which expels *T. spiralis* worms from the gut, some mouse strains expel worms more quickly than others. Mouse strains also differ in their ability to limit the fecundity of female worms. Naturally, strains of mice which expel worms quickly and effectively limit the reproduction of female worms will have fewer larvae in their muscle following the intestinal phase of the infection. Table 1 shows results of a study where 28 different inbred mouse strains were compared for their ability to resist *T. spiralis* infection. Mice were infected with 150 $L^1$ larvae of *T. spiralis*, and 30 days later the mice were killed and the numbers of larvae in the muscle were determined. To facilitate ease of comparison, results are expressed as a percentage of the count for C3HeB/FeJ mice which harbored 31,740 $\pm$ 2,141 worms (Mean $\pm$ SEM n=116). Several points can be drawn from these data. First, a wide range of

**TABLE 1. Strains of Mice Ranked in Order of Decreasing Susceptibility to *T. spiralis*[a]**

| Strain | H-2 haplotype | % C3H control ± SEM[b] |
|---|---|---|
| B10.BR | k | 104 ± 12.71 |
| BRVR | k | 102 ± 8.45 |
| C3HeB/FeJ | k | 100 |
| CBA/J | k | 93 ± 11.23 |
| C58/J | k | 85 ± 8.39 |
| C57L/J | b | 81 ± 5.05 |
| DBA/2J | d | 74 ± 5.44 |
| C3H.SW | b | 73 ± 6.46 |
| B10/Sn | b | 67 ± 7.49 |
| B6-K1 | b | 65 ± 3.75 |
| B10.S | s | 64 ± 6.31 |
| SJL | s | 64 ± 2.77 |
| B6 | b | 63 ± 3.95 |
| AKR/J | k | 62 ± 2.32 |
| SEA/GnJ | d | 62 ± 4.11 |
| C57BL/KsJ | d | 62 ± 5.44 |
| LG/J | d | 60 ± 3.33 |
| Balb/cByJ | d | 57 ± 5.23 |
| SEC/ReJ | d | 57 ± 3.26 |
| A.By | b | 51 ± 3.94 |
| AU/J | q | 50 ± 3.95 |
| 129/J | b | 48 ± 4.27 |
| B10.Q | q | 47 ± 5.42 |
| LP/J | b | 45 ± 2.86 |
| SWR/J | q | 45 ± 4.10 |
| A.SW | s | 41 ± 3.28 |
| DBA/1J | q | 39 ± 4.14 |
| BUB/BnJ | q | 31 ± 1.41 |

[a]From Wassom et al. (1983a); reproduced with permission of the publisher.
[b]Worm count for C3HeB/FeJ was 31,740 ± 2,141, see text.

host phenotypes exist within the sample tested. Second, mice which share common sets of H-2 genes, as designated by their H-2 haplotypes, may differ markedly in their worm count values. Thus, genes mapping outside the H-2 complex must influence the host's ability to resist infection. Third, there is a preponderance of mouse strains expressing the *H-2[k]* haplotype at the top of the list, while mice expressing *H-2[q]* alleles are clumped at the bottom. This suggests that H-2 genes might also influence levels of susceptibility to infection. A role for H-2 genes is confirmed by comparing counts from the H-2 congenic strains of mice also included in this experiment. Since B10.BR mice (*H-2[k]*) are significantly more susceptible than B10 (*H-2[b]*), B10.S(*H-2[s]*), or B10.Q (*H-2[q]*) mice, it can be concluded that H-2 genes also influence the host response to *Trichinella* infection. In the experiment described above, counts of larvae in the muscle were used as a criterion to assess levels of susceptibility to infection. However, other criteria may also be used. Factors to consider when selecting appropriate criteria are as follows:

## Counts of Larvae in the Muscle

This assay provides a comprehensive assessment of the host's ability to resist infection. The number of larvae in the muscle after the intestinal worms have been expelled represents the cumulative effects of all antiparasite responses expressed by the host. It must be considered, however, that strains of mice may have equal numbers of worms in the muscle, yet respond quite differently to the intestinal worms. For example, one strain may expel worms quickly but fail to limit the fecundity of female worms. Another strain may limit fecundity, yet allow worms to persist in the gut for a longer time. The net result, as measured counts of larvae in the muscle, could be similar but for different reasons. Nonetheless, if a single assay must be used, total body larval counts is the assay of choice; any other single assay could be very misleading, as discussed below.

## Counts of Worms in the Gut to Determine Rates of Expulsion

To determine rates of worm expulsion, assays must be performed at frequent intervals throughout the period when worms are present in the gut. A count very early in the infection determines how many worms have established from the original inoculum, and counts on subsequent days provide an estimate of how quickly the worms are expelled. The assay is a good index of host resistance as mice which expel worms quickly will usually have fewer worms in the muscle than mice which expel worms slowly. The rate of worm expulsion, however, can be markedly influenced by the number of worms present in the gut; expulsion is markedly delayed in heavy infections. It has been shown that expulsion rates may differ among strains of mice at one infective dose but not at another (Wassom et al., 1984b). An additional factor which must be considered is that mice which expel worms quickly may still accumulate high muscle burdens if worm fecundity is high.

## Measurements of Worm Fecundity

This assay is performed by harvesting female worms from the small intestine and culturing them in vitro in a highly enriched medium. The worms produce newborn larvae in culture and these larvae are counted 24 h later. Levels of fecundity are expressed as newborn larvae per female worm per day. As described for the expulsion assay, the fecundity assay must be performed every few days throughout the intestinal phase of the infection. Ideally, the fecundity and the expulsion assays are performed together using the same population of worms. If the number of female worms is known, and their levels of fecundity have been determined, it is possible to estimate the total numbers of newborn larvae produced each day throughout the course of infection. In general, female worms become less fecund as the intestinal phase of the infection progresses. This effect is mediated by immune factors and is most pronounced in resistant strains of mice.

Other assays may be useful in assessing levels of host resistance to *T. spiralis* infection, providing that they correlate with the biologically meaningful assays discussed above. For example, mesenteric lymph node cells harvested from *Trichinella*-infected mice will proliferate in vitro when cultured with *Trichinella* antigen. It has been shown that the strength and the kinetics of this proliferative response parallels the kinetics of the expulsion and antifecundity responses (Dougherty et al., 1985). An experiment demonstrating this relationship is shown in Figure 1. B10.BR mice were slow to expel worms when compared to B10.Q mice, and worms harvested from B10.BR mice produced more newborn larvae when cultured in vitro. The net result

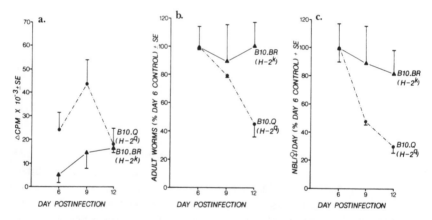

Fig. 1. Anti-*Trichinella* responses of mice 6, 9, and 12 days following infection with 150–200 L₁ of *T. spiralis*: B10.BR (▲ - ▲), B10.Q (● - - ●). **a:** Antigen-specific proliferation of mesenteric lymph node cells when cultured in vitro with 200 μg/ml *T. spiralis* antigens. Δ CPM = CPM in antigen-stimulated culture minus CPM in medium control. **b:** Adult worms in the small intestine. Results are expressed as a percentage of the worm count for day 6 controls. **c:** Number of newborn larvae produced per day per female worm expressed as a percentage of day 6 controls. Total body larval counts for B10.BR and B10.Q mice were 44,770 (± 3980) and 28,670 (±1970) respectively. (Reproduced from Dougherty et al., 1985, with permission of the publisher.)

of these responses is that B10.BR mice end up with more larvae in their muscle than do B10.Q mice. When mesenteric lymph node cells were harvested from B10.BR mice and cultured in vitro with *T. spiralis* antigen, the strength of the proliferative response was weak when compared to the response of cells from B10.Q. Thus, the kinetics and strength of lymphocyte responsiveness correlates directly with levels of resistance as measured by each of the other parameters. The data in Figure 1 demonstrate the value of using several criteria when attempting to determine the response phenotype of a given mouse strain and confirm that H-2 genes influence each of the responses observed since the B10.BR and B10.Q mice differ only at H-2 genes.

## IDENTIFICATION OF SPECIFIC GENES WHICH INFLUENCE THE ANTI-*TRICHINELLA* RESPONSE

Once it is known that H-2 genes influence a particular immune response, it is often possible to identify the specific gene that regulates the observed effect. This is possible because many H-2 congenic mouse strains have been established which express recombinant H-2 haplotypes. In the H-2 recombinant strains, alleles from two or more parental strains become fixed within the H-2 complex as the result of recombination. If the alleles expressed are from a susceptible and a resistant parent, and if recombinant strains are available where the crossover has occurred in different places, than the specific genes which regulate the anti-parasite response may be precisely mapped.

### I-Region Genes and Immunity to *T. spiralis*

Data in Table 2 show results from a comparison between recombinant strains expressing alleles from a susceptible $H-2^k$ parent and a resistant $H-2^s$ parent. Expression of s alleles at I region loci is associated with resistance to infection, even when genes from the susceptible $H-2^k$ parent are expressed elsewhere. These data implicate the I-A genes as important because mice expressing the $H-2^s$ haplotype have a defect in the $E_\alpha$ gene and do not express an I-E molecule. Thus, alleles at $E_\beta$ and $E_\alpha$ are irrelevant in this case. However, a curious relationship between expression of I-E alleles and susceptibility to *T. spiralis* infection emerges when more strains of mice are tested.

Many strains of mice possess defects in the $E_\beta$ and/or $E_\alpha$ gene(s) and do not express a cell surface I-E molecule (Murphy et al., 1980; Jones et al., 1981; Mathis et al., 1983). Such strains are immunologically normal in most

TABLE 2. The I-A Molecule Appears to Influence the Host Response to *T. spiralis*

| Experiment No. | Mouse strain | H-2 haplotype | H-2 loci | | | | | | | | Worm count (±SEM)[a] | Phenotype |
|---|---|---|---|---|---|---|---|---|---|---|---|---|
| | | | K | $A_\beta$ | $A_\alpha$ | $E_\beta$ | $E_\alpha$ | S | D | L | | |
| 1[b] | B10.BR | k | k | k | k | k | k | k | k | k | 26,640 (±3,236) | sus |
| | B10.S | s | s | s | s | s | s | s | s | s | 12,600 (±646) | res |
| | B10.S(8R) | as¹ | k | k | k | k | s | s | s | s | 29,120 (±2,760) | sus |
| | B10.HTT | t³ | s | s | s | s | k | k | d | d | 14,000 (±1,050) | res |
| 2[c] | B10.BR | k | k | k | k | k | k | k | k | k | 30,286 ± 1,291 | sus |
| | B10.A | a | k | k | k | k | k | d | d | d | 28,680 ± 2,485 | sus |
| | B10.TL | t¹ | s | k | k | k | k | k | d | d | 29,300 ± 2,564 | sus |
| | B10.S | s | s | s | s | s | s | s | s | s | 15,050 ± 1,024 | res |

[a]Total body larval counts 30 days following infection with 150 $L_1$ larvae of *T. spiralis*.
[b]Reinterpreted from Wassom et al. (1979); reproduced with permission of the publisher.
[c]Reinterpreted from Wassom et al. (1980); reproduced with permission of the publisher.

respects, although they are genetic nonresponders to antigens which must be presented to T cells in association with I-E. However, as shown in Table 3, strains of mice which do express I-E are susceptible to *Trichinella* infection when compared to mice which express only I-A. The association of I-E expression with susceptibility to infection is strengthened by the finding that offspring from resistant (I-E neg) and susceptible (I-E pos) parents are susceptible to *T. spiralis* infection (Table 4); I-E is expressed in these $F_1$ animals. These observations are not without precedent; the expression of I-E molecules has been associated with the induction of suppressor cells in certain experimental systems. For example, the antigens $LDH_B$ and Liver antigen F induce suppression when presented to T cells in association with I-E molecules (Baxevanis et al., 1981; Oliveira et al., 1985). In each case, the *I-E*$^k$ expressing host is a nonresponder to the antigen, and T cells harvested from the "immunized" mice fail to undergo antigen-driven proliferation in vitro. However, if I-E-restricted presentation of antigen is blocked by adding monoclonal anti-*I-E*$^k$ antibodies to such in vitro cultures, the I-A-restricted T cells proliferate strongly. Thus, in each of the above examples the mice were able to respond to antigen, but this response was suppressed by cells which recognized antigen in the context of *I-E*$^k$. It is possible that a similar

TABLE 3. Association of H-2 Haplotype, Expression of Cell Surface I-E Molecules, and Susceptibility to Infection With *T. spiralis*[a]

| Mouse strain | H-2 haplotype | I-E expression | Resistance index[b] | Resistance phenotype |
|---|---|---|---|---|
| B10.BR | k | + | 0 | sus |
| B10.P | p | + | −22 | sus |
| B10.RIII | r | + | 33 | sus |
| B10 | b | − | 63 | res/int |
| B10.S | s | − | 100 | res |
| B10.M | f | − | 104 | res |
| B10.Q | q | − | 105 | res |

[a]Reproduced from Wasson et al. (1987b) with permission of the publisher.

$$^b RI = \frac{\text{count for B10.BR} - \text{count for B10.?} \times 100}{\text{count for B10.BR} - \text{count for B10.S}}$$

RI for *T. spiralis* is calculated on basis of total body larval counts 30 days following per os infection with 150 $L_1$ larvae as described by Wassom et al. (1983a,b). In a typical experiment, $H$-$2^k$ mice harbor approximately twofold more worms than $H$-$2^s$ mice. The calculated RI, when greater than 100 indicates lower average worm counts than demonstrated for $H$-$2^s$ mice. A negative index value denotes susceptibility greater than that expressed by $H$-$2^k$.

**TABLE 4. Association of Susceptibility to *T. spiralis* Infection in $F_1$ Mice With Expression of *I-EK*[a]**

| Strain | No. tested | H-2 haplo-type | *I-E*[k] *expression* | Worm count $(\pm SEM)$[b] |
|---|---|---|---|---|
| B10.BR | 17 | k | + | 51,820 ± 5,620[c,d] |
| B10.Q | 22 | q | − | 25,370 ± 1,560 |
| B10.M | 20 | f | − | 28,050 ± 1,650 |
| B10.BR × B10.Q | 23 | k/q | + | 41,560 ± 3,060[c] |
| B10.BR × B10.M | 22 | k/f | + | 45,950 ± 2,800[d] |

[a]Reproduced from Wassom et al. (1987b) with permission of the publisher.
[b]Total body larval counts 30 days following infection per os with 150 $L_1$ larvae of *T. spiralis* as described by Wassom et al. (1983a,b).
[c]Significantly more susceptible than B10.Q ($P<.05$, Student's t-test).
[d]Significantly more susceptible than B10.M ($P<.05$, Student's t-test).

mechanism might operate in *Trichinella*-infected mice. For example, freshly explanted lymph node cells from *Trichinella* antigen-injected mice contain populations of I-A- and I-E-restricted T cells. However, when these cultures are passaged in vitro in the presence of *T. spiralis* antigen, only the I-E-restricted clones persist (Krco et al., 1983). It is possible therefore that the I-E-restricted cells have inhibited expansion of the I-A-restricted clones.

The association between I-E expression and susceptibility to T. *spiralis* infection is not absolute. It is not yet clear whether this phenomenon is associated with expression of all configurations of the I-E molecule, or if some configurations such as *I-E*[k] are involved to a greater extent than others. As discussed below, mice which express *I-E*[k] may resist infection when a specific gene on chromosome 4 is not expressed, and mice which fail to express I-E may still be susceptible to infection when a d allele is expressed at a gene between the S and D loci of the mouse H-2 complex.

## A Chromosome 4 Gene and Immunity to *T. spiralis*

Identification of the chromosome 4 gene which influences levels of susceptibility to *T. spiralis* infection came about quite by accident. As shown in Table 1, AKR/J mice are resistant to *T. spiralis* infection when compared to other strains which share the *H-2*[k] haplotype. All *H-2*[k] mice express the I-E molecule in the $E_\beta^k E_\alpha^k$ configuration. The relative resistance of AKR/J mice to infection with *T. spiralis* was an early clue that non-H-2 genes influenced the immune response to infection. Comparative studies between AKR/J and C3HeB/FeJ mice (both *H-2*[k]) have confirmed that non-H-2 genes influenced each of the relevant host responses for which there was an assay (Wassom et al., 1983a). However, it was difficult to determine how many non-H-2 genes were involved, and we had no means whereby the gene or genes could be mapped within the mouse genome. Then in 1984, Hayes et al. published data showing that AKR/J mice, unlike most $H$-$2^k$ strains, failed to express $I$-$J^k$ molecules on subpopulations of T cells. (I-J molecules were thought to be expressed on populations of suppressor T cells). The AKR/J gene defect was mapped to chromosome 4 using a series of recombinant inbred strains and a critical comparison between the strains AKR/J and AKR-Fv-1$^b$, which differ only at genes linked to the Fv-1 locus on chromosome 4. Although the two strains had identical H-2 genes, the AKR-Fv-1$^b$ mice expressed $I$-$J^k$ molecules and the AKR/J mice did not. Hayes et al. (1984) suggested that the I-J gene was erroneously mapped within the H-2 complex because I-E expression (at least $E_\alpha$) was requisite for the expression of I-J; thus, expression of $I$-$J^k$ was dependent on the concurrent expression of the $I$-$E^k$ genes and on the chromosome 4 gene designated Jt. Flood et al. (1986), studying transgenic mice, showed that when the $E_\alpha^k$ gene is expressed in (SJL×B6)F$_1$ mice, I-E molecules ( $E_\beta^s E_\alpha^k$ and $E_\beta^b E_\alpha^k$), and $I$-$J^k$ molecules are both expressed. Neither SJL, B6, nor (SJL×B6)F$_1$ mice express I-E or $I$-$J^k$ molecules in the absence of the inserted $E_\alpha^k$ gene. Thus, it is clear that I-E expression influences either the expression or the polymorphism of $I$-$J^k$ molecules. Because our studies of the *Trichinella*-mouse system had associated I-E expression with susceptibility to infection, because I-J expression has been associated with populations of suppressor T cells, and because we had characterized AKR/J mice to be resistant to *T. spiralis* infection, we were curious to determine whether or not the chromosome 4 gene identified by Hayes et al. (1984) influenced the immune response of mice to *Trichinella* infection. We compared AKR/J to AKR-Fv-1$^b$ mice in three separate experiments. In each case, the $I$-$J^k$ positive AKR-Fv-1$^b$ mice were dramatically more susceptible to infection (Table 5). Thus, a gene on chromosome 4 must influence the anti *Trichinella* response. We have speculated that the I-E-associated susceptibility to *T. spiralis* infection might be mediated by suppressor T cells which are regulated by the combined influence of the chromosome 4 and the $E_\beta$ and $E_\alpha$ genes (Wassom et al., 1987b).

## Mapping of a New Gene Between the S and D Regions of the H-2 Complex

Mice which fail to express I-E molecules may still be susceptible to *T. spiralis* infection if they possess a d allele at a new locus between the S and D regions of the H-2 complex (Wassom et al., 1983b). It was first observed that recombinant strains of mice which expressed s, b, f, or q alleles at I-region loci but d alleles at the D end of the H-2 complex were markedly

TABLE 5. Comparison of AKR/J and AKR-Fv-1$^b$ Mice for Susceptibility to *T. spiralis* Infection[a]

| Strain | H-2 haplo-type | *I-E*$^k$ expression | *I-J*$^k$ expression[b] | Worm count ($\pm$SEM)[c] |
|---|---|---|---|---|
| AKR/J | k | + | − | 13,800 ($\pm$4,690) |
| AKR-Fv-1$^b$ | k | + | + | 37,610 ($\pm$8,740)* |

[a]Reproduced from Wassom et al. (1987b) with permission of the publisher.
[b]Hayes et al. (1984).
[c]See Table 2.
*Significantly more susceptible than AKR/J ($P < .0001$, Student's t-test).

more susceptible to *Trichinella* infection than were mice expressing the independent s, b, f, or q haplotypes (Table 6). The observed effects were initially attributed to the D or L genes. However, examination of several $K^s$ $I^s$ $D^d$ or $K^b$ $I^b$ $D^d$ recombinant strains, which were believed to be genetically identical, showed that they differed in their ability to resist *Trichinella* infections (Table 7). It was concluded that although each of the strains tested shared common alleles at K, I, S, D, and L loci, a difference in the point of crossover between S and D resulted in the expression of a gene in some strains which was not expressed in others (Fig. 2). At the time of the initial observation, no genes of relevance to the immune response had been mapped between S and D in the H-2 complex. Recently however, the genes encoding tumor necrosis factor (TNF$_\alpha$ and TNF$\beta$) (Müller et al., 1987) and a gene which influences the immune response to TNP-Ficoll (Shapiro et al., 1985) have been mapped between S and D. It is known that TNF may influence the outcome of other host-parasite interactions (Taverne et al., 1987; Silberstein and David, 1986; Ruddle, 1987), but it remains to be elucidated whether the TNF genes and the gene which influences the anti*Trichinella* response are synonymous.

## CONCLUDING REMARKS

The immediate aim of the genetic studies outlined above was to identify relevant genes which influence the immune response to *T. spiralis* infections of mice. This aim was justified by the belief that once such genes had been identified, the mechanisms which underlie the functional immune response to this parasite could be more clearly discerned. This was to be achieved by studying strains of mice which differed in their ability to resist infection, yet which were genetically identical except for the allele expressed at a single

**TABLE 6. D-End Genes Influence Resistance to _T. spiralis_ Infection**

| Experiment no. | Mouse strain | H-2 haplotype | H-2 loci | | | | | | | | Worm count (±SEM)[a] |
|---|---|---|---|---|---|---|---|---|---|---|---|
| | | | K | Aβ | Aα | Eβ | Eα | S | D | L | |
| 1 | B10.S | s | s | s | s | s | s | s | s | s | 15,050 (±1,024)[b] |
| | B10.S(7R) | t2 | s | s | s | s | s | s | d | d | 24,044 (±4,836)[b] |
| 2 | B10.Q | q | q | q | q | q | q | q | q | q | 23,320 (±5,467) |
| | B10.T(6R) | y2 | q | q | q | q | q | q | d | d | 42,530 (±4,034) |
| 3 | B10.M | f | f | f | f | f | f | f | f | f | 15,750 (±1,394)[b] |
| | B10.M(11R) | ap1 | f | f | f | f | f | f | d | d | 20,301 (±1,793)[b] |

[a]See Table 2.
[b]Reproduced from Wassom et al. (1983a,b) with permission of the publisher.

**TABLE 7. Mouse Strains Sharing Alleles at K, I, S, and D Differ in Susceptibility to _T. spiralis_ Infection[a]**

| Mouse strain | H-2 haplotype | H-2 loci | | | | | | | | Worm count (±SEM)[b] |
|---|---|---|---|---|---|---|---|---|---|---|
| | | K | Aβ | Aα | Eβ | Eα | S | D | L | |
| B10.S | s | s | s | s | s | s | s | s | s | 15,050 (±1,024) |
| B10.S(7R) | t2 | s | s | s | s | s | s | d | d | 24,044 (±4,936) |
| B10.S(23R) | t11 | s | s | s | s | s | s | d | d | 17,260 (±1,961) |
| B10.S(24R) | t6 | s | s | s | s | s | s | d | d | 13,833 (±1,283) |
| B10.A(18R) | i18 | b | b | b | b | b | b | d | d | 25,033 (±2,301) |
| B10.S(21R) | i21 | b | b | b | b | b | b | d | d | 18,690 (±2,027) |
| B10.BAR-5 | i1 | b | b | b | b | b | b | d | d | 14,720 (±1,297) |
| B10.D2(R107) | i7 | b | b | b | b | b | b | d | d | 13,000 (±1,507) |

[a]Reinterpreted from Wassom et al. (1983a,b); reproduced with permission of the publisher.
[b]See Table 2.

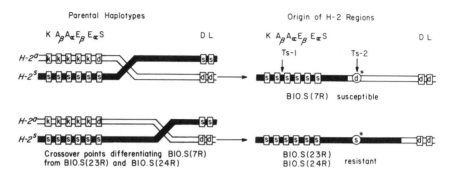

Fig. 2. Crossover points differentiating B10.S(7R) from B10.S(23R) and B10.S(24R). B10.S(7R) is more susceptible to *T. spiralis* infection than are the other two strains. The Ts-1 gene maps to the I-A region. The exact position of Ts-2 between S and D is not known. It is shown in close proximity to S for diagrammatic purposes only. (Reinterpreted from Wassom et al., 1983b, with permission of the publisher.)

relevant gene locus. Such experiments are now possible. For each of the three genes identified, pairs of mouse strains exist which are genetically identical, or nearly so, except for the gene of interest. For example, the demonstrated ability to construct and characterize I-E transgenic mice should allow a definitive test of the hypothesis that expression of I-E molecules is associated with the induction of T cells which down-regulate I-A-restricted responses. Comparisons between Fv-1 congenic strains such as AKR/J and AKR-Fv-1[b] should lead to identification of the cell(s) where the chromosome 4 gene is expressed, and should provide an opportunity to determine if the polymorphism of I-J molecules is associated with differential susceptibility to *T. spiralis* infection. Finally, comparisons among strains that differ only at genes between S and D should provide significant insights into how one H-2 gene may influence the expression of another. Of particular interest will be experiments to determine if the *Trichinella*-susceptible and the *Trichinella*-resistant strains of mice differ in the expression of TNF genes.

The preceding discussion on genetic control of the immune response to *T. spiralis* infections of mice was not intended to serve as a vehicle for extolling the merits of this experimental system or the analysis of it. To the contrary, it was intended to illustrate that genetic analyses can be conducted using such systems, however complex, and that if special care is paid to measuring biologically relevant parameters, the results obtained may be of genuine biological significance. Hopefully, the insights obtained from the *Trichinella* studies will be applied to other experimental systems of direct relevance to human and animal health.

## REFERENCES

Baxevanis CN, Nagy ZA, Klein J (1981): A novel type of T-T cell interaction removes the requirement of I-B region in the H-2 complex. Proc Natl Acad Sci USA 78:3809–3813.

Bell RG, McGregor DD, Despommier DD (1979): *Trichinella spiralis*. Mediation of the intestinal component of protective immunity in the rat by multiple, phase specific antiparasitic responses. Exp Parasitol 47:140–157.

Blackwell J (1985): Genetic control of discrete phases of complex infections: *Leishmania donovani* as a model. In E. Skamene (ed): Genetic Control of Host Resistance to Infection and Malignancy. New York: Alan R. Liss, Inc., pp 31–49.

Bradley DJ (1980): Genetic control of resistance to protozoal infections. In E. Skamene, P.A.L. Kongshavn, and M. Landy (eds): Genetic Control of Natural Resistance to Infection and Malignancy. New York: Academic Press, pp 9–28.

Dougherty DA, Gagliardo L, Krco CJ, David CS, Wassom DL (1985): H-2 genes influence the kinetics of lymphocyte responsiveness in *Trichinella spiralis*-infected mice. In E. Skamene (ed): Genetic Control of Host Resistance to Infection and Malignancy. New York: Alan R. Liss, Inc., pp 459–464.

Flood PM, Benoist C, Mathis D, Murphy DB (1986): Altered I-J phenotype in E$\alpha$ transgenic mice. Proc Natl Acad Sci USA 83:8308–8312.

Hayes CE, Klyczek KK, Krum DD, Whitcomb RM, Hullett DA, Cantor H (1984): Chromosome 4 Jt gene controls murine T cell surface I-J expression. Science 223:559–563.

Jones PP, Murphy DB, McDevitt HO (1981): Variable synthesis and expression of E$_\alpha$ and A$_e$ (E$_\beta$) Ia polypeptide chains in mice of different H-2 haplotypes. Immunogenetics 12:321–337.

Jungery M, Ogilvie BM (1982): Antibody response to stage specific *Trichinella spiralis* surface antigens in strong and weak responder mouse strains. J Immunol 129:839–843.

Krco CJ, David CS, Wassom DL (1982): Characterization of an in vitro proliferation response to solubilized *Trichinella spiralis* antigens. Role of Ia antigens and Ly-1 + T cells. Cell Immunol 68:359–367.

Krco CJ, Wassom DL, Abramson EJ, David CS (1983): Cloned T cells recognize *Trichinella spiralis* antigen in association with an E$_\beta^k$ E$_\alpha^k$ restriction element. Immunogenetics 18:435–444.

Mathis DJ, Benoist C, Williams VE, Kanter M, McDevitt HO (1983): Several mechanisms can account for defective E$_\alpha$ gene expression in different mouse haplotypes. Proc Natl Acad Sci USA 80:273–277.

Müller V, Jongeneel CV, Nedospasov SA, Fischer-Lindahl K, Steinmetz M (1987): Tumor Necrosis Factor and Lymphotoxin genes map close to H-2D in the mouse major histocompatibility complex. Nature 325:265–267.

Murphy DB, Jones PP, Loken MR, McDevitt HO (1980): Interaction between I region loci influences the expression of a cell surface Ia antigen. Proc Natl Acad Sci USA 77:5404–5408.

Oliveira DBG, Blackwell N, Virchis AE, Axelrod RA (1985): T helper and T suppressor cells are restricted by the A and E molecules respectively in the F antigen system. Immunogenetics 22:169–175.

Philipp M, Parkhouse RME, Ogilvie BM (1980): Changing proteins on the surface of a parasitic nematode. Nature 287:538–540.

Philipp M, Taylor PM, Parkhouse RME, Ogilvie BM (1981): Immune response to stage-specific surface antigens of the parasitic nematode *Trichinella spiralis*. J Exp Med 154:210–215.

Ruddle NH (1987): Tumor necrosis factor and related cytotoxins. Immunol Today 8:129–130.

Shapiro LH, Dugan ES, Neiderhuber JE (1985): Monoclonal antibody characterization of a unique immune response control locus between H-2 S and D. J Exp Med 162: 1477–1493.

Silberstein DS, David JR (1986): Tumor necrosis factor enhances eosinophil toxicity to *Schistosoma mansoni* larvae. Proc Natl Acad Sci USA 83:1055–1059.

Skamene E, Gros P, Furget A, Kongshavn PAL, St-Charles C, Taylor A (1982): Genetic regulation of resistance to intracellular pathogens. Nature 297:506–510.

Taverne J, Tavernier J, Fiers W, Playfair JHL (1987): Recombinant tumor necrosis factor inhibits malaria parasites *in vivo* but not *in vitro*. Clin Exp Immunol 67:1–4.

Wakelin D (1978): Genetic Control of Susceptibility and Resistance to Parasitic Infection. In W.H.R. Lumsden, R. Muller, and J.R. Baker (eds): Advances in Parasitology, Vol 16. New York: Academic Press, pp 219–308.

Wakelin D, Donachie AM (1980): Genetic control of immunity to parasites: Adoptive transfer of immunity between inbred strains of mice characterized by rapid and slow immune expulsion of *Trichinella spiralis*. Parasite Immunol 2:249–260.

Wassom DL, DeWitt CW, Grundmann AW (1974) Immunity to *Hymenolepis citelli* by *Peromyscus maniculatus*: Genetic control and ecological implications. J Parasit 60:47–52.

Wassom DL, David CS, Gleich GJ (1979): Genes within the major histocompatibility complex influence susceptibility to *Trichinella spiralis* in the mouse. Immunogenetics 9: 491–496.

Wassom DL, David CS, Gleich GJ (1980): MHC-linked genetic control of the immune response to parasites: *Trichinella spiralis* in the mouse. In E. Skamene, P.A.L. Kongshavn, and M. Landy (eds): Genetic Control of Natural Resistance to Infection and Malignancy. New York: Academic Press, pp 75–82.

Wassom DL, Brooks BO, Cypess RH (1983a): *Trichinella spiralis*: Role of non H-2 genes in resistance to primary infection. Exp Parasitol 55:153–158.

Wassom DL, Brooks BO, Babish JG, David CS (1983b): A gene mapping between the S and D regions of the H-2 complex influences resistance to *Trichinella spiralis* infections of mice. J Immunogenet 10:371–378.

Wassom DL, Wakelin D, Brooks BO, Krco CJ, David CS (1984a): Genetic control of immunity to *Trichinella spiralis* infections of mice: Hypothesis to explain the role of H-2 genes in primary and challenge infections. Immunology 51:625–631.

Wassom DL, Dougherty DA, Krco CJ, David CS (1984b): H-2-controlled dose dependent suppression of the response which expels *Trichinella spiralis* from the small intestine of mice. Immunology 53:811–818.

Wassom DL, Dougherty DA, Dick TA (1987a): *Trichinella spiralis* infections of inbred mice: Immunologically specific responses induced by different *Trichinella* isolates. J Parasitol, Volume 74: p. 283–287.

Wassom DL, Krco CJ, David CS (1987b): I-E expression and susceptibility to parasite infection. Immunol Today 8:39–43.

# PARASITE MOLECULAR BIOLOGY, BIOCHEMISTRY, AND GENETICS

The Biology of Parasitism, pages 349–369
© 1988 Alan R. Liss, Inc.

# Trypanosome Variant Surface Glycoprotein

**Mervyn J. Turner**

*Biochemical Parasitology, Merck Sharp & Dohme Research Laboratories, Rahway, New Jersey 07065*

## ANTIGENIC VARIATION

The parasitic protozoa have adopted a variety of strategies to evade elimination by the immune system of the host, but the most spectacular is surely the process of antigenic variation adopted by the African trypanosomes. These parasites are generally transmitted to mammals by the bite of an infected tsetse fly and multiply within the bloodstream of the vertebrate host to cause sleeping sickness in man, or the wasting disease Nagana in cattle. Tsetse flies are endemic within a 10-million-square-mile area of central Africa, rendering the breeding of domestic stock impossible due to the high incidence of trypanosomiasis.

Infection with the parasite is characterized by a relapsing parasitemia which can persist for weeks or months, generally until the death of the host. Peaks in the parasite population arise at roughly weekly intervals. The first reference to a mechanism to explain such behavior is contained within a paper published in 1905 by Franke (1905), who at the time was working in Ehrlich's laboratory. He was looking at the trypanocidal effects of arsenates in a monkey and made the chance observation that the blood of the monkey, removed 2 weeks after infection, contained a substance which was capable of lysing the initial inoculum of trypanosomes. However, trypanosomes collected later in the course of the infection were not lysed by this substance. Franke concluded "the trypanosomes must therefore have acquired other biological characters during their stay in his (the monkey's) semi-immune body that rendered them resistant to the defensive substances of the host." This observation provided the genesis of the concept of antigenic variation, which was later amply verified. The waves of parasites in the bloodstream are generated by the rapid multiplication of parasites until the host's immune system produces sufficient circulating antibody to kill most of the trypanosomes. Some parasites survive, however, because during the multiplication phase, they have switched to the expression of a different antigenic type, which is not recognized by the circulating antibody. These survivors are the progenitors of a second peak in parasite numbers, until these too are destroyed. The process goes on, literally ad nauseum.

The limits of antigenic variation have still been mapped only imprecisely. Capbern et al. (1977), in a classic experiment, performed with the equine parasite *Trypanosoma equiperdum*, identified 101 antigenically distinct variants as having arisen from a single cloned parasite used to infect a rabbit. Gene-counting techniques (Van der Ploeg et al., 1982) have been used to estimate that the trypanosome may contain genes for up to 1,000 different antigens. That would imply that the parasite devotes perhaps 10% of its genome to the storage of antigen genes. Studies on the genetic mechanism of antigenic variation (reviewed elsewhere in this volume) have led to the suggestion that the trypanosome may employ recombination between different antigen genes to construct new hybrid genes. Potentially, this would vastly increase the antigen repertoire. Truly, this is the protozoan response to the mechanisms by which vertebrates generate antibody diversity.

## THE SURFACE COAT

The structure responsible for antigenic variation was first identified by Vickerman (1969). Transmission electron microscopy of trypanosomes from infected blood revealed an electron-dense layer, some 12–15 nm thick, which covered the whole surface of the parasite. Vickerman christened the structure "the surface coat." He showed that ferritin-conjugated variant-specific-antibodies bound to the surface coat of trypanosomes of the homologous variant but not to trypanosomes of a heterologous antigenic type (Vickerman and Luckins, 1969). Furthermore, trypanosomes which had been cultured under conditions in which differentiation to a form equivalent to that found in the midgut of an infected tsetse occurs, no longer had a surface coat (Brown et al., 1973), nor did they bind variant-specific antibodies. Such trypanosomes are not infective to vertebrates. Examination of the different forms found in the tsetse fly revealed that the surface coat was present only on the metacyclic trypomastigotes (Vickerman, 1969), found in the salivary glands of the fly. This is the only form infective to vertebrates, and Vickerman therefore concluded that the surface coat was a necessary adaptation to life in the bloodstream of the host.

## THE VARIANT SURFACE GLYCOPROTEIN

The composition of the surface coat was first made clear as the result of studies by Le Page (1968) and by Cross (1975). The publication by Cross in

1975 of a simple protocol to allow the rapid purification of the glycoprotein molecules which form the surface coat was probably the single greatest impetus in propelling the trypanosome to the stellar status it has enjoyed ever since amongst molecular biologists. Cross started with an antigenically homogeneous population of *Trypanosoma brucei*. This can be acquired by the biological cloning of an individual parasite, and maintained by frequent passage in rats or mice, before an immune response can be generated. A pure suspension of trypanosomes can then be obtained by fractionation of cells from infected blood on diethylaminoethyl (DEAE) cellulose—red cells adhere, but trypanosomes do not. The key observation then was that disrupting the plasma membrane led to the release of variant antigens in a water-soluble form. We now know that this release is enzyme-mediated (Cardoso de Almeida and Turner, 1983). It is a consequence of activation of an endogenous glycophosphatidylinositol-specific phospholipase C, which cleaves dimyristylglycerol from a covalently bound glycolipid anchoring the variant surface glycoprotein (VSG) to the plasma membrane (Ferguson et al., 1985a,b). The structure and biosynthesis of the glycolipid and the nature of the enzyme are both discussed elsewhere in this volume. Suffice it to say that a better understanding of this event led to modifications in the isolation protocol, with greater control over coat release. The released molecules represent a high proportion of the total protein and can be readily purified to homogeneity by a very simple ion-exchange step and preparative isoelectric focusing. Indeed, in the modified protocol of Cross (1984a), passage over DEAE cellulose, under conditions in which the surface antigens do not bind, usually leads to material of sufficient purity for most requirements. Antibodies raised against the purified antigens were variant specific, and immunized mice were protected against challenge with the homologous but not the heterologous variant (Cross, 1975). Cross originally called the purified antigen the variant-specific surface antigen (VSSA), but it is now usually referred to as the variant surface glycoprotein (VSG). The yields are such that starting with $10^{11}$ trypanosomes, 50–80 mg of pure VSG can be obtained within 3 days. This has allowed the biochemistry of the VSGs to be explored in great detail.

## BIOCHEMISTRY OF VSGs

Because *Trypanosoma brucei brucei* is noninfective to humans yet produces very high levels of parasitemia in rats and mice, variants from this species have been studied in greatest detail. Unless specifically stated otherwise, all of the biochemistry discussed relates to VSGs of *T. brucei*.

Vsgs are glycoproteins with a molecular weight of around 60,000. The carbohydrate accounts for 7–17% by weight (Johnson and Cross, 1977), and the polypeptide contains 450–500 amino acids. Although the carbohydrates show heterogeneity in structure, it is diversity in the amino acid sequences that is responsible for antigenic variation. The carbohydrate is present in two different forms. All VSGs contain at least one N-linked oligosaccharide, sometimes two or more (Holder and Cross, 1981). The N-linked glycans appear to be similar to other eukaryotic high-mannose oligosaccharides insofar as they contain N-acetylglucosamine and mannose. Further, attachnent of the glycan to the nascent polypeptide chain is blocked in the presence of tunicamycin (Strickler and Patton, 1980; McConnell et al., 1983). At the time of writing, no complete structures for the N-linked glycans of *T. brucei* have been published (some partial structures for the equivalent glycan of a VSG of *T. congolense* are available, Savage et al. 1984). The second class of carbohydrate is contained within the polar head of the glycolipid membrane anchor, illustrated schematically in Figure 1. The polar head group is responsible for the only immunological cross-reaction seen between most VSGs (Holder and Cross, 1981). Removal of the diacylglycerol exposes an epitope, the cross-reacting determinant, present in all VSGs, and indeed on many other membrane glycoproteins isolated from species as diverse as *Homo sapiens* and *Paramecium aurelii,* which share a similar mode of membrane attachment (reviewed in Low, 1987). This glycolipid is discussed in more detail elsewhere in this volume.

Vsgs form dimers and occasionally higher oligomers in solution (Auffret and Turner, 1981; Strickler and Patton, 1982). Whether this is a true reflection of the form in which they are found on the cell surface is uncertain. VSGs form a very densely packed monolayer containing about $10^7$ molecules on the plasma membrane. Surface cross-linking experiments produce dimers—and other oligomers of virtually all sizes, presumably reflecting the very close packing that this surface density requires (Strickler and Patton, 1982). Hence, the surface cross-linking cannot be taken as definitive proof that dimers represent the natural association of VSGs on the cell surface. However, the solution dimers are very stable, both in sVSG (the released form, lacking dimyristylglycerol) and in mfVSG (the membrane form of the VSG, from which diacylglycerol has not been removed) (Gurnett et al., 1986). Indeed, as we shall see, VSGs can be crystallized as dimers.

Vsgs have a two-domain structure. Treatment with trypsin, or with a variety of other proteolytic enzymes, produces an N-terminal domain, which contains two-thirds to three-quarters of the sequence, and a C-terminal domain containing the remainder (Johnson and Cross, 1979). The N-terminal

## STRUCTURE OF VSG OF T.BRUCEI

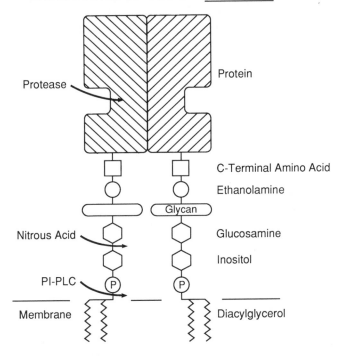

Fig. 1. Structure of a VSG of *T. brucei*. The structure shows the VSG as a dimer in the plasma membrane of the trypanosome. The shaded portion represents the protein portion of the molecule, which can be cleaved into two domains by protease. The rudiments of the structure of the C-terminal glycolipid are shown. The diacyl glycerol may be cleaved by a glycophosphatidylinositol-specific phospholipase C (PI-PLC). The phosphatidylinositol can be removed from purified mfVSG by diazotization of the amino group of glucosamine with nitrous acid.

domain also exists as a dimer in solution (Freymann et al., 1984). It is not clear whether the C-terminal domain associates into a dimer, too. The size of the two domains (as determined by SDS-PAGE) is constant for any one VSG, although there may be one or more intermediate forms before a limit digestion product is reached. However, there is quite a lot of size variation in comparing domains from different VSGs and in the degree to which different VSGs are resistant to proteolysis. This is one of many pieces of evidence implying that different VSGs have similar but not identical structures.

The C-terminal domain is, of course, glycosylated because of the presence of the C-terminal glycolipid. The N-terminal domain, on the other hand, does not invariably contain N-linked oligosaccharide. The best characterized VSG, MITat 1.2 (221), which is the subject of an x-ray crystallography study, has a glycosylated N-terminal domain, whereas a second crystallized VSG (ILTat 1.24) apparently has not. In both cases, the N-terminal domain crystallizes as a dimer, and the two domains share common structural motifs (vide infra); hence, it seems unlikely that carbohydrate has any essential role in structural organization at this level. The N-terminal domain is sometimes referred to as the hypervariable domain; the C-terminal, the homology domain, reflecting information obtained from sequence analysis.

## VSG SEQUENCES

Complete or partial sequences have been published for about ten VSGs, forming a useful data base. The existence of this data base is often taken as evidence that no further VSG sequence information is needed, or of any utility. I would strongly disagree with this viewpoint. The large number of immunoglobulin sequences on hand, even before the advent of DNA sequencing, proved of immense value in eventually unravelling the means by which diversity is generated in the immune system. Comparison of sequence with three-dimensional structure was of special value in understanding how changes in a very few portions of the sequence could profoundly affect antibody specificity. As we shall see, the problem of relating sequence and structure in VSGs is of a different kind, but there is little doubt in my mind that an expansion of the VSG sequence data base would be extremely useful.

That being said, what have the available sequences taught us? Firstly, that there is tremendous sequence diversity amongst VSGs. Secondly, that even amongst this diversity, certain homologies can be detected. Of the 450 or so amino acids, the N-terminal 350 or so are extraordinarily diverse in sequence (Rice-Ficht et al., 1981). There may be some conservation in cysteine placement, as we shall see, which could be very significant, and perhaps some conservative amino acid changes within the N-terminal 30 residues (Olafson et al., 1984); but other than that, the sequences show few resemblances to each other. The most obvious homologies are found with the C-terminal 50–100 residues, and more particularly at the C-terminus (Rice-Ficht et al., 1981; Holder and Cross, 1981). These homologies are present both in sequences found within the mature, purified VSG, and within a C-terminal sequence detected by cDNA sequence analysis, but absent from purified VSG (Boothroyd et al., 1980, 1981). This sequence is largely hydrophobic, and functions as a transient anchor for the VSG until its replacement in a transamidation reaction with nascent glycolipid. Again, both the nature and the significance of the C-terminal sequence homologies are discussed in great detail elsewhere in this volume, and so I will not dwell on this point, other than to say that C-terminal homologies have allowed VSGs to be classified into two groups, group I and group II. To date, all group I VSGs can be characterized by the presence of an aspartic acid or asparagine residue as the C-terminal amino acid to which the glycolipid is bound; within group II VSGs, the corresponding residue is a serine.

All group I VSGs have a site (that is, the sequence contains the glycosylation signal Asn-X-Ser/Thr) for at least one N-linked oligosaccharide within the C-terminal domain, some 50 residues from the C-terminus. The potential for additional N-linked oligosaccharides is present within several of the group I sequences. The group II VSGs all have an N-linked oligosaccharide five or six residues from the C-terminus. Again, the potential for other N-linked glycans is present, frequently within both domains.

The most commonly cited evidence for structural conservation amongst proteins of highly divergent sequences is the conservation of cysteine residues. In assigning new members to the fast-growing immunoglobulin superfamily, placement of cysteines within an immunoglobulin-like domain is virtually diagnostic. How do VSGs compare in this respect? The distribution of intrachain disulphide bonds has been determined for one VSG of group I type (Allen and Gurnett, 1983). As shown in Figure 2, this VSG contains eight cysteines in the C-terminal domain, which form four intra- chain disulphide bonds. In the N-terminal domain, there are four cysteines in two intramolecular disulphides. Comparison with other group I VSGs (Fig. 2) indicates that the presence of eight cysteines in the C-terminal domain is a strongly conserved feature of the family, and it is assumed that intrachain disulphides are formed between cysteines occupying similar positions relative to variant MITat 1.4. The situation in the N-terminal domain is more complex. At first, comparison of MITat 1.4 with members of the ANTAR 1 serodeme suggests that the cysteines are equally well conserved in this domain. However, this is misleading as the AnTat 1.1, 1.1b, and 1.10 are genetically closely related through a set of partial gene conversion events (Pays et al., 1983a,b). As other group I variants have been sequenced, more diversity in the number and placement of cysteines has been found. For example, the ILTat 1.25 VSG has seven cysteines in the N-terminal domain, in addition to the eight which are the conserved feature of the C-terminal domain.

Very few complete group II sequences are available, but from those that are, a couple of features stand out. Firstly, four or five cysteines in the C-

DISTRIBUTION OF CYSTEINES AND OF N-GLYCOSYLATION SITES
WITHIN VSGS OF T. BRUCEI

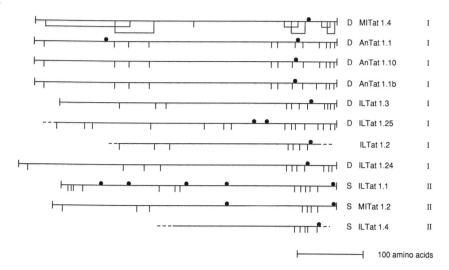

|  | | 100 amino acids |

Fig. 2. Distribution of cysteines and glycosylation sites within VSGs of *T. brucei*. Amino acid sequences for the mature glycoprotein are shown as scaled horizontal bars. Other than for variant MITat 1.4, the protein sequences were deduced from cDNA sequencing studies. The positions of cysteines are indicated by vertical bars. The placement of disulphide bonds is known for MITat 1.4. The presence within the sequence of the glycosylation triplet (Asn-X-Ser/Thr) is shown by the closed circles. The C-terminal amino acid is either aspartate (D) or serine (S). Sequences within the C-terminal region allow classification into two groups (I or II). Sequences are from Pays et al. (1983a,b), Rice-Ficht et al. (1981), Boothroyd et al. (1982), or are previously unpublished (M. Carrington, I. Roditi and M.J. Turner).

terminal domain, rather than eight, seems to be the rule. In the group I VSG, MITat 1.4, in which the placement of disulphide bonds has been established, the eight C-terminal cysteines form two clusters, each containing two intra-molecular disulphides. Cross (1984b) has suggested that group II VSGs are survivors of a deletion event, in which one of these clusters was lost. Clearly, the loss of cysteines has no effect on the proper functioning of the surface coat. A second feature of the group II VSGs is the lack of any clear pattern in N-terminal cysteine distribution, although the database is very small.

It has been pointed out that one cysteine residue is very strongly conserved amongst all VSG (Olafson et al., 1984). This cysteine is found at about amino acid 15 from the N-terminus and forms part of a segment of about 30 amino acids at the N-terminus in which conservative replacements are apparent in comparing different VSGs. In VSG MITat 1.2, this cysteine forms one end of a 131 amino acid loop, which can be seen, in part, in the crystal structure. In the structure, which will be described in more detail below, amino acid 15 falls within an $\alpha$-helical segment which may be conserved in a second VSG for which some structural information is available, ILTat 1.24. The two VSGs belong to different sequence classes. Sequence conservation in the first 30 residues of all VSGs may, therefore, be reflected in the placement of one end of a intrachain loop within a conserved structural element. All VSGs have the potential to form such a loop of about 140 amino acids, since in addition to a cysteine at residue 15, a cysteine at around residues 140–150 is also apparently conserved. In MITat 1.4, at least, the presence of a disulphide bond between these two residues has been established. Perhaps, in the N-terminal domain, the only structural constraint on cysteine distribution is imposed by these two residues.

## THREE-DIMENSIONAL VSG STRUCTURES

To me, one of the most interesting areas of VSG biochemistry is the extent to which sequence diversity in VSGs is reflected in structural diversity. It is reasonable to suppose that VSGs have all arisen from some common ancestral gene encoding a membrane protein of defined function. By amplifying the expression of that gene, a trypanosome may have gained a selective advantage through the generation of a crude nonvariant surface coat which may have provided protection against the lytic effects of complement, through either the classical or alternate pathways (Cross, 1977; Turner and Cordingley, 1981). Antigenic variation may have then developed through the accumulation of gene duplications, mutations, and the imposition of a control mechanism. If this scheme is correct, then today's VSGs might contain echos of the structure of the ancestral gene product. Presumably, the original function has been lost, hence sequence constraints imposed by that function no longer provide any evolutionary pressure. To the best of our knowledge, the only functions of the surface coat are to vary and to provide an impermeable barrier to antibody and complement, preventing access to underlying membrane structures common to all trypanosome variants. All variants are presumably produced through the same biosynthetic pathway, which must provide additional constraints; but all in all, the capacity for change would seem to be large, as indeed is reflected in VSG sequence diversity.

Two lines of argument have been used to suggest that VSGs have diverged to such an extent that there are no longer remnants of structural similarity.

The first is based on the use of algorithims such as those of Chou and Fassman, Kyte and Doolittle, or Hopp and Woods, to predict the existence and extent of secondary structure ($\alpha$-helix, $\beta$-sheet, and $\beta$-turn; or hydropathicity) within different VSG sequences. Lalor et al. (1984) conducted such an analysis and concluded that so diverse is the frequency and distribution of the different elements of secondary structure amongst the sequences analyzed that VSGs must be divergent in structure as well as in sequence. They suggested furthermore that such structural diversity is necessary to avoid the generation of cross-reacting antibodies which could neutralize otherwise antigenically distinct populations. The drawback with such analyses is that only the *potential* for secondary structure formation is measured. Frequently, there is very little difference in the potential for forming one kind of structure over another, and selection of the highest probability structure can be arbitrary.

The second argument is based on calculations of the content of $\alpha$-helix, $\beta$-sheet, and random coil derived from spectral analysis of different VSGs. For VSGs of *Trypanosoma equiperdum,* such an analysis showed a variation in the content of $\alpha$-helix of between 28% and 69%, of $\beta$-sheet, 25% to 49%; and of random coil plus $\beta$-turn of 17% to 43% (Duvillier et al., 1983). Even within such variation, there could, however, be a conserved structural core. Parenthetically, in an unpublished study from my own laboratory amongst the VSGs of *T. b. brucei* on which we were working, $\alpha$-helical content was consistently measured at around 50%.

Obviously, the only way to get at the answer is by determination of three-dimensional structure through x-ray crystallography. For the past several years, my colleagues and I have been collaborating with Don Wiley and his group to accomplish just that (Freymann et al., 1984; Metcalf et al., 1987). The crystallographic problems have been difficult, but interesting results have emerged. Ten VSGs were originally supplied for crystallographic analysis. Five have crystallized in some form. For two of them, solutions at 6 Å resolution have been obtained, and for one of these, a partial structure at 3 Å is available. In each case, the so-called soluble-form VSG lacking the dimyristylglycerol membrane anchor was supplied and used for crystallization experiments. However, in each case where crystals formed, analysis of the crystals showed that a VSG fragment, rather than intact VSG, had crystallized. Contamination with an endogenous protease is apparently responsible. If protease activity is eliminated, however, crystallization does not occur. The endogenous protease clips the VSG into its two component domains, one of which then selectively crystallizes. In four cases where the N-terminus of the crystallized protein has been sequenced, it is apparent that it is the N-terminal domain (about two-thirds of the molecule) that crystallizes. It is not known why intact VSG is refractory to crystallization—it could reflect flexibility around the "hinge" connecting the two domains. The VSG for which most structural information is available is MiTat 1.2, the sequence of which is shown in Figure 3. The mature glycoprotein contains 433 amino acids, after removal of an N-terminal signal peptide, and the C-terminal transient anchor sequence. It is a group II VSG, with a C-

Amino acid sequence of MiTat 1.2 VSG.

|  |  |  |  |  |  | 1 | 2 | 3 | 4 | 5 | 6 | 7 | 8 | 9 | 10 | 11 | 12 | 13 | 14 |  |
|---|---|---|---|---|---|---|---|---|---|---|---|---|---|---|---|---|---|---|---|---|
| *M* | *P* | *S* | *N* | *Q* | *E* | *A* | *R* | *L* | *F* | *L* | *A* | *V* | *L* | *V* | *L* | *A* | *Q* | *V* | *L* |  |
| *P* | *I* | *L* | *V* | *D* | *S* | A | A | E | K | G | F | K | Q | A | F | W | Q | P | L | 14 |
| C | Q | V | S | E | E | L | D | D | Q | P | K | G | A | L | F | T | L | Q | A | 34 |
| A | A | S | K | I | Q | K | M | R | D | A | A | L | R | A | S | I | Y | A | E | 54 |
| I | N | H | G | T | N | R | A | K | A | A | V | I | V | A | N | H | Y | A | M | 74 |
| K | A | D | S | G | L | E | A | L | K | Q | T | L | S | S | Q | E | V | T | A | 94 |
| T | A | T | A | S | Y | L | K | G | R | I | D | E | Y | L | N | L | L | L | Q | 114 |
| T | K | E | S | G | T | S | G | C | M | M | D | T | S | G | T | N | T | V | T | 134 |
| K | A | G | G | T | I | G | G | V | P | C | K | L | Q | L | S | P | I | Q | P | 154 |
| K | R | P | A | A | T | Y | L | G | K | A | G | Y | V | G | L | T | R | Q | A | 174 |
| D | A | A | N | N | F | P | I | T | T | T | P | N | A | G | Y | H | G | H | N | 194 |
| T | N | G | L | G | K | S | G | Q | L | S | A | A | V | T | M | A | A | G | Y | 214 |
| V | T | V | A | N | S | Q | T | A | V | T | V | Q | A | L | D | A | L | Q | E | 234 |
| A | S | G | A | A | H | Q | P | W | I | D | A | W | K | A | K | K | A | L | T | 254 |
| G | A | E | T | A | E | F | R | N | E | T | A | G | I | A | G | K | T | G | V | 274 |
| T | K | L | V | E | E | A | L | L | K | K | K | D | S | E | A | S | E | I | Q | 294 |
| T | E | L | K | K | Y | F | S | G | H | E | N | E | Q | W | T | A | I | E | K | 314 |
| L | I | S | E | Q | P | V | A | Q | N | L | V | D | G | N | Q | P | T | K | L | 334 |
| G | E | L | E | G | N | A | K | L | T | T | I | L | A | Y | Y | R | M | E | T | 354 |
| A | G | K | F | E | V | L | T | Q | K | H | K | P | A | E | S | Q | Q | Q | A | 374 |
| A | E | T | E | G | S | C | N | K | K | D | Q | N | E | C | K | S | P | C | K | 394 |
| W | H | N | D | A | E | N | K | K | C | T | L | D | K | E | E | A | K | K | V | 414 |
| A | D | E | T | A | K | D | G | K | T | G | N | T | N | T | N | T | T | G | S | S | N | 433 |
| *S* | *F* | *V* | *I* | *S* | *K* | *T* | *P* | *L* | *W* | *L* | *A* | *V* | *L* | *L* | *F* |  |  |  |  |  |

Fig. 3. Amino acid sequence of MiTat 1.2 VSG (determined by M. Carrington and I. Roditi, University of Cambridge). The sequence is presented in the form of the single-letter amino acid code. Italicized residues at the N- and C-terminii represent the hydrophobic cleavable signal sequence and transient membrane anchoring sequence respectively. Numbering starts from the first amino acid of the mature glycoprotein and terminates with the C-terminal amino acid carrying the membrane anchoring glycolipid. Underlined asparagine (N) and serine (S) residues are glycosylated, the serine with the glycolipid.

terminal serine, and a glycosylated asparagine, six residues from the C-terminus. There is a second N-linked oligosaccharide attached to asparagine 263. MITat 1.2 VSG has seven cysteines, four within the C-terminal domain and three within the crystallized N-terminal domain. The N-terminal cysteines are at residues 15, 123, and 146. Both intact MITat 1.2 VSG and the crystallized N-terminal domain form dimers, but there are no interchain disulphide bonds. Cysteine 123 apparently contains a free sulphydryl group.

The extent of our understanding of the MITat 1.2 VSG structure is schematized in Figure 4. The protease cuts at, or close to, Lys 357. Of the 360 or so amino acids in the domain, about 270 have been identified on the structure. All the unidentified residues fall within the same spatial area, at the "top" of the molecule. It is likely, but not proven, that these residues will delimit the boundary line between the trypanosome and the environment within the host. These residues have proved extremely refractory to identification for unknown reasons. The best guess for now is that they are present largely in the form of $\beta$-sheet.

The most striking feature of the visible 270 amino acids is the long $\alpha$-helical motif. Starting around amino acid 9, each monomeric subunit contains an $\alpha$-helix which extends to residue 56 with a "kink" at around residue 30. The helix turns and extends back to the top of structure, about 90 Å to residue 112, before the $\alpha$-carbon trace is lost. Since the N-terminal domain is dimeric, this forms a fourfold $\alpha$-helical bundle, with stabilization of the dimer occurring through Van de Waals interactions between $\alpha$-helices of each subunit. The existence of such long, interacting helices had been postulated by Cohen and her colleagues (1984), who analyzed VSG sequences for so-called heptad repeats in which, within sequences of seven amino acids, hydrophobic amino acids are present at positions 1 and 4. Such a motif places hydrophobic residues on one face of a helix, so that two helices can interact to form a coiled coil. There are other $\alpha$-helical segments in the structure. One is found in residues 340–355. This helix is about 30 Å long, parallel to the long helices, and in the dimer structure these two helices plus the four "core" helices account for the sixfold bundle of $\alpha$-helices visible in the early 6 Å electron density map (Freymann et al., 1984).

Another interesting feature is the Cys 15–Cys 146 disulphide bond. It will be interesting to see, in the future, whether this disulphide loop is structurally conserved, for reasons discussed earlier. The N-linked oligosaccharide can be seen at Asn 263. Since this glycosylation site is not conserved, little significance can be attached to this.

Although completion of the structure of a single VSG is obviously a priority, comparison of the structure of two VSGs is likely to prove more

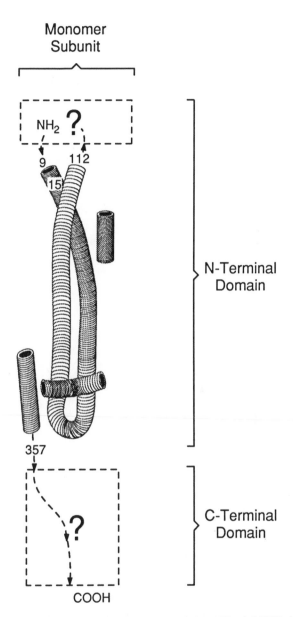

Fig. 4. Sketch of the structure of a monomeric subunit of the MITat 1.2 VSG, based on crystallographic evidence. Only the $\alpha$-helical segments, marked as cylinders, are shown for simplicity, although much of the $\alpha$-carbon trace is known. Untraced regions of the chain lie mainly at the "top" of the molecule, indicated by the boxed region. Nothing is known about the three-dimensional structure of the C-terminal domain. (Based on information from D. Freymann, M. Blum, and D.C. Wiley.)

interesting. A start in that direction has been made with the solution of a second VSG structure to 6 Å resolution (Metcalf et al., 1987). The ILTat 1.24 VSG was isolated from a different trypanosome serodeme than MITat 1.2. That is, the ILTat 1.24 trypanosome and the MITat 1.2 trypanosome have the capacity to express entirely different VSG repertoires. They come from different stocks, isolated from different vertebrate hosts, at different times and at different geographical locales. MITat 1.2 VSG is a group II variant. ILTat 1.24 is a member of group I containing 473 amino acids. The cDNA sequence predicts a C-terminal aspartate. There are 12 cysteines, 8 in the C-terminal domain and 4 in the N-terminal domain. A single glycosylation site is present, at Asn 425 within the C-terminal domain. The sequence is shown in Figure 5. Simple inspection of the two sequences shows few similarities. The two sequences have been aligned in Figure 6, using an algorithim called align, in which alignment of identical amino acids scores 3

Amino acid sequence of ILTat 1.24 VSG.

| | 1 | 2 | 3 | 4 | 5 | 6 | 7 | 8 | 9 | 10 | 11 | 12 | 13 | 14 | 15 | 16 | 17 | |
|---|---|---|---|---|---|---|---|---|---|---|---|---|---|---|---|---|---|---|
| *M* | *V* | *Y* | *R* | *N* | *I* | *L* | *Q* | *L* | *S* | *V* | *L* | *K* | *V* | *L* | *L* | *I* | *V* | *L* | *I* |
| *V* | *S* | *A* | T | H | F | G | V | K | Y | E | L | W | Q | P | E | C | E | L | T | 37 |
| A | E | L | R | K | T | A | G | V | A | K | M | K | V | N | S | D | L | N | S | 57 |
| F | K | T | L | E | L | T | K | M | K | L | L | T | F | A | A | K | F | P | E | 77 |
| S | K | E | A | L | T | L | R | A | L | E | A | A | L | N | T | D | L | R | A | 97 |
| L | R | D | N | I | A | N | G | I | D | R | A | V | R | A | T | A | Y | A | S | 117 |
| E | A | A | G | A | L | F | S | G | I | Q | T | L | H | D | A | M | T | P | P | 137 |
| T | A | R | P | I | A | L | A | Q | R | G | Q | G | S | N | G | N | A | A | M | 157 |
| A | S | Q | G | C | K | P | L | A | L | P | E | L | L | T | E | D | S | Y | F | 177 |
| S | D | V | I | S | D | K | G | S | E | D | F | A | T | N | K | C | P | R | T | 197 |
| G | Q | K | R | R | M | R | L | F | Q | A | A | S | ? | S | G | A | R | H | | 217 |
| R | C | A | V | L | R | G | Q | Q | D | K | L | R | P | W | R | H | S | S | K | 237 |
| R | S | Q | Q | P | T | R | P | D | L | S | D | F | S | G | T | A | R | N | Q | 257 |
| A | D | T | L | Y | G | K | A | H | A | F | I | T | E | L | L | Q | L | A | Q | 277 |
| Q | G | P | K | P | G | Q | T | E | V | E | T | M | K | L | L | A | Q | K | T | 297 |
| A | A | L | D | S | I | K | F | Q | L | A | A | S | T | G | K | K | T | F | R | 317 |
| L | Q | R | R | R | K | L | E | N | G | I | L | W | K | D | R | K | Q | Y | R | 337 |
| S | T | L | E | Q | V | K | E | E | K | V | K | G | A | D | P | E | D | P | S | 357 |
| K | E | S | K | I | S | D | L | N | T | ? | E | Q | L | Q | R | V | L | D | Y | 377 |
| Y | A | V | A | T | M | L | K | L | A | K | Q | A | E | D | I | A | K | L | E | 397 |
| T | E | I | A | D | Q | R | G | K | S | P | E | A | E | C | N | K | I | T | E | 417 |
| E | P | K | C | S | E | E | K | I | C | S | W | H | K | E | V | K | A | E | E | 437 |
| K | N | C | Q | F | N̲ | S | T | K | A | S | K | S | G | V | P | V | T | Q | T | 457 |
| Q | T | A | G | A | D̲ | T | T | A | E | K | C | K | G | K | G | E | K | D | C | 477 |
| K | S | P | D | C | K | W | E | G | G | T | C | K | D̲ | S | S | I | L | A | N | 491 |
| *N* | *K* | *Q* | *F* | *A* | *L* | *S* | *V* | *A* | *S* | *A* | *A* | *F* | *V* | *A* | *L* | *L* | *F* | | |

Fig. 5. Amino acid sequence of ILTat 1.24 VSG (determined by M. Carrington, University of Cambridge). Lettering and numbering as in Figure 3.

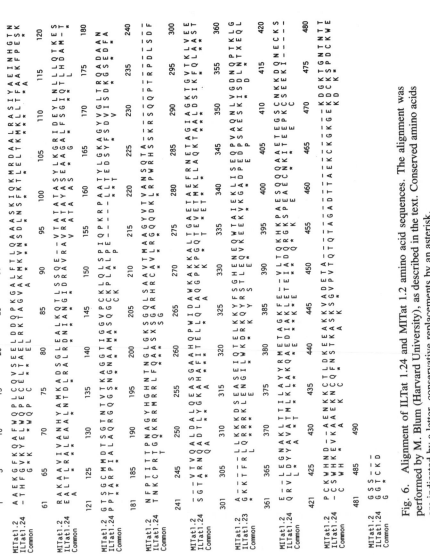

Fig. 6. Alignment of ILTat 1.24 and MITat 1.2 amino acid sequences. The alignment was performed by M. Blum (Harvard University), as described in the text. Conserved amino acids are indicated by a letter, conservative replacements by an asterisk.

points (except cysteines, 9 points). Alignment of conservative replacements scores 2 points. Total aligned scores are optimized through the introduction of gaps in each sequence. However, each gap costs 9 points. The score produced is compared with the mean score produced by randomizing the sequences and aligning the random sequences. Ten randomization cycles are conducted to give a mean and standard deviation for random alignment, which for the MITat 1.2 and ILTat 1.24 sequences gave values of 366 for the mean with a standard deviation of 9. The score for alignment of the two real nonrandomized VSG sequences was 396—just outside three standard deviations from the mean. In comparison, the alignment score between MITat 1.4 and ILTat 1.24, two group I VSGs, was 641.

Such statistical manipulations are fun—but what does the structural comparison show us? As shown in Figure 7, there is very considerable overlap between the two 6 Å structures, which identify the $\alpha$-helical regions in the structure. This suggests the exciting possibility of an underlying structural motif, common to these VSGs—and perhaps, by extension, to all VSGs. If that is correct, then 1,000 different VSG sequences might correspond to a single protein fold. That, in turn, would have important implications for the development of new rules relating structure and sequence.

## AN ANTIGENIC CONUNDRUM

It would seem self-evident that the driving force for the generation of new VSG sequences is the production of antigenically distinct variants. However, this is apparently not entirely true. Several labs have reported on the production of variant-specific monoclonal antibodies (Pearson et al., 1980; Miller et al., 1984a,b; Hall and Esser, 1984; Pinder et al., 1987; Clarke et al., 1987). Only a fraction of the monoclonal antibodies bind to living trypanosomes or, more critically, can neutralize infectivity. Many monoclonals can only be shown to bind to trypanosomes after fixation, although all are capable of binding to VSGs. The interpretation is that there are "cryptic" antigenic determinants—variant-specific sequences which are not exposed on the living trypanosome. Since there is no apparent selective advantage in the evolution of such sequences, how are they selected and maintained within the population? The corollary is, where on the structure are the sequences that define epitopes recognized by neutralizing antibodies? Presumably, changes in this region of the molecule define the practical limits of antigenic variation. Surface-exposed antigenic determinants are located within the N-terminal domain, but no epitope has yet been properly mapped. An analysis performed in my lab with a set of 16 monoclonal antibodies specific for the MITat 1.2

ILTat 1.24          MITat 1.2

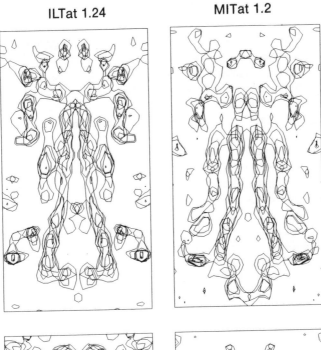

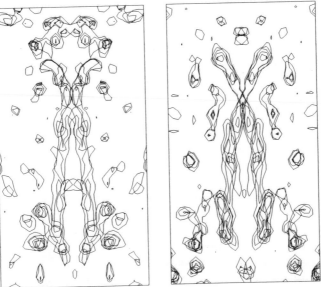

Fig. 7.   Sections of the 6 Å electron-density maps for ILTat 1.24 and MITat 1.2. For each variant, the crystal contains only the N-terminal two-thirds of the molecules. Sections are parallel to the long axis of the molecule and show the dimeric nature of the molecule. Top and bottom sections were taken at 90° to each other, and each diagram represents an area of 10 Å × 60 Å. At this resolution, the observable structures are α-helices. About 50% of the sequence in this portion of each molecule can be accounted for by this α-helical core.(Reproduced with permission from Metcalf et al., 1987.)

VSG was at least consistent with the view that neutralizing antibodies recognize sequences at the "top" of the structure shown in Figure 4 (Masterson et al., 1988). Such is the talent of the trypanosome to tantalize and frustrate the investigator that I am confident that neutralizing epitopes will be present in the "missing" elements of the structure! Much more detailed analyses of the antigenic determinants will be needed before that slur can be justified or refuted.

What of the cryptic variant-specific epitopes? One possibility is that sequence changes within exposed segments require changes in cryptic sites as a structural counterpoise. I prefer an alternative view, which is that the primary selection pressure for new VSG sequences is the ability to conform to the constraints of a conserved structure. As discussed elsewhere in this volume, the mechanism of antigenic variation involves gene conversion. This conversion can be complete, resulting in the total replacement of one VSG sequence by another, or it can be partial, producing a new VSG sequence containing elements of both the donor and recipient genes involved in the conversion. It has been proposed (Longacre and Eisen, 1986; Roth et al., 1986) that multiple, partial gene conversions can occur within a single VSG gene, leading to the extensive "scrambling" of two VSG sequences throughout their entire length. Many such scrambled sequences may be formed which are incapable of adopting the "VSG fold." Incorrect folding would either produce a VSG which could not be expressed at the surface at all, or which is expressed but does not pack to form a compact surface coat. Such sequences would be selected against. Only sequences containing the information to form the "VSG fold" would be maintained. Selection from such sequences of those which, in addition, contain antigenically distinct surface epitopes would be the function of the immune system.

## CONCLUSIONS

The study of antigenic variation in the African trypanosomes has provided the biggest incentive for the recruitment of molecular biologists into parasitology over the last decade. It is easy to see why. Francis Crick said that molecular biology was the study of that which interested molecular biologists. Whether you are a molecular biologist interested in DNA, RNA, protein, or carbohydrate, the biochemistry and genetics of antigenic variation hold something special for you. Personally, I see no end in sight. Analysis of the events involved in glycolipid addition is going to take years. We still do not know how expression of a single VSG gene is controlled. The analysis of relationships between VSG structure and sequence should prove fertile ground for a long while to come. In each of those three examples, the findings will be of great interest, not just to molecular biologists working on trypanosomes, but to the wider community of molecular biologists working on all forms of eukaryotic life. Thanks to the principles of biochemical economy, trypanosomes can tell us almost as much about their hosts as they can about themselves.

## REFERENCES

Allen G, Gurnett LP (1983): Locations of the six disulphide bonds in a variant surface glycoprotein (VSG 117) of *Trypanosoma brucei*. Biochem J 209:481–487.

Auffret CA, Turner MJ (1981): Variant specific antigens of *Trypanosoma brucei* exist in solution as glycoprotein dimers. Biochem J 193:647–650.

Boothroyd JC, Cross GAM, Hoeijmakers JH, Borst P (1980): A variant surface glycoprotein of *Trypanosoma brucei* synthesized with a C-terminal hydrophobic "tail" absent from purified glycoprotein. Nature 288:624–626.

Boothroyd JG, Paynter CA, Cross GA, Bernards A, Borst P (1981): Variant surface glycoproteins of *Trypanosoma brucei* are synthesized with cleavable hydrophobic sequences at the carboxy and amino termini. Nucleic Acids Res 9:4735–4743.

Boothroyd JC, Paynter CA, Coleman SL, Cross GA (1982): Complete nucleotide sequence of complementary DNA coding for a variant surface glycoprotein from *Trypanosoma brucei*. J Mol Biol 157:547–556.

Brown RC, Evans DA, Vickerman K (1973): Changes in oxidative metabolism and ultrastructure accompanying differentiation of the mitochondrion in *Trypanosoma brucei*. Int J Parasitol 3:691–704.

Capbern A, Giroud C, Baltz T, Mattern P (1977): *Trypanosoma equiperdum:* Etude des variation antigenique au cours de la trypanosomose experimentale du lapin. Exp Parasitol 42:6–13.

Cardoso de Almeida ML, Turner MJ (1983): The membrane form of variant surface glycoproteins of *Trypanosoma brucei*. Nature 302:349–352.

Clarke MW, Barbet AF, Pearson UC (1987): Structural features of antigenic determinants on variant surface glycoproteins from *Trypanosoma brucei*. Mol Immunol 7:707–713.

Cohen C, Reinhardt B, Parry DA, Roelants GE, Hirsch W, Kanwe B (1984): Alpha-helical coiled coil structures of *Trypanosoma brucei* variable surface glycoproteins. Nature 311:169–171.

Cross GAM (1975): Identification, purification and properties of clone-specific glycoprotein antigens constituting the surface coat of *Trypanosoma brucei*. Parasitology 71:393–417.

Cross GAM (1977): Isolation, structure and function of variant specific surface antigens. Ann Soc Belge Med Trop 57:389–399.

Cross GAM (1984a): Release and purification of *Trypanosoma brucei* VSG. J Cell Biochem 24:79–90.

Cross GAM (1984b): Structure of the variant glycoproteins and surface coat of *Trypanosoma brucei*. Philos Trans R Soc Lond [Biol] 307:3–12.

Duvillier G, Aubert JP, Baltz T, Richet C, Degand P (1983): Variant specific surface antigens from *Trypanosoma equiperdum:* Chemical and physical studies. Biochem Biophys Res Commun 110:491–498.

Ferguson MAJ, Halder K, Cross JAM (1985a): *Trypanosoma brucei* variant surface glycoprotein has a *sn*-1,2-dimyristylglycerol membrane anchor at its COOH terminus. J Biol Chem 260:4963–4968.

Ferguson MAJ, Low MG, Cross GAM (1985b): Glycosyl-*sn*-1,2-dimyristyl-phosphatidylinositol is covalently linked to *Trypanosoma brucei* variant surface glycoprotein. J Biol Chem 260:14547–14555.

Franke E (1905): Therapeutische versuche bei Trypanosomen-enkrankung. Munch Med Wochenschr 52(11):2059. Abstracted in Bulletin of Sleeping Sickness (1909) 1:218.

Freymann D, Metcalf P, Turner MJ, Wiley DC (1984): The 6A resolution X-ray structure of a variable surface glycoprotein from *Trypanosoma brucei*. Nature 311:167–169.

Gurnett AM, Raper J, Turner MJ (1986): Solution properties of the variant surface glycoprotein of *Trypanosoma brucei*. Mol Biochem Parasitol 18:141–153.

Hall T, Esser K (1984): Topologic mapping of protective and non-protective epitopes on the variant surface glycoprotein of the WRATat 1 clone of *Trypanosoma brucei rhodesiense*. J Immunol 132:2059–2063.

Holder AA, Cross GAM (1981): Glycopeptides from variant surface glycoproteins of *Trypanosoma brucei*. C-terminal location of antigenically cross-reacting carbohydrate moieties. Mol Biochem Parasitol 2:135–150.

Johnson JG, Cross GAM (1977): Carbohydrate composition of variant specific surface antigen glycoproteins from *Trypanosoma brucei*. J Protozool 24:587–591.

Johnson JG, Cross GAM (1979): Selective cleavage of variant surface glycoproteins from *Trypanosoma brucei*. Biochem J 178:689–697.

Lalor TM, Kjeldgaard M, Shimamato GT, Strickler JE, Konigsberg WH, Richards FF (1984): Trypanosome variant-specific glycoproteins: A polygene protein family with multiple folding patterns. Proc Natl Acad Sci USA 81:998–1002.

LePage RWF (1968): Antigenic Variation in *Trypanosoma brucei*. Ph.D. thesis, University of Cambridge.

Longacre S, Eisen H (1986): Expression of whole and hybrid genes in *Trypanosoma equiperdum* antigenic variation. EMBO J 5:1057–1063.

Low MG (1987): Biochemistry of the glycosyl-phosphatidylinositol membrane protein anchors. Biochem J 244:1–13.

Masterson WJ, Taylor D, Turner MJ (1988): Topologic analysis of the epitopes of a variant surface glycoprotein of *Trypanosoma brucei*. J Immunol 140:3194–3199.

McConnell J, Turner MJ, Rovis L (1983): Biosynthesis of *Trypanosona brucei* variant surface glycoproteins—analysis of carbohydrate heterogeneity and timing of post-translational modifications. Mol Biochem Parasitol 8:119–135.

Metcalf P, Blum M, Freymann D, Turner MJ, Wiley DC (1987): Two variant surface glycoproteins of *Trypanosoma brucei* of different sequence classes have similar 6Å resolution X-ray structure. Nature 325:84–86.

Miller EN, Allan LM, Turner MW (1984a): Topological analysis of antigenic determinants of a variant surface glycoprotein of *Trypanosoma brucei*. Mol Biochem Parasitol 13:67–81.

Miller EN, Allan LM, Turner MJ (1984b): Relationship of antigenic determinants to structure within a variant surface glycoprotein of *Trypanosoma brucei*. Mol Biochem Parasitol 13:309–322.

Olafson RW, Clarke MW, Kielland SO, Barbet AF, McGuire TD (1984): Amino terminal sequence homologies amongst variant surface glycoproteins of the African trypanosomes. Mol Biochem Parasitol 2:287–298.

Pays E, Delauw HF, Van Assel S, Laurent M, Vervoort T, Van Meirvenne N, Steinert M (1983a): Modifications of a *Trypanosoma b. brucei* antigen gene repertoire by different DNA recombinational mechanisms. Cell 35:721–731.

Pays E, Van Assel S, Laurent M, Darville M, Vervoort T, Van Meirvenne N, Steinert M (1983b): Gene conversion as a mechanism for antigenic variation in trypanosomes. Cell 34:371–381.

Pearson TW, Pinder M, Roelants GE, Kar SK, Lundin LB, Mayor-Withey KS, Hewett RS (1980): Methods for the derivation and detection of anti-parasite monoclonal antibodies. J Immunol Methods 34:141–154.

Pinder M, Van Melick A, Vernet G (1987): Analysis of protective epitopes on the variant surface glycoprotein of a *Trypanosoma brucei brucei* (DITat 1.3) using monoclonal antibodies. Parasite Immunol 9:395–400.

Rice-Ficht AC, Chen KK, Donelson JE (1981): Sequence homologies near the C-termini of the variable surface glycoproteins of *Trypanosoma brucei*. Nature 294:53–57.

Roth CW, Longacre S, Raibaud A, Baltz T, Eisen H (1986): The use of incomplete genes for the construction of a *Trypanosoma equiperdum* variant surface glycoprotein gene. EMBO J 5:1065–1070.

Savage A, Geyer R, Stirm S, Reinwald E, Risse HJ (1984): Structural studies on the major oligosaccharides in a variant surface glycoprotein of *Trypanosoma congolense*. Mol Biochem Parasitol 11:309–328.

Strickler JE, Patton CL (1980): *Trypanosoma brucei brucei:* Inhibition of glycosylation of the major variable surface glycoprotein by tunicamycin. Proc Natl Acad Sci USA 77:1529–1533.

Strickler JE, Patton CL (1982*): Trypanosoma brucei*: Nearest neighbor analyses on the major variable surface coat glycoprotein—crosslinking patterns with intact cells. Exp Parasitol 53:117–131.

Turner MJ, Cordingley JS (1981): Evolution of antigenic variation in the salivarian trypanosomes. In M.J. Carlile, B.F. Collins, and B.E.B. Moseley (eds): Molecular and Cellular Aspects of Microbial Evolution. Society for General Microbiology Symposium 32. Oxford: Cambridge University Press, pp 313–347.

Van der Ploeg LHT, Valerio D, DeLange T, Bernards A, Borst P, Grosveld FG (1982): An analysis of cosmid clones of nuclear DNA from *Trypanosoma brucei* shows that the genes for VSGs are clustered in the genome. Nucleic Acids Res 10:5905–5923.

Vickerman K (1969): On the surface coat and flagellar adhesion in trypanosomes. J Cell Sci 5:163–193.

Vickerman K, Luckins AG (1969): Localization of variable antigens in the surface of *Trypanosoma brucei* using ferritin-conjugated antibody. Nature 224:1125–1126.

The Biology of Parasitism, pages 371–400
© 1988 Alan R. Liss, Inc.

# Unsolved Mysteries of Trypanosome Antigenic Variation

## John E. Donelson

*Department of Biochemistry, University of Iowa, Iowa City, Iowa 52442*

## INTRODUCTION

African trypanosomes are protozoan parasites that cause sleeping sickness in humans and nagana in domestic livestock throughout much of equatorial Africa. These organisms spend part of their life cycle in their insect vector, the tsetse fly, and the remainder of the life cycle in the bloodstream of their mammalian host. During their time in the bloodstream each trypanosome is covered by about $10^7$ copies of a single surface protein called the variant surface glycoprotein (VSG). The parasites evade the immune system of their mammalian host by periodically switching from the production of one VSG on their surface to another VSG, a phenomenon called antigenic variation. This process enables a population of trypanosomes in the bloodstream of an infected animal to keep "one step ahead" of the antibodies that are raised against their VSGs.

The crucial event in antigenic variation is the step at which an individual trypanosome switches from the synthesis of one VSG to another. The molecular basis for this switch, and for the remarkable phenomenon of antigenic variation, is now partially understood as the result of much effort by several research groups during the past few years. This work has demonstrated that complex rearrangements of the several hundred VSG genes in the trypanosome genome are involved in the selection of a specific VSG to be expressed. However, many questions about how these gene rearrangements relate to the actual switch mechanism itself remain to be answered.

This chapter will begin by summarizing the biology of African trypanosomes and what has been learned about the molecular basis for antigenic variation. It will then focus on those aspects of antigenic variation that are not understood and, in some cases, indicate how these mysteries can be further evaluated experimentally.

## THE BIOLOGY OF AFRICAN TRYPANOSOMES

African trypanosomes can potentially occur wherever one of several species of tsetse flies in the genus *Glossina* are located. This includes an area of Africa south of the Sahara Desert approximately the size of the United States that supports a human population of over 200 million people. These parasites belong to the genus *Trypanosoma* within the order Kinetoplastida. Their species classification is based on their morphology, geographical location, the hosts they infect, and occasionally, their virulence. *Trypanosoma brucei brucei* (Fig. 1) has been used for most experimental investigations because it is easy to maintain in laboratory animals and does not survive in human serum, reportedly because it is lyzed by the high-density lipoprotein (HDL) fraction (Rifkin, 1978). The two subspecies of *T. brucei brucei* that infect humans are *Trypanosoma brucei rhodesiense* (acute infections) and *Trypanosoma brucei gambiense* (chronic infections). Two other important species that have a tremendous impact on the African cattle industry are *Trypanosoma congolense* and *Trypanosoma vivax*. In Africa today the species that

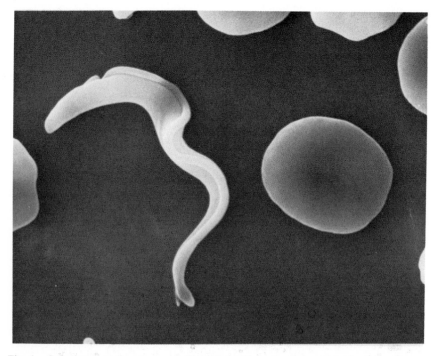

Fig. 1.  Scanning electron micrograph of a *T. brucei brucei* among red blood cells. ×5500.

infect domestic animals are much more prevalent than the species that infect humans and have a far greater impact. This has not always been the case. It has been estimated that in the late 1890s two-thirds of the population living near Lake Victoria in Uganda, about 200,000 people, died of the disease. Today the threat of human epidemics on this scale is reduced by constant surveillance and treatment with antiparasitic drugs. However, the disease remains a severe impediment to the development of the livestock industry in the endemic areas.

Compared to many parasites trypanosomes have a relatively simple life cycle (Vickerman, 1974). When a tsetse fly bites an infected mammal, trypomastigotes are ingested with the blood meal. Once in the midgut of the fly, the organisms change from anaerobic to aerobic growth, lose their surface coat containing the VSG, and begin to multiply. After about 3 weeks in the midgut, the parasites move to the salivary glands or proboscis (depending on the trypanosome species) and differentiate into the metacyclic form. Metacyclic trypanosomes reacquire the VSG coat but do not divide. They are injected into the mammal during the fly bite and are the first parasites to be exposed to the host immune response. An infected tsetse fly can harbor as many as 50,000 metacyclic trypanosomes.

Shortly after entering the bloodstream, trypanosomes begin to multiply by binary fission and undergo a variety of additional morphological and metabolic changes (returning to anaerobic glycolysis as the main source of ATP). They continue, however, to express the metacyclic VSGs for about 5 days after the infection (Hadjuk and Vickerman, 1981; Esser et al., 1982). Between days 5 and 7 of infection the parasites switch from the expression of metacyclic VSGs to the first bloodstream VSGs, one of which is often the same VSG that was expressed by the trypanosomes ingested by the fly (Hadjuk et al., 1981; Hajduk, 1984). The bloodstream forms of the parasite persist in the bloodstream and eventually invade the central nervous system. The infected individual ultimately becomes comatose and dies—hence, the name sleeping sickness.

In the bloodstream two distinct forms of the parasite occur. The long, slender forms actively divide and undergo antigenic variation. The short, stumpy forms do not divide but have a more developed mitochondria and are thought to be the form that infects the insect. The life cycle is completed when a tsetse fly takes up the bloodstream forms while feeding on an infected mammal.

## THE PHENOMENON OF ANTIGENIC VARIATION

During the early investigations of trypanosomiasis at the beginning of this century, it was noticed that the number of trypanosomes in the bloodstream of an infected individual fluctuated dramatically with time. The most complete characterization of these successive waves of parasitemia was conducted by Ross and Thomson (1910). They systematically examined the blood of an infected patient and demonstrated that a peak of parasitemia was followed a few days later by a low level of parasites and then by another peak about 2 weeks later (Fig. 2A). In a remarkable burst of insight for the time, they and other investigators speculated that a new wave of parasitemia was composed of trypanosomes that had somehow escaped the action of antibodies directed against the previous parasites. It would be nearly 60 years before the relationship between trypanosomes and the immune system would be better understood.

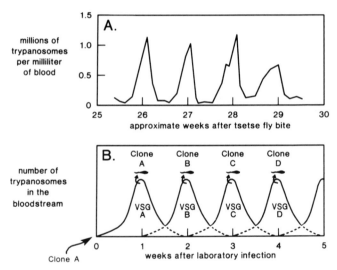

Fig. 2. **A:** Demonstration of the successive waves of parasitemia during a *T. rhodesiense* infection of a human patient (redrawn from Ross and Thompson, 1910). **B:** An example of a carefully controlled infection of *T. rhodesiense* in an immunocompetent laboratory animal. A single trypanosome expressing VSG$_A$ (clone A) is injected into the blood and gives rise to the first peak of parasitemia in which all of the parasites are expressing VSG$_A$. The host immune system kills most of the parasites in the next few days, but one or more trypanosomes switch to the expression of VSG$_B$ which is not recognized by the anti-VSG$_A$ antibodies. These trypanosomes begin the next wave of parasitemia. VSG$_B$ then elicits a second set of antibodies to start the process again. From each peak of parasitemia, individual trypanosomes can be cloned to investigate the VSG structure and the molecular mechanism of antigenic variation. Since the switch to a new VSG occurs spontaneously and individual trypanosomes can have slightly different growth rates, the synchrony of the parasitemia peaks is lost later in the infection.

By the mid-1970s it had been shown that the trypanosome, including its flagellum, was covered by a single major protein (Vickerman, 1974). Then in 1976 another revealing study was reported (Bridgen et al., 1976). These investigators prepared four cloned trypanosome populations from a single rabbit infected with a cloned strain of *T. b. brucei* and isolated the different VSGs from these four populations. N-terminal degradation of the four VSGs demonstrated that the first 25–30 amino acids of each were remarkably different and not related by potential frame shift changes or point mutations. This information provided a more detailed explanation of the antigenic variation observed much earlier by Ross and Thompson.

As depicted in Figure 2B, each peak of parasitemia early in the infection of a laboratory animal by a single trypanosome clone is composed of a population of trypanosomes expressing the same immunologically distinct VSG. In one often-cited case a trypanosome infection of an experimental rabbit resulted in the successive expression of over 100 different antigenic types (Capbern et al., 1977). It has been estimated that well-adapted laboratory strains of *T. brucei* spontaneously switch VSGs at a rate of $10^{-4}$ to $10^{-6}$ per division (Miller and Turner, 1981; Lamont et al., 1986). The immune response of the host does not appear to induce the switch (Doyle et al., 1980; Vickerman et al., 1980) but it does indirectly select new variants (because it does not immediately recognize their VSGs) that will give rise to the next parasitemia.

The order of appearance of new bloodstream variants remains a somewhat controversial point. In most cases the VSGs appear more or less at random (Gray, 1965; Campbell et al., 1979), although some VSGs occur more frequently than others. In *T. equiperdum*, which is transmitted venereally in horses (without an insect vector), a subset of VSGs appears early in the infection, while other VSGs are more likely to occur later in the infection (Longacre et al., 1983a). In still other cases a given VSG has a propensity to follow another (Miller and Turner, 1981). It is possible that the genomic location of a VSG gene influences the frequency with which it expressed and whether it is likely to follow another specific VSG. We will return to a consideration of the order of appearance of VSGs later in the chapter.

## THE STRUCTURE OF VSGs

The amino-acid and higher-order structure of the different VSGs and their attachment to the outer trypanosome membrane constitute fascinating aspects of antigenic variation that are discussed in other chapters, so they will not be presented here except as they relate to the molecular biology.

The approximately $10^7$ VSG molecules on the surface of a trypanosome probably compose about 5% of the total protein of the organism (Cross, 1975; Turner, 1982, 1985). Likewise, about 5% of the total polyadenylated RNA of the parasite codes for the VSG. This large percentage contributes to the ease with which VSG cDNAs can be identified in a trypanosome cDNA library by plus/minus screening. The complete or partial nucleotide sequences of the cDNAs for about 20 different VSGs from at least five different trypanosome species have been determined (see Donelson and Rice-Ficht, 1985, for a partial summary). Comparison of these different VSG cDNA nucleotide sequences reveals that the VSGs of most trypanosome species can be placed into two groups based on sequence similarities among the coding regions for the last 50 amino acids of the nascent VSGs, a region that includes a C-terminal hydrophobic tail of 17 or 23 amino acids. In addition, the 80–100 nucleotide 3'-nontranslated region of the VSG mRNAs display extensive similarity. As we shall see, these nucleotide homologies at the 3'-ends of the VSG genes often form the downstream boundary of a transposed VSG gene.

## CHROMOSOMAL DISTRIBUTION OF VSG GENES

One of the ironies in the study of the genetic basis for trypanosome antigenic variation is that, to date, it has not used the conventional tools of genetics. For example, no mutants that affect the process of antigenic variation exist; and, in fact, no nuclear mutations at all are known. This means that virtually nothing is known about the genetics of trypanosomes except that DNA exchange may occur during an insect stage (Tait, 1980; Jenni et al., 1986). Like other lower eukaryotes, trypanosome chromosomes do not sufficiently condense during the cell cycle for cytological staining techniques to be useful in determining the chromosome number. Furthermore, without mutations, genetic linkage groups cannot be determined. Even the ploidy of the parasite remains somewhat uncertain, although most of the evidence suggests that bloodstream trypanosomes are diploid while metacyclic parasites may be haploid (Tait, 1980; Gibson et al., 1985; Zampetti et al., 1986; Jenni et al., 1986; Vickerman, 1986). DNA complexity measurements ($C_0t$ analyses) indicate that the trypanosome genome contains about $3.7 \times 10^7$ base pairs while sensitive colorimetric measurements suggest that the nuclear DNA content is twice that (Borst et al., 1982). This haploid genome size is about ten times that of *Escherichia coli* and two to three times the haploid value for yeast. It is 1% of the size of the mammalian genome.

Although conventional approaches have not been useful in determining the number of chromosomes, it is now clear that the trypanosome genome is distributed among a large number of DNA molecules, some of which are quite small by eukaryotic standards. The first evidence for these small chromosomes was the identification by sucrose gradients of a specific VSG gene on an 80-kilobase (kb) "minichromosome," i.e., a DNA molecule only a little larger than the 50-kb DNA of bacteriophage lambda (Williams et al., 1982). These smaller chromosomes can also be resolved by the technique of pulsed-field gradient (PFG) gel electrophoresis (Schwartz and Cantor, 1984; Van der Ploeg et al., 1984c; Gibson and Borst, 1986). By pulsing a direct current through an agarose gel at different angles, this technique separates DNA molecules whose sizes are between 100 kb and several thousand kilobases. Trypanosome nuclear DNA resolves into four general size classes using this approach (Fig. 3A). Densitometer tracings of ethidium-bromide-stained gels indicate that there are about 100 minichromosomes of 50–100 kb, 5–7 DNA molecules of 200–700 kb, several molecules in the 2,000-kb

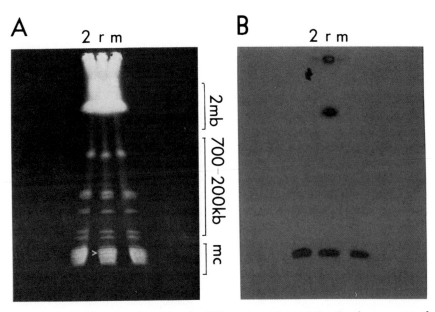

Fig. 3.   **A:** Ethidium bromide stain of a PFG agarose gel containing the chromosomes of three closely related *T. brucei* clones (2, r, and m). Minichromosomes of about 100 kb (mc); the 200–700-kb chromosomes; and the 2,000-kb, or 2-megabase (2 mb), chromosomes can be observed. **B:** Autoradiogram of a Southern blot of the same gel probed with $VSG_r$ cDNA. This probe hybridizes to the basic copy $VSG_r$ gene on a minichromosome in all three genomes and to an $ELC_r$ gene on a 2-mb chromosome in the genome of trypanosome clone r. Thus, the expression site is on a different chromosome than the basic copy gene.

range, and some molecules that are too big to measure; i.e., they do not effectively enter the gel. Furthermore, individual trypanosomes derived from the same original trypanosome clone sometimes display variation in the size of one or more chromosomes as well as a difference in the number of resolved chromosomes (Bernards et al., 1986). This demonstrates that chromosome size and number is not constant in trypanosomes; rather, chromosomes increase or decrease in size, perhaps by occasionally breaking into pieces that integrate into or fuse onto other chromosomes. The significance of these frequent chromosomal changes is not clear. They are not strictly confined to parasites that undergo antigenic variation because they can also be detected in *Trypanosoma cruzi* and *Leishmania* which don't display antigenic variation (Engman et al., 1987; Gibson and Borst, 1986; Spithill and Samaras, 1985). Minichromosomes, however, are less prominent in these latter kinetoplastids, which might suggest that their presence does correlate in some way with the ability to undergo antigenic variation (Gibson and Borst, 1986).

When the trypanosome DNA molecules are transferred from PFG gels for Southern blots under low-hybridization stringency, DNA probes coding for the VSG C-terminal similarity regions hybridize to all of the size classes (Van der Ploeg et al., 1984c). Therefore, VSG genes appear to be scattered among all of the chromosomes, independent of the size of the chromosome.

When the same C-terminal similarity probes are used in conventional one-dimensional Southern blots of restricted genomic DNA, they hybridize (under low stringency) to many fragments. From these experiments it has been estimated that there are several hundred VSG genes in the genome and perhaps as many as a thousand (Van der Ploeg, 1982a). Since VSG mRNAs are about 1,600 nucleotides, the coding regions for a thousand VSG genes would consume $1.6 \times 10^6$ base pairs or about 4% of the genome. As described below, the actual transposition unit in the VSG gene rearrangements is between 2,500 and 3,500 base pairs. Thus, as much as 9% of the genome is consumed in specifying the VSGs and accommodating the rearrangements of their genes. Indeed, recombinant DNA cosmid clones have been reported in which several unrelated VSG genes were present (Van der Ploeg et al., 1982a). A commitment of nearly one-tenth of the genome seems a large genetic price to pay for the ability to undergo antigenic variation but perhaps it indicates that a VSG gene repertoire of this size is required for antigenic variation to be effective under natural conditions.

## THE BASIC GENE REARRANGEMENT: FORMATION OF AN EXPRESSION-LINKED COPY GENE

Most of the approximately 1,000 *basic copy* VSG genes contains the complete, or nearly complete, coding region for that VSG. These genes are situated at one of two general chromosomal locations, either at an internal DNA site or near a double-strand DNA break thought to be a chromosomal end or telomere. The internal genes are never transcribed in their intrachromosomal location. They must be duplicated and translocated to a telomere-linked site before their transcription can occur. This duplicative translocation is a *gene conversion* event—a term used to describe the nonreciprocal transfer of nucleotide sequences from one DNA region to another (Fink and Petes, 1984). The telomere-linked duplicated VSG gene was named the "expression-linked copy" (ELC) gene by Hoeijmakers et al. (1980), who first detected it in genomic Southern blots. Pays et al. (1981) subsequently showed that the ELC gene is more susceptible to DNase I digestion than the corresponding internal basic copy gene and, therefore, is probably the transcribed gene.

Telomere-linked basic copy VSG genes, on the other hand, need not undergo gene conversion to be transcribed. They seemingly have the potential to be expressed in situ even though they sometimes do undergo duplication prior to their transcription (Buck et al., 1984; Laurent et al., 1983; Pays et al., 1983b; Myler et al., 1984a,b; Young et al., 1984). Since there may be 100 distinct DNA molecules in the trypanosome nucleus, as many as 200 VSG genes could be telomere linked. It is possible that many, if not all, of these genes are poised for transcription because, on the basis of Southern blots, it is not essential for them to move from their telomere-linked site to be transcriptionally activated.

Experimental data for the two basic chromosomal locations of VSG genes and their modes of activation are shown in Figure 4. The left panel shows the autoradiogram of a Southern blot (Murphy et al., 1984; Donelson and Rice-Ficht, 1985) in which the radioactive probe is a VSG cDNA from trypanosome clone IATat 1.2. This cDNA hybridizes under *high* stringency to two restriction fragments in the genomes of trypanosome clones that are not expressing the IATat 1.2 VSG and to a third fragment in the IATat 1.2 genome. Thus, there are two similar internal basic copy genes, or isogenes, for this VSG, one of which is duplicated and expressed as an ELC gene in the IATat 1.2 clone.

The right panel of Figure 4 shows a similar blot in which the radioactive probe contains the coding sequence for another VSG gene (Donelson et al., 1983; Donelson and Rice-Ficht, 1985). In this case the probe hybridizes to the same number of fragments (two) in each genome but one fragment is a different size in each clone. The VSG gene on the constant-sized fragment is an internal gene and the gene on the variable-sized fragment is telomere-linked. As we shall see, the size of this fragment is variable because the

## DNA from Trypanosome Clones:

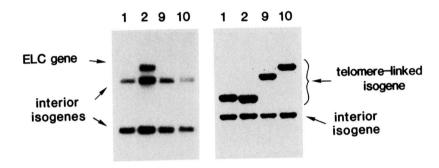

Fig. 4. Autoradiograms of two Southern blots showing the duplicative and non-duplicative modes of VSG gene rearrangements (Murphy et al., 1984). **Left:** Genomic DNAs from trypanosome clones IATat 1.1, 1.2, 1.9, and 1.10 **(lanes 1, 2, 9, 10)** were restricted and probed with a cDNA for the IATat 1.2 VSG. Note the presence of two isogenes for this VSG in all four genomes and an additional ELC gene in the IATat 1.2 genome. **Right:** The same four genomes were probed with the coding region for a different VSG. In this case, the probe hybridizes to two isogenes, one of which is on a variable-sized restriction fragment containing a chromosomal telomere at one end. This telomere-linked VSG gene does not need to be duplicated to be expressed.

number of base pairs between the end of the gene and the telomere end of the fragment constantly changes.

Figure 4 also shows that each of the VSG genes used as a probe belongs to a family of two closely related basic copy isogenes. Similar Southern blots under high stringency have revealed that other VSG genes occur within families of one to ten different isogenes (Donelson et al., 1982; Parsons et al., 1983; Pays et al., 1983b; Young et al., 1983b). These isogenes have extensive sequence similarities throughout their entire coding regions rather than just within the C-terminal 50 codons and the 3' nontranslated regions. This suggests that they evolved by duplications followed by point mutations. In some cases these isogenes may arise from ELCs that are not deleted from the genome during the switch to another VSG but "linger" as unexpressed telomere-linked VSG genes; i.e., they remain in the genome to become basic copy genes (Buck et al., 1984; Laurent et al., 1983; Young et al., 1983b). In other cases, a silent gene conversion of an internal basic copy gene to another internal site may occur. Tandem duplication as a mechanism to generate isogenes may also occur although, to date, isogenes have been found scattered about the genome rather than adjacent to each other (Van der Ploeg, 1982a). Some evidence exists to suggest that different isogenes accumulate point mutations at different rates for unknown reasons (Frasch et al., 1982).

## THE ANATOMY OF TELOMERE-LINKED VSG GENES

The proximity of the ELC and other VSG genes to a naturally occurring double-strand DNA break (a telomere) was first detected by careful mapping of restriction sites surrounding the gene (De Lange and Borst, 1982; Donelson et al., 1982; Borst et al., 1983; Van der Ploeg et al., 1982b; Young et al., 1982). These initial experiments suggested that virtually every restriction enzyme tested cleaved at a specific site 5–20 kb downstream from these VSG genes (usually with no restriction sites in between). The only rational explanation for such a "universal" restriction site was that a double-strand break in the DNA already existed at that position. This was demonstrated directly using Bal31 nuclease digestion of the genomic DNA prior to restriction enzyme analysis (De Lange et al., 1982). The Bal31 progressively degrades linear DNA from the ends with time, thereby reducing the sizes of restriction fragments that contain a naturally occurring double-strand break at one end. Bal31 susceptibility has now been used extensively to establish telomere linkage of both expressed and nonexpressed VSG genes (Parsons et al., 1983; Pays et al., 1983c; Lenardo et al., 1984; Young et al., 1984).

Most of what is known about telomere-linked VSG genes has come from sequence determinations of recombinant DNA clones containing these genes and their flanking regions. Initially, such clones were difficult to obtain because of the lack of flanking restriction sites at which ligation to a cloning vector could be conducted. Now a few recombinant DNA clones of the ELC or other telomere-linked VSG genes have been acquired by cloning fragments of sheared DNA or by identifying rare restriction sites in the flanking "barren" regions (Campbell et al., 1984a; Blackburn and Challoner, 1984; Murphy et al., 1984; Shea et al., 1987).

The information obtained from sequence analyses of these cloned telomere-linked VSG genes is summarized in Figure 5. Downstream from these genes are tandem repeats of the hexanucleotide 5' CCCTAA 3' that are studded with related sequences and AT-rich segments (Blackburn and Challoner, 1984; Van der Ploeg, 1984b). Although not unambiguously demonstrated, it seems likely that the entire downstream barren region of 5–20 kb is composed of these hexameric repeats and AT-rich regions or their derivatives. Bal31 digestion of total genomic DNA preferentially degrades DNA

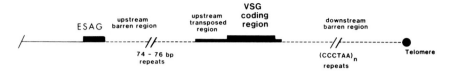

Fig. 5. The general structure of telomere-linked basic copy VSG genes and ELC genes. The solid lines indicate sequences containing a conventional distribution of restriction sites. The dotted lines indicate "barren regions" of 5–30 kb that are composed, at least in part, of repetitive sequences lacking restriction sites. Transcription of the ESAG and the adjacent telomere-linked VSG gene is coordinately regulated.

regions that hybridize to these hexamer repreats, suggesting that the repeat sequence occurs at virtually all natural DNA termini of the cell.

The presence of VSG genes near telomeres permits a given VSG cDNA to be used as a probe in Southern blots for a specific telomere. This has been used to demonstrate that the distance between a specific VSG gene and its telomere steadily increases with time, occasionally undergoing large dramatic deletions (Bernards et al., 1983; Pays et al., 1983c; Van der Ploeg et al., 1984b). Estimates of this growth rate are between 6 and 10 base pairs per generation, suggesting that one hexamer repeat is added during each replication event. Presumably, periodic deletions of many repeats then bring the number of telomeric repeats back down to a manageable size for the cell, after which the telomere growth begins again. This continual change in the number of hexamer repeats adjacent to the telomere is probably a function of telomere replication and is responsible for the variable size of the restriction fragments observed in Figure 4. It is likely that similar telomere growth and contraction occurs in other organisms as well but in these cases a probe for a specific telomere is not as readily available to detect it.

The downstream (3') boundary of the ELC transposition unit usually occurs somewhere within the region that includes the coding sequence for the C-terminal hydrophobic tail and the 3' nontranslated region of the mRNA (Bernards et al., 1981; Majumder et al., 1981; Donelson et al., 1983). This means that the expression site into which the new ELC moves provides the sequences for the C-terminus of the nascent VSG or at least the polyadenylation signal for the mRNA. In a least one case (Donelson et al., 1983), the internal basic copy gene did not possess a termination codon in the correct location—the proper termination codon was contributed to the incoming ELC gene by the expression site. This finding suggests that excision of an ELC from an expression site often does not occur at exactly the same boundary nucleotide as its insertion. Thus, remnants of a previous ELC may show up at the end of a new ELC. Such variability in the transposition boundary might be expected if switching from one ELC to another in the same expression site occurred via a gene conversion mechanism employing similar C-terminal similarity sequences (Michels et al., 1982; Pays et al., 1982; Van der Ploeg et al., 1984a). Therefore, not only are the C-terminal amino acid similarities required for the folding and deposition of the VSG on the parasite surface, but the corresponding DNA sequence similarities are also required for the ELC transposition.

The upstream (5′) boundary of the ELC transposition unit typically occurs 1–2 kb in front of the initiator methionine codon of the VSG. Since the average VSG coding sequence is about 1.6 kb, this means that the transposition unit is between 2.5 and 3.5 kb. In front of this transposition unit is another "barren" region of 5–30 kb containing few, if any, restriction sites. In the cases that have been studied, this upstream "barren" region contains tandem repeats of a 74–76 base pair sequence (some of which are interrupted by TAA repeats) that are unrelated to the downstream telomeric hexamer repeats (Campbell et al., 1984a; Liu et al., 1983). A few copies of this 74–76 base pair repeat are also upstream of many internal basic copy VSG genes (Van der Ploeg et al., 1982b). These findings suggest that these upstream repeats serve as homologous "nucleation sites" for the gene conversion event that replaces one ELC with another in an expression site. The conserved coding sequences for the C-terminal 50 amino acids and the 3′ nontranslated region could then serve as the downstream "nucleation site" for homologous recombination (Bernards et al., 1981; Donelson et al., 1983; Michels et al., 1983). One reservation about this model is that in separate characterizations of two unrelated ELCs, these 74–76 base pair repeats were not observed at the 5′ transposition boundary (Murphy et al., 1984; Pays et al., 1983d). In contrast, another repetitive sequence appeared to occur at this boundary. Therefore, other sequences and/or factors could also be involved in the gene conversion events. In addition, it should be emphasized that the 5′ barren region is sometimes as large as 40 kb (Pays et al., 1983d), and there is no evidence that this entire region is composed of the 74–76 base pair repeats.

Immediately upstream of the 5′ barren regions of expressed telomere-linked VSG genes are other genes that are transcribed. The best studied of these are a family of novel isogenes called the expression-site-associated genes (ESAGs) (Cully et al., 1985, 1986). Transcription of these genes is coordinately regulated with the downstream VSG gene in the expression site. If the VSG gene is transcribed, then the corresponding ESAG is also transcribed. If the VSG gene is not transcribed, then its neighboring ESAG is likewise not expressed. In addition, neither ESAG mRNA nor VSG mRNA is present in procyclic trypanosomes, the culture form of the parasite that does not possess a VSG coat. Southern blots under low stringency indicate that the genome contains 14–25 ESAG isogenes. The nucleotide sequences of two different ESAG genes possess 60% identity and encode similar, but not identical, glycoproteins. These proteins do not resemble VSGs in sequence but do appear to have an N-terminal signal peptide and a hydrophobic component that might act as a membrane anchor. Their presence has been detected directly by immunoprecipitation of $^{35}$S-labeled trypanosome extracts using antisera against a recombinant fusion protein containing an ESAG coding region. In contrast to the VSG which represents 5% of the total protein, the corresponding ESAG protein is probably only about 0.01% of the total protein. The subcellular location of this small amount of ESAG protein and its role in antigenic variation remain a major mystery to be discussed later in the chapter.

The number of potential expression sites in the genome that are available to a given VSG gene has been the subject of much debate. The simplest model is one in which the genome contains a unique expression site that a VSG gene must occupy before it can be transcribed (Borst and Cross, 1982). Support for this model comes from the demonstration that some VSG genes are repeatedly expressed from a site that has the same restriction pattern (Michels et al., 1983). However, the situation is much more complex, and several lines of evidence indicate that there are at least several different telomeres at which a VSG gene can be transcribed. In the BoTAR serodeme of *T. equiperdum* the BoTat 1 VSG gene appears to be capable of undergoing transcription from three different telomere-linked sites as determined by restriction mapping (Longacre et al., 1983a; Buck et al., 1984). In the *T. brucei* AnTAR serodeme the AnTat 1.1C VSG gene is transcribed from a site with a different restriction pattern than the site from which previous VSG genes are expressed just before it (Pays et al., 1983a). In addition, Southern blots of PFG gels clearly indicate that ELCs of different VSG genes can be located on different-sized chromosomes. For example, Figure 3B shows a case in which the basic copy VSG gene is on a minichromosome while its corresponding ELC is on a 2,000-kb chromosome. Another example is the *T. brucei* 221 VSG gene, which is telomere-linked and can be expressed from either its own site on a large chromosome (which does not migrate into a PFG gel) or as an ELG that is located on a 2,000-kb chromosome (Van der Ploeg et al., 1984c). These examples, and others, clearly show that the gene conversion leading to an ELC can be a trans-chromosomal event. It is tempting to think that the estimate of 14–25 ESAGs means there are a similar number of expression sites. There is no evidence for this, however.

Thus, we are left with a complex and incomplete picture of the expression sites for VSG genes that will be taken up again at the end of the chapter.

## ALTERNATIVES TO ELC FORMATION FOR VSG GENE ACTIVATION

The generation of an ELC appears to be essential for an internal basic copy VSG gene to be transcribed. It is, however, just one of several alternatives available to a telomere-linked VSG gene for activation.

Sometimes when a telomeric VSG gene is duplicatively transposed to another chromosome, enormous chunks of DNA are transferred—from more than 40 kb upstream of the gene all of the way down, presumably, to the telomere (Pays et al., 1983a; Bernards et al., 1984a). This event appears to be a nonreciprocal duplication of a large telomeric region onto another chromosome. It is equivalent to a gene conversion process in which there is an upstream boundary to the transposition but the telomere itself forms the downstream boundary. As a result, the process has been called a *telomere conversion* (Bernards et al., 1984a). One scenario for telomere conversion might be that it starts out as a simple ELC formation of an internal basic copy gene in which the upstream nucleation of the 74–76 base pair repeats takes place, but the downstream nucleation of the C-terminal similarity doesn't occur, swinging a new telomere onto the chromosome and activating its linked VSG gene. In contrast, the transposed unit of internal basic copy genes seems to be primarily confined to the 1.6-kb coding segment and 1–2 kb upstream.

There is also evidence that a reciprocal, rather than a nonreciprocal, recombination of telomeres between two chromosomes can be accompanied by activation of a VSG gene (Pays et al., 1985). In the example reported, there appeared to be a reciprocal crossover within the upstream (5') barren regions of two telomere-linked VSG genes such that the two telomeres involved simply exchanged positions. The result was inactivation of the expressed VSG gene and activation of the previously silent VSG gene. The inactivated gene remained in the genome, presumably now suppressed by the silent telomere environment of one chromosome while the other gene inherited the active environment of the second chromosome. Since the two telomeres appeared to just exchange positions on their respective chromosomes, the process was called *telomere exchange*.

It should be pointed out that in all three mechanisms, *ELC formation*, *telomere conversion*, and *telomere exchange*, there is no evidence that the DNA rearrangements are accompanied by an *obligatory* switch to the expression of a new VSG. Quite the contrary, many rearrangements may result in no change in the VSG being expressed, *or* they may shut off one VSG gene without activating another with fatal consequences for the organism. The detection of these silent rearrangements would have to be fortuitous, and nonlethal, since there is no way to select for their occurrence. Nevertheless, some evidence for their presence exists (Young et al., 1984; Liu et al., 1985; Aline and Stuart, 1985).

It is also likely that none of the above rearrangements is essential for telomere-linked VSG gene activation since in many cases these genes can be activated in situ with no *detectable* change in the vicinity of the gene (Myler et al., 1984a,b; Lenardo et al., 1984; Young et al., 1984; Longacre et al., 1983a,b; Van der Ploeg, 1984c). However, just because rearrangements cannot be detected does not mean they don't occur, so one has to be careful with this interpretation.

## FORMATION OF COMPOSITE VSG GENES

Perhaps the most interesting aspect of VSG gene rearrangements, and one with stimulating implications for the evolution of antigenic variation, is partial gene conversion. ELC formation is usually a complete gene conversion event in which all of the coding sequence (except, sometimes, the final few codons) of a basic copy gene is transposed into an expression site. However, as mentioned earlier, the crossover boundaries of the ELC transposition unit are not precise. Sometimes, a flanking region of a previous ELC remains in place after the new ELC has come in. In most cases this imprecision of the insertion/excision process does not interfere significantly with the amino acid sequence of the new VSG. Occasionally however, during the switch a crossover event takes place *within* the coding sequence itself.

The best-characterized of these partial gene conversion events has been reported by Pays et al. (1983b). This study examined the expressed ELC genes in three successively derived trypanosome clones which for the purpose of discussion we will call A, B, and C. The ELC genes for $VSG_A$ and $VSG_B$ are closely related and are faithful duplicates of their telomere-linked basic copy gene equivalents. About 80% of their nucleotide positions are identical. However, the two VSGs, which are 67% identical, are immunologically distinct. The third VSG, $VSG_C$, is serologically the same as $VSG_A$ but its ELC is a composite of the two previous ELCs. The front two-thirds of $ELC_C$ is the same as $ELC_A$, the back one-third is the same as $ELC_B$ and 133 base pairs (bp) in the middle are of unknown origin. This suggested to the investigators that $ELC_B$ was only partially displaced by an attempt to re-express $ELC_A$ with, perhaps, the participation of a third gene that contributed the interior 133 bp-segment.

The sequential steps and overall mechanism for these particular recombinations within the coding regions are mysterious, but their potential biological significance is not. Since similar VSGs, such as $VSG_A$ and $VSG_B$, can be immunologically distinct, these intermediate isogene recombinations during ELC formation could create new VSGs with different antigenic properties than any of the encoded VSGs. Thus, a trypanosome has the potential to produce more VSGs than are simultaneously encoded in its gene pool. It seems likely that this additional capacity for making new VSGs is a natural consequence of the gene rearrangements and expands the ability of trypanosomes for antigenic variation to even further limits.

Figure 6 summarizes the various gene rearrangements discussed above. We now turn to the questions that remain about the molecular biology of antigenic variation.

## WHAT REGULATES THE MUTUALLY EXCLUSIVE TRANSCRIPTION OF A SINGLE VSG GENE?

The fundamental question about the molecular biology of trypanosome antigenic variation continues to be, How does a trypanosome turn on one VSG gene at a given time and turn off the hundreds of others? Despite nearly 10 years of investigation since the first published data related to this question (Williams et al., 1979), and hundreds of researcher-years consumed, we still do not have a definitive answer to this question. During this time, however, our knowledge of VSG gene regulation has increased to nearly the same level of sophistication as understanding of other eukaryotic gene regulation systems. Its further elucidation is also stymied by the same bottleneck as research on these other systems. In other words, VSG gene regulation is known to occur at the transcriptional level, the structures of several VSG genes and their flanking regions have been determined, and experimental conditions for achieving differential expression of those genes have been found. Nevertheless, we know very little about the actual mechanisms and molecules that control transcription initiation of a VSG gene. Nor do we know what triggers the switch from the transcription of one telomere-linked gene to another. Correspondingly, a great deal is also known about the structures of the genes for insulin, globin, growth hormone, and many other eukaryotic proteins, but this information has not yet elucidated the mechanism that ensures insulin mRNA is synthesized only in pancreatic cells, globin only in erythrocytes, growth hormone only in the pituitary, etc.

The trouble with both trypanosomes and mammalian cells as experimental systems is that genetic mutants affecting gene regulation are frequently

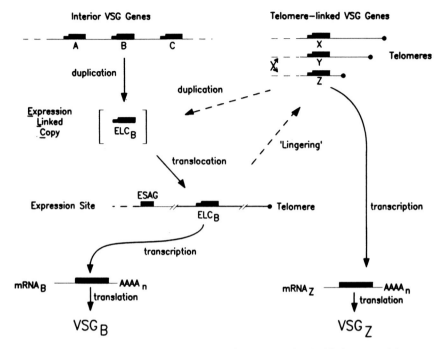

Fig. 6. Summary of the DNA rearrangements that are associated with the sequential expression of individual VSG genes. Many basic copy VSG genes are located at interior chromosomal positions as represented by genes A, B, and C. These interior genes are only transcribed after undergoing a duplicative transposition to form an ELC gene in a telomere-linked expression site. Other basic copy VSG genes, represented by genes X, Y, and Z, are already situated near telomeres. A duplication/transposition event is not essential for their expression, although ELCs of telomere-linked basic copy genes sometimes occurs. ELC genes usually disappear from the genome after the switch to the expression of another VSG gene, although they occasionally "linger" to become nonexpressed, telomere-linked, basic copy genes. Expression of telomere-linked genes has also been correlated with exchange of telomeres between two chromosomes and with displacement of one telomere by a duplicated copy of another telomere region. However, the precise molecular event that triggers a switch to the transcription of a new ELC or another telomere-linked gene is not known.

difficult to obtain. Without defined genetic mutants that affect the regulation of a specific gene, it is difficult to address specific questions about that regulation. Furthermore, the problem often is not in generating general mutations but in identifying the desired ones against the background of other mutations. For example, one can mutagenize trypanosomes with any of the common mutagens, but, to date, a selection technique has not been devised for identifying mutants with a defect in, or enhancement of, antigenic variation. That means that molecular biology techniques have to substitute for the tools of genetics.

The lack of firm data on the molecular mechanisms that regulate VSG gene expression has not inhibited speculation, however. For example, a single "floating" promoter or mobile enhancer element that bounces among telomeres to activate the linked VSG gene would explain much of the data. This model is probably eliminated by those rare instances in which more than one expression site appears to be simultaneously active for a least a brief time (Cornelissen et al., 1985; Baltz et al., 1986). Another possibility is the formation of one complex per cell of proteins and/or other factors that act in *trans* to initiate transcription of a telomere-linked gene. Production of a "double dose" of this complex would account for simultaneous activation of two VSG genes, but it is difficult to design tests for the presence of such a complex.

One interesting model for the mutually exclusive transcription of VSG genes is based on the possibility of nuclear compartmentalization (Shea et al., 1987). These authors detected a VSG transposition unit with an unusually large upstream region. Within this upstream region is a sequence resembling an RNA polymerase I promoter that maps just in front of the 5′-end of the initial transcript. This is of interest because of earlier observations that, unlike other protein-coding genes, the VSG genes are transcribed by RNA polymerases that are insensitive to $\alpha$-amanitin (Kooter and Borst, 1984). Since RNA polymerase I, which transcribes the ribosomal RNA genes, is also insensitive to $\alpha$-amanitin, the data suggest the same polymerase activity may be responsible for the transcription of both sets of genes. This, in turn, indicates that the nucleolus might participate in VSG gene regulation. If a telomere were confined to the nucleolus with the ribosomal RNA genes and RNA polymerase I, then its linked VSG gene would be transcribed. All of the other telomere-linked VSG genes would be outside the nucleolus and not exposed to the RNA polymerase I activity.

Although very attractive from several standpoints, this model raises as many questions as it answers. What directs a specific telomere to the nucleolus? What excludes all other telomeres from associating with the nucleolus? What tosses the expressed telomere out of the nucleolus and admits another one when expression switches telomeres? Why are most transposition units lacking a cotransposed promoter for RNA polymerase I? Why is the primary transcript of some VSG genes greater than 40 kb while for others it is only a few kilobases?

## WHY IS THE EXPRESSED VSG GENE LINKED TO A TELOMERE?

One of the few initial observations about VSG gene expression that has stood the test of continued experimentation is that a VSG gene must be linked to a telomere to be transcribed. Yet, it remains unclear what, if anything, the telomere has to do with the transcription process itself. The proximity of the telomere is clearly *necessary* but not *sufficient* for transcription since many unexpressed basic copy VSG genes are also telomere-linked. Two features about telomeric regions in other organisms may provide clues to this question. The first is that telomeric regions tend to be very recombinagenic, probably as a result of the dual properties of possessing free DNA ends for strand invasion events and containing multiple repeats of short sequences for homologous base pairing. Since antigenic variation relies on recombination to generate its diversity, it is perhaps not surprising to find a region of the chromosome involved whose physical structure enhances its recombinational capacity. The second consideration is that ribosomal RNA genes in the macronucleus of tetrahymena, a ciliated protozoan, are also telomere-linked and on minichromosomes (Karrer and Gall, 1976). In trypanosomes there is no evidence for either a micro/macronucleus arrangement or for the presence of a telomere adjacent to the ribosomal RNA genes (Dorfman et al., 1985). Nevertheless, if, indeed, VSG genes are transcribed by an RNA polymerase I-like activity, the coincidence of both sets of genes' being adjacent to telomeres in the different protozoons is probably too great to be dismissed.

## WHAT IS THE PLOIDY OF VSG GENES?

As mentioned earlier, there is evidence that the genes for housekeeping enzymes are diploid in bloodstream trypanosomes. This raises a major question about Southern blots probed with a VSG cDNA in which an ELC gene is present. In these autoradiograms the band containing the ELC gene is about the same intensity as the band containing the corresponding basic copy VSG gene (see Figs. 3B, 4). This equivalent intensity means that if the basic copy VSG gene is diploid, then the ELC must also be diploid; alternatively, if only one ELC occurs per nucleus, then, likewise, there must be a single copy of its basic copy gene. Neither possibility is very satisfactory.

Diploid basic copy VSG genes would mean that two ELCs are formed during an antigenic switch and that they are located at sites precisely the same distance in from two different telomeres—an unlikely scenario. A haploid ELC would imply that the corresponding basic copy gene is haploid in a nucleus full of diploid housekeeping genes. This could be explained by the presence of "sites" on sister chromosomes that are occupied by different basic copy VSG genes, i.e, diploid sites for haploid genes. This would seem to create difficulties in the inheritance by progeny trypanosomes of a full parental complement of VSG genes.

For the moment there is no obvious way around this dilemma. It should be noted that only a few housekeeping genes have been cloned, but most of these appear in the genome as tandem duplicates. As with most features of trypanosomes, there is at least one exception to this, however, so gene duplications are unlikely to be related to the ploidy question. In addition, if genetic exchange between trypanosomes occurs in the tsetse fly, it would seem necessary that the full complement of basic copy VSG genes be maintained for transfer to the progeny.

## WHY DO MULTIPLE EXPRESSION SITES EXIST?

As discussed above, the experimental evidence that VSG genes are transcribed from multiple expression sites seems irrefutable. Yet the need for more than one expression site is not obvious; a single expression site would seem to be sufficient and in fact would reduce the chances for complications in regulation of both the transcription and the switching. No limitation appears to be imposed on where a VSG gene may occur in the genome so this is unlikely to be the reason for multiple sites. ELCs do not have to be on the same chromosome as their corresponding basic copy gene (see Fig. 3B) and are not confined to a specific chromosomal size class. No definitive evidence exists to suggest that different subclasses of VSG genes must be expressed from a given expression site, although this possibility constitutes the most likely explanation for the presence of multiple expression sites. The fact that the ELCs of at least some internal basic copy genes are repeatedly directed to the same expression site would suggest the existence of favored expression sites for specific VSG genes. This could be explained by a propensity of the regions flanking a given transposition unit to undergo homologous recombination with sequences in one expression site rather than in other expression sites. Likewise, the tendency of some VSGs to appear early or late in infection could correlate with the ease in which the expression sites for these genes can be activated. Perhaps differential activation of specific expression sites influences the overall timing of VSG appearance, rather like the activation of early and late viral genes during virus infections. As with some of the other unanswered mysteries about antigenic variation, it is difficult to design experiments that directly address this question.

## WHAT IS THE FUNCTION OF ESAGs?

As discussed above, ESAGs are neighboring upstream genes whose transcription occurs in concert with the expressed VSG gene. The need for such *cis*-transcribed genes remains completely obscure. The sequences of several ESAGs suggest that their protein products have a membrane location since they have a signal peptide, a hydrophobic C-terminus, and amphiphilic helices. Yet, considerable diversity is tolerated in this family of 14–25 genes even though only one ESAG is usually transcribed at a time. In addition, there is only one ESAG protein molecule for every 1,000 or so VSG molecules.

One possibility is that the ESAG protein assists VSGs to their surface location and/or associates with some copies of the VSG on the membrane. Indeed, the sequence diversity of ESAGs may suggest that their proteins undergo a reduced form of antigenic variation. If so, one is left to ponder what the nature of this assistance or association is and why it is required. In addition, why must the ESAG be transcribed in *cis* with the VSG gene? Since the different ESAGs produce diffusible products, why must their transcription occur from the same telomere as the VSG? Why must a gene be transcribed in *cis* with the VSG gene to produce a product that acts in *trans*? Why is variation in the ESAG protein sequence tolerated if they act as specific carriers or scaffolding for other cellular constituents?

## WHY IS DISCONTINUOUS TRANSCRIPTION NECESSARY?

A distinctive feature of all kinetoplastids that was not discussed above because it seems to have nothing to do with antigenic variation is that the mRNAs of these organisms all begin with the same 35 nucleotides (Boothroyd and Cross, 1982; Van der Ploeg et al., 1982c; Walder et al., 1986). These 35 nucleotides at the 5'-ends of the mRNAs have been called the mini-exon or the spliced leader sequence (Van der Ploeg et al., 1982c; Parsons et al., 1984). In *T. brucei*, where they have been best studied, they are encoded within about 200 copies of a 1.35-kb tandem DNA repeat whose primary transcript of about 140 nucleotides contains the 35 nucleotides at its 5'-end (Nelson et al., 1983; Dorfman and Donelson, 1984; DeLange et al., 1983; Campbell et al., 1984b). They are donated from this 140 nucleotide transcript to the initial transcript of all protein-coding genes via a *trans*-splicing mechanism (Murphy et al., 1986; Sutton and Boothroyd, 1986). Since the resultant mRNAs are derived from two independent transcription events, the process is called discontinuous transcription.

The reason for discontinuous transcription is not clear. The mini-exon must play a general role of some sort for the cell, since all mRNAs have it. The initial 140-nucleotide transcript possesses a CAP structure so one function of the mini-exon is to donate the CAP structure to the protein-coding mRNAs (Lenardo et al., 1985). It seems unlikely, however, that trypanosomes would replace the normal enzymatic steps for modifying 5' termini with such an involved process, especially since it must still cap the 5' terminus of the 140-nucleotide transcript. Other possible functions are that the mini-exon is required for i) transport across the nuclear membrane, ii) stability of the mRNA, iii) translation initiation by trypanosome ribosome, iv) cytoplasmic compartmentalization, v) some other unknown function, or vi) no function at all. Of these possibilities, the fifth and sixth have to be entertained in the absence of any evidence for the first four.

Stimulating the issue even more are recent reports that a few, but certainly not all, mRNAs in other organisms also possess mini-exon-like sequences. In the free-living nematode, *Caenorhabditis elegans*, three of the four actin mRNAs contain a 5' terminal sequence of 22 nucleotides that is encoded within a DNA repeat far removed from the actin genes (Krause and Hirsh, 1987). Since some genes of *C. elegans* have introns, the organism appears to accommodate both *cis-* and *trans-*splicing of RNA. In vaccina virus, an animal DNA virus, a similar discontinuous synthesis of mRNA appears to occur during the late stage of transcription (Bertholet et al., 1987; Schwer et al., 1987). Clearly the potential relationship between *cis-* and *trans-*splicing are going to be active areas of research in the future.

## WHY DO TRYPANOSOMES HAVE SO MANY UNUSUAL FEATURES?

The nucleic acids of kinetoplastids have still other features that are peculiar. For example, the basis for their classification in the order Kinetoplastida is the kinetoplast, an appendix of the mitochondrion that contains as much as 25% of the total DNA content of the cell (Simpson, 1986). This kinetoplast DNA is a enormous network of concatenated maxicircles and minicircles. The maxicircles of about 20 kb are the trypanosome equivalent of mitochondrial DNA. The 1–2-kb minicircles, of which there may be over 10,000 per kinetoplast, have no known genetic function. Collectively, these DNAs have some very bizarre properties. Maxicircles have ribosomal RNA genes that are among the smallest in nature (Eperon et al., 1983) and protein-coding genes that require U-residue insertions into the mRNA (Benne et al., 1986; Feagin et al., 1987). Minicircles present enormous topological problems during replication/cell division and possess heterogeneous sequences that "bend" (Kitchin et al., 1985; Ntambi et al., 1984). None of these properties suggest why the minicircles consume so much of the trypanosome's DNA.

Another curious feature is the presence of nucleotide modifications in the nuclear DNA that increase its buoyant density, rather than decrease it as do most base modifications (Raibaud et al., 1983). The chemical structure of these modifications are unknown. Furthermore, when these, or other, modifications occur within telomere-linked VSG genes, they prevent cleavage by restriction enzymes, such as PstI, at specific sites in some molecules of a chromosome but not in other molecules of the same chromosome (Bernards et al., 1984b; Pays et al., 1984). Thus, the modifications are not uniform among seemingly identical cells. Again, this does not conform to conventional wisdom about DNA modifications.

A final thought about all of these so-called unusual features of trypanosomes is that, perhaps, they are not so unique after all. Maybe they are (or were, or will be) common properties of other organisms as well; it is just that trypanosomes are the first experimental system in which they have been detected. This was certainly the case with the distinctive attachment of the VSG to the membrane that was originally thought to be unique to trypanosomes and has now been found in a number of other cell types (Seki et al., 1985; Hemperly et al., 1986). Indeed, variations on the theme of antigenic variation also occur in a number of other infectious agents (Cruse and Lewis, 1987). Therefore, the uniqueness of trypanosomes depends on one's frame of reference and they are best taken for just what they are: an excellent experimental system for exploring such diverse biological phenomena as antigenic variation, discontinuous transcription, mitochondrial function, and membrane structure.

"Sometimes the whole world seems strange except for thee and me, and sometimes I wonder about thee."

—old Quaker maxim

### ACKNOWLEDGMENTS

Research described here that was performed in the author's laboratory was supported by grants from U.S. Public Health Service, the U.S. Army, and an award from the Burroughs-Wellcome Fund.

# REFERENCES

Aline R, Stuart K (1985): The two mechanisms for antigenic variation in *Trypanosoma brucei* are independent processes. Mol Biochem Parasitol 16:11–20.

Baltz T, Giroud C, Baltz D, Roth C, Raibaud A, Eisen H (1986): Stable expression of two variable surface glycoproteins by cloned *Trypanosoma equiperdum*. Nature 319:602–604.

Benne R, Van Den Borg J, Brakenhoff J, Sloof P, Van Boom J, Tromp M (1986): The major transcript of the frameshifted coxII gene from trypanosome mitochondria contain four nucleotides that are not encoded in the DNA. Cell 46:819–826.

Bernards A, Van der Ploeg L, Frasch A, Borst P, Boothroyd J, Coleman S, Cross G (1981): Activation of trypanosome surface glycoprotein genes involves a duplication-transposition leading to an altered 3' end. Cell 27:497–505.

Bernards A, Michels P, Lincke C, Borst P (1983): Growth of chromosome ends in multiplying trypanosomes. Nature 303:592–597.

Bernards A, De Lange T, Michels P, Liu A, Huisman M, Borst P (1984a): Two modes of activation of a single surface antigen gene of *Trypanosoma brucei*. Cell 36:163–170.

Bernards A, van Harten-Loosbroek N, Borst P (1984b): Modification of telomeric DNA in *Trypanosoma brucei:* A role in antigenic variation. Nucleic Acids Res 12:4153–4170.

Bernards A, Kooter J, Michels P, Moberts R, Borst P (1986): Pulsed field gradient electrophoresis of DNA digested in agarose allows the sizing of the large duplication unit of a surface antigen gene in trypanosomes. Gene 42(3):313–322.

Bertholet C, Van Meir E, ten Heggeler-Bordier B, Wittek R (1987): Vaccina virus produces late mRNAs by discontinuous synthesis. Cell 50:153–162.

Blackburn E, Challoner P (1984): Identification of a telomeric DNA sequence in *Trypanosoma brucei*. Cell 36:447–457.

Boothroyd J, Cross G (1982): Transcripts coding for different variant surface glycoproteins in *Trypanosoma brucei* have a short identical exon at their 5'-end. Gene 20:279–287.

Borst P, Cross G (1982): Molecular basis for trypanosome antigenic variation. Cell 29:291–303.

Borst P, Van der Ploeg L, van Hoek J, Tas J, James T (1982): On the DNA content and ploidy of trypanosomes. Mol Biochem Parasitol 6:13–23.

Borst P, Bernards A, Van der Ploeg L, Michels P, Liu A, De Lange T, Sloof P, Veeneman G, Tromp M, Van Boom J (1983): DNA rearrangements controlling the expression of genes for variant surface antigens in trypanosomes. In C. Chater, D. Cullis, D. Hopwood, A. Johnston, and H. Woolhouse (eds.): Genetic Rearrangement. London: Croom Helm, pp 207–233.

Bridgen P, Cross G, Bridgen J (1976): N-terminal amino acid sequences of variant-specific antigens from *Trypanosoma brucei*. Nature 263:613–614.

Buck G, Longacre S, Raibaud A, Hibner U, Giroud C, Baltz T, Baltz D, Eisen H (1984): Stability of expression-linked surface antigen genes in *Trypansoma equiperdum*. Nature 307:563–566.

Campbell G, Esser K, Wellde B, Diggs C (1979): Isolation and characterization of a serodeme of *Trypanosoma rhodesiense*. Am J Trop Med Hyg 28:974–983.

Campbell D, Thornton D, Boothroyd J (1984a): The 5'-limit of transposition and upstream barren region of a trypanosome VSG gene. Tandem 76 base-pair repeats flanking (TAA)$_{90}$. Nature 12:2759–2774.

Campbell D, Thornton D, Boothroyd J (1984b): Apparent discontinuous transcription of *Trypanosoma brucei* surface antigen genes. Nature 311:350–355.

Capbern A, Biroud C, Baltz T, Maltern P (1977): *Trypanosoma equiperdum:* Etude des variations antigeniques au cours de la trypanosome experimentale du lapin. Exp Parasitol 42:6–13.

Cornelissen A, Johnson P, Kooter J, Van der Ploeg L, Borst P (1985): Two simultaneously active VSG gene transcription units in a single *Trypanosoma brucei* variant. Cell 41:825–832.

Cross G (1975): Identification, purification and properties of clone-specific glycoprotein antigens constituting the surface coat of *Trypanosoma brucei*. Parasitology 71:393–417.

Cruse J, Lewis R (1987): Contemporary concepts of antigenic variation. In J Cruse and R. Lewis (eds): Antigenic Variation: Molecular and Genetic Mechanisms of Relapsing Disease. Contributions to Microbiology and Immunology, Volume 8. Basel: S. Karger AG, pp 1–19.

Cully D, Ip H, Cross G (1985): Coordinate transcription of variant surface glycoprotein genes and an expression site associated gene family in *Trypanosoma brucei*. Cell 42:173–182.

Cully D, Gibbs C, Cross M (1986): Identification of proteins encoded by variant surface glycoprotein expression site-associated genes in *Trypanosoma brucei*. Mol Biochem Parasitol 21:189–197.

De Lange T, Borst P (1982): Genomic environment of the expression-linked extra copies of genes for surface antigens of *Trypanosoma brucei* resembles the end of a chromosome. Nature 299:451–453.

De Lange T, Liu A, Van der Ploeg L, Borst P, Tromp M, Van Boom J (1983): Tandem repetition of the 5' mini-exon of variant surface glycoprotein genes. A multiple promoter for VSG gene transcription? Cell 34:891–900.

Donelson J, Rice-Ficht A (1985): Molecular biology of trypanosome antigenic variation. Microbiol Rev 49:107–125.

Donelson J, Young J, Dorfman D, Majiwa P, Williams R (1982): The ILTat 1.4 surface antigen gene family of *Trypanosoma brucei*. Nucleic Acids Res 10:6581–6595.

Donelson J, Murphy W, Brentano S, Rice-Ficht A, Cain G (1983): Comparison of the expression-linked extra copy (ELC) and basic copy (BC) genes of a trypanosome surface antigen. J Cell Biochem 23:1–12.

Dorfman D, Donelson J (1984): Characterization of the 1.35 kb DNA repeat until containing the conserved 35 nucleotides at the 5'-termini of VSG mRNAs in *Trypanosoma brucei*. Nucleic Acids Res 12:4907–4920.

Dorfman D, Lenardo M, Reddy L, Van der Ploeg L, Donelson J (1985): The 5.8S ribosomal RNA gene of *Trypanosoma brucei:* Structural and transcriptional studies. Nucleic Acids Res 13:3533–3549.

Doyle J, Hirumi H, Hirumi H, Lupton E, Cross G (1980): Antigenic variation in clones of animal-infective *Trypanosoma brucei* derived and maintained in vitro. Parasitology 80:359–369.

Engman D, Reddy L, Donelson J, Kirchhoff L (1987): *Trypanosoma cruzi* exhibits inter- and intra-strain heterogeneity in molecular karyotype and chromosomal gene location. Mol Biochem Parasitol 22:115-123.

Eperon I, Janssen J, Hoeijmakers J, Borst P (1983): The major transcripts of the kinetoplast DNA of *Trypanosoma brucei* are very small ribosomal RNAs. Nucleic Acids Res 11:105-125.

Esser K, Schoenbechler M, Gingerich J (1982): *Trypanosoma rhodesiense* blood forms express all antigen specificities relevant to protection against metacyclic (insect form) challenge. J Immunol 129:1715-1718.

Feagin J, Jasmer D, Stuart K (1987): Developmentally regulated addition of nucleotides within apocytochrome b transcripts in *Trypanosoma brucei*. Cell 49:337-345.

Fink G, Petes T (1984): Gene conversion in the absence of reciprocal recombination. Nature 310:728-729.

Frasch A, Borst P, Van den Burg J (1982): Rapid evolution of genes coding for variant surface glycoproteins in trypanosomes. Gene 17:197-211.

Gibson W, Borst P (1986) Size-fractionation of the small chromosomes of Trypanozoon and Nannomonas trypanosomes by pulsed field gradient gel electrophoresis. Mol Biochem Parasitol 18(2):127-140.

Gibson W, Osinga K, Michels P, Borst P (1985): Trypanosomes of subgenus Trypanozoon are diploid for housekeeping genes. Mol Biochem Parasitol 16(3):231-242.

Gray A (1965): Antigenic variation in a strain of *Trypanosoma brucei* transmitted by *Glossina morsitans* and *G. palpalis*. J Gen Microbiol 41:195-214.

Hadjuk S (1984): Antigenic variation during the developmental cycle of *Trypanosoma brucei*. J Protozool 31:41-42.

Hajduk S, Vickerman K (1981): Antigenic variation in cyclically transmitted *Trypanosoma brucei*. Variable antigen type composition of the first parasitaemia in mice bitten by trypanosome-infected *Glossina morsitans*. Parasitology 83:609-621.

Hajduk S, Cameron C, Barry J, Vickerman K (1981): Antigenic variation in cyclically-transmitted *Trypanosoma brucei*. I. Variable antigen type composition of metacyclic trypanosome populations from the salivary glands of *Glossina morsitans*. Parasitology 83:595-607.

Hemperly J, Edelman G, Cunningham B (1986): cDNA clones of the neural adhesion molecule (N-CAM) lacking a membrane-spanning region consistent with evidence for membrane attachment via a phosphatidylinositol intermediate. Proc Natl Acad Sci USA 53:9822-9826.

Hoeijmakers J, Frasch A, Bernards A, Borst P, Cross G (1980): Novel expression-linked copies of the genes for variant surface antigens in trypanosomes. Nature 284:78-80.

Jenni L, Marti S, Schweizer J, Betschart B, Le Page R, Wells J, Tait A, Paindovoine P, Pays E, Steinert M (1986): Hybrid formation between African trypanosomes during cyclical transmission. Nature 322:173-175.

Karrer K, Gall J (1976): The macronuclear ribosomal DNA of tetrahymena priformis is a palindrome. J Mol Biol 104:421-454.

Kitchin P, Klein V, Englund P (1985): Intermediates in the replication of kinetoplast DNA minicircles. J Biol Chem 260:3844-3851.

Kooter J, Borst P (1984): Alpha-amanitin-insensitive transcription of variant surface glycoprotein genes provides further evidence for discontinuous transcription in trypanosomes. Nucleic Acids Res 12:9457-9472.

Krause M, Hirsch D (1987): A trans-spliced leader sequence on actin mRNA in *C. elegans*. Cell 49:753-761.

Lamont G, Tucker R, Cross G (1986): Analysis of antigen switching rates in *Trypanosoma brucei*. Parasitology (Pt 2):355–367.

Laurent M, Pays E, Magnus E, Van Meirvenne N, Natthijssens G, Williams R, Steinert M (1983): DNA rearrangements linked to expression of a predominant surface antigen gene of trypanosomes. Nature 302:263–266.

Lenardo M, Rice-Ficht A, Kelly G, Esser K, Donelson J (1984): Characterization of the genes specifying two metacyclic variable antigen types in *Trypanosoma brucei rhodesiense*. Proc Natl Acad Sci USA 81:6642–6646.

Lenardo, M, Dorfman D, Donelson J (1985): The spliced leader sequence of *Trypanosoma brucei* has a potential role as a cap donor structure. Mol Cell Biol 9:2487–2490.

Liu A, Van der Ploeg L, Rijsewijk F, Borst P (1983): The transposition unit of VSG gene 118 of *Trypanosoma brucei*. Presence of repeated elements at its border and absence of promoter-associated sequences. J Mol Biol 167:57–75.

Liu A, Michels P, Bernards A, Borst P (1985): Trypanosome variant surface glycoprotein genes expressed early in infection. J Mol Biol 182:383–396.

Longacre S, Hibner U, Raibaud A, Eisen H, Baltz T, Giroud C, Baltz D (1983a): DNA rearrangement and antigenic variation of *Trypanosoma equiperdum*. Multiple expression linked sites in independent isolates of trypanosomes expressing the same antigen. Mol Cell Biol 3:399–409.

Longacre S, Raibaud A, Hibner U, Buck G, Eisen H (1983b): DNA rearrangements and antigenic variation in *Trypanosoma equiperdum*. Expression-independent DNA rearrangements in the basic copy of a variant surface glycoprotein gene. Mol Cell Biol 3:410–414.

Majumder H, Boothroyd J, Weber H (1981): Homologous 3'-terminal regions of mRNAs for surface antigens of different antigenic variants of *Trypanosoma brucei*. Nucleic Acids Res 9:4745–4753.

Michels P, Bernards A, Van der Ploeg L, Borst P (1982): Characterization of the expression-linked copies of variant surface glycoprotein 118 in two independently isolated clones of *Trypanosoma brucei*. Nucleic Acids Res 10:2353–2366.

Michels P, Liu A, Bernards A, Sloof P, Van der Bijl M, Schinkel A, Menke H, Borst P, Veeneman G, Tromp M, Van Boom J (1983): Activation of the genes for variant surface glycoproteins 117 and 118 in *Trypanosoma brucei*. J Mol Biol 66:537–556.

Miller E, Turner J (1981): Analysis of antigenic types appearing in first relapse populations of clones in *Trypanosoma brucei*. Parasitology 82:63–80.

Murphy W, Brentano S, Rice-Ficht A, Dorfman D, Donelson J (1984): DNA rearrangements of the variable surface antigen genes of trypanosomes. J Protozool 31:65–73.

Murphy W, Watkins K, Agabian N (1986): Identification of a novel Y branch structure as an intermediate in trypanosome mRNA splicing: Evidence for trans splicing. Cell 47:517–525.

Myler P, Allison J, Agabian N, Stuart K (1984a): Antigenic variation in African trypanosomes by replacement or activation of alternate telomeres. Cell 39:203–211.

Myler P, Nelson R, Agabian N, Stuart K (1984b): Two mechanisms of expression of a predominant variant antigen gene of *Trypanosoma brucei*. Nature 369:282–284.

Nelson R, Parsons M, Barr P, Stuart K, Selkirk M, Agabian N (1983): Sequences homologous to the variant antigen mRNA spliced leader are located in tandem repeats and variable orphons in *Trypanosoma brucei*. Cell 34:901–909.

Ntambi J, Marini J, Bangs J, Hajduk S, Jimenez H, Kitchin P, Klein V, Ryan K, Englund P (1984): Presence of a bent helix in fragments of kinetoplast DNA minicircles from several trypanosomatid species. Mol Biochem Parasitol 12:273–286.

Parsons M, Nelson R, Newport G, Milhausen M, Stuart K, Agabian N (1983): Genomic organization of *Trypanosoma brucei* variant antigen gene families in sequential parasitemias. Mol Biochem Parasitol 9:255–269.

Parsons M, Nelson R, Watkins J, Agabian N (1984): Trypanosome mRNAs share a common 5' spliced leader sequence. Cell 38:309–316.

Pays E, L'Heureux M, Steinert M (1981): The expression-linked copy of surface antigen gene in *Trypanosoma* is probably the one transcribed. Nature 292:365–367.

Pays E, L'Heureux M, Steinert M (1982): Structure and expression of a *Trypanosoma brucei gambiense* variant specific antigen gene. Nucleic Acids Res 10:3149–3163.

Pays E, Van Assel S, Laurent M, Dero B, Michiels F, Kronenberger P, Matthyssens G, Van Meirvenne N, Le Ray D, Steinert M (1983a): At least two transposed sequences are associated in the expression site of a surface antigen gene in different trypanosome clones. Cell 34:359–369.

Pays E, Van Assel S, Laurent M, Darville M, Vervoort T, Van Meirveene N, Steinert M (1983b): Gene conversion as a mechanism for antigenic variation in trypanosomes. Cell 34:371–381.

Pays E, Laurent M, Delinte K, Van Meirvenne N, Steinert M (1983c): Differential size variations between transcriptionally active and inactive telomeres in *Trypanosoma brucei*. Nucleic Acids Res 11:8137–8147.

Pays E, Delauw M, Van Assel S, Laurent M, Vervoort T, Van Meirvenne N, Steinert M (1983d): Modifications of a *Trypanosoma b. brucei* antigen gene repertoire by different DNA recombinational mechanisms. Cell 35:721–731.

Pays E, Delauw M, Laurent M, Steinert M (1984): Possible DNA modification in GC dinucleotides of *Trypanosoma brucei* telomeric sequences; relationship with antigenic variation. Nucleic Acids Res 12:5235–5247.

Pays E, Gayaux M, Aerts D, Van Meirvenne N, Steinert M (1985): Telomeric reciprocal recombination as a possible mechanism for antigenic variation in trypanosomes. Nature 316:562–564.

Raibaud A, Gaillard C, Longacre S, Hibner U, Buck G, Bernardi G, Eisen H (1983): Genomic environment of variant surface antigen genes of *Trypanosoma equiperdum*. Proc Natl Acad Sci USA 80:4306–4310.

Rifkin M (1978): Identification of the trypanocidal factor in normal human serum high density lipoprotein. Proc Natl Acad Sci USA 75:3450–3454.

Ross R, Thomson D (1910): A case of sleeping sickness studied by precise enumerative methods. Regular periodic increase in the parasites disclosed. Proc R Soc Lond [Biol] 82:411–415.

Schwartz D, Cantor C (1984): Separation of yeast chromosome-sized DNA by pulsed field gradient gel electrophoresis. Cell 37:67–75.

Schwer B, Visca P, Vos J, Stunnenberg H (1987): Discontinuous transcription or RNA processing of vaccina virus late messenger RNAs results in S' poly(A) leader. Cell 50:163–169.

Seki T, Moriuchi T, Chang H, Denome R, Silver J (1985): Structural organization of the rat thy-1 gene. Nature 313:485–487.

Shea C, Lee M, Van der Ploeg L (1987): VSG gene 118 is transcribed from a cotransposed Pol-I like promoter. Cell 50:603–612.

Simpson L (1986) Kinetoplast DNA in trypanosomatid flagellates. Int Rev Cytol 99:119–179.

Spithill T, Samaras N (1985): The molecular karyotype of *Leishania major* and mapping of α- and β-tubulin gene families to multiple unlinked chromosomal loci. Nucleic Acids Res 13:4155–4169.

Sutton R, Boothroyd J (1986): Evidence for trans splicing in trypanosomes. Cell 47:527–535.

Tait A (1980): Evidence for diploidy and mating in trypanosomes. Nature 287:536–538.

Turner M (1982): Biochemistry of the variant surface glycoproteins of salivarian trypanosomes. Adv Parasitol 21:69–153.

Turner M (1985): The biochemistry of the surface antigens of the African trypanosomes. Br Med Bull 41(2):137–143.

Van der Ploeg L, Valerio D, De Lange T, Bernards A, Borst P, Groveld F (1982a): An analysis of cosmid clones of nuclear DNA from *Trypanosoma brucei* shows that the genes for variant surface glycoproteins are clustered in the genome. Nucleic Acids Res 10:5905–5923.

Van der Ploeg L, Bernards A, Rijsewijk F, Borst P (1982b): Characterization of the DNA duplication-transposition that controls the expression of two genes for variant surface glycoproteins in *Trypanosoma brucei*. Nucleic Acids Res 10:593–609.

Van der Ploeg L, Liu A, Michels P, De Lange T, Borst P, Majumder H, Weber H, Veeneman G, Van Boom J (1982c): RNA splicing is required to make the messenger RNA for a variant surface antigen in trypanosomes. Nucleic Acids Res 10:3591–3604.

Van der Ploeg L, Cornelissen A, Michels P, Borst P (1984a): Chromosomes rearrangement in *Trypanosoma brucei*. Cell 39:213–221.

Van der Ploeg L, Liu A, Borst P (1984b): The structure of the growing telomeres of trypanosomes. Cell 36:459–465.

Van der Ploeg L, Schwartz D, Cantor C, Borst P (1984c): Antigenic variation in *Trypanosoma brucei* analyzed by electrophoretic separation of chromosome-sized DNA molecules. Cell 37:77–84.

Vickerman K (1974): Antigenic variation in African trypanosomes. In Parasites in the Immunized Host: Mechanism of Survival. CIBA Foundation Symposium 25. Amsterdam: Elsevier/North Holland, pp 53–80.

Vickerman K (1986): Parasitology. Clandestine sex in trypanosomes. Nature 322:113–114.

Vickerman K, Barry J, Hajduk S, Tetley L (1980): Antigenic variation in trypanosomes. In The Host Invader Interplay. Amsterdam: Elsevier/North Holland/Biomedical Press, pp 179–190.

Walder J, Eder P, Engman D, Brentano S, Walder R, Knutzon D, Dorfman D, Donelson J (1986): The 35 nucleotide spliced leader sequence is common to all trypanosome messenger RNAs. Science 233:569–571.

Williams R, Young J, Majiwa P (1979): Genomic rearrangements correlated with antigenic variation in *Trypanosoma brucei*. Nature 282:847–849.

Williams R, Young J, Majiwa P (1982): Genomic environment of *Trypanosoma brucei* VSG genes: Presence of a minichromosome. Nature 299:417–421.

Young J, Donelson J, Majiwa P, Shapiro S, Williams R (1982): Analysis of genomic rearrangements associated with two variable antigen genes of *Trypanosoma brucei*. Nucleic Acids Res 10:803–819.

Young J, Miller E, Williams R, Turner M (1983a): Are there two classes of VSG gene in *Trypanosoma brucei?* Nature 306:196–198.

Young J, Shah J, Matthyssens G, Williams R (1983b): Relationship between multiple copies of a *T. brucei* variable surface glycoprotein gene whose expression is not controlled by duplication. Cell 32:1149–1159.

Young J, Turner M, Williams R (1984): The role of duplication in the expression of a variable surface glycoprotein gene of *Trypanosoma brucei*. J Cell Biochem 24:287–295.

Zampetti-Bosseler F, Schweizer J, Pays E, Jenni L, Steinert M (1986): Evidence for haploidy in metacyclic forms of *Trypanosoma brucei*. Proc Natl Acad Sci USA 83(16):6063–6064.

The Biology of Parasitism, pages 401–412
© 1988 Alan R. Liss, Inc.

# Glycolipid Membrane Anchor of the Trypanosome Variant Surface Glycoprotein: Its Biosynthesis and Cleavage

Paul T. Englund, Dale Hereld, Jessica L. Krakow,
Tamara L. Doering, Wayne J. Masterson, and Gerald W. Hart
*Department of Biological Chemistry, Johns Hopkins School of Medicine,
Baltimore, Maryland 21205*

## INTRODUCTION

African trypanosomiasis, a disease of humans and cattle in equatorial Africa, is caused by protozoan parasites of the genus *Trypanosoma*. Trypanosomes inhabit the bloodstream of their mammalian host and are transmitted between mammals by tsetse flies. Even today chemotherapy for trypanosomiasis is unsatisfactory, and attempts at vaccination have been thwarted by the trypanosome's remarkable property of antigenic variation.

Antigenic variation is the process by which the trypanosome periodically switches its surface antigens and thereby evades its host's immune defenses. Ten million identical molecules of a variant surface glycoprotein (VSG) form a densely packed surface coat that covers the entire cell and presumably conceals underlying structures from the immune system. The VSG coat itself elicits a specific immune response which eliminates most of the trypanosomes. However, a few parasites in the population survive because they had replaced their coat with a new one composed of an antigenically distinct VSG molecule. These trypanosomes proliferate until they too are destroyed by an immune response specific for this new antigenic variant. The infection persists, however, because new variants continue to appear. Trypanosomes have the genetic potential for synthesis of hundreds of VSGs with different amino acid sequences; therefore, they can switch surface coats and maintain an infection almost indefinitely (see Englund et al., 1982; Borst and Cross,

1982; Boothroyd, 1985; Donelson and Rice-Ficht, 1985, for reviews of antigenic variation).

## THE VARIANT SURFACE GLYCOPROTEIN MOLECULE

In *Trypanosoma brucei* VSGs have between 500 and 600 amino acid residues; those from different trypanosome variants differ strikingly in amino acid sequence (see Turner, 1982, for review of VSG structure). Each VSG molecule contains two different kinds of carbohydrate modifications. First, there are one or more asparagine-linked oligosaccharides of either the high mannose or complex type (the latter lack sialic acid) (Holder, 1985). Second, there is a phosphatidylinositol-containing glycolipid which is covalently linked, through ethanolamine, to the $\alpha$-carboxyl group of the C-terminal amino acid residue (Ferguson et al., 1985b). The only fatty acids present in this glycolipid are myristates (Ferguson and Cross, 1984), and these anchor the VSG to the trypanosome surface by inserting into the lipid bilayer of the plasma membrane. Ferguson and co-workers (1987) have recently deduced the complete structure of this glycolipid, which is shown in Figure 1. A slightly different structure has been proposed by Schmitz and co-workers (1987).

Phosphatidylinositol-containing glycolipid anchors are not unique to VSGs. It is now clear that many membrane proteins, including some from mammals, are anchored by glycolipids similar to those on VSG (Low et al., 1986; Low, 1987; Cross, 1987). These proteins include acetylcholinesterase, alkaline phosphatase, 5'-nucleotidase, decay accelerating factor (DAF), and Thy-1. Since VSG is by far the most abundant of these proteins, it offers a convenient system for studying the structure, biosynthesis, enzymatic cleavage, and function of these glycolipid anchors. In this essay we describe some of our experimental results and discuss our ideas on the biosynthesis of the VSG glycolipid and its specific cleavage by an endogenous phospholipase C known as VSG lipase.

## VSG BIOSYNTHESIS

The VSG polypeptide is synthesized on a membrane-bound ribosome and then is processed in several ways (see Turner et al., 1985, for review). Two of the processing events, the removal of an N-terminal signal sequence and the glycosylation of specific asparagine residues, commonly occur during the biosynthesis of membrane or secretory proteins in other eukaryotic cells. The other two events, the removal of a hydrophobic sequence of about 20 amino

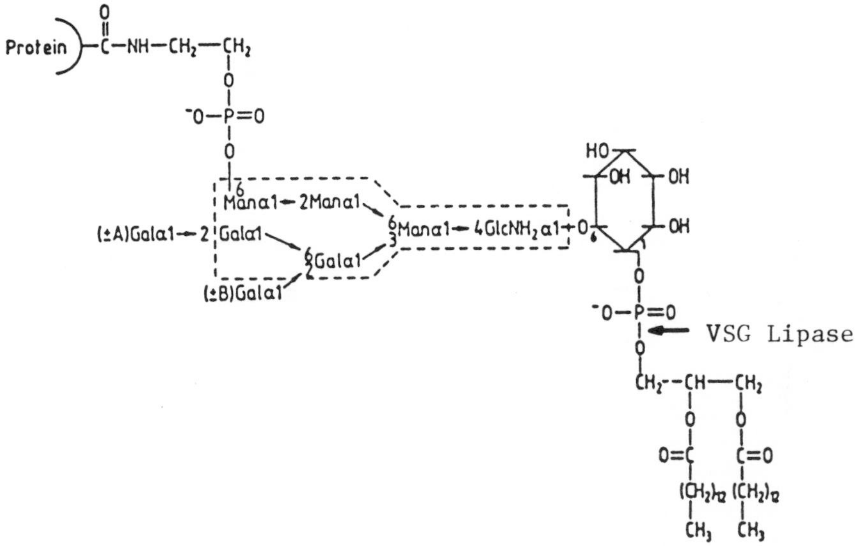

Fig. 1. Structure of the glycolipid from VSG 117 (MITat 1.4) (Ferguson et al., 1987). The structure is reprinted with permission. The bond cleaved by VSG lipase is indicated.

acid residues from the VSG's C-terminus (Boothroyd et al., 1980) and the attachment of the glycolipid to the new C-terminus, are still poorly understood. These two unusual processing events are discussed below.

## A VSG GLYCOLIPID PRECURSOR

Early studies in our laboratory (Bangs et al., 1985) and elsewhere (Ferguson et al., 1986) indicated that the glycolipid is attached to the VSG within 1 min after synthesis of the polypeptide. This surprisingly rapid processing suggested that the glycolipid may be preconstructed and situated in the endoplasmic reticulum (ER) membrane. A hypothetical scheme for its attachment to VSG is shown in Figure 2. In this scheme we assume that the newly synthesized VSG polypeptide, after translocation through the ER membrane, is transiently anchored in that membrane by the hydrophobic sequence at its C-terminus. Rapid nucleophilic attack by the preconstructed glycolipid on the polypeptide could result in transfer of the VSG to the glycolipid, with concomitant displacement of the C-terminal peptide. The glycolipid-contain-

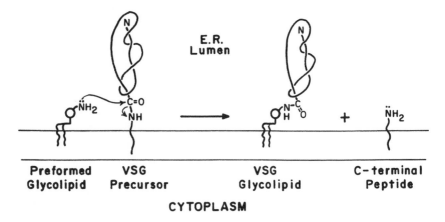

Fig. 2. Hypothetical scheme for attachment of the glycolipid to the VSG in the endoplasmic reticulum. See text for details.

ing VSG could then proceed to the cell surface. An alternative two-step mechanism, involving proteolytic removal of the C-terminal peptide followed by independent glycolipid attachment, is less likely than the concerted reaction proposed in Figure 2; proteolysis would not only release the VSG intermediate from the ER membrane but would also waste the peptide bond energy which could have been conserved for addition of the glycolipid.

Isolation of the putative preformed glycolipid precursor has provided strong support for the model shown in Figure 2 (Krakow et al., 1986). Since the glycolipid on VSG can be efficiently labeled in vivo with [³H]-myristate (Ferguson and Cross, 1984), we expected that the precursor would be labeled as well. We incubated *Trypanosoma brucei* cells with [³H]-myristate, extracted the lipids with organic solvent, and chromatographed them on silica thin layer plates. About 10–15 radioactive lipids were separated by this system; many of these are probably common phospholipids. One of the most polar of these lipids, designated lipid A, is likely to be a VSG glycolipid precursor (Krakow et al., 1986).

The findings implicating lipid A as a precursor are as follows: a) During continuous labeling, incorporation of [³H]-myristate into lipid A reaches a steady state as expected of a metabolic intermediate; the radioactivity disappears during a chase. b) Like the VSG glycolipid, lipid A can be labeled with [³H]-myristate, [³H]-mannose, and [³H]-ethanolamine, but not with [³H]-palmitate. c) Like the VSG glycolipid, lipid A is cleaved by nitrous acid to generate dimyristoyl phosphatidylinositol, suggesting that this moiety is linked glycosidically to non-acetylated glucosamine. d) Like the VSG glycolipid, lipid A is cleaved by the highly specific trypanosomal enzyme VSG lipase liberating dimyristoylglycerol. e) VSG lipase treatment of lipid A yields a hydrophilic molecule which binds to antibodies ("anti-cross-reacting determinant") specific for VSG lipase-treated VSG glycolipid, suggesting structural similarities between the two molecules (Krakow et al., 1986).

The glycolipid appears to be processed after its addition to VSG. This conclusion is based primarily on size measurements by gel filtration of the VSG lipase-treated glycolipids derived from protease digests of mature VSG (59 kilodaltons [kD]) and of an immature intracellular form of VSG (58 kD) (unpublished observations and Bangs et al., 1986). An apparent increase in size and heterogeneity of the glycolipid as it progresses to its mature form could be due to addition of galactose residues. Galactosidase treatment of the mature glycolipid decreased its apparent size, whereas this enzyme had no effect on the immature form. Galactose is present in varying amounts (ranging from zero to eight residues) in different VSG glycolipids (Holder, 1985); this variation could be due to variable processing of the glycolipid after it is incorporated into the VSG. The variant we study, ILTat 1.3, has approximately four galactose residues in its glycolipid.

Based on the kinetic and structural studies summarized above, we propose that lipid A is a VSG glycolipid precursor. Its existence is consistent with the scheme depicted in Figure 2. Whether lipid A is the molecule which is directly incorporated into VSG or an intermediate on the pathway to that ultimate precursor is not yet certain. We are currently studying the biosynthesis of the precursor, its structure, the mechanism by which it is incorporated into VSG, and its modification after attachment to VSG.

## VSG TRANSFERASE

According to the model in Figure 2, the VSG polypeptide in the ER membrane is transferred from its C-terminal hydrophobic peptide to the glycolipid precursor residing in the same membrane. We refer to the hypothetical enzyme which catalyzes this reaction as VSG transferase, and in the following paragraphs we speculate on some of its properties.

Because many different VSG genes are expressed during a trypanosome infection, VSG transferase probably is able to recognize and process many different VSG sequences. Therefore, we inspected the C-terminal sequences of numerous VSGs, derived from cDNA sequencing, for common features which might be recognized by this enzyme. In Figure 3 these sequences are aligned so that the C-terminus of the mature sequence (either aspartic acid,

```
                      ···WEGETCKD │ SSILVTKKFALTVVSAAFVALLF  MITat 1.6ᵃ
 TTDKCKDKKKDDCKSPDCKWEGETCKD │ SSFILNKQFALSVVSAAFAALLF  ILTat 1.3ᵇ
 TDKCKGKLEDTCKKESNCKWENNACKD │ SSILVTKKFALTVVSAAFVALLF  MITat 1.4aᶜ
 TDKCKGKLEDTCKKESNCKWENNACKD │ SSILATKKFALSMVSAAFVTLLF  MITat 1.4bᵈ
 TDKCKGKLEDTCKKESNCKWEGETCKD │ SSILVNKQLALSVVSAAFAALLF  ANTat 1.8ᵉ
 AEKCTGKKKDDCK  DGCKWEAETCKD │ SSILLTKNFALSVVSAALVALLF  ANTat 1.1ᶠ
 TDKCKDKTKDECKSPN CKWEGETCKD │ SSILVTKKFALSLVSAAFASLLF  ANTat 1.10ᶠ
 EEKCKGKLEPECTKAPECKWEGETCKD │ SSILVNKQFTLSMISAAFM...    MVAT 4ᵍ
 TECEGVKGTPPTGKAKVCGWIEGKCQD │ SSFLLSKQFALSVVSAAFAALLF  MVAT 7ᵍ
 PGKSADCGFRKGKDGETDEPDKEKCRN │ GSFLTSKQFAFSVVSAAFVALLF  MITat 1.5aʰ
 ASTCAGKKQGECEKENGCKWDGKECKD │ SSILVNKQFAFSVVSAAFMALLF  IATat 1.3Mⁱ
 TTEEKCKGKKKDDCKDGCKWEGETCKD │ SSFLLTKKFALSVVSATFVALLF  IATat 1.3Rⁱ
 TAEKCKGKGEKDCKSPDCKWEGGTCKD │ SSILANKQFALSVASAAFVALLF  ILTat 1.24ʲ

                      NTNTTGSS │ NSFVISKTPLWLAVLLF  MITat 1.2aᵏ
            EEAAENQEGKKEKTSNTTAS │ NSFVINKAPLLLGFLLF  TXTat 1ˡ
     AKRVAEQAATNQETEGKDGKTTNTTGS │ NSFVIHKAPLFLAFLLF  ILTat 1.1ᵇ
     DKEEAKKLEEKTEQNDSKTVTTNTTGS │ HSFVINKTPLLLAFLLF  IATat 1.2ᵐ
   PKAGTEAATTGPGERDAGATANTTGSS │ NSFVIKTSPLLFAFLLF  MVAT 5-2ⁿ
   KLSEEGKQAEKENQEGKDGKANTTGSS │ NSFVIKTSPLLLAVLLL  MVAT 5-3ⁿ
```

Fig. 3. C-terminal amino acid sequences of *T. brucei* VSGs determined by cDNA sequencing. The amino acid immediately to the left of the vertical line (aspartic acid, asparagine, or serine) is at the C-terminus of the mature VSG and its alpha carboxyl group is linked to the glycolipid. (This fact has been proven only for MITats 1.2, 1.4, 1.5, and 1.6 [Holder, 1985]; in other VSGs it has been inferred from sequence similarities.) The underlined sequence may be involved in recognition by VSG transferase; see text for discussion. Original references for sequences are a, Holder and Cross (1981) and Cross (1984); b, Rice-Ficht et al. (1981); c, Boothroyd et al. (1980); d, Michels et al. (1983); e, Matthyssens et al. (1981); f, Pays et al. (1983); g, Lenardo et al. (1984); h, Majumder et al. (1981); i, S. Brentano and J. Donelson, personal communication; j, M. Turner, personal communication; k, Boothroyd et al. (1981); l, Merritt et al. (1983); m, Murphy et al. (1984); n, J. Donelson, personal communication. The sequences designated MVAT are for VSGs from metacyclic trypanosomes of the WRATat 1 serodeme.

asparagine, or serine, to which the glycolipid is ultimately attached) is just to the left of the vertical line. The C-terminal tail, either 23 or 17 residues, is to the right of the line.

We expect that the sequence recognized by VSG transferase would be near the residue to which the glycolipid will be attached. As shown in Figure 3, the sequences to the left of that residue are highly variable and therefore less likely to be involved in recognition. In contrast, the sequences immediately to the right of that residue share characteristics. Most VSGs have these common features: the first residue to the right of the line is hydrophilic (asparagine, glycine, serine, or histidine); the second is serine; the next three are hydrophobic (phenylalanine, valine, isoleucine, leucine, or alanine); the sixth is hydrophilic (asparagine, serine, histidine, or threonine); and the seventh is lysine. There are exceptions, however. In the case of MITat 1.5a, one of the three hydrophobic residues is substituted by threonine; with MVATs 5-2 and 5-3 the sixth residue (lysine) and the seventh residue (threonine) would have to be reversed to conform to the proposed sequence. Rigorous experiments will be essential to determine which residues are recognized by VSG transferase. It would not be surprising if other hydrophobic residues, such as those conserved near the C-terminus of the tail, are also involved.

The hydrophobic tail could act as a temporary anchor, allowing the VSG to diffuse laterally in the ER membrane until attack by the glycolipid precursor. Inspection of the sequences of the hydrophobic tails, however, indicates that they differ in some respects from other membrane spanning sequences (Eisenberg, 1984). Although rich in hydrophobic residues, they do contain charged residues (one or two lysines) and hydrophilic residues (several serines or threonines). Furthermore, several sequences include proline which would probably disrupt a membrane-spanning $\alpha$-helix. For these reasons, it seems possible that the C-terminal sequence may not serve as a conventional membrane anchor. Instead, the VSG transferase may act while its VSG substrate is leaving the ribosome with the C-terminal sequence still undergoing translocation through the endoplasmic reticulum membrane. Perhaps the VSG transferase, with a bound glycolipid precursor, waits until the part of the VSG containing the "recognition sequence" passes through the membrane. As it interacts with this sequence it would transfer the VSG to the glycolipid. This hypothesis is depicted in Figure 4.

The postulated VSG transferase reaction is energetically feasible: the reaction involves breakage of one amide bond (e.g., between aspartic acid and the C-terminal peptide) and the formation of another (between aspartic acid and the ethanolamine residue of the glycolipid). Nevertheless, it is

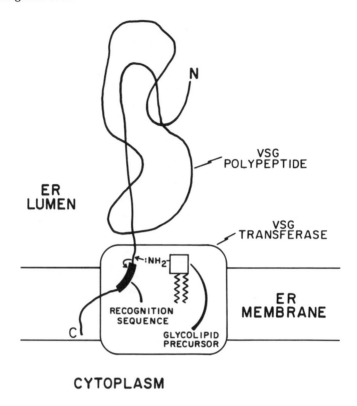

Fig. 4. Hypothetical mechanism for VSG transferase. See text for discussion.

possible that the mechanism is more complex than indicated in Figure 4 and that energy is introduced into the system to make this reaction irreversible. It is also possible that rapid galactosylation of the glycolipid or degradation of the displaced C-terminal peptide drives the reaction forward.

**VSG LIPASE**

Early studies on VSG, prior to the discovery of its glycolipid anchor, indicated that this protein is hydrophilic with no apparent affinity for membranes. These results were puzzling because in live trypanosomes the VSG appears to be tightly bound to the plasma membrane. Cardoso de Almeida and Turner (1983) resolved this paradox when they discovered the "membrane form" of VSG (mfVSG). mfVSG is an amphiphilic protein which contains the intact glycolipid. In the soluble form of VSG (sVSG), the glycolipid has been cleaved to release dimyristoylglycerol (Ferguson et al., 1985a). Lysis of trypanosomes under nondenaturing conditions results in an exceedingly rapid cleavage of the VSG glycolipid, catalyzed by an endogenous trypanosome enzyme we designate "VSG lipase." VSG lipase has the specificity of a phospholipase C; the bond which it cleaves is indicated on the glycolipid structure in Figure 1. mfVSG may be purified only under conditions in which VSG lipase has been inhibited (e.g., by SDS or mercurials).

VSG lipase is a membrane-associated enzyme. It is present in a particulate fraction of the cell but may be solubilized in active form by mild detergents (Cardoso de Almeida et al., 1983, 1984). It has been purified to homogeneity and studied in our laboratory (Hereld et al., 1986) and elsewhere (Bulow and Overath, 1986; Fox et al., 1986). The purified enzyme in its native form has a single polypeptide chain of 37–39 kD. About 30,000 copies are present in each cell, representing about 0.04% of the cell's protein. Under our standard assay conditions in vitro, one molecule of VSG lipase converts about 100 molecules of mfVSG to sVSG per minute.

VSG lipase differs strikingly from other known phospholipase C's (Bulow and Overath, 1986; Hereld et al., 1986; Fox et al., 1986). In contrast to many of these enzymes it does not require $Ca^{2+}$. It also demonstrates remarkably high specificity: it does not cleave other phospholipids including 1-stearoyl-2-arachidonoyl-$sn$-phosphatidylinositol and it has barely detectable activity on dimyristoyl phosphatidylinositol (a component of the VSG glycolipid). Known substrates include VSG and other glycolipid-anchored proteins and the VSG glycolipid precursor, lipid A. Therefore, this enzyme is not a phosphatidylinositol-specific phospholipase C (PI-PLC), but instead is a glycan-phosphatidylinositol-specific phospholipase C (GPI-PLC). VSG lipase's high degree of specificity for the VSG glycolipid suggests that it plays a role in VSG metabolism.

The biological function of VSG lipase is not known, but it seems likely that it serves to release VSG from the plasma membrane under certain circumstances. The presence of a common glycolipid membrane anchor on VSGs of different amino acid sequence provides a target for a single releasing activity. If the VSGs were anchored by transmembrane peptide sequences, they could be released by proteolytic cleavage. However, a phospholipase acting on a glycolipid common to all VSGs may achieve higher specificity for release than a protease acting on peptide anchors which are continually subject to mutation.

VSG lipase appears quiescent on living trypanosomes, as there is little or no release of VSG from these cells (Black et al., 1982; Shapiro, 1986). Its apparent inactivity could be explained by some regulatory mechanism or by compartmentalization away from the VSG in the plasma membrane. Activation of VSG lipase and/or translocation of the enzyme to the plasma membrane to allow VSG cleavage could occur under special physiological conditions. One such situation, when the VSG coat is totally lost, is during transformation of the trypanosome from its bloodstream form to its procyclic form either in the midgut of the tsetse fly or in vitro in 27°C culture (Vickerman, 1969; Overath et al., 1983). VSG lipase could also function in the release of VSG during antigenic variation in the animal bloodstream. In any case, it is likely that VSG lipase plays an important role in the life of trypanosomes, and one of our laboratory's objectives in the next several years is to discover that function.

## ACKNOWLEDGMENTS

This research was supported by grants from NIH (AI21334) and the MacArthur Foundation. P.T.E. is a Burroughs Wellcome Scholar in Molecular Parasitology and G.W.H. is an Established Investigator of the American Heart Association. D.H. and T.L.D. are supported by Medical Scientist Training Grant GM7309. We thank Viiu Klein and Shirley Metzger for invaluable help.

## REFERENCES

Bangs JD, Hereld D, Krakow JL, Hart GW, Englund PT (1985): Rapid processing of the carboxyl terminus of a trypanosome variant surface glycoprotein. Proc Natl Acad Sci USA 82:3207–3211.

Bangs JD, Andrews NW, Hart GW, Englund PT (1986): Posttranslational modification and intracellular transport of a trypanosome variant surface glycoprotein. J Cell Biol 103:255–263.

Black SJ, Hewett RS, Sendashonga CN (1982): *Trypanosoma brucei* variable surface antigen is released by degenerating parasites but not by actively dividing parasites. Parasite Immunol 4:233–244.

Boothroyd JC (1985): Antigenic variation in African trypanosomes. Annu Rev Microbiol 39:475–502.

Boothroyd JC, Cross GAM, Hoeijmakers JHJ, Borst P (1980): A variant surface glycoprotein of *Trypanosoma brucei* synthesized with a C-terminal hydrophobic "tail" absent from purified glycoprotein. Nature 288:624–626.

Boothroyd JC, Paynter CA, Cross GAM, Bernards A, Borst P (1981): Variant surface glycoproteins of *Trypanosoma brucei* are synthesized with cleavable hydrophobic sequences at the carboxy and amino termini. Nucleic Acids Res 9:4735–4743.

Borst P, Cross GAM (1982): Molecular basis for trypanosome antigenic variation. Cell 29:291–303.

Bulow R, Overath P (1986): Purification and characterization of the membrane-form variant surface glycoprotein hydrolase of *Trypanosoma brucei*. J Biol Chem 261:11918–11923.

Cardoso de Almeida ML, Turner MJ (1983): The membrane form of variant surface glycoproteins of *Trypanosoma brucei*. Nature 302:349–352.

Cardoso de Almeida ML, Le Page RWF, Turner MJ (1984): The release of variant surface glycoproteins of *Trypanosoma brucei*. In J.T. August (ed): Proceedings of the 3rd John Jacob Abel Symposium on Drug Development: Molecular Parasitology. New York: Academic Press, pp 19–31.

Cross GAM (1984) Structure of the variant glycoproteins and surface coat of *Trypanosoma brucei*. Philos Trans R Soc Lond [Biol] 307:3–12.

Cross GAM (1987): Eukaryotic protein modification and membrane attachment via phosphatidylinositol. Cell 48:179–181.

Donelson JE, Rice-Ficht AC (1985): Molecular biology of trypanosome antigenic variation. Microbiol Rev 49:107–125.

Eisenberg D (1984): Three-dimensional structure of membrane and surface proteins. Annu Rev Biochem 53:595–623.

Englund PT, Hajduk SL, Marini JC (1982): The molecular biology of trypanosomes. Annu Rev Biochem 51:695–726.

Ferguson MAJ, Cross GAM (1984): Myristylation of the membrane form of a *Trypanosoma brucei* variant surface glycoprotein. J Biol Chem 259:3011–3015.

Ferguson MAJ, Haldar K, Cross GAM (1985a): *Trypanosoma brucei* variant surface glycoprotein has a sn-1,2-dimyristyl glycerol membrane anchor at its COOH terminus. J Biol Chem 260:4963–4968.

Ferguson MAJ, Low MG, Cross GAM (1985b): Glycosyl-sn-1,2-dimyristylphosphatidylinositol is covalently linked to *Trypanosoma brucei* variant surface glycoprotein. J Biol Chem 260:14547–14555.

Ferguson MAJ, Duszenko M, Lamont GS, Overath P, Cross GAM (1986): Biosynthesis of *Trypanosoma brucei* variant surface glycoproteins: N-glycosylation and addition of a phosphatidylinositol membrane anchor. J Biol Chem 261:356–362.

Ferguson MAJ, Homans SW, Dwek RA, and Rademacher TW (1988): Glycosylphosphatidyl linositol moiety that anchors *Trypanosoma brucei* variant surface glycoprotein to the membrane. Science 239:753–759.

Fox JA, Duszenko M, Ferguson MAJ, Low MG, Cross GAM (1986): Purification and characterization of a novel glycan-phosphatidylinositol-specific phospholipase C from *Trypanosoma brucei*. J Biol Chem 261:15767–15771.

Hereld D, Krakow JL, Bangs JD, Hart GW, Englund PT (1986): A phospholipase C from *Trypanosoma brucei* which selectively cleaves the glycolipid on the variant surface glycoprotein. J Biol Chem 261:13813–13819.

Holder AA (1985): Glycosylation of the variant surface antigens of *Trypanosoma brucei*. Curr Top Microbiol Immunol 117:57–74.

Holder AA, Cross GAM (1981): Glycopeptides from variant surface glycoproteins of *Trypanosoma brucei*. C-terminal location of antigenically cross reacting carbohydrate moieties. Mol Biochem Parasitol 2:135–150.

Krakow JL, Hereld D, Bangs JD, Hart GW, Englund PT (1986): Identification of a glycolipid precursor of the *Trypanosoma brucei* variant surface glycoprotein. J Biol Chem 261:12147–12153.

Lenardo MJ, Rice-Ficht AC, Kelly G, Esser KM, Donelson JE (1984): Characterization of the genes specifying two metacyclic variable antigen types in *Trypanosoma brucei rhodesiense*. Proc Natl Acad Sci USA 81:6642–6646.

Low MG (1987): Biochemistry of the glycosyl phosphatidyl-inositol membrane protein anchors. Biochem J 244:1–13.

Low MG, Ferguson MAJ, Futerman AH, Silman I (1986): Covalently attached phosphatidylinositol as a hydrophobic anchor for membrane proteins. Trends Biochem Sci 11:212–215.

Majumder HK, Boothroyd JC, Weber H (1981): Homologous 3' terminal regions of mRNAs for surface antigens of different antigenic variants of *Trypanosoma brucei*. Nucleic Acids Res 9:4745–4753.

Matthyssens G, Michiels F, Hamers R, Pays E, Steinert M (1981): Two variant surface glycoproteins of *Trypanosoma brucei* have a conserved C-terminus. Nature 293:230–233.

Merritt SC, Tschudi C, Konigsberg WH, and Richards FF (1983): Reverse transcription of trypanosome variable antigen mRNAs initiated by a specific oligonucleotide primer. Proc Natl Acad Sci USA 80:1536–1540.

Michels PAM, Liu AYC, Bernards A, Sloof P, Van Der Bijl MMW, Schinkel AH, Menke HH, Borst P, Veeneman GH, Tromp MC, Van Boom JH (1983): Activation of the genes for variant surface glycoproteins 117 and 118 in *Trypanosoma brucei*. J Mol Biol 166:537–556.

Murphy WJ, Brentano ST, Rice-Ficht AC, Dorfman DM, Donelson JE (1984): DNA rearrangements of the variable surface antigen genes of the trypanosomes. J Protozool 31:65–73.

Overath P, Czichos S, Stock U, Nonnengaesser C (1983): Repression of glycoprotein synthesis and release of surface coat during transformation of *Trypanosoma brucei*. EMBO J 2:1721–1728.

Pays E, Van Assel S, Laurent M, Darville M, Vervoort T, Van Meirvenne N, Steinert M (1983): Gene conversion as a mechanism for antigenic variation in trypanosomes. Cell 34:371–381.

Rice-Ficht AC, Chen KK, Donelson JE (1981): Sequence homologies near the C-termini of the variable surface glycoproteins of *Trypanosoma brucei*. Nature 294:53–57.

Schmitz B, Klein RA, Duncan IA, Egge H, Gunawan J, Peter-Katalinic J, Dabrowski U, Dabrowski J (1987): MS and NMR analysis of the cross-reacting determinant glycan from *Trypanosoma brucei brucei* MITat 1.6 variant specific glycoprotein. Biochem Biophys Res Commun 146:1055–1063.

Shapiro SZ (1986): *Trypanosoma brucei*: Release of variant surface glycoprotein during the parasite life cycle. Exp Parasitol 61:432–437.

Turner MJ (1982): Biochemistry of the variant surface glycoproteins of salivarian trypanosomes. Adv Parasitol 21:69–153.

Turner MJ, Cardoso de Almeida ML, Gurnett AM, Raper J, Ward J (1985): Biosynthesis, attachment and release of variant surface glycoproteins of the African trypanosome. Curr Top Microbiol Immunol 117:23–55.

Vickerman K (1969): On the surface coat and flagellar adhesion in trypanosomes. J Cell Sci 5:163–193.

The Biology of Parasitism, pages 413–429
© 1988 Alan R. Liss, Inc.

# Parasite Pharmacology

Ching C. Wang

*Department of Pharmaceutical Chemistry, University of California*
*San Francisco, San Francisco, California 94143*

## INTRODUCTION

Parasitic infections belong to a class of infectious diseases that have protozoa or helminths as the infectious agents. Although this classification lies on rather shaky scientific ground owing to the distant relationships among the many species of protozoa and worms, it does specify that the diseases are caused by well-defined external agents. They are thus basically similar to bacterial or fungal infections. In principle, chemotherapeutic control of parasitic infections should exploit metabolic differences between the parasite and the host, such as the antibacterial action of $\beta$-lactams on bacterial cell wall synthesis.

Parasites are, by definition, organisms incapable of growing, differentiating, or even surviving in a natural environment like soil or water. They must live in highly nutritionally enriched surroundings, e.g., the intestinal lumen, the bloodstream or the cytoplasm of another living cell. This unusual requirement for "luxurious" living conditions reflects highly deficient metabolic activities in parasites, which necessitate a reliance on supplies of a large variety of exogenous nutrients from the host. These metabolic deficiencies in parasites make them vulnerable to simple chemotherapeutic intervention on nutrient supplies. Indeed, the practice of chemotherapeutic control of parasitic infections can be traced back to the beginnings of human civilization. On the other hand, there has not yet been a single successful vaccine against any parasitic disease. Parasites must have highly elaborate means of evading host immunity in order to survive inside the host body for prolonged periods. Chemotherapy is aimed at the weak points of parasites and will thus remain the most effective way of controlling parasitic infections within the foreseeable future.

Although the search for new agents for antiparasitic chemotherapy should have, in theory, a fairly good chance of success, experience in the recent past has indicated the opposite. Until a few years ago, there had not been a new antileishmanial or antitrypanosomal agent developed for clinical use for 40 years. No cure is yet available for the Chaga's disease. In some cases, such as malaria, additional difficulties have been encountered owing to development of drug resistance among the parasites. The root of the problem, however, lies in our lack of knowledge of parasite metabolism, which is the source of chemotherapeutic opportunities. It was not until the past two decades, when many parasites began to be cultivated in vitro successfully, that the myth of a certain mysterious parasite dependence on the living host started to fade away. The in vitro growth of many parasites without contamination of bacteria or host tissues has made intensive and focused biochemical investigations of these parasites possible (Wang, 1984). The ability to clone single protozoan parasites has made it possible to study a genetically homogeneous population of a parasite species. The capability of some parasites to grow in a defined medium has made metabolic studies even easier. Biochemical investigations of parasites performed in the past few years have yielded a wealth of knowledge, some of which has already led to the identification of new therapeutic targets and, in a couple of cases, new antiparasitic agents (Wang, 1984). Advanced studies of the mechanism of action of some newly discovered antiparasitic agents have also contributed to the identification of new opportunities for antiparasitic chemotherapy (Wang and Pong, 1982). A few successful examples of antiparasitic drug development based on biochemical knowledge instead of empirical screens have recently become available. These positive experiences may provide some guidance for our future attempts at antiparasitic chemotherapy as well as our ability to recognize an opportunity for selective attack on parasite metabolism.

### DL-α-DIFLUOROMETHYLORNITHINE

Among eukaryotes, the formation of polyamines, required for cellular proliferation, is controlled by ornithine decarboxylase (Pegg and McCann, 1982). In the long, slender bloodstream trypomastigotes of African trypanosomes, putrescine and spermidine constitute the main pool of polyamines (Bacchi, 1980). They are mainly synthesized from ornithine by *Trypanosoma brucei* and are taken up only very slowly from extracellular sources (Bacchi et al., 1979). DL-α-difluoromethylornithine (DFMO), a catalytic, irreversible inhibitor of ornithine decarboxylase with some antitumor activities (Sjoerdsma and Schechter, 1984), is effective against African trypanosomes both in vitro and in vivo (Bacchi et al., 1980). Because of its remarkable lack of toxicity, it is currently undergoing clinical trials with encouraging results, especially on *Trypanosoma gambiense* infections (van Nieuwenhove et al., 1985). DFMO has been shown to deplete the intracellular polyamine pool of *T. brucei* (Giffin et al., 1986), and its in vitro activity against *T. brucei* is reversible by putrescine (Phillips et al., 1987). Its antitrypanosomal activity in the experimental murine model is also reversible by intraperitoneal injections of polyamines (Bacchi et al., 1982). It appears thus that DFMO acts on *T. brucei* by inhibiting the ornithine decarboxylase of the parasite. The physiological consequence of such enzyme inhibition and subsequent polyamine depletion turned out to be quite interesting. The long-slender form of *T. brucei* is simply transformed to the nondividing short-stumpy form by DFMO treatment either in vivo or in vitro (Giffin et al., 1986). The nondividing form is apparently incapable of changing its variant surface glycoprotein coat and is eventually caught up by the host immune response. This is verified by experiments indicating that the antitrypanosomal activity of DFMO is much reduced in immunodeficient rats (Bitoni et al., 1986). This drug-induced transformation bears a close resemblance to the natural transformation of *T. brucei* from long-slender to short-stumpy forms, during which the mitochondrial genes encoding cytochrome b, cytochrome oxidase, and NADH dehydrogenase are transcribed at high levels in both cases (Feagin et al., 1986). Preliminary studies in our laboratory indicated also that during the natural transformation of pleomorphic strains of *T. brucei* from long-slender to short-stumpy forms, the specific ornithine decarboxylase activity in the parasite decreased rapidly to a level beyond detection (unpublished observation), which was presumably accompanied by polyamine depletion in *T. brucei*. These observations, together with the previous finding by Mancini and Patton (1981) that cyclic 3′,5′-AMP increased in *T. brucei* during its early phase of transformation from long-slender to short-stumpy forms, suggests the involvement of a typical mechanism of eukaryotic cell differentiation in the *T. brucei* transformation. α-Monofluoromethylornithine, another mechanism-based suicide inhibitor of ornithine decarboxylase, and dibutyryl 3′,5′-cyclic AMP were found to induce differentiation of NB-15 mouse neuroblastoma cells (Chen et al., 1983). The ornithine decarboxylase activity decreased and the polyamine content was depleted in the neuroblastoma cells during differentiation (Chen et al., 1982). The spermine deficiency elicits a specific activation of nuclear protein kinase NI which phosphorylates two proteins with subunit molecular weights of 11 kilodaltons (kd) and 10 kd in the nuclear matrix (Verma and Chen, 1986), a process usually related to the regulation of gene expression (Song and Adolph, 1983). It is possible that similar events occur during *T. brucei* transformation from the long-slender to the short-stumpy form and that *T. brucei* ornithine decarboxylase could be a strictly regulated enzyme in the parasite.

Understanding the mechanism of antitrypanosomal action of DFMO cannot, however, explain why the trypanosomes are much more susceptible to DFMO than the mammalian host. Apparently, the trypanosomal ornithine decarboxylase has a higher $K_i$ value of 130 $\mu$M for DFMO (Bitonti et al., 1985) than does the mammalian enzyme ($K_i$ = 39 $\mu$M) (Bey, 1978), and should be, in theory, less sensitive to DFMO inhibition. Detailed comparisons between the mammalian and the parasite ornithine decarboxylase are thus necessary for understanding the basis of therapeutic action in DFMO. This has been proved to be difficult owing to the unavailability of *T. brucei* enzyme ($\sim$ 10 nmol/h/mg protein) for extensive purifications and characterizations. Fortunately, the close homology between the two enzymes allowed the use of a mouse lymphoma S49 cDNA of ornithine decarboxylase as a heterologous probe to identify and clone the *T. brucei* gene encoding ornithine decarboxylase under relatively stringent conditions (Phillips et al., 1987). The sequence of the parasite enzyme gene revealed two major discrepancies from that of the Balb/c mouse cDNA when translated into open reading frames (Fig. 1). The parasite enzyme has an extra 20-amino-acid peptide at the N-terminus but lacks the 36-amino-acid peptide at the C-terminus of the Balb/c mouse ornithine decarboxylase. There is 61.5% homology between residues 21-445 of *T. brucei* ornithine decarboxylase and residues 1-425 of Balb/c mouse ornithine decarboxylase. When the homologous amino acid substitutions are discounted, the percentage of homology reaches 90%. Clearly, these two enzymes are very similar, which explains their similar susceptibilities toward DFMO, similar subunit molecular weights, and ready hybridizations between the two encoding DNAs, but fails to explain the selective susceptibility of trypanosomes toward DFMO.

The biological significance of the additional N-terminal peptide in the parasite enzyme remains unclear, but the similar N-terminal three amino acids in the two enzymes (Met-Thr-Thr- vs. Met-Ser-Ser-; see Fig. 1) suggest similar N-terminal functions. The deletion of mouse C-terminal 36-amino-acid peptide in the *T. brucei* enzyme is of considerable interest because MaCrae and Coffino (1987) were able to make C-terminal deletions of the mouse ornithine decarboxylase by cDNA excision, cloning, and expression in an *Escherichia coli* ornithine decarboxylase deletion mutant HT289, and found that the mouse enzyme C-terminus could be deleted up to 38 amino acid residues without an appreciable decrease in enzymatic activity. Thus, the *T. brucei* enzyme with fewer C-terminal deletions must be a functional enzyme. Furthermore, the mouse enzyme C-terminal 423-449 is classified as

Tb   1                                   MTTKSTPSSLSVNCLVAQTE

```
              21              40                  60                  80
Tb    21  KSMDIVVNDDLSCRFLEGFNTRDALCKKISMNTCDEGDPFFVADLGDIVRKHETWKKCLPRVTPFYAVKCND
          *          *          *     * *  *******  **   * *  **************
M      1  MSSFTKDEFDCHILDEGFTAKDILDQKINEVSSSDDKDAFYVADLGDILKKHLRWLKALPRVTPFYAVKCND

              100             120                 140                160
Tb    93  DWRVLGTLAALGTGFDCASNTEIQRVRGIGVPPEKIIYANPCKQISHIRYARDSGVDVMTFDCVDELEKVAK
          ****  ********  **** * * *** *  ******** *  *  **   **  ****   ** ***
M      73 SRAIVSTLAAIGTGFDCASKTEIQLVQGLGVPAERVIYANPCKQVSQIKYAASNGVQMMTFDSEIELMKVAR

              180             200                 220
Tb    165 THPKAKMVLRISTDDDSLARCRLSVKFGAKVEDCRFILEQAKKLNIDVTGVSFHVGSGSTDASTFAQAISDSR
          *****  ****  **** *  ********       * ** ** ***** ********* **  ** ** ** *
M      145 AHPKAKLVLRIATDDSKAVCRLSVKFGATLKTSRLLLERAKELNIDVIGVSFHVGSGCTDPDTFVQAVSDAR

              240             260                 280                300
Tb    237 FVFDMGTELGFNMHILDIGGGFPGTRDAPLKFEEIAGVINNALEKHFPPDLKLTIVAEPGRYYVASAFTLAV
          ****  **  **  ** ********* *  ******  *** ** * ** *  * **************
M      217 CVFDMATEVGFSMHLLDIGGGFPGSEDTKLKFEEITSVINPALDKYFPSDSGVRIIAEPGRYYVASAFTLAV

              320             340                 360
Tb    309 NVIAKKVTPGVQTDVGAHAESNAQSFMYYVNDGVYGSFNCILYDHAVVRPLPQREPIPNEKLYPSSVWGPTC
          *  ****       *       *** * ********************** *   * * ** * ** *****
M      289 NIIAKKTVWKEQPGSDDEDESNEQTFMYYVNDGVYGSFNCILYDHAHVKALLQKRPKPDEKYYSSSIWGPTC

              381             400                 420                440
Tb    381 DGLDQIVERYYLPEMQVGEWLLFEDMGAYTVVGTSSFNGFQSPTIYYVVSGLPDHVVRELKSQKS
          ****  ****   ****  **  * *** * *** ****** * ***** *  **** *  *
M      361 DGLDRIVERCNLPEMHVGDWMLFENMGAYTVAAASTFNGFQRPNIYYVMSRPMWQLMKQIQSHGFPPEVEEQ
```

M   433   DDGTLPMSCAQESGMDRHPAACASARINV

Fig. 1. Comparison of the amino acid sequences between *Trypanosoma brucei* ornithine decarboxylase and Balb/c mouse lymphoma S49 ornithine decarboxylase. The identical segments between the *T. brucei* sequence (Tb) and the mouse sequence (M) are marked by asterisks (*).

a PEST region of the protein with a PEST score of 5.2 (Rogers et al., 1986). It means that the peptide is very rich in proline, glutamic acid, serine, and threonine, and is flanked by basic amino acids. These sequences are commonly found in many eukaryotic proteins with half-lives less than 2 h but not in those with longer half-lives (Rogers et al., 1986). According to the PEST hypothesis, phosphorylation of the serine residues in the PEST region may generate $Ca^{2+}$-binding sites which may elicit intracellular protein degradation by $Ca^{2+}$-activated cytoplasmic proteases such as calpain. Since mammalian ornithine decarboxylase is one of the most rapidly turned-over eukaryotic proteins, with an estimated half-life of 20 min (Tabor and Tabor, 1984), it is assigned with two PEST regions at the C-terminal 423-449 and the middle 295-307. *T. brucei* ornithine decarboxylase does not have the mouse enzyme C-terminus 426-461, and its sequence corresponding to the mouse sequence 295-307 contains no PEST (see Fig. 1). It is thus likely that, by the PEST theory, *T. brucei* ornithine decarboxylase should have a relatively long half-life. This elucidation was confirmed by our recent observations that *T. brucei* procyclic forms, kept under an inhibited state of protein synthesis, maintained a constant level of ornithine decarboxylase for at least 6 h (Phillips et al., 1987). This lack of turning-over of the parasite enzyme, which may be necessary for well-controlled *T. brucei* differentiation at a specific developmental stage by polyamine deficiency, may have constituted the basis of the therapeutic usefulness of DFMO as an antitrypanosomal agent. While DFMO remains a potent, irreversible inhibitor of mammalian ornithine decarboxylase, the in vivo drug metabolism coupled with continuous, rapid synthesis of new enzyme molecules in the mammalian host may quickly nullify the drug effect. But the same enzyme activity in *T. brucei* will remain inhibited even after the removal of free drug molecules owing to the absence of newly synthesized ornithine decarboxylase. This inhibited state, reflected by polyamine deficiency in DFMO-treated *T. brucei* both in vitro and in vivo (Giffin et al., 1986), pushes the parasite into the dormant short -stumpy form to be killed by the host immune response. It is thus ironic that by rigidly controlling the turnover of a key enzyme, the parasite achieves specific differentiations at crucial stages of its development, but it also provides an opportunity for chemotherapeutic attack.

Since the successful elucidation of the mechanism of DFMO action, we have further pursued the investigation of *T. brucei* ornithine decarboxylase. The genomic DNA encoding this enzyme was engineered by site-directed mutagenesis techniques to convert the 5'-initiating open reading sequence from 5'-ATGACC-3' to 5'-AGTACT-3' so that a Sca I restriction site was generated between 5'-AGT and ACT-3'. The Sca I fragment of the modified *T. brucei* ornithine decarboxylase gene, missing the initiating codon ATG but retaining the second codon ACT for threonine (ref. Fig. 1), was ligated to the BamHI site of a PL vector, pCQV2 (MacRae and Coffino, 1987). The ligated genome regained the ATG initiation codon at its 5'-end downstream from a strong lambda $P_R$ promoter and a mutant lambda cI gene encoding a thermolabile repressor of the $P_R$ promoter. This recombinant construct, termed pQtODC, was used to transform the *E. coli* HT289 ornithine decarboxylase deletion mutant. The *E. coli* transformants were cultivated and shocked at a higher temperature of 42°C to inactivate the lambda repressor. Enzyme activity assays and gel electrophoretic analysis of the [$^{14}$C]-DFMO-labeled protein indicated that about 1% of the total *E. coli* protein was

*T. brucei* ornithine decarboxylase (unpublished results). This protein was purified from *E. coli* to total homogeneity by ammonium sulfate fractionation (25–75%), pyridoxal phosphate agarose affinity column chromatography, and fast performance liquid chromatography on a Mono Q column. The purified recombinant enzyme has been characterized and found to be identical with the partially purified native *T. brucei* ornithine decarboxylase in all aspects examined. This may be the first time a therapeutic target enzyme in a parasite has been produced in pure form in significant quantities by means of genetic engineering means. Future efforts will be concentrated on enzyme crystallization, x-ray crystallography, and inhibitor design by computer graphics.

## AGENTS AFFECTING FUNCTIONS OF GLYCOSOMES

In the long-slender bloodstream forms of *T. brucei*, each trypomastigote consists of a single mitochondrion devoid of a functioning tricarboxylic acid cycle and a cytochrome chain of electron transport (Bowman and Flynn, 1976). The parasite is entirely dependent on glycolysis for energy production, which generates, from every glucose molecule, two ATP molecules and two pyruvate molecules under aerobic conditions (Grant and Sargent, 1960). The relatively inefficient energy production is compensated for by having the glycolysis proceed at an exceedingly high rate in order to support the rapid cell divisions of *T. brucei* every 6–8 h (Brohn and Clarkson, 1980). This is apparently made possible by aggregating the first seven glycolytic enzymes and two glycerol-metabolizing enzymes in membrane-bound, microbody-like organelles, termed the glycosomes (Opperdoes and Borst, 1977). The glycosome, having a protein concentration of 340 mg/ml and steady concentrations of glycolytic intermediates in the millimolar range (Visser et al., 1981), regenerates $NAD^+$ from NADH by a dihydroxyacetone phosphate: glycerol-3-phosphate shuttle plus a glycerol-3-phosphate oxidase in the mitochondrion. During anaerobiosis, the oxidase cannot function. Glycerol-3-phosphate becomes accumulated in the glycosome, and is accompanied by the accumulation of NADH and ADP, which eventually revert the glycerol kinase catalyzed reaction to generate glycerol and ATP to enable the parasite to survive (Visser and Opperdoes, 1980).

This delicate and indispensable glycolytic system is an obvious and attractive target for antitrypanosomal chemotherapy. Recent evidence suggested that the oxidase could be inhibited by salicyl hydroxamic acid (SHAM) to bring *T. brucei* under an anaerobic-equivalent condition; glycolysis could then be stopped by inhibiting glycerol kinase with glycerol (Clarkson and Brohn, 1976; Fairlamb et al., 1977). The SHAM and glycerol combination lyses African trypanosome bloodstream forms readily in vitro (Brohn and Clarkson, 1978) and suppresses parasitemia in animal models (Evans et al., 1977). However, while glycerol presented few problems, the curative dose of SHAM caused 44% animal mortality owing to acute toxicity (van der Meer et al., 1979). A replacement for SHAM with a less toxic but more potent inhibition of trypanosomal glycerol-3-phosphate oxidase is thus necessary if the combination therapy is to have clinical values. A recent finding that p-n-tetradecyloxybenzhydroxamic acid was a 70 times more potent inhibitor of the oxidase and a 10–20-fold better trypanocidal agent in vitro than SHAM made this direction seem promising (Grady et al., 1986) but success may ultimately have to rely on further understanding of *T. brucei* glycerol-3-phosphate oxidase.

The value of glycosome as a potential target for antitrypanosomal chemotherapy can be further illustrated by the apparently unique mechanism of its biogenesis. It has been assumed that the genes encoding glycosomal proteins are located in the nucleus, and the products of these genes are synthesized in the cytoplasm on free polysomes (Hart et al., 1987). Thus, glycosomal assembly would require insertion of proteins across the organelle membrane post-translationally. Thus far, in vitro translation of mRNA yields glycosomal proteins with the same molecular weights as the mature products inside the glycosome (Hart et al., 1987), thus suggesting that the import may not involve proteolytic processing. Recent studies on the genomic sequences of four *T. brucei* glycosomal enzymes—aldolase (Clayton, 1985), glyceraldehyde-3-phosphate dehydrogenase (Michels et al., 1986), 3-phosphoglycerate kinase (Osinga et al., 1985), and triosephosphate isomerase (Swinkels et al., 1986)—indicated that none of them have a leader sequence, but all have very high isoelectric points (Misset et al., 1986). The high pI's are the result of basic amino acids interspersed along the molecules. Three-dimensional structural analyses of these enzymes by superimposing their deduced sequences on the crystalline structures of the same mammalian enzymes suggested that some of the extra basic amino acids are clustered in two areas on the surface of the molecule, 40 Å apart. It was postulated that these "hot spots" may be involved in the import of these proteins into the glycosome (Wierenga et al., 1987). We have recently established an in vitro protein import assay with the purified intact glycosomes and [$^{35}$S]-methionine-labeled in vitro translational products (Dovey et al., 1988). Specificity of the protein import was verified using translational products derived from cloned genes encoding *T. brucei* glycosomal 3-phosphoglycerate kinase and its 95% homologous cytosolic isozyme. Glycosomal 3-phosphoglycerate kinase was inserted into the glycosome with 27.6% efficiency, but no imported cytosolic 3-phosphoglycerate kinase was detectable. Preliminary data suggest that certain sequences between the N-terminus and residue 123 may be important for import of glycosomal 3-phosphoglycerate kinase. Future studies with the potential use of genetically altered substrate proteins may reveal the mechanism and the recognition systems involved in glycosome biogenesis. These findings are all potential directions for future antitrypanosomal chemotherapy. Meanwhile, an interesting correlation between the basic "hot spots" of a glycosomal protein and the chemical structure of suramin, a well-known antitrypanosomal agent, was noticed by Misset and Opperdoes (1987). Suramin contains six negatively charged sulfonyl groups with a molecular weight of 1,429. It is entirely possible that these sulfonyl groups may facilitate suramin binding to some of the basic "hot spots" on a glycosomal protein. Misset and Opperdoes (1987) indicated that suramin is a more potent inhibitor of glycosomal dehydrogenases and kinases when compared with the cytoplasmic enzymes. But these higher inhibitory potencies may not be attributed to better binding of suramin to glycosomal enzymes because binding to the peripheral basic "hot spots" does not necessarily inhibit the enzyme activity. Nor is it at all clear that the mechanism of antitrypanosomal action of suramin is its preferential inhibition of glycosomal enzymes, because there is some doubt as to whether suramin can get inside the glycosome at all. Suramin, trypan blue, and trypan red are very similar chemical compounds. They are all incapable of penetrating the plasma membrane of a living cell because of their negatively charged sulfonyl groups, but they are all highly effective antitrypanosomal agents. They are apparently taken up by the trypanosomes through endocytosis of drug-bound serum proteins (Fairlamb and Bowman, 1980). The glycosomes are simple microbody organelles devoid of any cytoskeletons and are thus most unlikely to perform endocytosis. Suramin and its analogs will probably remain in the cytoplasm of the trypanosome and have no chance to inhibit the enzyme activities inside glycosome.

On the other hand, however, the cytoplasmic suramin, trypan blue or trypan red should have the chance of binding to the basic "hot spots" of newly synthesized glycosomal enzymes before their import into the glycosome. These drug bindings should block glycosome biogenesis if the "hot spots" are essential for glycosomal protein import. Thus, the mechanism of action of this type of drug could simply be blocking glycosomal biogenesis in trypanosomes. Preliminary studies in our in vitro glycosomal protein import assays suggested that suramin may indeed inhibit protein imports into glycosomes (unpublished observation). Similar speculation may be made regarding the mechanism of action of the antitrypanosomal diamidines such as pentamide, stilbamidine, and berenil. These compounds are highly positively charged at physiological pH. They may compete with the basic "hot spots" of newly synthesized glycosomal enzymes and result in blocking the glycosomal import of proteins. Further studies should be able to clarify these points.

## ALLOPURINOL RIBOSIDE

One of the major metabolic deficiencies of all the parasitic protozoa studied to date is their deficient in *de novo* synthesis of purine nucleotides and their dependence on purine salvage for survival (Wang, 1984). Effective exploitation of this weakness for chemotherapy has not, however, been overwhelmingly successful. The reason may be attributable to the presence in most parasites of multiple purine salvage pathways, which are difficult to block with a single compound. The other difficulty may be due to the unavailability of potent inhibitors of purine phosphoribosyltransferases which are the major salvage enzymes in many of the protozoan parasites (Wang and Simashkevich, 1981; Aldritt and Wang, 1986). Other parasites relying on purine nucleoside kinases (Heyworth et al., 1982) or phosphotransferases (Wang and Cheng, 1984b) as main salvage enzymes face a similar dilemma of lacking specific inhibitors. Again, the problem may result from a lack of in-depth understanding of these target enzymes. Hypoxanthine-guanine-xanthine phosphoribosyltransferase of *Eimeria tenella* (Wang and Simashkevich, 1981) and guanine phosphoribosyltransferase of *Giardia lamblia* (Aldritt and Wang, 1986) have both been purified to apparent homogeneity and shown to have unique substrate specificities suitable as potential therapeutic targets. But further characterizations have not been performed because of technical difficulties in preparing enough of the pure enzymes. This problem will have to be resolved by gene cloning and expression in the future.

Meanwhile, some of the purine salvage enzymes in protozoan parasites appear to have more relaxed substrate specificities, which allow incorporation of certain purine analogs into the nucleotide pools and nucleic acids of parasites that do not occur in their mammalian hosts (Marr and Berens, 1983; Rainey and Santi, 1983). Some of the purine nucleotide analogs thus formed by the parasites are harmful to the parasites, and so achieve the antiparasitic purpose (Looker et al., 1986). One of the most successful examples in this case has been the antileishmanial activity of allopurinol, a hypoxanthine analog and a relatively nontoxic antigout drug for human use (Hitchings, 1975), found actively incorporated into the purine nucleotide pool and the RNA of *Leishmania donovani*. Apparently, the hypoxanthine-guanine phosphoribosyltransferase of *L. donovani*, as well as the rest of the parasite enzymes involved in converting hypoxanthine to ATP and incorporating ATP into RNA all recognize allopurinol as another hypoxanthine (Looker et al., 1986), a mistake not committed by humans. Allopurinol ribonucleoside monophosphate is an inhibitor of GMP reductase and IMP dehydrogenase, whereas the aminopurinol ribonucleotides increased the catabolism of RNA in *L. donovani*, all of which could have contributed to the antileishmanial activity of allopurinol (Looker et al., 1986). Allopurinol riboside was found an even more potent antileishmanial agent (Nelson et al., 1979). It is incorporated into the leishmanial nucleotide pool by the action of a unique parasite enzyme, purine nucleoside phosphotransferase. Encouraging results have been obtained from recent clinical trials of allopurinol and allopurinol riboside on leishmaniasis patients (Kager et al., 1981; Saenz et al., 1986). The compound may become the first new antileishmanial agent discovered in 40 years. Along the same lines, other inosine analogs such as formycin B (Carson and Chang, 1981) and 9-deazainosine (Fish et al., 1985) have been also discovered as effective antileishmanial as well as antitrypanosomal agents. The eventual future of these drug candidates will be determined by their antiparasitic efficacy, host toxicity, ease of synthesis, patent coverage, etc.

Thus, the exploitation of deficient purine metabolism in parasitic protozoa has not succeeded by inhibiting the action of some crucial purine salvage enzymes to block the purine supply for the parasites. Rather, it has worked by allowing some false substrates to be converted to something poisonous to the parasite through the action of a parasite enzyme with a relaxed substrate specificity. The enzyme, such as phosphotransferase in leishmania, does not have to be of crucial importance for the survival of parasite. Its relaxed substrate specificity is not necessarily related to the deficient *de novo* purine nucleotide synthesis of leishmania. It is simply an accident, i.e., the recent discoveries of antileishmanial purine nucleosides may be no more than chance.

This rather discouraging conclusion is further fortified by our recent observations on *Schistosoma mansoni*. *S. mansoni* is also incapable of *de novo* synthesis of purine nucleotides. Its main pathways for purine salvage are by a hypoxanthine-guanine phosphoribosyltransferase and adenine phosphoribosyltransferase with little interconversion between adenine and guanine nucleotides (Dovey et al., 1984). The hypoxanthine-guanine phosphoribosyltransferase has been purified to apparent homogeneity; partial characterization of the pure enzyme revealed significant differences from the mammalian enzyme (Dovey et al., 1986). Since tubercidin (7-deazaadenosine) has long been recognized as a powerful antischistosomal drug (Dovey et al., 1985), it was assumed that the active form of the drug could be either 7-deazaadenine, which may be incorporated by the action of adenine phosphoribosyltransferase, or 7-deazahypoxanthine which may be incorporated by the action of hypoxanthine-guanine phosphoribosyltransferase. The experimental results suggested that neither purine base has antischistosomal activity and 7-deazaadenine is not a substrate of *S. mansoni* adenine phosphoribosyltransferase (Dovey et al., 1985). Tubercidin is incorporated into the schistosomal nucleotide pool by the action of a very low level of adenosine kinase in the parasite. The enzyme apparently plays apparently no major role in schistosomal purine salvage; its presence has no apparent correlation with the absence of *de novo* purine synthesis—i.e., it is not needed for the survival of *S. mansoni*. The formation of tubercidin mono-, di-, and triphosphate and its incorporation into schistosomal nucleic acids have been demonstrated. Similar incorporation of formycin A into *S. mansoni* was also observed, but the compound has no antischistosomal activity even when enough formycin A, equal to the lethal dose of tubercidin, was found in the parasite nucleic acids. In fact, formycin A can revert the antischistosomal activity of tubercidin, suggesting that the presence of formycin A nucleotides and their incorporation into nucleic acids are not harmful at all to the parasite (Dovey et al., 1985).

In spite of all the complications resulting from the complex nature of purine salvage networks in some of the parasites, the original concept in exploiting metabolic deficiencies for chemotherapy remains valid. When the metabolism in a parasite becomes simplified to the point of depending on a single enzyme to fulfill a vital need, an effective inhibition of such an enzymic function should kill the parasite. We have found, in recent years, many such examples among the anaerobic protozoan parasites *Trichomonas foetus*, *Trichomonas vaginalis* and *Giardia lamblia*. These organisms are totally incapable of *de novo* synthesis of either purine nucleotides or pyrimidine nucleotides (Wang et al., 1983a; Wang and Cheng, 1984a; Aldritt et al., 1985). Their means of obtaining a particular purine or pyrimidine nucleotide often rely on a single, major salvage enzyme. For instance, *G. lamblia* acquires essentially all its adenine nucleotides by a functioning adenine phosphoribosyltransferase and all the guanine nucleotides by guanine phosphoribosyltransferase (Wang and Aldritt, 1983). *T. vaginalis* has adenosine kinase and guanosine kinase to fulfill its needs (Heyworth et al., 1982). *T. foetus* relies on one major enzyme, hypoxanthine-guanin-exanthine phosphoribosyltransferase, to supply most of the purine nucleotides (Wang et al., 1983b). *T. vaginalis* cannot convert ribonucleotides to deoxyribonucleotides. It depends on a membrane-bound enzyme, deoxyribonucleoside phosphotransferase, to provide all four deoxyribonucleotides for DNA synthesis (Wang and Cheng, 1984b). All these enzymes can be classified as excellent targets for antiparasitic controls. Design of specific inhibitors of these enzymes relies on their extensive characterization, which in turn must wait for cloning and expression of their genes.

## CONCLUSION

We have, in the present chapter, discussed only three potential antiparasitic drugs which are either in the very preliminary stage of investigation or in the early phases of clinical trials. It is uncertain if any of them will eventually succeed and be approved for general clinical uses. But they share the distinction of being the first group of drugs discovered through some prior understanding of the metabolism of the parasites. By means of continuing studies of their mechanisms of action we become increasingly aware of the identities of some of the valuable chemotherapeutic targets in parasites. Perhaps even more importantly, the information has helped us to learn how to search for a therapeutic target in a parasite. There is certainly more than one way to approach such a problem. The metabolic deficiencies ordinarily associated with a parasite can provide a good starting point to look for an indispensable and unique enzyme for inhibition studies. But such an ideal target does not always exist. Instead, there may be several alternative enzymes involved with the simplified metabolism, which means that there is no special advantage in exploiting such metabolic defects. The antileishmanial purines and purine nucleosides thus do not owe their activities to the lack of de novo purine nucleotide synthesis in the parasite. On the other hand, however, the action of SHAM and glycerol on African trypanosomes has provided a good example of successful exploitation of simple parasite metabolism without demanding a distinctive substrate specificity from the inhibited enzymes. The simple glycolysis in trypanosomes makes it necessary to aggregate the glycolytic enzymes within membrane-bound microbodies in order to accelerate the glycolytic rate. The state of aggregation can be treated as an opportunity for therapeutic attack without demonstrated differences in enzyme substrate specificities between parasite and host. The trypanosomal ornithine decarboxylase is another example that a parasite enzyme does not have to have distinctive catalytic properties in order to be a therapeutic target. The qualification of this particular enzyme is its lack of turnover when compared with the mammalian ornithine decarboxylase. An extension of lesson from this finding would be to identify all the mammalian enzymes known to have relatively short half lives in vivo, and to search for these same enzymes in the parasites to examine their half-lives. Those with very long half-lives may be potential targets for irreversible inhibition.

In spite of all the therapeutic opportunities unrelated to unique, essential enzymes in parasites, the latter remain the most interesting and perhaps the most challenging targets for antiparasitic chemotherapy. As I have discussed previously, specific inhibitor designs have to rely on thorough characterizations of the target enzyme, X-ray crystallography and eventually, design of inhibitors by computer graphics. Gene cloning and gene expression for yielding large quantities of pure, native enzyme appear to be the only means of approaching the goal. Since the original thinking is to look into possible differences between the host and parasite enzyme, the mammalian cDNA encoding the enzyme is often unsuitable as a heterologous probe for identifying the parasite gene. The parasite enzyme must be purified to homegeneity in modest quantities in order to provide some peptide sequence data or specific antibodies before the identification of the gene can be attempted. Future approaches to antiparasitic chemotherapy must thus be a combination of intensive biochemical and molecular biological studies.

## REFERENCES

Aldritt SM, Tien P, Wang CC (1985): Pyrimidine salvage in *Giardia lamblia*. J Exp Med 161:437–445.
Aldritt SM, Wang CC (1986): Purification and characterization of guanine phosphoribosyltransferase from *Giardia lamblia*. J Biol Chem 261:8528–8533.
Bacchi CJ (1980): Content, synthesis and function of polyamines in trypanosomatids: Relationships to chemotherapy. J Protozool 28:2027.
Bacchi GJ, Vergara C, Garofalo J, Lipschik GY, Hutner SH (1979): Synthesis and content of polyamines in bloodstream *Trypanosoma brucei*. J Protozool 26:484–488.
Bacchi CJ, Nathan HG, Hutner SH, McCann PP, Sjoersma A (1980): Polyamine metabolism: A potential therapeutic target in trypanosomes. Science 210:332–334.
Bacchi CJ, Nathan HC, Hutner SH, McCann PP, Sjoersma A (1982): Novel combination chemotherapy of experimental trypanosomiasis by using bleomycin and DL-α-difluoromethylornithine: Reversal by polyamines. Biochem Pharmacol 31:2833–2836.
Bey P (1978): Substrate-induced irreversible inhibition of α-amino acid decarboxylase: Application to glutamate, aromatic-L-amino acid and ornithine decarboxylases. In N Seiler, MJ Jung, J Koch-Weser (eds): Enzyme-Activated Irreversible Inhibitors. Amsterdam: Elsevier/North Holland Biomedical Press, pp 27–41.
Bitonti AJ, Bacchi CJ, McCann PP, Sjoersma A (1985): Catalytic irreversible inhibition of *Trypanosoma brucei brucei* ornithine decarboxylase by substrate and product analogs and their effects on murine trypanosomiasis. Biochem Pharmacol 34: 1773–1777.
Bitonti AJ, McCann PP, Sjoersma A (1986): Necessity of antibody response in the treatment of African trypanosomiasis with α-difluoromethylornithine. Biochem Pharmacol 35:331–334.

Bowman IBR and Flynn IW (1976): Oxidative metabolism of trypanosomes. In Lumsden W.H.R., DA Evans (eds): The Biology of the Kinetoplastida. New York: Academic Press, pp 436–467.

Brohn FH, Clarkson AB (1980): *Trypanosoma brucei brucei*: Patterns of glycolysis at 37°C *in vitro*. Mol Biochem Parasitol 1:291–305.

Brohn FH, Clarkson AB (1978): Quantitative effects of salicyl hydroxamic acid and glycerol on *Trypanosoma brucei* glycolysis *in vitro* and *in vivo*. Acta Trop 35:23–33.

Carson DA, Chang K-P (1981): Phosphorylation and antileishmanial activity of formycin B. Biochem Biophys Res Commun 100:1377–1383.

Chen KY, Nau D, Liu AY-C (1983): Effects of inhibitors of ornithine decarboxylase on the differentiation of mouse neuroblastoma cells. Cancer Res 43:2812–2818.

Chen KY, Presepe V, Parken N, Liu AY-C (1982): Changes of ornithine decarboxylase activity and polyamine content upon differentiation of mouse NB-15 neuroblastoma cells. J Cell Physiol 110:285–290.

Clarkson AB, Brohn FH (1976): Trypanosomiasis: An approach to chemotherapy by the inhibition of carbohydrate metabolism. Science 194:204–206.

Clayton CE (1985): Structure and regulated expression of genes encoding fructose bisphophate aldolase in *Trypanosoma brucei*. EMBO J 4:2997–3003.

Dovey HF, McKerrow JH, Wang CC (1984): Purine salvage in *Schistosoma mansoni* schistosomules. Mol Biochem Parasitol 11:157–167.

Dovey HF, McKerrow JH, Wang CC (1985): Action of tubercidin and other adenosine analogs on *Schistosoma mansoni* schistosomules. Mol Biochem Parasitol 16:185–198.

Dovey HF, McKerrow JH, Wang CC (1986): Purification and characterization of hypoxanthine-guanine phosphoribosyltransferase from *Schistosoma mansoni*: A potential target for chemotherapy. J Biol Chem 261:944–948.

Dovey HF, Parsons M, Wang CC (1988): Biogenesis of glycosomes of *Trypanosoma brucei*: An *in vitro* model of 3-phosphoglycerate kinase import. Proc Natl Acad Sci USA (in press).

Evans DA, Brightman CJ, Holland MF (1977): Salicylhydroxamic acid/glycerol in experimental trypanosomiasis. Lancet ii:769.

Fairlamb AH, Bowman IBR (1980): Uptake of the trypanocidal drug suramin by bloodstream forms of *Trypanosoma brucei* and its effect on respiration and growth rate *in vivo*. Mol Biochem Parasitol 1:315333.

Fairlamb AH, Opperdoes FR, Borst P (1977): New approach to screening drugs for activity against African trypanosomes. Nature (London) 265:270–271.

Fish WR, Marr JJ, Berens RL, Looker DL, Nelson DJ, LaFon SW, Balber AE (1985): Inosine analogs as chemotherapeutic agents for African trypanosomes: Metabolism in trypanosomes and efficacy in tissue culture. Microbiol Agents Chemother 27:33–36.

Giffin BF, McCann PP, Bitonti AJ, Bacchi CJ (1986): Polyamine depletion following exposure to DL-α-difluoromethylornithine both *in vivo* and *in vitro* initiates morphological alterations and mitochondrial activation in a monomorphic strain of *Trypanosoma brucei brucei*. J Protozool 33:238–243.

Grady RW, Bienen EJ, Clarkson Jr. AB (1986): p-Alkyloxybenzhydroxamic acids, effective inhibitors of the trypanosome glycerol-3-phosphate oxidase. Mol Biochem Parasitol 19:231–240.

Grant PT, Sargent JR (1960): Properties of L-glycerol-3-phosphate oxidase and its role in respiration of *Trypanosoma rhodesiense*. Biochem J 76:229–237.

Hart DT, Baudhuin P, Opperdoes FR, DeDuve C (1987): Biogenesis of the glycosome in *Trypanosoma brucei*: The synthesis, translocation and turnover of glycosomal polypeptides. EMBO J 6:1403–1411.

Heyworth PG, Gutteridge WE, Ginger CD (1982): Purine metabolism in *Trichomonas vaginalis*. FEBS Lett 141:106–110.

Hitchings GH (1975): Pharmacology of allopurinol. Arthritis Rheum 18:863–870.

Kager PA, Rees PH, Wellde PT, Hockmeyer WT, Lyerly WH (1981): Allopurinol in the treatment of visceral leishmaniasis. Trans Roy Soc Trop Med Hyg 75:556–559.

Looker DL, Marr JJ, Berens RL (1986): Mechanisms of action of pyrazolopyrimidines in *Leishmania donovani*. J Biol Chem 261:9412–9415.

MaCrae M, Coffino P (1987): Complimentation of a polyamine-deficient *Escherichia coli* mutant by expression of mouse ornithine decarboxylase. Mol Cell Biol 7:564–567.

Mancini PE, Patton CL (1981): Cyclic 3':5'-adenosine monophosphate levels during the developmental cycle of *Trypanosoma brucei brucei* in the rat. Mol Biochem Parasitol 3:19–31.

Marr JJ, Berens RL (1983): Mini Review: Pyrazolopyrimidine, metabolism in the pathogenic trypanosomatidae. Mol Biochem Parasitol 7:339–356.

Michels PAM, Poliszczak A, Osinga KA, Misset O, Van Beeumen J, Wierenga RK, Borst P, Opperdoes FR (1986). Two tandemly linked identical genes code for the glycosomal glyceraldehyde-phosphate dehydrogenase in *Trypanosoma brucei*. EMBO J 5:1049–1056.

Misset O, Opperdoes FR (1987): The phosphoglycerate kinases from *Trypanosoma brucei*: A comparison of the glycosomal and the cytosolic isoenzyme and their sensitivity towards suramin. Eur J Biochem 162:493–500.

Misset O, Bos OJM, Opperdoes FR (1986): Glycolytic enzymes of *Trypanosoma brucei*: Simultaneous purification, intraglycosomal concentrations and physical properties. Eur J Biochem 157:441–453.

Nelson DJ, LaFon SW, Tuttle JV, Miller WH, Miller RL, Krenitsky TA, Elion GB, Berens RL, Marr JJ (1979): Allopurinol ribonucleoside as an antileishmanial agent. J Biol Chem 254:11544–11549.

Opperdoes FR, Borst P (1977): Localization of nine glycolytic enzymes in a microbody-like organelle in *Trypanosoma brucei*: the glycosome. FEBS Lett 80:360–364.

Osinga KA, Swinkels BW, Gibson WC, Borst P, Veeneman GH, Van Boom JJ, Michels PAM, Opperdoes FR (1985): Topogenesis of microbody enzymes: A sequence comparison of the genes for the glycosome (microbody) and cytosolic phosphoglycerate kinases of *Trypanosoma brucei*. EMBO J 4:3811–3817.

Pegg AE, McCann PP (1982): Polyamine metabolism and function. Am J Physiol 243:212–221.

Phillips MA, Coffino P, Wang CC (1987): Cloning and sequencing of the ornithine decarboxylase gene from *Trypanosoma brucei*: Implications for enzyme turnover and selective difluoromethylornithine inhibition. J Biol Chem 262:8721–8727.

Rainey P, Santi DV (1983): Metabolism and mechanism of action of formycin B in *Leishmania*. Proc Natl Acad Sci USA 80:288–292.

Rogers S, Wells R, Rechsteiner M (1986): Amino acid sequences common to rapidly degraded proteins: The PEST hypothesis. Science 234:364–368.

Saenz RE, Paz H, Johnson CM, Rogers MD, Nelson DJ, Pattishall KH (1986): Efficacy and safety of allopurinol riboside in the treatment of American cutaneous leishmaniasis. Nineth Internatl Cong Infect and Parasit Dis, Münich, July 20–26.

Sjoerdsma A, Schechter PJ (1984): Commentary: Chemotherapeutic implications of polyamine biosynthesis inhibition. Clin Pharmacol Therap 35:287–300.

Song MH, Adolph KW (1983): Phosphorylation of nonhistone proteins during the HeLa cell cycle: Relationship to DNA synthesis and mitotic chromosome condensation. J Biol Chem 258:3309–3318.

Swinkels BW, Gibson WG, Osinga KA, Kramer R, Veeneman GH, Van Boom JH, Borst P (1986). Characterization of the gene for the microbody (glycosomal) triosephosphate isomerase of *Trypanosoma brucei*. EMBO J 5:1291–1298.

Tabor CW, Tabor H (1984): Polyamines. Annu Rev Biochem 53:749–790. van der Meer C, Versluijs-Broers JAM, Opperdoes FR (1979): *Trypanosoma brucei*: Trypanocidal effect of salicylhydroxamic acid plus glycerol in infected rats. Exp Parasitol 48:126–134.

van Nieuwenhove S, Schechter PJ, Declercq J, Bone G, Burke J, Sjoerdsma A (1985): Treatment of gambiense sleeping sickness in the Sudan with oral DFMO (DL-α-difluoromethylornithine), an inhibitor of ornithine decarboxylase; first field trial. Trans R Soc Trop Med Hyg 79:692–698.

Verma R, Chen KY (1986); Spermine inhibits the phosphorylation of the 11,000- and 10,000-dalton nuclear proteins catalyzed by nuclear protein kinase NI in NB-15 mouse neuroblastoma cells. J Biol Chem 261:2890–2896.

Visser N, Opperdoes FR (1980): Glycolysis in *Trypanosoma brucei*. Eur Biochem 103:623–632.

Visser N, Opperdoes FR, Borst P (1981): Subcellular compartmentation of glycolytic intermediates in *Trypanosoma brucei*. Eur J Biochem 118:521–526.

Wang CC (1984): Enzymes as potential targets for antiparasitic chemotherapy. J Med Chem Perspective 27:1–9.

Wang CC, Aldritt SM (1983): Purine salvage networks in *Giardia lamblia*. J Exp Med 158:1703–1712.

Wang CC, Cheng H-W (1984a): Salvage of pyrimidine nucleosides by *Trichomonas vaginalis*. Mol Biochem Parasitol 10:171–184.

Wang CC, Cheng H-W (1984b): The deoxyribonucleoside phosphotransferase of *Trichomonas vaginalis*. J Exp Med 160:987–1000.

Wang CC, Pong S-S (1982): Actions of avermectin $B_{1a}$ on GABA nerves. In J.R. Sheppard, V.E. Anderson, J.W. Eaton (eds): "Membranes and Genetic Disease. New York: Alan R. Liss, Inc., pp. 373–395.

Wang CC, Simashkevich PM (1981): Purine metabolism in a protozoan parasite *Eimeria tenella*. Proc Natl Acad Sci USA 78:6618–6622.

Wang CC, Verham R, Tzeng S-F, Aldritt SM, Cheng H-W (1983a): Pyrimidine metabolism in *Tritrichomonas foetus*. Proc Natl Acad Sci USA 80:2564–2568.

Wang CC, Verham R, Rice A, Tzeng S-F (1983b): Purine salvage by *Tritrichomonas foetus*. Mol Biochem Parasitol 8:325–337.

Wierenga RK, Swinkels B, Michels PAM, Osinga K, Misset O, Van Beeumen J., Gibson WC, Postma JPM, Borst P, Opperdoes FR, Hol, WGJ (1987): Common elements on the surface of glycolytic enzymes from *Trypanosoma brucei* may serve as topogenic signals for import into glycosomes. EMBO J 6:215–221.

The Biology of Parasitism, pages 431–448
© 1988 Alan R. Liss, Inc.

# Gene Amplification in *Leishmania*

Stephen M. Beverley, Thomas E. Ellenberger, David M. Iovannisci,
Geoffrey M. Kapler, Maria Petrillo-Peixoto, and Barbara J. Sina

*Department of Biological Chemistry and Molecular Pharmacology, Harvard
Medical School, Boston, Massachusetts 02115*

## INTRODUCTION

Drug resistance in human parasites has been studied since the work of
Paul Ehrlich at the beginning of this century. One area of emphasis has been
the mechanisms that parasites employ in becoming resistant, both in laboratory and clinical settings. Although progress in identifying the precise genetic
alterations has been limited, the addition of molecular, genetic, and immunological techniques to our repertoire in recent years has made it possible to
contemplate the identification and isolation of the genes responsible for drug
resistance. A resistance mechanism that has proven particularly amenable to
study is that mediated by specific gene amplification in the protozoan parasite
*Leishmania*. We showed that promastigotes of *L. major*, when selected in a
stepwise manner for high levels of resistance to the folate analog methotrexate (MTX), exhibited enzyme overproduction and gene amplification reminiscent of those observed in MTX-resistant mammalian cells (Coderre et al.,
1983; note that the *Leishmania* line utilized was incorrectly named *L. tropica*
at the time, its true identity being *L. major*; Beverley et al., 1987). As
described in this paper, these studies have led to the isolation of the gene
encoding the target of MTX, dihydrofolate reductase-thymidylate synthase
(DHFR-TS), a novel bifunctional enzyme found in all protists characterized
to date (Ferone and Roland, 1980; Garrett et al., 1984; Beverley et al., 1986;
Grumont et al., 1986). Molecular and biochemical studies of other drug-
resistant lines have revealed other mutations mediating drug resistance,
including additional gene amplifications (Beverley et al., 1984; Petrillo-
Peixoto and Beverley, 1988; Ellenberger and Beverley, 1988). Moreover,
from studies of the amplified genes we have been able to garner information
concerning the genetic ploidy of *Leishmania*, the mechanism of genomic
rearrangement, and the segregation of genetic information.

The term *specific gene amplification* is usually applied to changes in relative copy number of a limited portion of the genome, and excludes genomic changes such as alterations in chromosomal ploidy or polytenization. Gene amplification has been most intensively studied in drug-resistant cultured mammalian cells, obtained by serial stepwise selection to high levels of resistance (Schimke, 1982; Stark and Wahl, 1984). Current data from studies of mammalian cells suggests that a large number of loci mediating a wide spectrum of enzymatic reactions can be amplified under appropriate conditions, suggesting that no particular class of enzymes or loci are especially prone to amplify. Moreover, a significant fraction of tumors have been shown to contain amplifications of oncogenes, and the role of amplification in the process of tumorigenesis and progression is under intensive study in many laboratories (Schimke, 1984; Stark and Wahl, 1984). Many of the properties of gene amplification in *Leishmania* are reminiscent of amplification in other eukaryotes, and the study of amplification in *Leishmania* offers certain advantages as a model system for analyzing basic mechanisms of amplification as described below.

## DETECTION AND ISOLATION OF AMPLIFIED DNAs

There have been a number of methods utilized for the detection and isolation of amplified DNA sequences in cultured mammalian cells; generally, these exploit in some manner the kinetically favored hybridization of the amplified sequences relative to nonamplified sequences (Alt et al., 1978; Brison et al., 1982; Roninson, 1983). In *Leishmania* and other organisms of small genome size amplified DNAs can be easily visualized following electrophoresis of restriction enzyme-digested DNAs (Coderre et al., 1983) (Fig. 1A). The sensitivity of this method allows amplifications of only 5–10-fold to be visualized, depending on the specific enzyme used. Once visualized, the amplified DNAs can be isolated from the gel and employed as hybridization probes, either directly or after molecular cloning (Coderre et al., 1983; Petrillo-Peixoto and Beverley, 1988). Since amplified DNAs in *Leishmania* appear to exist primarily as extrachromosomal circular forms (Fig. 2 and below) (Beverley et al., 1984; Petrillo-Peixoto and Beverley, 1988; Ellenberger and Beverley, 1988), another approach relies upon the unique properties of circular DNAs relative to linear DNAs. Circular molecules can easily be isolated in CsCl/ethidium bromide density gradients or following electrophoresis under appropriate conditions (Petrillo-Peixoto and Beverley, 1988; Beverley et al., 1984); pulsed field gradient electrophoresis can be especially useful in isolating very pure amplified DNA (Beverley, 1988a). These ap-

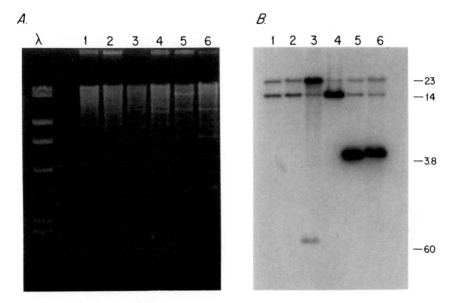

Fig. 1. Detection of amplified DNA by ethidium bromide staining and blot hybridization. **A:** DNAs from wild-type *L. major* or lines selected for resistance to MTX (**Lanes 2–5**) or 10-propargyl-5,8-dideazafolate (**Lane 6**) were digested with BglII, electrophoresed on 0.8% agarose gels, and stained with ethidium bromide. **B:** The gel shown in A was blotted to a membrane and hybridized with a 3.2-kb EcoRI-BglI fragment located at the left rearrangement site of many cell lines, indicated by the leftmost black box on the wild-type map of the R region shown in Figure 3. This fragment bears a direct repeat, which is also found in the rightmost rearrangement site, as indicated by the black box in Figure 3, which accounts for the hybridization to two bands in the wild-type line. The DNAs are as follows: **lane 1**, Wild-type *L. major*, strain LT252 (Beverley et al., 1987); **lane 2**, C5A-R1000; **lane 3**, D4A-R1000; **lane 4**, D7B-R1000 (C5, D4A, and D7B are clonal derivatives of the LT252 line, and the DNAs shown are those from MTX-resistant lines selected for resistance to 1 mM MTX; Beverley, Coderre, Schimke, and Santi, unpublished data); **lane 5**, R1000 DNA (Coderre et al., 1983); **lane 6**, CB-50 DNA (Garvey et al., 1985).

proaches have the advantage that the entire amplified region is isolated in a single step, in a form suitable for molecular cloning or hybridization analysis.

## OCCURRENCE OF SPECIFIC GENE AMPLIFICATION IN *LEISHMANIA*

Currently our knowledge of the ability of different *Leishmania* genes to amplify is limited, though evidence is rapidly accumulating. In *Leishmania* four different loci have been amplified in three different species, using five different drugs. The dihydrofolate reductase-thymidylate synthase (DHFR-

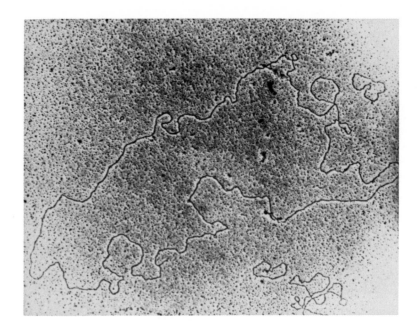

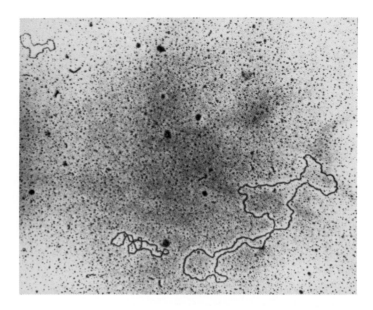

Fig. 2.   Visualization of circular amplified DNAs by electron microscopy. **Top**: The 30-kb R region. **Bottom**: The 85-kb H region. Small circular DNAs are phiX174 RF, included as a size marker (5.4 kb).

TS) gene has been amplified in *L. major* after selection with methotrexate, a specific inhibitor of dihydrofolate reductase (Beverley et al., 1984; Beverley et al., 1986), and 10-propargyl-5,8-dideaza-folate, an inhibitor of both DHFR and TS activities (Garvey et al., 1985; Grumont et al., 1986). The H region, whose function is currently unknown, was first observed in *L. major* in MTX-resistant lines and subsequently in lines selected with terbinafine and primaquine (representatives of two classes of potential antileishmanials; Ellenberger and Beverley, 1988), and in two natural isolates of *L. tarentolae* (Petrillo-Peixoto and Beverley, 1988). The amplification of H region in these diverse settings suggests that amplification may confer simultaneous resistance to many drugs, as reported for amplifications in certain multi-drug-resistant mammalian cells (Borst, 1984). However, in contrast to mammalian cells in which P-glycoprotein amplification mediates alterations in drug accumulation, our data indicate that the H-region amplification does not alter drug influx, efflux, or accumulation in *Leishmania*, and thus another mechanism must be sought (Ellenberger and Beverley, 1987b, 1988).Another natural isolate of *L. tarentolae* exhibits amplification of a different locus, the T region (Petrillo-Peixoto and Beverley, 1988). Unlike the H amplifications in this species, the naturally occurring T-amplified line does not exhibit resistance to MTX or any drug tested thus far. The potential role of these amplifications in *L. tarentolae*, which is found naturally in lizards, is discussed elsewhere (Petrillo-Peixoto and Beverley, 1988). Finally, tunicamycin-resistant *L. mexicana* exhibit overproduction of N-acetylglucosaminyl transferase and amplification of DNA which presumably encodes the structural gene (Kink and Chang, 1987).

These data suggest that gene amplification is certainly not infrequent in *Leishmania*. However, it should be emphasized that the majority of drug-resistance mutations characterized thus far appear to involve mechanisms other than amplification (Iovannisci et al., 1984; Scott et al., 1987). For example, about 50% of the lines of *L. major* selected for MTX resistance fail to exhibit any detectable amplification (unpublished data), yet every line examined thus far exhibits decreased MTX uptake (Ellenberger and Beverley, 1987b). Moreover, various combinations of R- and H- region amplification are observed in different lines and at different selective steps. Apparently, parasites are capable of elaborating several different mechanisms of resistance (Ellenberger and Beverley, 1987b). The specific occurrence of a given form of resistance appears to be the result of a stochastic process, dependent upon factors such as relative mutation rates, the degree of resistance attained, and the number of selective steps employed.

## STRUCTURE OF AMPLIFIED DNAs

Amplified DNAs in *Leishmania* characterized thus far exist primarily as extrachromosomal circular DNAs (Beverley et al., 1984; Garvey et al., 1985; Petrillo-Peixoto and Beverley, 1988). The R1000 line of *L. major*, for example, contains two separate regions of amplification, the R (DHFR-TS) and H regions, and represents the first example of a single eukaryotic cell bearing two amplifications (Beverley et al., 1984). This line also exhibits at least one other mutation, mediating decreased influx of MTX (Ellenberger and Beverley, 1987b). Two basic structural types of extrachromosomal amplification have been observed (Beverley et al., 1984): direct, as observed for the amplified R region (Fig. 3; the directly repeating structure can be visualized by tracing around the map), or inverted, as for the amplified H region in *L. major* and *L. tarentolae* (Fig. 4). Following the description of inverted amplifications in *Leishmania*, Ford and Fried (1986) have shown that amplifications of the *myc* oncogene and CAD gene (encoding the first three enzymatic activities of de novo pyrimidine biosynthesis) also exist as large inverted duplications in cultured mammalian cells.

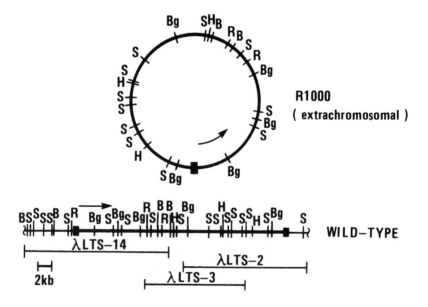

Fig. 3. Structure of the amplified R region. For further description see Beverley et al. (1984).

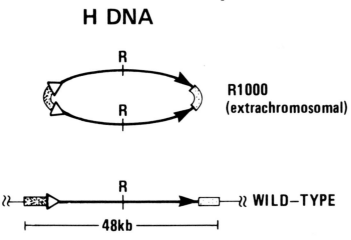

Fig. 4. Structure of the amplified H region. For further description see Beverley et al. (1984).

In *Leishmania*, the circular amplifications are limited to a discrete region of chromosomal DNA, which has become excised from the chromosome and converted into the extrachromosomal circular form. Multiple cell lines bearing amplification of the R region have been derived, and the location of the amplified segments of chromosomal DNA determined (Fig. 5; Beverley et al., 1984) (Beverley, Coderre, Schimke and Santi, unpublished data). The region amplified always includes the coding region for the DHFR-TS, as well as substantial flanking DNA, ranging in size from 30 to 90 kilobases (kb). Within a given line, there appears to be only one structural rearrangement associated with R amplification (Fig. 5), which is the minimum number of rearrangements required to give a direct repeat structure. Multiple H-region amplifications have also been observed in two species, and within these lines there appear to be two separate sites of DNA rearrangement in the amplified DNA, again the minimum number necessary to generate an inverted repeat type of structure (Fig. 4; Beverley et al., 1984; Ellenberger and Beverley, 1988; Petrillo-Peixoto and Beverley, 1988). As cells are selected in increasing concentrations of the drug the rearranged structure first observed is identical with that seen in more highly resistant cells, i.e., there are no additional rearrangements observed other than the primary one. In total, these observations suggest that the rearrangements observed were those

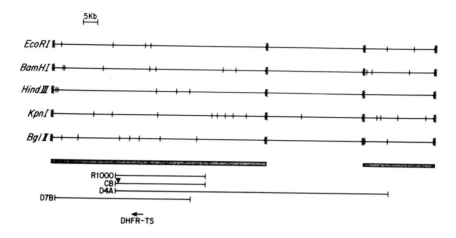

Fig. 5. Map of a 135-kb region surrounding the DHFR-TS gene of *Leishmania major* and amplifications thereof. The location of cleavage sites for five restriction enzymes are shown. Much of this DNA has been isolated in molecular recombinants in bacteriophage lambda, indicated by the gray bars. The joining of the two cloned regions was inferred by blot hybridization of gels of restriction enzyme-cleaved DNAs and pulsed field gradient electrophoresis of chromosomes. The location of the rearrangement sites joined during amplification for the four different lines shown in Figure 1 are shown; additional lines with amplifications similar to the R1000 line structure are not shown (unpublished data). The small triangle above the CB50 line indicates the location of a small insertion of DNA. The location and orientation of the DHFR-TS mRNA is also shown (Beverley et al., 1986; Grumont et al., 1986; Kapler et al., 1987).

which actually arose from the primary amplification event, and do not represent secondary rearrangements.

## STABILITY OF AMPLIFIED DNAs

In the absence of continued drug pressure, amplified genes in *Leishmania* are maintained for varying periods, depending upon the cell line. The phenotypic extremes are termed unstable amplifications, which are rapidly lost with a "half-life" of about 25 cell doublings, or stable amplifications, which are retained for at least 200 cell doublings (Beverley et al., 1984). Lines with intermediate stabilities are also observed (Garvey et al., 1985; Kink and Chang, 1987; Sina et al., 1988). When amplified DNAs from the R1000 line were examined by CsCl/ethidium bromide density-gradient centrifugation, it was evident that the unstable lines exhibited a 30-kb supercoiled circular R region plasmid (unit length), whereas in the stable lines the amplified DNA banded with chromosomal DNA (Beverley et al., 1984). By analogy with mammalian cells, it was proposed that the stable amplified DNA had integrated into the chromosome, thus acquiring the determinants necessary for stability. However, it was also noted that other molecular configurations, such as large extrachromosomal tandem arrays, were consistent with the data.

Recent work employing pulsed field gradient electrophoresis methodology for separating chromosomal-sized DNAs has provided a new approach for examining the structure of the amplified DNA. These experiments have shown that the amplified DNA present within the stably amplified lines can be extrachromosomal and circular (Garvey and Santi, 1986; Petrillo-Peixoto and Beverley, 1988; Beverley, 1988a,b). Moreover, integration of the amplified DNA into the chromosomes is not observed, either at the site of the resident gene or at anyb other location (Beverley et al., 1984; Beverley, 1988b). The R region in the stably amplified lines appears as a set of extrachromosomal circular multimers, in contrast to the predominant unit length form in the unstable line. Interestingly, the H region is maintained as the dimeric inverted repeat and does not show oligomerization, although stable tetrameric forms are observed in *L. tarentolae* (Petrillo-Peixoto and Beverley, 1988). These data suggest that while increased size may be an important factor in acquiring stability (as is known for artificial yeast chromosomes; Hieter et al., 1985; Murray et al., 1986; Zakian et al., 1986), other factors must play a role. By restriction enzyme mapping the structure of the amplified DNA in the unstable and stable lines appears to be identical, suggesting that the increased stability is not due to the acquisition of sequences such as centromeres. Thus, we must consider a possible role for trans-acting factors encoded outside the amplified DNA in the development of stable amplifications. In yeast it is known that trans-acting loci contribute to the maintenance of chromosomes (Meeks-Wagner et al., 1986), and perhaps similar loci are involved in some aspect of stabilization in *Leishmania* (Beverley, 1988b).

## MECHANISM OF AMPLIFICATION

Analysis of the structures of amplified DNAs, such as those shown in Figures 3–5 and many others, have provided some clues concerning the mechanism of gene amplification. The first clue came from the observation that six cell lines, including the R1000 and CB50 lines shown in Figures 1 and 5, have rearranged the R region in identical fashion (Beverley et al., 1984; Garvey et al., 1985) (unpublished data), while four lines have rearranged the H region in the same manner (Ellenberger and Beverley, 1988).

This indicated that certain sequences may preferentially be located at the site of rearrangements leading to amplification. Current data suggest that sites which are joined by amplification-linked rearrangements contain homologous repetitive sequences, suggesting that homologous recombination may be responsible for directing and creating the rearrangement site.

## Deletional Gene Amplification and the Ploidy of the DHFR-TS Chromosome

Reciprocal homologous recombination within the chromosome would be expected to yield an extrachromosomal circle, and simultaneously, deletion of the locus from the chromosome, in a manner similar to the excision of the lambda prophage from the bacterial chromosome. In the cell line shown in Figure 6 this prediction has been fulfilled, as a new and nonamplified DNA fragment appears following amplification which has precisely the size predicted for the deletional rearrangement (Beverley, Coderre, Schimke and Santi, unpublished data). Analysis of revertant clones of this line reveals the presence of equal amounts of the normal and deletion chromosome, showing that each chromosomal type is found within the cell (Fig. 6). This observation constitutes direct molecular evidence that the ploidy of the DHFR-TS chromosome in *L. major* is even, and most likely diploid, with the amplified line and its clonal descendants now being heterozygous for the wild-type and deletion chromosome. Thus, in this line amplification appears to be "deletional," in which one wild-type chromosomal copy of the R region is lost during amplification.

## Conservative Gene Amplification

In most other lines both copies of the normal chromosomal locus appear to be retained. This was shown previously for the R1000 line (Beverley et al., 1984), and an example of this is shown for another line in Figure 7 (Beverley, Coderre, Schimke and Santi, unpublished data). It is evident that no new fragments corresponding to the predicted deletion fragment are observed; moreover, the normal level of hybridization to the wild-type fragment is found. In lines such as these amplification is "conservative," retaining the wild-type ploidy of unrearranged genes. One explanation for this result would be that the recombinational event occurred on extra copies of the R region, created as a result of localized over-replication. Similar models have been proposed for the excision of integrated viral genomes (Botchan et al., 1979) and amplification in cultured mammalian cells (Schimke, 1982; Roberts et al., 1983; Stark and Wahl, 1984; Schimke et al., 1986). The prevalence of conservative amplification (about 90% of all events) may be due to the abundance of the over-replicated molecules relative to the

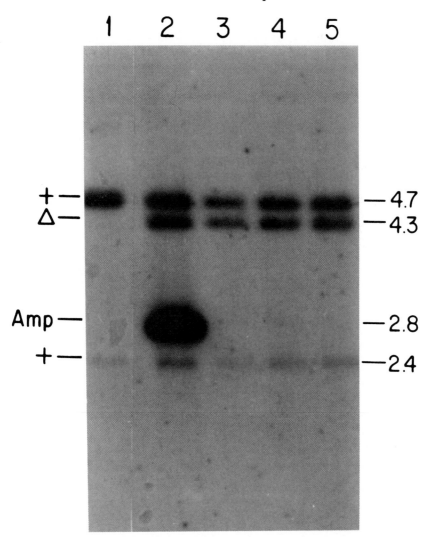

Fig. 6. Deletional gene amplification. DNA from wild-type DNA **(lane 1)**, the CB50 line **(lane 2)**, and three clonal revertant lines **(lanes 3–5)** are shown. DNAs were digested with PstI, electrophoresed, blotted, and hybridized with the same probe described in the legend to Figure 1B; this probe overlaps the 4.7- and 4.3-kb fragments to the same extent and can be used to assess the relative copy number of each. The 4.3-kb fragment has the size expected for a precise deletion of the R region from the chromosome, based upon the size of the 2.8-kb amplified rearrangement fragment. Digests and blots using four other enzymes also give the predicted deletion fragment (not shown).

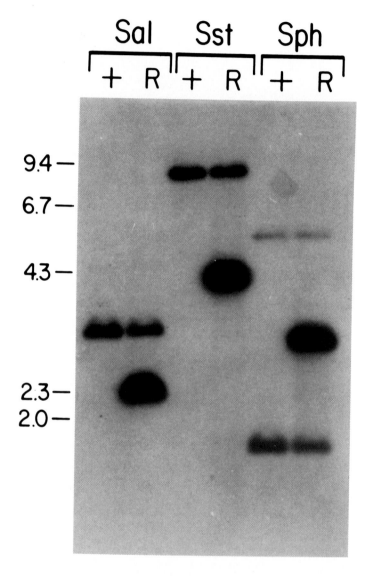

Fig. 7. Conservative gene amplification. DNA from the clonal D7B line (+) and the highly methotrexate-resistant derivative D7B-R1000 (R; see the legend to Fig. 1) were digested with restriction enzymes as indicated, electrophoresed, blotted to membranes, and then hybridized with a fragment corresponding to the leftmost rearrangement site of this line (Fig. 5). Other than the wild-type fragment(s) and the rearrangement fragment in each digest, no other hybridizing fragments are observed, in contrast to "deletional" gene amplification shown in Figure 6.

chromosomal copies, or an enhanced ability of the hypothetical short linear over-replicated copies to recombine into the circular form.

Other structural types of gene amplification are possible. For example, "integrational" amplification similar to that found in stably amplified cultured mammalian cells, or extrachromosomal linear amplifications, could occur. As more amplified lines and loci are examined, the diversity of amplified structures will no doubt increase.

Ultimately, many copies of the rearranged circular DNA must accumulate, either as the result of a rapid single-step amplification, or accretionally as the selective pressure is increased during stepwise selection. Either way, the amplified DNAs must be maintained at high levels. We have obtained evidence suggesting that the circular DNAs may replicate autonomously (unpublished work). Additionally, the amplified DNAs are located within the nucleus and appear to be packaged within nucleosomes, indicating that the amplified DNAs now constitute true minichromosomes.

## AMPLIFICATION IN THE REAL WORLD

To play a role in the biology of the leishmania parasite, amplified DNAs must be maintained in the natural cycle. This includes the promastigote stage, normally found in the midgut of the sandfly vector (the culture form closely resembles this stage) as well as the amastigote stage, present during infection within the phagolysosome of macrophages within the vertebrate host. We have successfully passed both unstably and stably amplified lines through the amastigote stage in mice, and recovered cells bearing amplified DNA (Sina et al., 1988). No effect upon infectivity was observed between the wild-type and unstably and stably amplified lines, as monitored by the time required to develop a lesion. This suggests that cells bearing amplified DNA can be effectively maintained in both stages of the parasite life cycle. Kink and Chang (1987) have also shown that amplified DNAs present in tunicamycin-resistant *L. mexicana* may also be maintained during the amastigote stage in cultured macrophages.

Interestingly, the stability of the amplified DNAs during animal infection are comparable to those observed in vitro, with unstable lines exhibiting loss and stable lines retention of the amplified DNA during propagation through the amastigote stage (Sina et al., 1988). This observation is relevant to the maintenance of gene amplifications and other drug-resistance mutations in the natural infectious cycle. Normally, the parasite spends the majority of its time in the insect vector or reservoir vertebrate animals (usually wild) in the absence of drug pressure, as leishmaniasis is primarily a zoonosis (WHO,

1984). In such a natural cycle unstable forms of resistance would be expected to be disfavored due to the long periods without drug pressure, and ultimately stable forms of resistance would be expected to predominate. In support of this expectation, the tetrameric H amplification found in natural isolates of *L. tarentolae* appears to be extremely stable in the absence of selective pressure (Petrillo-Peixoto and Beverley, 1988).

## PARALLELS BETWEEN GENE AMPLIFICATION IN *LEISHMANIA* AND DRUG RESISTANCE IN PROKARYOTES AND EUKARYOTES

There are many similarities between the amplified DNAs mediating drug resistance in *Leishmania*, gene amplification in cultured mammalian cells, and drug-resistance factors in prokaryotes. In many ways, the properties of amplifications in *Leishmania* bridge the resistance mechanisms of both these kingdoms, and incorporates features of each. Bacterial drug resistance factors usually are found as part of small extrachromosomal circular DNAs, or plasmids. These plasmids can be found at varying copy numbers per cell, and replicate autonomously. Moreover, bacterial plasmids frequently contain gene functions which mediate horizontal transfer among different cells, strains, and species, thus facilitating the spread of drug resistance. The amplified DNAs of *Leishmania* resemble these in every way except for transfer functions, as these DNAs consist of normal chromosomal DNA in a rearranged form rather than an evolutionarily adapted structure.

Gene amplification in cultured mammallian cells has been extensively studied in drug-resistant lines and in tumor lines bearing amplified oncogenes. Similarities with *Leishmania* include the progression from unstable to stable amplifications, extrachromosomal amplification and replication (as double-minute chromosomes in cultured cells), and the amplification of substantial flanking regions of DNA besides the selected target gene (Schimke, 1984; Stark and Wahl, 1984). Though stable amplifications in cultured mammalian cells are usually integrated into chromosomal DNA (forming homogeneously staining regions), examples of extrachromosomal amplifications which are stable have been reported (Hamkalo et al., 1985). Differences in the two systems include the extent of DNA amplified, which is much smaller in *Leishmania* (less than 100 kb vs. upwards of 1,000 kb in mammalian cells), and the homogeneous nature of amplifications in *Leishmania* (vs. heterogeneity in mammalian systems; Roberts et al., 1983; Federspiel et al., 1984; Schimke, 1984; Stark and Wahl, 1984). These latter two features in combination with the ability to detect amplified DNA easily and resolve parasite chromosomes by pulsed field gradient electrophoresis methodology make the parasite amplification system an ideal one for studying the chromosomal structure of amplified DNAs.

## USES OF AMPLIFIED DNA

Gene amplification leads to greatly increased levels of the protein(s) encoded by the amplified region. This can be of great use in purification of normally nonabundant proteins, such as DHFR-TS (Meek et al., 1985). This feature would be especially useful in obtaining quantities of parasite proteins which undergo unusual metabolic modifications not performed in bacterial or eukaryotic expression systems. For example, Kavathas and Herzenberg (1983) have shown that by using specific antisera and the fluorescent-activated cell sorter it is possible to select for overexpression of membrane antigens (such a protocol could also be successful even if amplification was not the mechanism leading to overexpression). This might prove exceedingly beneficial in the study of leishmanial surface antigens of interest as possible vaccines, and could possibly be successful with nonprotein antigens as well. Accompanying overexpression of protein is elevation of mRNA levels, facilitating the analysis of their structure and expression (Kapler et al., 1987).

Amplification in response to drug pressure may constitute a genetic identification of a drug-resistance locus. Moreover, because amplified DNA may be readily isolated and subjected to molecular analysis, this form of genetic mutation may rapidly yield the drug resistance gene(s); this has been rarely possible with other drug-resistance mutations in parasites thus far. In the case of drugs such as antifolates, whose targets can be assumed to include the DHFR-TS enzyme, amplification of the DHFR-TS gene can be observed (Beverley et al., 1986; Grumont et al., 1986). However, amplification of resistance genes unanticipated by pharmacological or biochemical studies can also be observed, as possibly exemplified by the resistance determinant encoded by the H region. Analysis of this region may lead to an understanding of the biochemical mechanism of resistance conferred by its amplification, and possibly, a prospective new target for chemotherapy (Ellenberger and Beverley, 1988).

Another potential use of amplified DNA is as a tool for studying basic genetic processes. Because amplified DNA is transcribed, replicated, and inherited, it must contain all required cis-acting information. Currently, our knowledge of all three of these processes is limited in the trypanosomatids, though rapid progress is being made in the study of transcription (Borst, 1986). Analysis of mutants of the R region such as those shown in Figure 5 provide some clue as to the extent of DNA surrounding the DHFR-TS coding region necessary for these functions. Ultimately, we would like to be able to manipulate the structure of the DNA and reintroduce it back into the parasite via genetic transformation, and thereby develop a functional test of the role(s) of amplified sequences. The power of this approach to many lines of investigation is obvious, and work on genetic transformation is being actively pursued in many laboratories, including our own.

## THE FUTURE

It is relevant to consider the role of gene amplification in the treatment of leishmaniasis. Currently, drug resistance in this parasite is not widely recognized to be a clinical problem; instead, the most pressing problem is to identify drugs which are more satisfactory than the current agents of choice, the organic antimonials. We have selected parasites resistant to antifolates and two classes of experimental antileishmanials which bear amplified DNA, and shown that amplified DNA can be maintained during both stages of the parasite life cycle. This suggests that gene amplification may potentially play a role in future clinical drug resistance as better antileishmanials are discovered and introduced into clinical usage.

The study of gene amplification has provided important new insights into the basic molecular genetics of *Leishmania*, including gene structure, mechanisms of gene rearrangement, replication, and genetic ploidy. These studies have also increased our understanding of the diverse repertoire of drug-resistance mechanisms available to the parasite, and may prove useful in identifying new chemotherapeutic targets. Gene amplification thus simultaneously offers exciting new insights into both basic and applied questions of parasitology.

## ACKNOWLEDGMENTS

This work was supported by grants from the National Institute of Health, the March of Dimes, and the Pharmaceutical Manufacturers Association. We thank D. Dobson and J. Coderre for discussions. S.M.B. is a Burroughs-Wellcome Scholar in Molecular Parasitology.

## REFERENCES

Alt FW, Kellems RE, Bertino JR, Schimke RT (1978): Selective multiplication of dihydrofolate reductase genes in methotrexate-resistant variants of cultured murine cells. J Biol Chem 253:1357–1370.

Beverley SM (1988a): Circular DNAs in pulsed-field-gradient electrophoresis. Nucleic Acids Res 16:925–939.

Beverley SM (1988b): manuscripts submitted or in preparation:

Beverley SM, Coderre JA, Santi DV, Schimke RT (1984): Unstable DNA amplifications in methotrexate-resistant *Leishmania* consist of extra-chromosomal circles which relocalize during stabilization. Cell 38:431–439.

Beverley SM, Ellenberger TE, Cordingley JS (1986): Primary structure of the gene encoding the bifunctional dihydrofolate reductase-thymidylate synthase of *Leishmania major*. Proc Natl Acad Sci USA 83:2584–2588.

Beverley SM, Ismach RB, McMahon-Pratt D (1987): Evolution of the genus *Leishmania* as revealed by comparisons of nuclear DNA restriction fragment patterns. Proc Natl Acad Sci USA 84:484–488.

Borst P (1984): DNA amplification and multi-drug resistance. Nature 309:580.

Borst P (1986): Discontinuous transcription and antigenic variation in trypanosomes. Annu Rev Biochem 55:701–32.

Botchan M, Topp W, Sambrook J (1979): Studies on SV40 excision from cellular chromosomes. Cold Spring Harbor Symp Quant Biol XLIII:709–719.

Brison O, Ardeshir F, Stark GR (1982): General method for cloning amplified DNA by differential screening with genomic probes. Mol Cell Biol 2:578–587.

Coderre JA, Beverley SM, Schimke RT, Santi DV (1983): Overproduction of a bifunctional thymidylate synthetase-dihydrofolate reductase and DNA amplification in methotrexate-resistant *Leishmania*. Proc Natl Acad Sci USA 80:2132–2136.

Ellenberger TE, Beverley SM (1987a): Biochemistry and regulation of folate and methotrexate transport in *Leishmania major*. J Biol Chem 262:10053–10058

Ellenberger TE, Beverley SM, (1987b): Reductions in Methotrexate and folate influx in methotrexate-resistant lines of *Leishmania major* are independent of R or H region amplification. J Biol Chem 262:13501–13506.

Ellenberger TE, Beverley SM (1988): manuscript in preparation.

Federspiel NA, Beverley SM, Schilling JW, Schimke RT (1984): Novel DNA rearrangements are associated with dihydrofolate reductase gene amplification. J Biol Chem 259:9127–9140.

Ferone R, Roland S (1980): Dihydrofolate reductase-thymidylate synthase, a bifunctional polypeptide from *Crithidia fasciculata*. Proc Natl Acad Sci USA 77:5802–5806.

Ford M, Fried M (1986): Large inverted duplications are associated with gene amplification. Cell 45:425–420.

Garrett CE, Coderre J, Meek TD, Garvey EP, Claman DM, Beverley SM, Santi DV (1984): A bifunctional thymidylate synthase-dihydrofolate reductase in protozoa. Mol Biochem Parasitol 11:257–265.

Garvey EP, Santi DV (1986): Stable amplified DNA in drug-resistant *Leishmania* exists as extra-chromosomal circles. Science 233:535–540.

Garvey EP, Coderre JA, Santi DV (1985): Selection and properties of *Leishmania* resistant to 10-propargyl-5,8-dideazafolate, an inhibitor of thymidylate synthetase. Mol Biochem Parasitol 17:79–91.

Grumont R, Washtien WL, Caput D, Santi DV (1986): Bifunctional thymidylate synthase-dihydrofolate reductase from *Leishmania*: sequence homology with the corresponding monofunctional proteins. Proc Natl Acad Sci USA 83:5387–5391.

Hamkalo B, Farnham P, Johnston RN, Schimke RT (1985): Ultrastructural features of minute chromosomes in a methotrexate-resistant mouse 3T3 cell line. Proc Natl Acad Sci USA 82:1126–1130.

Hieter P, Mann C, Snyder M, Davis RW (1985): Mitotic stability of yeast chromosomes: A colony color assay that measures nondisjunction and chromosome loss. Cell 40:381–392.

Iovannisci DM, Kaur K, Young L, Ullman B (1984): Genetic analysis of nucleoside transport in Leishmania donovani. Mol Cell Biol 4:1013–1019.

Kapler GM, Zhang K, Beverley SM (1987): Sequence and S1 nuclease mapping of the 5' region of the dihydrofolate reductase-thymidylate synthase gene of Leishmania major. Nucleic Acids Res 15:3369–3383.

Kavathas P, Herzenberg LA (1983): Amplification of a gene coding for human T-cell differentiation antigen. Nature 306:385–387.

Kink JA, Chang K-P (1987): Tunicamycin-resistant Leishmania mexicana amazonensis: expression of virulence associated with an increased activity of N-acetylglucosaminyltransferase and amplification of its presumptive gene. Proc Natl Acad Sci USA 84:1253–1257.

Meek TD, Garvey EP, Santi DV (1985): Purification and characterization of the bifunctional thymidylate synthetase-dihydrofolate reductase from methotrexate-resistant Leishmania. Biochemistry 24:678–686.

Meeks-Wagner D, Wood JS, Garvik B, Hartwell LH (1986): Isolation of two genes that affect mitotic chromosome transmission in S. cerevisiae. Cell 44:53–63.

Murray AW, Schultes NP, Szostak JW (1986): Chromosome length controls mitotic chromosome segregation in yeast. Cell 45:529–536.

Petrillo-Peixoto M, Beverley SM (1988): manuscripts submitted or in preparation.

Roberts JM, Buck L, Axel R (1983): A structure for amplified DNA. Cell 33:53–63.

Roninson IB (1983): Detection and mapping of homologous, repeated and amplified DNA sequences by DNA renaturation in agarose gels. Nucleic Acids Res 11:5413–5431.

Schimke RT (1982): Gene Amplification. Cold Spring Harbor, NY: Cold Spring Harbor Press.

Schimke RT (1984): Gene amplification in cultured animal cells. Cell 37:705–713.

Schimke RT, Sherwood SW, Hill AB, Johnston RN (1986): Overreplication and recombination of DNA in higher eukaryotes: Potential consequences and biological mechanisms. Proc Natl Acad Sci USA 83:2157–2161.

Scott DA, Coombs GH, Sanderson, BE (1987): Effects of methotrexate and other antifolates on the growth and dihydrofolate reductase activity of leishmania promastigotes. Biochem Pharmacol, in press.

Sina BJ, Kapler GM, Beverley SM (1988): manuscript in preparation.

Stark GR, Wahl GM (1984): Gene Amplification. Annu Rev. Biochem 53:447–91.

Wahl GM, Padgett RA, Stark GR (1979): Gene amplification causes overproduction of the first three enzymes of UMP synthesis in N-(phosphonacetyl)-L-aspartate-resistant hamster cells. J Biol Chem 254:8679–8689.

WHO Expert Committee (1984): The Leishmaniasis. New York: WHO.

Zakian VA, Blanton HM, Wetzel L, Dani G (1986): Size threshold for Saccharomyces cerevisiae chromosomes: Generation of telocentric chromosomes from an unstable minichromosome. Mol Cell Biol 6:925–932.

The Biology of Parasitism, pages 449–465
© 1988 Alan R. Liss, Inc.

# Plasma Membrane Functions: A Biochemical Approach to Understanding the Biology of *Leishmania*

## Michael Gottlieb and Dennis M. Dwyer

*Department of Immunology and Infectious Diseases, The Johns Hopkins University School of Hygiene and Public Health, Baltimore, Maryland 21205 (M.G.) and Cell Biology and Immunology Section, Laboratory of Parasitic Diseases, National Institutes of Health, Bethesda, Maryland 20892*

## PLASMA MEMBRANES AND THE HOST-PARASITE INTERFACE

The plasma membrane plays numerous essential roles in the maintenance and survival of all cells. Such membranes and their specific constituents maintain differences between the intracellular milieu and the external environment as well as respond to signals from the environment. For parasites the external environment is that provided by the infected host organism. Thus, for protozoan parasites the plasma membrane is the interface between them and their hosts and in that position must carry out all requisite physiologic functions, such as nutrient acquisition, secretion, and excretion, in addition to providing protection against a variety of host defense mechanisms.

In the case of protozoa of the genus *Leishmania,* adaptations involving the plasma membrane must have evolved for resistance to host hydrolytic environments. These organisms develop and multiply as extracellular promastigotes within the alimentary tract of their sandfly vectors and as obligate intracellular amastigotes within phagolysosomes of mammalian macrophages (for review of the biology of *Leishmania* see Chang et al., 1985). Presumably, such membrane adaptations have enabled the various *Leishmania* spp. to survive in the mammalian macrophage, a cell which is central to resistance and immunity to infectious agents. This survival leads to the development of the various clinically distinct disease entities observed in infected individuals. Leishmaniases can range from simple and self-healing cutaneous lesions to visceral disease with significant mortality (for review of the clinical aspects see Marsden and Jones, 1985). This clinical spectrum reflects the predilection of the infecting species for specific macrophage populations as well as factors inherent in the host.

Considering the overall significance of the leishmanial surface membrane many studies have been undertaken to investigate its structure and composition (for review see Dwyer and Gottlieb, 1983, 1985). In particular, the application of recently developed cellular and molecular biological techniques has led to the identification of a number of leishmanial surface membrane antigens. These investigations have taken advantage of hybridoma-derived monoclonal antibodies in conjunction with specific labeling reagents to detect surface proteins and other molecules on intact parasites. Of special interest are antigens which appear to be developmentally regulated (e.g., Kahl and McMahon-Pratt, 1987) and antibodies which are useful in parasite typing and clinical diagnosis (McMahon-Pratt and David, 1981; Jaffe et al., 1984). These methods have facilitated the isolation of some of these molecules and their subsequent physicochemical characterization. This approach has permitted rapid progress to be made in the study of leishmanial surface membrane antigens which should prove useful in the development of molecularly defined, candidate vaccines against the diseases caused by this group of human pathogens. Further, it has facilitated the study of molecular events involved in the recognition of the leishmanial parasites by macrophages (reviewed in Chang, 1983; Alexander and Russell, 1985; Blackwell, 1985).

## "Function-First" Approach to the Study of Parasite Membranes

Despite the aforementioned advances, the biological activities of such characterized antigens remain generally unknown. To learn more of the physiological functions of components of the leishmanial plasma membrane we and others have approached the problem in an alternative manner. In this "function-first" approach, biological activities are first identified, for example, enzymatic catalyses, and subsequently the proteins responsible are purified and characterized with respect to their subcellular distribution, as well as their kinetic, physicochemical, and regulatory properties. A variation on this approach, which has been used to only a limited extent with *Leishmania*, is the generation and characterization of mutants in specific biological functions. It is assumed that the knowledge gained from these approaches of the physiological functions essential to the parasite will be useful in not only understanding the biology of the host-parasite interaction but also in identifying new means of controlling infection and disease.

This overview will focus on catalytic activities, which have been clearly associated with leishmanial surface membrane proteins, as well as transport activities. In the course of discussing each activity, the known structure, function, and regulatory properties of the membrane constituent will be considered in the context of its presumptive adaptive advantage to the parasite. In many instances data regarding specific membrane functions have been obtained from a single leishmanial species; therefore, the general applicability of such results to other species in this genus must await comparative studies. Similarly, most results described are derived from studies using promastigotes from axenic cultures, as significantly less is known regarding the membrane constituents of amastigotes.

## LEISHMANIAL PLASMA-MEMBRANE-ASSOCIATED ACTIVITIES

### Purine Acquisition

**Significance.** *Leishmania*, as well as other trypanosomatids, require an exogenous source of preformed purines for continuous growth because of their inability to synthesize these molecules de novo (for review see Hammond and Gutteridge, 1984). The host environment must provide these nutrients. In contrast, most mammalian cells are capable of de novo purine synthesis. This biochemical difference between host and parasite has implications for the development of more effective chemotherapeutic agents for diseases caused by these organisms. Thus, attention has been given to the effects of purine analogs on the growth of these parasites. Specifically, pyrazolopyrimidines, such as allopurinol, have been advanced as potential antileishmanial drugs (for review see Marr and Berens, 1983).

To date, investigations of purine salvage in trypanosomatids have focused on the analysis of the intracellular metabolism of purines, especially on the enzymatic machinery involved in the interconversion of purines from exogenously supplied nucleoside and nucleobase precursors (for review see Marr and Berens, 1985). Considerably less attention has been given to adaptations of the parasite surface membrane which are involved in the scavenging of these essential nutrients and little is known of the specific membrane constituents.

In this regard, it is generally assumed that *Leishmania* rely on the salvage of purines from bases and nucleosides present in its hosts. However, in these host environments, nucleotides and nucleic acids are likely to be present in large excess over nucleosides or bases. To serve as purine sources such molecules must first be hydrolyzed to membrane-permeable products. Parasite enzymes which can catalyze such nucleic acid and nucleotide hydrolysis would offer a significant advantage to the parasite.

**Nucleotidase and nuclease activities.** Distinct 5'- and 3'-nucleotidases have been localized to the surface membrane of promastigotes of *L. donovani* by a variety of methods (Gottlieb and Dwyer, 1983; Dwyer and Gottlieb, 1984). Recently confirmatory results were reported (Hassan and Coombs, 1987) for *L. mexicana mexicana*. The cumulative results indicate that the surface membrane is the predominant, if not exclusive, site of these activities. The relationship of a nucleotidase activity reported from *L. tropica* promastigote membranes by Pereira and Konigk (1981) to these other activities awaits further comparisons. Additional studies indicate that amastigotes of *L. donovani* and *L. m. mexicana* also exhibit both enzyme activities (Gottlieb and Dwyer, 1982a; Hassan and Coombs, 1987).

Nucleotidases specific for 5'-mononucleotides are characteristically associated with plasma membranes of various cell types (Drummond and Yamamoto, 1971). Although not all of its functions are known, the enzyme's position at the cell surface has implications for the initiation of purine salvage, particularly in cells, such as lymphocytes, with limited de novo synthetic abilities (Thompson, 1985). By analogy, the leishmanial 5'-nucleotidase (5'-Nase) enzyme can catalyze the release of nucleosides from both extracellular ribo- and deoxyribonucleotides (Gottlieb and Dwyer, 1983), making them available for transport and subsequent intracellular utilization.

The evolutionary homology between the leishmanial and mammalian 5'-Nases is suggested by preliminary studies (Bates, Luzio, and Dwyer, unpublished observations) using a cross-reactive antiserum prepared against purified rat liver 5'-Nase (supplied by P. Luzio of Cambridge University). The antiserum immunoprecipitates a 72,000 $M_r$ leishmanial surface membrane polypeptide. Similar results were obtained by Western blot analysis with that antiserum and isolated *L. donovani* membranes. Confirmation of this apparent homology awaits the purification of the leishmanial enzyme. Evidence acquired to date, based on lectin binding properties, indicates that the *L. donovani* 5'-Nase is a mannose-containing glycoprotein, analogous to the rat liver enzyme (Odera, Bates, and Dwyer, unpublished).

In contrast to 5'-Nase, 3'-nucleotidase (3'-Nase) has been less well studied as it has not been found in, or isolated from, mammalian cells and tissues. The presence of 3'-Nase in leishmanial surface membranes define another biochemical difference between host and parasite. A suggestion of the importance of the 3'-Nase to the parasite is the observation that the level of its activity is 20-fold greater than that of the 5'-Nase. Initial characterization (Gottlieb and Dwyer, 1983) of the 3'-Nase demonstrated its ability to hydrolyze 3'-ribo-, but not 3'-deoxyribo-, nucleotides, making ribonucleoside reaction products available for transport and intracellular metabolism.

Although initially described as a phosphomonoesterase specific for 3'-ribomononucleotides, more recent evidence indicates that the *L. donovani*

3'-Nase hydrolyzes nucleic acids (Gottlieb and Zlotnick, 1987). The purified enzyme is active with both DNA and RNA substrates and liberates 5'-nucleotides from them. The demonstration that both nucleotidase and nuclease activities reside in the same polypeptide was possible because the protein can be renatured following sodium dodecyl sulfate-polyacrylamide gel electrophoresis (SDS-PAGE) as shown by Zlotnick et al. (1987). Both activities were displayed by a 43,000 $M_r$ protein. However, a catalytically active doublet, with an $M_r$ of 35,000 and 37,000, has been observed in some preparations of purified leishmanial 3'-Nase. These fragments are presumably generated from the action of an endogenous protease which copurifies with 3'-Nase. Preliminary evidence (Gottlieb, Pastakia, and Dwyer, unpublished observations) indicates that this protease is the promastigote surface protease (Etges et al., 1986; Bouvier et al., 1987) which is considered in more detail below. The susceptibility of the native 3'-Nase to proteolysis is indicated by the observation that trypsin treatment of leishmanial membranes, but not intact cells, yields a 37,000 molecular weight fragment which is capable of hydrolyzing 3'-AMP. Of special interest is the demonstration that this putative proteolytic fragment, in contrast to the parent molecule, is soluble in the absence of detergents.

The cumulative results indicate that the surface 3'-Nase and surface 5'-Nase function together in the liberation of nucleosides from available nucleic acids and/or mononucleotides. Thus these enzymes fulfill the parasite's obligatory requirement for the salvage of purines from the host environment. In support of this, *L. donovani* can be grown continuously in chemically defined culture medium, containing either nucleic acids or nucleotides as the sole source of purine (Gottlieb, unpublished).

**3'-Nucleotidase/nucleases from other organisms.** Activities homologous to the leishmanial 3'-Nase have recently been described in related trypanosomatids including a species of *Crithidia* (Gottlieb, 1985) and African trypanosomes of the *Trypanosoma brucei* complex (Gottlieb et al., 1986; Gbenle et al., 1986). Interestingly, these trypanosomes lack 5'-Nase activity and apparently, therefore, cannot take advantage of the 5'-nucleotide products of nuclease hydrolysis generated by the 3'-Nase. Curiously, the stercorarian trypanosome *T. cruzi* lacks 3'-Nase entirely (Gottlieb, unpublished observations).

Enzymes having both 3'-nucleotidase and nuclease activities are not novel, having also been described from plant and fungal sources (Shishido and Ando, 1985). So-called nucleases from mung bean and wheat germ seedlings, although soluble proteins, share many kinetic properties, including pH optima and substrate specificity, with the trypanosomatid 3'-Nase. Despite the widespread use of these enzymes, especially the mung bean nuclease, in molecular biology, little is known of their biological role. Their differential expression in germinating seedlings has led to the suggestion (Hanson and Fairley, 1969) that they salvage purines which are used during this period of rapid growth and cell division.

**Regulation of nucleotidase and nuclease activities.** Studies (Gottlieb, 1985) of the 3'-Nase in *Crithidia luciliae* (incorrectly identified in that report as *C. fasciculata*) have implications for the physiological role of this activity in purine acquisition. The results of these studies indicate that *C. luciliae,* grown in purine-containing media, exhibit very low levels of 3'-Nase activity, whereas those organisms from purine-depleted medium exhibit 100–1000-fold greater levels of 3'-Nase activity. The activation of enzyme activity appears to be a specific adaptive response to conditions in which purines are growth limiting. The increased activity of the surface 3'-Nase under purine starvation conditions would allow these cells to utilize purine nucleotide and/ or nucleic acid sources readily, should these substrates become available. Studies of the regulatory process by which the enzyme is activated are in progress. Results to date (Gottlieb et al., 1986), from studies using inhibitors of macromolecular synthesis, indicate that enzyme activation requires the synthesis of new RNA transcripts and protein products. In contrast, hydroxyurea, an inhibitor of DNA synthesis, has no effect on the increase in enzyme activity. The activation of the 3'-Nase appears to be specific as there is only a modest 4–5-fold increase in the level of the crithidial 5'-nucleotidase and a membrane-associated acid phosphatase in cells which have been purine-starved.

Although it was initially assumed that the leishmanial 3'-Nase was expressed at constitutively high levels in purine-rich medium, further studies (Gottlieb and Wyman, unpublished) have revealed that this basal level can be elevated, in some cases by as much as tenfold, by purine starvation of the promastigotes in defined medium.

**Purine transport.** The nature of purine nucleoside transport in leishmanial promastigotes has been addressed in several studies. However, the difficulties in interpreting the results obtained are due to the inability to separate transmembrane events from subsequent intracellular metabolism. Despite these difficulties, Aronow et al. (1987) have concluded that there are two independent nucleoside transporters in *L. donovani* promastigotes. One transports inosine, guanosine, and their analogs, whereas the other transports adenosine, its analogs as well as the pyrimidine nucleosides. In contrast to mammalian cells, nucleoside transport in *L. donovani* is of higher affinity and is resistant to dipyridamole and 4-nitrobenzylthioinosine. Hansen et al.

(1982) have also provided evidence for multiple purine transport pathways in *L. braziliensis panamensis*. The further dissection of the leishmanial transport process will be aided by the availability of mutants to tubericidin and formycin B (Iovannisci et al., 1984). These mutant phenotypes appear to result from the loss of functional purine transport components rather than from changes in intracellular metabolism. It is anticipated that these studies will lead to the identification, specific labeling, and isolation of proteins responsible for transport functions as well as their functional reconstitution in artificial membranes.

Preliminary studies (Gottlieb, unpublished) indicate that purine-starved *L. donovani* and *C. luciliae* take up purines at a much faster rate than do purine-replete cells. This up-regulation of purine acquisition may reflect differential expression of membrane transporters or changes in intracellular utilization events. Alterations in 3'-Nase and purine transport levels presumably reflect the parasite's ability to adaptively respond to transient environmental fluctuations.

**Perspectives.** From the foregoing it is clear that further studies are needed to define the relationships between the proteins and activities described. In particular, the structural relationships between the membrane nucleotidases and purine transporters remain to be elucidated. Studies with *L. donovani* suggest that in this organism the rate-limiting step in purine salvage, from nucleotide or nucleic acid precursors, is that of purine transport. In support of this conclusion is the observation that when promastigotes are grown in chemically defined medium with nucleic acids or nucleotides, as the sole source of purines, nucleosides are generated and detected extracellularly. If nucleotide hydrolysis was rate limiting, such free extracellular products would not be detected. Moreover, the conclusion that the transport is rate limiting is supported by results with intact promastigotes which demonstrate that the rate of hydrolysis of 3'-AMP is faster than the uptake rate for free adenosine. Paradoxically, the apparent $K_m$ of the nucleoside transporter is very low ($< 1$ $\mu$M for adenosine) compared to the $K_m$ of the 3'-Nase for 3'-AMP (0.2 mM).

The further study of the regulated changes in purine salvage should lead to the identification of the underlying molecular events in such adaptive responses. Although nutrient-induced changes have been the subject of many investigations in other microorganisms (for review see Harder and Dijkhuizen, 1983), few studies on nutrient depletion of the trypanosomatid protozoa have been carried out. Such studies have implications for the characterization of surface antigens since changes in culture composition may be responsible for their differential expression. More importantly, such studies may provide clues to environmental signals which trigger specific responses in these organisms. For example, stationary phase promastigotes are much more infectious for macrophages and animals than log phase organisms (Sacks and Perkins, 1984). The increased infectiousness of L. major promastigotes has been correlated with altered expression of a surface membrane carbohydrate component (Sacks et al., 1985). The environmental stimulus for such a differentiation event may be changes in nutrient availability. Indeed, nutritional stimuli may be responsible for parasite development in the sandfly, e.g., leading to the migration of promastigotes to the proboscis, where they are capable of infecting a suitable mammalian host. Purine starvation can serve as a model to characterize such environmentally regulated changes. The availability of specific antibodies to the surface membrane enzymes and of molecular probes to the nucleic acids encoding these molecules will facilitate studies concerning the regulated changes of their expression and activities.

While the foregoing discussion has considered the role of membrane constituents in purine salvage, it is possible that the 5'-Nase and the 3'-Nase may serve other physiological needs of the parasite. The ability of the Nases to generate extracellular nucleosides, in excess of that required for transport and utilization, suggests their role in influencing host responses. For example, it is known that adenosine exhibits hormonelike and immunoregulatory properties.

### Sugar and Amino Acid Acquisition

The leishmanial parasite must access its sugar and amino acid requirements from its hosts for its energy yielding and biosynthetic processes. These compounds are presumably acquired by specific membrane transport systems which are considered here.

**Hexose transport.** Several studies of hexose (primarily glucose) transport by promastigotes of various leishmanial species have been reported (Schaefer et al., 1974; Mukkada et al., 1974; Schaefer and Mukkada, 1976; Zilberstein and Dwyer, 1984a, 1985a). L. donovani promastigotes possess a very-high-affinity, carrier-mediated, stereospecific surface membrane active transport system for the accumulation of D-glucose (Zilberstein and Dwyer, 1984a). Such a high-affinity transporter is unusual for a eukaryotic cell. However, in this instance, it would permit the parasite to maintain high levels of intracellular glucose and thus normal glucose metabolism throughout its development, even under conditions of limited glucose availability. Recently, the L. donovani surface membrane glucose transport protein was specifically labeled, identified, and characterized (Zilberstein et al., 1986). This transporter appears to be a mannose-containing glycoprotein with an $M_r$ of 19,000, which is significantly smaller than mammalian glucose transport proteins. Isolation of the parasite molecule and its reconstitution into artificial lipid membranes should permit further in-depth studies of its biochemical and physiological properties. Generation of specific antibodies would permit further investigations of this transport protein per se as well as facilitate the screening of parasite expression libraries for its gene.

*Leishmania* presumably encounter a variety of other hexose sugars in their host. However, among those that have been tested only fructose was accumulated by *L. donovani* promastigotes (Pastakia and Dwyer, 1987), apparently via the glucose carrier as reported previously for *L. tropica* (Schaefer and Mukkada, 1976). As sandflies ingest plant juice meals between blood feedings, fructose would be available to the parasite and could be accessed by this mechanism.

**Pentose transport.** In addition to hexoses, the parasite is presumably exposed to a variety of pentose sugars. Among those tested only ribose was shown to be accumulated by a variety of promastigotes of *Leishmania* spp., including those of *L. donovani* (Pastakia and Dwyer, 1987). Ribose transport appears to be constitutive and to occur via a facilitated diffusion mechanism. Among eukaryotic organisms, ribose transport is not prevalent. However, this pentose carrier could confer a selective advantage to the parasite by allowing the parasite to utilize this sugar as an alternative energy source via the pentose shunt pathway (Marr, 1980) or as a precursor for essential nucleotide biosynthesis in conjunction with carriers specific for purines.

**Amino acid transport.** In addition to carbohydrates, *Leishmania* accumulate a variety of amino acids. Carrier-mediated transport systems, characterized by typical saturation kinetics, have been demonstrated in several different species for the uptake of L-proline (Law and Mukkada, 1979; Zilberstein and Dwyer, 1985a), L-methionine (Mukkada and Simon, 1977; Simon and Mukkada, 1977), L-alanine (Bonay and Cohen, 1983) and the neutral amino acid analog $\alpha$-aminoisobutyric acid (Lepley and Mukkada, 1983). To date, although defined physiologically with intact cells, none of these transporters has been specifically labeled, isolated, or biochemically characterized.

**Membrane-associated ATPase.** Recently, it was demonstrated that the active transport of both D-glucose and L-proline, in *L. donovani* promastigotes, is accomplished via a proton-motive force (pmf)-driven mechanism (Zilberstein and Dwyer, 1985a). The pmf is a proton electrochemical gradient which is created across the parasite surface membrane and which couples symport translocation of specific substrates with protons. Proton pumps play critical roles in regulating both the magnitude of the pmf and in maintaining

intracellular pH homeostasis. Since *Leishmania* encounter extreme environmental changes during development, its surface membrane proton pump
would be essential in maintaining active transport of nutrients and in stabiliz-
ing intracellular pH levels. Membrane bound proton-translocating ATPases
($H^+$-ATPase) have been demonstrated in a variety of systems to be energy
transducers which couple the energy released from ATP hydrolysis to generate such proton-electrochemical gradients. Recently, a plasma membrane
ATPase, which satisfies the criteria for a primary proton pump, has been
identified, localized cytochemically, and characterized in *L. donovani* promastigotes (Zilberstein et al., 1987) (Zilberstein and Dwyer, in preparation).
Although characterized physiologically in surface membranes and membrane
vesicles derived from promastigotes, this enzyme remains to be physically
isolated from such sources. Its reconstitution into liposomes should allow for
the demonstration of its proton pumping activity and provide evidence for its
biochemical and physiological functions. In a practical sense, the $H^+$-ATPase
should be a logical target for chemotherapeutic attack. In that regard, results
with both pro- and amastigotes of *L. donovani* and *L. major* have demonstrated the rationale of this approach by using drugs, such as imipramine
derivatives, which disrupt the proton pumping function (Zilberstein and
Dwyer, 1984b).

## Other Plasma Membrane Enzymes

In addition to enzymatic and transport activities involved in the acquisition
of purines, sugars, and amino acids, a number of other enzymatic activities
have been identified and associated with the leishmanial surface membrane.
To date, the physiological roles of these activities have been less well
established; however, some intriguing possible associations between these
activities and the interactions between host and parasite have been advanced.
These are discussed briefly in this section.

**Acid phosphatase.** The surface membrane of *L. donovani* promastigotes
possesses a nonspecific acid phosphomonoesterase, resistant to tartrate ions,
with an extracellularly oriented active site (Gottlieb and Dwyer, 1981a,b).
The demonstration of a very active surface acid hydrolase in an organism
which resides in hydrolytic environments rich in such enzymatic activities
prompted further studies of this activity and the enzyme(s) responsible. Glew
and his collaborators have purified the predominant acid phosphatase activity
from a crude membrane fraction of *L. donovani* (Glew et al., 1982; Remaley
et al., 1985a). Based primarily upon its resistance to tartrate ions, it was
assumed that the purified enzyme is of surface origin. A variety of biological
roles have been proposed for this enzyme, including one based on the

following observation. Preincubation of isolated human neutrophils with the purified leishmanial acid phosphatase leads to the inhibition of the respiratory burst stimulated upon the addition of the chemotactic peptide fmet-leu-phe or concanavalin A, but not phorbol ester (Remaley et al., 1984, 1985b). The inhibition of neutrophil superoxide production by the acid phosphatase apparently requires the catalytic activity of the enzyme as demonstrated by control studies in which such catalysis was specifically abolished. Although leishmanial parasites do not survive in neutrophils, it has been argued that the ability of the acid phosphatase to inhibit a similar respiratory burst in macrophages may account for the ability of these organisms to survive in these cells. The mechanism by which the acid phosphatase prevents the superoxide production by the neutrophils has not been established, although it is unlikely that a phosphoprotein phosphatase activity of the leishmanial enzyme is involved (Das et al., 1986). Additional work is needed to understand more fully the role of the acid phosphatase in the host-parasite interactions. In particular, the relationship of the major tartrate-resistant enzyme to other membrane-associated acid phosphatase activities, identified in this organism, needs to be addressed.

In addition to the surface acid phosphatase, promastigotes of *L. donovani* and most other leishmanial species, but not *L. major*, secrete a soluble acid phosphatase (Gottlieb and Dwyer, 1982b; Lovelace et al., 1986; Lovelace and Gottlieb, 1986). The role of this enzyme, whose activity is not able to inhibit the neutrophil respiratory burst (Das et al., 1986), also remains to be established. However, the ability of such an extracellular protein to act at sites distal to the parasite is consistent with a role in the modification of host activities. Further, the study of this enzyme will help to elucidate the secretory pathways, including the role of various membrane compartments, involved in the routing of such exoenzymes in these parasites. Such studies are underway (Bates and Dwyer, 1987).

**Protease.** In contrast to the aforementioned activities, the protein responsible for a surface-membrane-associated proteolytic activity was identified prior to the establishment of its catalytic properties.

The major surface protein of promastigotes of human leishmanias, as identified initially by surface iodination, is an evolutionarily conserved glycoprotein of approximately 63,000 molecular weight (Lepay et al., 1983; Etges et al., 1985; Bouvier et al., 1987). This molecule is anchored into the membrane by means of a glycophospholipid anchor, homologous to that of the variant specific antigens of African trypanosomes (Bordier et al., 1986). The membrane form of the enzyme can be solubilized by treatment with phospholipase C from bloodstream forms of *T. brucei* (Bordier et al., 1986).

Etges et al. (1986) have provided evidence for the protease activity for both the hydrophilic and amphiphilic forms of this protein. Various physiologicals roles have been suggested for this enzyme activity, including nutritional and protective; as yet, however, there is no direct evidence to support any of these. Further, although it has been demonstrated that antibodies from infected individuals recognize this protein, the localization of the enzyme protein and its activity in the amastigote stage of the parasite has not been established.

Despite the lack of evidence for a role based upon the enzyme's activity, there is data which indicate that the protein has a very important role in the recognition of the parasite by the macrophage. Russell and Wilhelm (1986) have shown that antibodies to this protein block, at least partially, the internalization of promastigotes by macrophages. Moreover, liposome vesicles containing this glycoprotein are readily phagocytosed by macrophages. Thus, this surface protein may be directly involved in the receptor mediated uptake of the parasite by its target cell.

The highly conserved nature of this surface protein/protease, common to all human leishmanias, and its abundance at the parasite interface with the host suggest its importance and therefore require its further investigation. Of special interest will be the role of the lipid-linkage with the membrane and of the catalytic activity.

**Lipolytic activities.** Efforts directed at characterizing the leishmanial surface membranes have focused primarily on their protein constituents. However, it should be noted that lipids account for approximately one-half of the dry weight of surface membranes isolated from *L. donovani* promastigotes (Wassef et al., 1985). To maintain membrane fluidity and integrity, these lipid components presumably undergo restructuring in adapting to the altered environments encountered by the parasite during the course of its development. To date, little is known of either the functional roles of individual lipid constituents or of their regulated changes. However, three lipolytic enzymes, phospholipase-A1, -A2, and -C, which are presumably involved in such membrane restructuring events as well as other aspects of lipid metabolism, have been partially characterized from surface membranes isolated from *L. donovani* promastigotes (Dwyer, 1987) (Wassef, Fioretti, and Dwyer, unpublished observations). The three phospholipases demonstrated their highest activities with phosphatidylethanolamine, the major leishmanial surface membrane phospholipid. The physiological roles in which these phospholipases are involved have yet to be established. In other systems, phospholipase-A1 and -A2 have been shown to be involved in deacylation and reacylation cycling events, in the remodeling of fatty acid components esterified into membrane phospholipids, and in affecting the fluidity of the membrane. By further analogy, phospholipase-C has been implicated in transmembrane signal recognition events and in exocytotic processes. It is conceivable that the leishmanial phospholipase-C is involved in expression, i.e., turnover or release, of lipid-linked membrane proteins, including the surface membrane protease considered above. In addition these lipase activities may be involved in the release of the hydrophilic form ("excreted factor") of surface membrane glycoconjugates (Kaneshiro et al., 1982; Handman et al., 1984). Further, these enzymes may be involved in nutrient acquisition and in contributing to parasite survival by the modification of host membrane components (Dwyer, 1987).

## CONCLUSIONS AND FUTURE DIRECTIONS

From the foregoing discussion it is evident that a number of *Leishmania* surface membrane enzyme and transport activities have been identified. These have been partially characterized; however, with the exception of the surface protease, the dual-functioning 3'nucleotidase/nuclease, and the tartrate-resistant acid phosphatase, the specific proteins responsible for these activities have not been defined. Thus, it seems obvious that to understand more fully these activities and their physiological relevance in parasite development and survival, additional knowledge of these constituents is required. It will be necessary to characterize these proteins more fully with regard to their physicochemical composition/structure, their biosynthesis, and their regulation/expression in response to environmental stimuli and their interactions with other membrane components. The availability of specific antibody and molecular probes will greatly facilitate such studies. This function and structure approach has already generated significant information concerning leishmanial biology. We anticipate that it will continue to provide insights into the mechanisms involved in the host-parasite relationship.

## ACKNOWLEDGMENTS

The studies of M.G. were supported in part by awards from the USPHS (grant AI-16530), from the John D. and Catherine T. MacArthur Program in Molecular Parasitology, and from the UNDP/World Bank/WHO Special Programme for Tropical Disease Research.

## REFERENCES

Alexander J, Russell DG (1985): Parasite antigens, their role in protection, diagnosis and escape: The Leishmaniases. Curr Top Microbiol Immunol 120:43–67.

Aronow B, Kaur K, McCartan K, Ullman B (1987): Two high affinity nucleoside transporters in *Leishmania donovani*. Mol Biochem Parasitol 22:29–37.

Bates PA, Dwyer DM (1987): Biosynthesis and secretion of acid phosphatase by *Leishmania donovani* promastigotes. Mol Biochem Parasitol 26:289–296.

Blackwell JM (1985): Receptors and recognition mechanisms of *Leishmania* species. Trans R Soc Trop Med Hyg 79:606–612.

Bonay P, Cohen BE (1983): Neutral amino acid transport in *Leishmania* promastigotes. Biochim Biophys Acta 731:222–228.

Bordier C, Etges RJ, Ward J, Turner MJ, Cardoso De Almeida ML (1986): *Leishmania* and *Trypanosoma* surface glycoproteins have a common glycophospholipid membrane anchor. Proc Natl Acad Sci USA 83:5988–5991.

Bouvier J, Etges R, Bordier C (1987): Identification of the promastigote surface protease in seven species of *Leishmania*. Mol Biochem Parasitol 24:73–79.

Chang K-P (1983): Cellular and molecular mechanisms of intracellular symbiosis in leishmaniasis. Int Rev Cytol [Suppl] 14:267–302.

Chang K-P, Fong D, Bray RS (1985): Biology of *Leishmania* and leishmaniasis. In K.-P. Chang, and R.S. Bray (eds): Leishmaniasis. Amsterdam: Elsevier, pp 1–30.

Das S, Saha AK, Remaley AT, Glew RH, Dowling JN, Kajiyoshi M, Gottlieb M (1986): Hydrolysis of phosphoproteins and inositol phosphates by cell surface acid phosphatase of *Leishmania donovani*. Mol Biochem Parasitol 20:143–153.

Drummond GI, Yamamoto M (1971): Nucleotide phosphomonoesterases. In P.D. Boyer (ed): The Enzymes, Vol IV. New York: Academic Press, pp 337–354.

Dwyer DM (1987): The roles of surface membrane enzymes and transporters in the survival of *Leishmania*. In K.-P. Chang and D. Snary (eds): Host-Parasite Cellular and Molecular Interactions in Protozoal Infections. NATO ASI Series. Heidelberg: Springer-Verlag, pp 175–182.

Dwyer DM, Gottlieb M (1983): The surface membrane chemistry of *Leishmania*: Its possible role in parasite sequestration and survival. J Cell Biochem 23:35–45.

Dwyer DM, Gottlieb M (1984): Surface membrane localization of 3'- and 5'-nucleotidase activities in *Leishmania donovani* promastigotes. Mol Biochem Parasitol 10:139–150.

Dwyer DM, Gottlieb M (1985): The biochemistry of *Leishmania* surface membranes. In K.-P. Chang and R.S. Bray (eds): Leishmaniasis. Amsterdam: Elsevier, pp 32–47.

Etges RJ, Bouvier J, Hoffman R, Bordier C (1985): Evidence that the major surface proteins of three Leishmania species are structurally related. Mol Biochem Parasitol 14:141–149.

Etges RJ, Bouvier J, Bordier C (1986): The major surface protein of *Leishmania* is a protease. J Biol Chem 261:9091–9098.

Gbenle GO, Opperdoes FR, Van Roy J (1986): Localization of 3'-nucleotidase and calcium dependent endoribonuclease in the plasma-membrane of *Trypanosoma brucei*. Acta Trop (Basel) 43:295–305.

Glew RH, Czuczman MS, Diven WF, Berens RL, Pope MT, Katsoulis DE (1982): Partial purification and characterization of particulate acid phosphatase of *Leishmania donovani* promastigotes. Comp Biochem Physiol [B] 72:581–590.

Gottlieb M (1985): Enzyme regulation in a trypanosomatid: Effect of purine starvation on levels of 3'-nucleotidase activity. Science 227:72–74.

Gottlieb M, Dwyer DM (1981a): Protozoan parasite of humans: Surface membrane with externally disposed acid phosphatase. Science 212:939–941.

Gottlieb M, Dwyer DM (1981b): *Leishmania donovani*: surface membrane acid phosphatase activity of promastigotes. Exp Parasitol 52:117–128.

Gottlieb M, Dwyer DM (1982a): Association of a membrane bound 3'-nucleotidase activity with both developmental stages of *Leishmania donovani*. Mol Biochem Parasitol [Suppl]:193.

Gottlieb M, Dwyer DM (1982b): Identification and characterization of an extracellular acid phosphatase activity of *Leishmania donovani* promastigotes. Mol Cell Biol 2:76-81.

Gottlieb M, Dwyer DM (1983): Evidence for distinct 5'- and 3'-nucleotidase activities in the surface membrane fraction of *Leishmania donovani* promastigotes. Mol Biochem Parasitol 7:303-317.

Gottlieb M, Zlotnick GW (1987): 3'-Nucleotidase of *Leishmania donovani*—evidence for exonuclease activity. In N. Agabian, H. Goodman, and N. Nogueira (eds): Molecular Strategies of Parasitic Invasion. UCLA Symposium on Molecular and Cellular Biology, New Series, Vol 42. New York: Alan R. Liss, Inc., pp 585-594.

Gottlieb M, Gardiner PR, Dwyer DM (1986): 3'-Nucleotidase activity in procyclic and bloodstream stages of *Trypanosoma rhodesiense*. Comp Biochem Physiol [B] 83:63-69.

Gottlieb M, Mackow MC, Neubert TA (1988): *Crithidia lociliae*: Factors affecting the expression of 3'-nucleotidase/nuclease activity. Exp Parasitol, in press.

Hammond DJ, Gutteridge WE (1984): Purine and pyrimidine metabolism in the trypanosomatidae. Mol Biochem Parasitol 13:243-261.

Handman E, Greenblatt CL, Goding JW (1984): An amphipathic sulphated glycoconjugate of *Leishmania*: Characterization with monoclonal antibodies. EMBO J 3:2301-2306.

Hansen BD, Perez-Arbelo J, Walkony JF, Hendricks LD (1982): The specificity of purine base and nucleoside uptake in promastigotes of *Leishmania braziliensis panamensis*. Parasitology 85:271-282.

Hanson DM, Fairley JL (1969): Enzymes of nucleic acid metabolism from wheat seedlings. J Biol Chem 244:2440-2449.

Harder W, Dijkhuizen L (1983): Physiological responses to nutrient limitation. Annu Rev Microbiol 37:1-23.

Hassan HF, Coombs GH (1987): Phosphomonoesterases of *Leishmania mexicana mexicana* and other flagellates. Mol Biochem Parasitol 23:285-296.

Iovannisci DM, Kaur K, Young L, Ullman B (1984): Genetic analysis of nucleoside transport in *Leishmania donovani*. Mol Cell Biol 4:1013-1019.

Jaffe CL, Bennett E, Grimaldi G, McMahon-Pratt D (1984): Production and characterization of species-specific monoclonal antibodies against *Leishmania donovani* for immunodiagnosis. J Immunol 133:440-447.

Kahl LP, McMahon-Pratt D (1987): Structural and antigenic characterization of a species- and promastigote-specific *Leishmania mexicana amazonensis* membrane protein. J Immunol 138:1587-1595.

Kaneshiro ES, Gottlieb M, Dwyer DM (1982): Cell surface origin of antigens shed by *Leishmania donovani* during growth in axenic culture. Infect Immun 37:558-567.

Law SS, Mukkada AJ (1979): Transport of L-proline and its regulation in *Leishmania tropica* promastigotes. J Protozool 26:295-301.

Lepay DA, Nogueira N, Cohn Z (1983): Surface antigens of *Leishmania donovani* promastigotes. J Exp Med 157:1562-1572.

Lepley PR, Mukkada AJ (1983): Characteristics of an uptake system for α-aminoisobutyric acid in *Leishmania tropica* promastigotes. J Protozool 30:41-46.

Lovelace JK, Gottlieb M (1986): Comparison of extracellular acid phosphatases from various isolates of *Leishmania*. Am J Trop Med Hyg 35:1121-1128.

Lovelace JK, Dwyer DM, Gottlieb M (1986): Purification and characterization of the extracellular acid phosphatase of *Leishmania donovani*. Mol Biochem Parasitol 20:243–251.

Marr JJ (1980): Carbohydrate metabolism in *Leishmania*. In M. Levandowsky and S.H. Hutner (eds): Biochemistry and Physiology of Protozoa. New York: Academic Press, Vol 3, pp 311–340.

Marr JJ, Berens R (1983): Pyrazolopyrimidine metabolism in the pathogenic trypanosomatidae. Mol Biochem Parasitol 7:339–356.

Marr JJ, Berens R (1985): Purine and pyrimidine metabolism in *Leishmania*. In K.-P. Chang and R.S. Bray (eds): Leishmaniasis. Amsterdam: Elsevier, pp 66–78.

Marsden PD, Jones TC (1985): Clinical manifestations, diagnosis and treatment of leishmaniasis. In K.-P. Chang and R.S. Bray (eds): Leishmaniasis. Amsterdam: Elsevier, pp 183–198.

McMahon-Pratt D, David JR (1981): Monoclonal antibodies that distinguish between New World species of *Leishmania*. Nature 291:581–583.

Mukkada AJ, Simon MW (1977): *Leishmania tropica*: uptake of methionine by promastigotes. Exp Parasitol 42:87–96.

Mukkada AJ, Schaefer FW III, Simon MW, Neu C (1974): Delayed *in vitro* utilization of glucose by *Leishmania tropica* promastigotes. J Protozool 21:393–397.

Pastakia KB, Dwyer DM (1987): Identification and characterization of a ribose transport system in *Leishmania donovani* promastigotes. Mol Biochem Parasitol 26:175–182.

Pereira NM, Konigk E (1981): A nucleotidase from *Leishmania tropica* promastigotes: Partial purification and properties. Trope Med Parasitol 32:209–214.

Remaley AT, Kuhns DB, Basford RE, Glew RH, Kaplan SS (1984): Leishmanial acid phosphatase blocks neutrophil $O_2$ production. J Biol Chem 259:11173–11175.

Remaley AT, Das S, Campbell PI, LaRocca GM, Pope MT, Glew RH (1985a): Characterization of *Leishmania donovani* acid phosphatases. J Biol Chem 260:880–886.

Remaley AT, Glew RH, Kuhns DB, Basford RE, Waggoner AS, Ernst LA, Pope MT (1985b): *Leishmania donovani*: Surface membrane acid phosphatase blocks neutrophils oxidative metabolite production. Exp Parasitol 60:331–341.

Russell DG, Wilhelm H (1986): The involvement of the major surface glycoprotein (gp63) of *Leishmania* promastigotes in attachment to macrophages. J Immunol 136:2613–2620.

Sacks DL, Perkins PV (1984): Identification of an infective stage of *Leishmania* promastigotes. Science 223:1417–1419.

Sacks DL, Hieny S, Sher A (1985): Identification of cell surface carbohydrate and antigenic changes between noninfective and infective developmental stages of *Leishmania major* promastigotes. J Immunol 135:564–569.

Schaefer FW III, Mukkada AJ (1976): Specificity of the glucose transport system in *Leishmania tropica* promastigotes. J Protozool 23:446–449.

Schaefer FW III, Martin E, Mukkada AJ (1974): The glucose transport system in *Leishmania tropica* promastigotes. J Protozool 21:592–596.

Shishido K, Ando T (1985): Single-strand-specific nucleases. In S.M. Linn and R.J. Roberts (eds): Nucleases. Cold Spring Harbor, NY: Cold Spring Harbor Laboratory Press, pp 155–185.

Simon MW, Mukkada AJ (1977): *Leishmania tropica*: regulation and specificity of the methionine transport system in promastigotes. Exp Parasitol 42:97–105.

Thompson LF (1985): Ecto-5′-nucleotidase can provide the total purine requirements of mitogen-stimulated human T cells and rapidly dividing human B lymphoblastoid cells. J Immunol 134:3794–3797.

Wassef MK, Fioretti TB, Dwyer DM (1985): Lipid analyses of isolated surface membranes of *Leishmania donovani* promastigotes. Lipids 20:108–115.

Zilberstein D, Dwyer DM (1984a): Glucose transport in *Leishmania donovani* promastigotes. Mol Biochem Parasitol 12:327–336.

Zilberstein D, Dwyer DM (1984b): Antidepressants cause lethal disruption of membrane function in the human protozoan parasite *Leishmania*. Science 226:977–979.

Zilberstein D, Dwyer DM (1985a): Protonmotive force driven active transport of D-glucose and L-proline in the protozoan parasite *Leishmania donovani*. Proc Natl Acad Sci USA 82:1716–1720.

Zilberstein D, Dwyer DM (1985b): Antiprotozoal activity of tricyclic compounds. Science 230:1064.

Zilberstein D, Dwyer DM, Matthaei S, Horuk R (1986): Identification and biochemical characterization of the plasma membrane glucose transporter of *Leishmania donovani*. J Biol Chem 261:15053–15057.

Zilberstein D, Sheppard HW, Dwyer DM (1987): The plasma membrane H+-ATPase of *Leishmania donovani* promastigotes. In K.-P. Chang and D. Snary (eds): Host-Parasite Cellular and Molecular Interactions in Protozoal Infections. NATO ASI Series. Heidelberg: Springer-Verlag, pp 183–188.

Zlotnick GW, Mackow MC, Gottlieb M (1987): Renaturation of *Leishmania donovani* 3'-nucleotidase following sodium dodecyl sulfate electrophoresis. Comp Biochem Physiol, [B] 87:629–635.

The Biology of Parasitism, pages 467–477
© 1988 Alan R. Liss, Inc.

# Genetics of Malaria Parasites

**David Walliker**

*Department of Genetics, University of Edinburgh, Edinburgh EH9 3JN, Scotland*

## INTRODUCTION

Most current research on malaria is aimed at finding new ways of treating and controlling the disease. This has become necessary largely because of the widespread occurrence of drug-resistant forms of the parasite; such resistance is especially common in *Plasmodium falciparum*, the most pathogenic species of malaria infecting man. In the past 10 years there has been a change of emphasis from research into new antimalarial drugs to the development of vaccines. This has resulted in the identification of parasite antigens thought to be important in stimulating protective immunity. However, considerable diversity in some antigens has been detected, and this is causing concern that vaccines may quickly become ineffective due to the selection and spread of parasites with novel antigenic forms.

It can be expected that variant forms of characters such as antigens and drug-sensitivity are determined by genes which undergo processes such as mutation, recombination, etc. The techniques of molecular biology have been used to isolate genes for many *P. falciparum* antigens, and this approach has shown the extent to which variant forms of a given antigen may differ in their amino acid sequences. Genetic analysis shows how the genes for variant characters are inherited and recombine with each other to produce parasites with novel genotypes. The subject is thus of crucial importance if we are to understand how parasites will respond to chemical treatments or vaccines. However, remarkably little genetic work has been carried out on these organisms despite the fact that some malaria vaccines are already being tested in man (Ballou et al., 1987; Herrington et al., 1987). The purpose of this paper is to illustrate the relevance of the subject to our understanding of the basic biology of the parasite.

## GENOME ORGANIZATION IN MALARIA PARASITES

Malaria parasites, being eukaryotic organisms, are likely to have their genetic material arranged as discrete chromosomes. Cytological studies using conventional light microscopy have provided disappointingly little information on the arrangement of the genome, probably because the chromosomes are too small to be detected. However, electron microscopy of both blood and mosquito stages has revealed the presence of kinetochores, which are the attachment points of chromosomes to spindle microtubules of dividing cells. In serial sections of schizonts, Prensier and Slomianny (1986) demonstrated 14 kinetochores, and they concluded that this is the haploid chromosome number. Sinden and Hartley (1985) detected synaptomemal complexes, characteristic of meiotically dividing chromosomes, during the early divisions of the zygote in the mosquito midgut. Genetic work on the inheritance of isoenzymes in the rodent malaria species *P. chabaudi* has established that the blood forms are haploid (Walliker et al., 1975). The parasite therefore undergoes a mainly haploid life-cycle, the diploid phase being probably restricted to the ookinete stage in mosquitoes.

A recent alternative approach to the study of chromosomes has been the use of pulsed field gradient gel electrophoresis (PFG) to separate large fragments of parasite DNA. Fourteen such fragments have been identified by this method (Wellems et al., 1987), and on the assumption that these represent intact chromosomes, the numbers are in agreement with the cytological evidence described above.

## GENETIC DIVERSITY IN MALARIA PARASITES

Evidence for variation between individual parasites of a species of *Plasmodium* was formerly based on characters such as blood form morphology and infection patterns (Garnham, 1966). In more recent times, major advances in our ability to differentiate parasites have come from the development of biochemical and immunological techniques for detecting variation in proteins, including a) enzyme electrophoresis (Sanderson et al., 1981), b) protein variation revealed by two-dimensional electrophoresis (2D-PAGE) (Fenton et al., 1985), and c) antigen diversity revealed by monoclonal antibodies (McBride et al., 1985). Variations in genomic structure between different parasites have been revealed by the application of recombinant DNA techniques to *Plasmodium*. For example, repetitive DNA sequences show marked differences in hybridization patterns in different isolates (Goman et al., 1982), chromosomes exhibit marked size variations (Kemp et al., 1985; Van der Ploeg et al., 1985), and the base sequences of some antigen genes show considerable diversity, notably in regions coding for repeated epitopes (Kemp et al., 1986).

The genetic diversity which occurs within a species is illustrated here by enzyme and antigen variants in *P. falciparum*. It is considered in two ways:

a) the extent of diversity in the parasite population, and b) the extent of diversity which may occur in a single host.

## Diversity in the Parasite Population

Table 1 illustrates the electrophoretic forms of two enzymes, glucose phosphate isomerase (GPI) and adenosine deaminase (ADA), occurring in blood forms of *P. falciparum* isolates of The Gambia, Thailand, and Brazil. Two allelic forms of each enzyme occur, GPI-1 and -2. Both forms of each enzyme are found in parasites of all three countries, but there are some differences in the frequencies with which they occur. Thus, in The Gambia and Thailand, parasites characterized by type 1 of each enzyme are more frequent than those of type 2. In Brazil, however, parasites characterised by type 2 of each enzyme are more common.

Table 2 illustrates the distribution of seven allelic variants (serotypes) of a major surface antigen of schizonts and merozoites called p190, distinguished by strain-specific monoclonal antibodies (McBride et al., 1985), in parasites

**TABLE 1. Enzyme Variation in *Plasmodium falciparum* of Three Countries[a]**

| | GPI | | | | ADA | | | |
|---|---|---|---|---|---|---|---|---|
| | No. examined | 1 (%) | 2 (%) | 1 + 2 (%) | No. examined | 1 (%) | 2 (%) | 1 + 2 (%) |
| The Gambia | 172 | 63 | 10 | 27 | 53 | 90 | 4 | 4 |
| Thailand | 176 | 62 | 22 | 16 | 135 | 94 | 3 | 3 |
| Brazil | 110 | 32 | 65 | 3 | 103 | 29 | 69 | 2 |

[a]Nos. represent electrophoretic forms of each enzyme. Percentages represent the proportion of isolates examined showing each enzyme form.

**TABLE 2. Frequencies of Serotypes I–VII of p190 Schizont Antigen of *Plasmodium falciparum* in Three Countries[a]**

| | Frequencies (% of isolates examined in each country) | | |
|---|---|---|---|
| Serotype | Thailand | Brazil | The Gambia |
| I | 28 | 4 | 2 |
| II | 25 | 41 | 26 |
| III | 8 | 29 | 16 |
| IV | 3 | 7 | 18 |
| V | 9 | — | 11 |
| VI | 3 | 12 | 16 |
| VII | 24 | 8 | 11 |

[a]Serotypes are defined by reactions of the antigen with a panel of monoclonal antibodies (McBride et al., 1985).

of the same three countries. As with the enzymes, variations occur in the frequencies of each form; serotype 1, for example, is rarely found in The Gambia and Brazil but is comparatively common in Thailand.

*P. falciparum* in different parts of the world, therefore, possesses similar enzyme and antigen forms, providing strong evidence that it is a single species worldwide. Variations in the frequencies of the forms of each character are presumably due to different selection pressures on the parasites in each region, the nature of which are unknown at present.

## Parasite Diversity in a Single Host

In the enzyme data shown in Table 1, it can be seen that a considerable number of isolates possess more than one form of a given enzyme. For example, in The Gambia, 27% of isolates are characterized by GPI-1 and -2. As the blood forms of *Plasmodium* are haploid and each parasite thus possesses only a single allele of each gene, such isolates represent mixed infections of at least two types of organism, some characterised by GPI-1 and some by GPI-2. This has been confirmed by establishing clones from certain isolates. For example, Thaithong et al. (1984) established clones from an isolate from Thailand called T9, which exhibited two forms of three enzymes, was positive for five monoclonal antibodies recognising the p190 schizont antigen, and exhibited resistance to both pyrimethamine and chloroquine. Each clone exhibited only single forms of each enzyme, and single serotypes of the p190 antigen. Certain clones exhibited sensitivity to pyrimethamine and to chloroquine. On cloning, five genetically distinct types of parasite were found (Table 3).

## THE GENETIC BASIS OF VARIANT CHARACTERS IN *PLASMODIUM*

For a conventional genetic analysis, crosses are made between cloned parasite lines differing by the characters under investigation, in order to examine their patterns of inheritance. During the malaria life cycle, gametocytes are produced in the blood of the vertebrate host, while fertilization between gametes occurs in the mosquito vector. To carry out a cross, therefore, mosquitoes are fed on blood containing a mixture of gametocytes of two clones to permit fertilization of gametes to occur (Fig. 1). Assuming that random fertilizations occur, equal numbers of parent-type and hybrid zygotes are formed. The resulting sporozoites are then used to infect a suitable new host, and the blood forms which develop are cloned and examined for the characters distinguishing the parent clones. The presence of parasites exhibiting nonparental characteristics shows that hybridization

TABLE 3. Enzymes, Drug-Sensitivities and Antigen Forms of Five Clones Derived From Isolate T9 of *P. falciparum* (Thaithong et al., 1984)

| Characters | Uncloned T9 | T9 clones | | | | |
|---|---|---|---|---|---|---|
| | | 94 | 96 | 98 | 101 | 102 |
| Enzymes[a] | | | | | | |
| GPI | 1 + 2 | 1 | 2 | 2 | 1 | 1 |
| ADA | 1 + 2 | 1 | 1 | 1 | 2 | 1 |
| PEP | 1 + 2 | 1 | 2 | 2 | 1 | 1 |
| Drug sensitivity (MIC) | | | | | | |
| Chloroquine | 1.0 | 0.5 | 0.1 | 0.1 | 1.0 | 1.0 |
| Pyrimethamine | 50 | 0.001 | 0.001 | 50 | 50 | 50 |
| P190 antigen | | | | | | |
| Monoclonal | | | | | | |
| antibodies 12.4 | + | + | + | + | + | + |
| 7.3 | + | + | − | − | − | − |
| 9.2 | + | − | + | + | + | + |
| 9.5 | + | − | − | − | + | − |
| 12.2 | + | − | + | + | − | − |
| 12.1 | + | − | + | + | − | + |

[a]GPI = gluclose phospate isomerase; ADA = adenosine deaminase; PEP = peptidase; MIC = minimum inhibitory concentration of drug $\times 10^{-6}$M. P190 monoclonal antibodies tested by immunofluorescence: +, positive; −, no fluorescence.

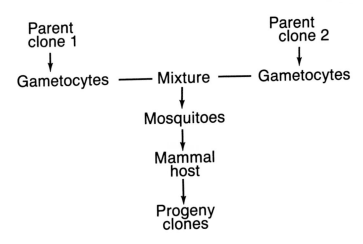

Fig. 1. Procedure for crossing mammalian malaria parasites.

has occurred. The technique has been used extensively to study the genetics of the rodent malaria species *Plasmodium yoelii* and *P. chabaudi* (Walliker, 1983), and more recently *P. falciparum* (Walliker et al., 1987). Examples are given here of genetic studies on variant proteins, antigens, and drug sensitivity.

## Genetics of Variant Proteins

Crossing experiments have demonstrated clearly that the electrophoretic forms of isoenzymes and of certain proteins detected by 2D-PAGE are determined by allelic variation of their respective genes. Table 4 illustrates the inheritance of forms of the enzyme adenosine deaminase and of four 2D-PAGE proteins in a cross between two *P. falciparum* clones (Walliker et al., 1987; B. Fenton, unpublished results). The parent clones possessed different forms of each character. Progeny clones derived from the cross exhibited only single forms of each protein, and in the majority of cases these were in different combinations from those of the parents. Thus, segregation of each form of each protein had occurred, as well as recombination between the genes determining them. These patterns of inheritance were those expected for allelic variants of unlinked genes in a haploid organism.

**TABLE 4. Enzyme (ADA) and 2D-PAGE Protein Characteristics of Parents and Nine Progeny Clones of *P. falciparum* Cross[a]**

| Parasites | ADA type | 2D-PAGE proteins | | | | |
|---|---|---|---|---|---|---|
| | | C | D | G | K | |
| Parent clones | | | | | | |
| 3D7 | 1 | 3 | 2 | 1 | 3 | |
| HB3 | 2 | 7 | 4 | 2 | 1 | |
| Progeny clones | | | | | | |
| X8 | 1 | 3 | 2 | 1 | 3 | Parent-type |
| X10 | 2 | 7 | 4 | 2 | 1 | Parent-type |
| X2 | 1 | 3 | 4 | 2 | 1 | Recombinant |
| X4 | 2 | 7 | 2 | 1 | 3 | Recombinant |
| X6 | 2 | 7 | 4 | 2 | 3 | Recombinant |
| X11 | 1 | 7 | 4 | 1 | 3 | Recombinant |
| X12 | 1 | 7 | 4 | 2 | 3 | Recombinant |
| XP1 | 2 | 7 | 2 | 2 | 3 | Recombinant |
| XP9 | 2 | 7 | 2 | 1 | 1 | Recombinant |

[a]2D-PAGE proteins are indicated by capital letters, and variant forms of each protein by Nos. (Fenton et al., 1985; and unpublished work).

## Genetics of Antigens

The variation seen in the p190 antigen of *P. falciparum* mentioned above has been shown to be due to allelic variation of a single gene, both by genetic and molecular studies (Walliker et al., 1987; Tanabe et al, 1987). In genetic work, two clones were used which exhibited different forms of the antigen detectable by two monoclonal antibodies—7.3 and 9.2. In clone 3D7, the antigen was 7.3− and 9.2+, while in clone HB3 it was 7.3+ and 9.2−. Following a cross between the clones, each progeny clone possessed the antigen type of one or other parent. No parasites were detected which were 7.3+,9.2+ or 7.3−,9.2−; such forms could have been produced by recombination had the genes determining the antigen been at different loci in eaph parent. The allelic nature of the various forms of p190 seen in *P. falciparum* has been confirmed by sequencing studies on the gene from four different clones (Tanabe et al., 1987). Only single copies of the gene were present in each clone. Regions of similarity as well as difference were seen in the sequence of each allele, and there was evidence that the diverse forms of the antigen seen in the parasite population could be generated by intragenic recombination.

Another group of highly diverse malaria antigens are the S-antigens, first described in the serum of patients infected with *P. falciparum* in The Gambia (Wilson et al., 1969). Studies on their molecular biology have shown that they vary considerably in size, and contain long sequences of tandemly repeated groups of amino acids, which differ markedly from one isolate to another (Cowman et al., 1985). No genetic studies on these antigens have yet been carried out, but it is possible that the repeated sequences provide sites for intragenic recombination, which could generate molecules of novel size.

## Genetics of Drug Resistance

An example of a phenotypic character under the control of genes at more than one locus is provided by resistance to the drug chloroquine. In the rodent species *P. chabaudi*, Padua (1981) crossed a cloned line exhibiting high-level resistance, obtained by prolonged drug selection in the laboratory, with a drug-sensitive clone. Among the progeny of the cross, she detected many clones which exhibited a level of resistance intermediate between that of the parents. This was most easily explained by assuming that a succession of mutations had occurred at different gene loci in the drug-selected line, each conferring a low level of resistance, but additively producing high resistance. During meiosis of hybrid zygotes, the various resistance genes had undergone recombination, thus producing the intermediate forms of resistance.

## GENETIC RECOMBINATION, LINKAGE, AND CHROMOSOMAL VARIATION

The results of the crossing experiments have demonstrated that genetic recombination occurs readily in malaria parasites, during meiosis in the mosquito phase of the life cycle. Recombination may be the result of simple reassortment of unlinked genes on separate chromosomes, or of crossing-over events between linked genes on homologous chromosomes. In the crossing experiments carried out so far, recombination has been detected between each character studied, and thus closely linked genes have not been identified.

An alternative way of studying linkage is that of chromosome mapping using the PFG technique. DNA probes for *Plasmodium* genes can be hybridised to blots of PFG gels to determine their chromosomal locations. This approach has shown that, for example, the genes for two antigens known as RESA and Ag169 are located on chromosome 1 (Kemp et al., 1986), and for the sporozoite antigen (circumsporozoite [CSP]) on chromosome 3 (Walliker et al., 1987) Considerable variations in the sizes of homologous chromosomes are found in different laboratory-adapted cultures of *P. falciparum* and in parasites isolated directly from infected patients; some of these size variations have been correlated with the presence or absence of certain genes (Corcoran et al., 1986). Following mosquito transmission of a mixture of two *P. falciparum* clones in which homologous chromosomes differed in size, many progeny contained chromosomes of novel sizes, which were probably produced by unequal crossing-over events at meiosis (Walliker et al., 1987) (Fig. 2). Thus, genomic rearrangements appear to occur readily during the mosquito phase of the life cycle, allowing recombination between genes on homologous chromosomes to occur frequently. This may explain the high frequencies of recombination detected in crossing experiments.

## CONCLUSIONS

The principal findings which are emerging from genetic studies on malaria parasites can be summarised as follows:-

i. Considerable diversity in characters such as enzymes and drug sensitivity occurs in *P. falciparum* populations.

ii. Mixed infections with more than one genetically distinct organism are common in naturally infected patients.

iii. When mosquitoes transmit mixtures of genetically distinct parasites, cross-fertilization of gametes occurs readily.

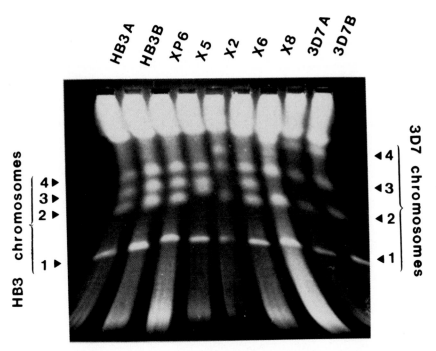

Fig. 2. Chromosomes 1–4 of parents and five progeny clones derived from cross between clones 3D7 and HB3 of *P. falciparum*, detected by pulsed field gradient gel electrophoresis. Note non-parental-type chromosomes in progeny X5 (chromosome 2) and X2 (chromosome 4). (Reprinted from Walliker et al., 1987, with the permission of the publisher; copyright 1987 by the AAAS.)

iv. During meiosis of the zygote in mosquitoes, recombination between different genes occurs frequently, and considerable rearrangements of the genome may take place. This can result in the production of parasites which are genetically very different from their parents. Thus, the resulting sporozoites which infect a new host may be of different genotypes from those which gave rise to the mosquito infection.

v. From these findings, it can be predicted that a given parasite clone exists in nature for only a short time. Such a clone is likely to coexist in a patient with other, genetically different, parasites which, on passage through mosquitoes, yield novel genotypes.

In the past, malaria has usually been considered as a collection of distinct "strains." It is customary, for example, to consider the spread of drug resistance in *P. falciparum* as the replacement of sensitive by resistant parasite "strains." In reality, we should consider malaria parasites as a population of organisms in which interbreeding by random mating of gametes in mosquitoes probably occurs at high frequency, resulting in the production of novel gene combinations. The spread of drug resistance, therefore, should be perceived as the spread of resistance genes rather than of resistant strains. Variations in the level of resistance among parasites are determined by different combinations of the genes which affect the resistance character.

Much of the current research into new antimalarial drugs and vaccines is carried out with cloned *P. falciparum* blood forms cultured in the laboratory over long periods. In future, it will be necessary to develop model systems which include the whole parasite life cycle, to take account of the parasite's capacity for genetic recombination.

## REFERENCES

Ballou RS, Hoffman SL, Sherwood JA, Hollingdale MR, Neva FA, Hockmeyer WT, Gordon DM, Schneider I, Wirtz RA, Young JF, Wasserman GF, Reeve P, Diggs CL, Chulay JF (1987): Safety and efficacy of a recombinant DNA Plasmodium falciparum sporozoite vaccine. Lancet i:1277–1281.

Corcoran LM, Forsyth KP, Bianco AE, Brown GV, Kemp DJ (1986): Chromosome size polymorphisms in Plasmodium falciparum can involve deletions and are frequent in natural parasite populations. Cell 44:87–95.

Cowman AF, Saint RB, Coppel RL, Brown GV, Anders RF, Kemp DJ (1985) Conserved sequences flank variable tandem repeats in two S-antigen genes of Plasmodium falciparum. Cell 40:775–783.

Fenton B, Walker A, Walliker D (1985): Protein variation in clones of Plasmodium falciparum detected by two dimensional electrophoresis. Mol Biochem Parasitol 16:173–183.

Garnham PCC (1966): Malaria Parasites and Other Haemosporidia. Oxford: Blackwell Scientific Publications.

Goman M, Langsley G, Hyde JE, Yankovsky NK, Zolg JW, Scaife JG (1982): The establishment of genomic DNA libraries for the human malaria parasite Plasmodium falciparum and identification of individual clones by hybridisation. Mol Biochem Parasitol 5:391–400.

Herrington DA, Clyde DF, Losonsky G, Cortesia M, Murphy JR, Davis J, Baqar S, Felix AM, Heimer EP, Gillessen D, Nardin E, Nussenzweig RS, Nussenzweig V, Hollingdale M, Levine MM (1987): Safety and immunogenicity in man of a synthetic peptide malaria vaccine against Plasmodium falciparum sporozoites. Nature 328:257–259.

Kemp DJ, Corcoran LM, Coppel RL, Stahl HD, Bianco AE, Brown GV, Anders RF (1985): Size variations in chromosomes from independent cultured isolates of Plasmodium falciparum. Nature 315:347–350.

Kemp DJ, Coppel RL, Stahl HD, Bianco AE, Corcoran LM, McIntyre P, Langford CJ, Favaloro JM, Crewther PE, Brown GV, Mitchell GF, Anders RF (1986): Genes for antigens of Plasmodium falciparum. Parasitology 92:S83–S108.

McBride JS, Newbold CI, Anand R (1985): Polymorphism of a high molecular weight schizont antigen of the human malaria parasite Plasmodium falciparum. J Exp Med 161:160–180.

Padua RA (1981): Plasmodium chabaudi: Genetics of resistance to chloroquine. Exp Parasitol 52:419–426.

Prensier G, Slomianny Ch (1986): The karyotype of Plasmodium falciparum determined by ultrastructural serial sectioning and 3D reconstruction. J Parasitol 72:731–736.

Sanderson A, Walliker D, Molez J-F (1981): Enzyme typing of Plasmodium falciparum from some African and other old world countries. Trans R Soc Trop Med Hyg 75:263–267.

Sinden RE, Hartley RH (1985): Identification of the meiotic division of malarial parasites. J Protozool 32:742–744.

Tanabe K, Mackay M, Goman M, Scaife JG (1987): Allelic dimorphism in a surface antigen gene of the malaria parasite Plasmodium falciparum. J Mol Biol 195:273–287.

Thaithong S, Beale GH, Fenton B, McBride JS, Rosario V, Walker A, Walliker D (1984): Clonal diversity in a single isolate of the malaria parasite Plasmodium falciparum. Trans R Soc Trop Med Hyg 78:242–245.

Van der Ploeg LHT, Smits M, Ponnudurai T, Vermeulen A, Meuwissen JHETh, Langsley G (1985): Chromosome-sized DNA molecules of Plasmodium falciparum. Science 229:658–661.

Walliker D (1983): The genetic basis of diversity in malaria parasites. Adv Parasitol 22:217–259.

Walliker D, Carter R, Sanderson A (1975): Genetic studies on Plasmodium chabaudi: Recombination between enzyme markers. Parasitology 70:19–24.

Walliker D, Quakyi IA, Wellems TE, McCutchan TF, Szarfman A, London WT, Corcoran LM, Burkot TM, Carter R (1987): Genetic analysis of the human malaria parasite Plasmodium falciparum. Science 236:1661–1666.

Wellems TE, Walliker D, Smith CL, Rosario VE, Maloy WL, Howard RJ, Carter R, McCutchan TF (1987): A histidine-rich protein gene marks a linkage group favored strongly in a genetic cross of Plasmodium falciparum. Cell 49:633–642.

Wilson RJM, McGregor IA, Hall P, Williams K, Bartholomew R (1969): Antigens associated with Plasmodium falciparum infections in man. Lancet ii:201–205.

The Biology of Parasitism, pages 479–501
© 1988 Alan R. Liss, Inc.

# *Toxoplasma gondii* Viewed From a Virological Perspective

**E.R. Pfefferkorn**

*Department of Microbiology, Dartmouth Medical School, Hanover,
New Hampshire 03756*

## BECOMING A PARASITOLOGIST

Just over a decade ago, I abandoned my research in animal virology and turned to the study of obligate intracellular protozoan parasites. This decision was based on the hope that much of the intellectual capital that had been accumulated by a generation of biochemically oriented animal virologists could be profitably invested in comparable studies of intracellular protozoa. This account of successes and failures will show that my experimental outlook is deeply rooted in animal virology. My choice of a parasite was determined by a chance event of three decades earlier. During an exceptionally well taught Harvard Medical School course in parasitology, Thomas Weller mentioned that he was growing *Toxoplasma gondii* in cultured cells. Since the key to much of the success of animal virology was the switch from in vivo experiments to tissue culture, *T. gondii* seemed an ideal protozoan parasite to study by the methodology of animal virology. In retrospect this choice, although based more upon hope than upon knowledge, proved to be reasonable. *Toxoplasma gondii* is undoubtedly the easiest obligate intracellular protozoan parasite to study in cultured cells.

## INTRODUCTION TO *TOXOPLASMA GONDII*

A brief review of the life cycle of *T. gondii* is essential to understanding some of its advantages as a model intracellular protozoan parasite. The stage that multiplies in cultured cells is the tachyzoite, the rapidly growing asexual form. The tachyzoite enters its host cell by invagination of the plasma membrane and comes to lie in the cytoplasm within a vacuole, the parasitophorous vacuole. Within this ever-expanding vacuole, the tachyzoites divide every 5 to 7 h. The parasitophorous vacuole and its membrane must play a critical role in the intracellular growth of *T. gondii*. Mitochondria of the host cell are generally found packed against the vacuolar membrane (Hirsh et al., 1974). All biochemical communication between host cell and parasite crosses this membrane, and the vacuolar fluid must provide a suitable environment for the growth of *T. gondii*. Unfortunately nothing is known of the composition of the vacuolar membrane, and apart from the important observation that it is not acidified (Sibley et al., 1985), little is known of the vacuolar fluid. By 2 days after infection, the host cell lyses and releases progeny parasites that go on to infect additional cells in the culture. Tachyzoites are also seen in acutely infected animals. As the in vivo infection progresses, some tachyzoites, particularly in muscle and brain cells, are transformed into the bradyzoite stage, which is characterized by essentially dormant parasites within a cyst, the wall of which is at least partially of host cell origin. An animal that survives infection with *T. gondii* probably retains encysted bradyzoites for life. When these bradyzoites are ingested by a nonfeline carnivore, they are transformed into tachyzoites to produce an acute infection that resolves into a chronic state, again characterized by encysted bradyzoites. However when bradyzoites are ingested by any member of the cat family, they differentiate, through several intermediate steps in intestinal epithelial cells, into microgametes (male) and macrogametes (female). The gametes fuse to form a zygote, which synthesizes for itself a rigid impermeable wall and is excreted in the feces as an unsporulated oocyst. At ambient temperature the single cell within this oocyst goes through three divisions to yield eight sporozoites. The resulting sporulated oocyst is infectious. A nonimmune animal accidentally eating sporulated oocysts from cat fecal contamination will become infected. The sporozoites are released from the oocyst in the intestine and multiply intracellularly as tachyzoites to produce a systemic acute infection. Thus the cycle is completed.

In tissue cultures, this complex life cycle does not take place. Under most conditions, only the rapidly multiplying tachyzoites are produced. Thus, only one stage of the life cycle of *T. gondii* can be studied in vitro. However, the tachyzoite provides an easily manipulated homogeneous population of obligate intracellular parasites in which to explore various aspects of the host-parasite relationship using some of the methodology originally developed by animal virologists. One of the key developments in the study of animal viruses was the plaque assay. Work in the laboratory of another animal virologist had already adapted the animal virus plaque assay for use with *T. gondii*. We simplified the procedure by omitting the agar overlay. This overlay is essential in the plaque assay of viruses because the progeny virions would otherwise diffuse through the liquid medium and destroy the entire culture. *Toxoplasma gondii*, however, produces discrete plaques even when the cultured cells are incubated in liquid medium. Slowly dividing contact-inhibited cultures yield the best plaques. In unstained human fibroblasts, plaques of *T. gondii* can easily be counted 1 week after infection (Pfefferkorn and Pfefferkorn, 1976). A simple plaque assay offers not only reliable quantitation of infectious parasites but also easy cloning, a critical step in the genetic experiments described below.

## MUTANT HOST CELLS

Animal virologists have long used mutant host cells to study specific problems. For example, the first conclusive evidence that herpesviruses encode their own thymidine kinase came from enzyme assays of infected mutant cells that lacked this enzyme. We have used mutant cells to examine various aspects of the host-parasite relationship of *T. gondii*. The main advantage of this approach is that it allows the examination of growth of the parasite under one additional constraint. Our first use of mutant host cells solved a simple biochemical problem. Protein and nucleic acid syntheses of many virulent animal viruses are easy to study because infection effectively shuts down these host cell functions. Unfortunately infection of human fibroblasts with *T. gondii* has no such effect. Infected cells even proceed through mitosis and cell division. Thus radioactive precursors supplied to parasitized cultures largely label host cell proteins and nucleic acids, especially during the early hours of infection. We achieved specific labeling of parasite nucleic acids by using fibroblasts from a patient with the Lesch-Nyhan syndrome (Pfefferkorn and Pfefferkorn, 1977a,b). Since these mutant cells lack the salvage enzyme hypoxanthine-guanine phosphoribosyltransferase, they cannot incorporate hypoxanthine. The autoradiographic patterns of [3H]-hypoxanthine labeling of infected normal and Lesch-Nyhan fibroblasts are compared in Figure 1. In the mutant cultures all of the precursor is incorporated by the intracellular parasites. This autoradiograph also shows that the purine nucleotides of the parasite are unavailable to the host cell. If they were available, the cytoplasm and nuclei of infected host cells would have been labeled by the [3H]-ATP made by the parasite from [3H]-hypoxanthine.

The only disadvantage of using Lesch-Nyhan cells for the specific labeling of *T. gondii* nucleic acids is that a mutant host cell is required. Two alternative nucleic acid precursors allow the use of wild-type cells. Radioactive uracil can be used specifically to label the pyrimidines in the nucleic acids of

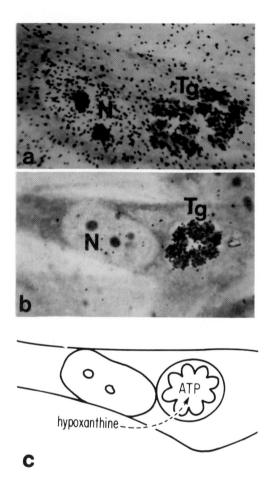

Fig. 1. Autoradiographic analysis of the incorporation of [³H]-hypoxanthine by fibroblasts labeled from 20 to 24 h after infection with wild-type *T. gondii*. **a:** Normal human fibroblasts. **b:** Lesch-Nyhan fibroblasts. **c:** Drawing based on the cell in b to emphasize that ATP of the parasites is unavailable to the host cell, which remains unlabeled. Tg = *T. gondii*; N = host cell nucleus. Bar is 10 μM. (from Pfefferkorn and Pfefferkorn, 1977a).

intracellular *T. gondii* because mammalian host cells lack the salvage enzyme uracil phosphoribosyltransferase. Alternatively, guanine of the parasite's nucleic acids can be specifically labeled with radioactive xanthine because the hypoxanthine-guanine phosphoribosyltransferase of the parasite recognizes xanthine as a substrate while that of the host cell does not (unpublished).

Mutant host cells are also valuable in determining the biosynthetic capability of an intracellular parasite. The ability to synthesize a given metabolite can readily be determined for a free-living protozoan. If the metabolite can be synthesized, the protozoan will grow in medium in which it is lacking. During growth the free-living protozoan will convert a labeled precursor to the metabolite. Neither criterion is valid for an obligate intracellular parasite, because the host cell could be synthesizing the metabolite in question and exporting it to the parasite. The easiest way to attack this question is selectively to blot out a biosynthetic pathway in the host cell by using an appropriate mutant. A good many such mammalian mutants are available, especially in Chinese hamster ovary (CHO) cells. We infected mutant CHO cells that were incapable of pyrimidine synthesis and labeled them with [$^{14}$C]-aspartate, a major precursor of the pyrimidine ring. Analyses showed that the growing parasites had incorporated the [$^{14}$C]-aspartate into their nucleic acid pyrimidines, and were thus capable of synthesizing them de novo (Schwartzman and Pfefferkorn, 1981).

The opposite result was obtained when we used a purine-requiring CHO mutant as the host cell. In these studies we used [$^{14}$C]-formate as the purine precursor. *Toxoplasma gondii* grew normally in these mutant cells in medium that contained no purines. But the purines of the parasites contained no radioactivity, showing that the parasite was incapable of de novo synthesis of purines and must have obtained these essential metabolites from its host cell (Schwartzman and Pfefferkorn, 1982). The form in which host cell purines are passed to intracellular *T. gondii* remains unknown. We have shown that freshly prepared extracellular parasites, which are briefly capable of limited nucleic acid synthesis, can incorporate exogenous ATP into RNA. Closer examination showed that all three phosphates were removed before incorporation. However, the fact that *T. gondii*, in contrast to mammalian cells, uses exogenous ATP efficiently suggested that host cell ATP might be the source of the purines that the parasite required. This suggestion led to speculation that the host cell mitochondria that cluster about the parasitophorous vacuole are there for a purpose: They might secrete into the vacuolar space the ATP that the parasite required as its source of purines. This theory was open to an experimental test through the use of a mutant CHO cell whose mitochondria cannot produce ATP because of a defect in oxidative phosphorylation. *Toxo-*

*plasma gondii* grew at the normal rate in these mutant cells. Thus synthesis of ATP by mitochondria of the host cell is not critical for the growth of *T. gondii* (Schwartzman and Pfefferkorn, 1982).

We have also used mutants to determine the importance of DNA, RNA, and protein synthesis by the host cell for the intracellular growth of *T. gondii.* The mutant host cells in this case were each temperature-sensitive mutants with the appropriate defect. Each mutant gave the same result: *Toxoplasma gondii* grew at the restrictive temperature and is thus not dependent on the synthesis of macromolecules by its host cell.

## MUTANTS OF *TOXOPLASMA GONDII*

Much of our understanding of the multiplication of animal viruses has come through the study of viral mutants. For a number of simple viruses, every gene has been identified through the study of mutants. The most valuable animal virus mutants have probably been temperature sensitive. These are mutants that have lost the ability to grow at a high (restrictive) temperature but retained the ability to grow at a low (permissive) tempera-ture. The biochemical basis of temperature sensitivity is an altered amino acid acid sequence. The mutant protein is either denatured at the restrictive temperature or is unable to assume its normal functional configuration when synthesized at this temperature. Our first mutants of *T. gondii* were temper-ature sensitive (Pfefferkorn and Pfefferkorn, 1976). Since there is no conve-nient selective pressure for these mutants, they must be isolated by brute-force methodology: parasites cloned by plaque formation at the permissive temperature are individually scored for the ability to grow at the restrictive temperature. With the aid of chemical mutagenesis, success is guaranteed since all of the many parasite genes that encode proteins are targets for temperature sensitive mutations. We find that about 1% of the survivors of mutagenesis are temperature sensitive. Although we have a large catalogue of temperature-sensitive mutants, they have not proven to be as useful as we had hoped. In viruses that have a small number of genes, assignment of a function to a gene bearing a temperature-sensitive mutation has proven to be feasible. However, generality of this type of mutation and the large genome size of *T. gondii* has made the determination of specific defects most difficult. In short, we have not determined the precise defect of any of our mutants.

Many of our temperature-sensitive mutants of *T. gondii* are avirulent in mice. This is remarkable because they were isolated from the hypervirulent RH strain. Presumably, this avirulence is a result of restriction of mutant parasite growth at core body temperature of the mouse. In collaboration with

Dr. J.K. Frenkel, we have used the avirulence of temperature-sensitive mutants to design a veterinary vaccine for toxoplasmosis that has the additional useful property of being incapable of producing a chronic infection because of a defect in bradyzoite formation.

There is little evidence that *T. gondii* exhibits the antigenic diversity displayed by some other parasitic protozoa, notably the African trypanosomes. Various isolates of the parasite do not appear to have marked antigenic differences and there is no evidence of successive waves of parasitemia in the course of experimental infections of animals. We selected an antigenic mutant of *T. gondii* by modifying a technique first used by animal virologists. These mutants of animal viruses can be isolated through the use of monoclonal antibodies that neutralize infectivity. Antigenic mutants are readily selected because they are unaffected by the antibody. Characteristically, these viral mutants have lost the epitope recognized by the neutralizing monoclonal antibody in the wild-type virus. We have never observed significant neutralization of *T. gondii* by monoclonal or polyclonal antisera, using a sensitive test that depends on the reduction of plaque titer. Thus we had to use an alternative lethal effect of monoclonal antibodies on the parasite. We chose to use complement, which has a marked lytic effect on *T. gondii* in the presence of antibodies directed against the surface antigens of the parasite. Suitably mutagenized extracellular parasites were treated briefly with a monoclonal antibody and complement. This treatment did not eliminate all of the wild-type parasites because some 1% of them were phenotypically resistant and failed to lyse. The basis for this resistance is unclear but is clearly not inherited because the progeny of the apparently resistant *T. gondii* were again 99% sensitive to complement-mediated lysis. In contrast, the rare antigenic mutants that were genotypically resistant yielded progeny that remained resistant. Thus when this selection was continued through several cycles of complement-mediated lysis followed by regrowth of the survivors, the antigenic mutants came to represent a larger and larger fraction of the population. When this resistant population came to predominate, an antigenic mutant was cloned and analyzed. Table 1 shows that this mutant was totally resistant to the complement-mediated lytic effect of monoclonal antibody G. The resistance was specific for a single monoclonal antibody and did not represent general resistance to complement since the mutant was as sensitive as the wild type to lysis in the presence of a different monoclonal antibody.

Based on the experience with animal viruses, we expected that our antigenic mutant would have an altered epitope that was not recognized by monoclonal antibody G. We were surprised to find that this antigenic mutant appeared completely to lack the protein to which monoclonal antibody bound

**TABLE 1. The Complement-Mediated Lytic Effect of Two Monoclonal Antibodies on Wild-Type *T. gondii* and on An Antibody-Resistant Mutant[a]**

| | Percent survival[b] | |
|---|---|---|
| Treatment | Wild-type *T. gondii* | *T. gondii* resistant to monoclonal G |
| Complement only[c] | 105 | 85 |
| Monoclonal antibody G only | 91 | 94 |
| Monoclonal antibody G plus complement | 2.2 | 94 |
| Monoclonal antibody B plus complement | 3.3 | 4.3 |

[a]From Kasper et al. (1982).
[b]Measured by a plaque assay.
[c]Supplied by normal human serum.

in the wild type. Purified mutant and wild-type parasites were surface iodinated with [125]I and analyzed by acrylamide gel electrophoresis in the presence of sodium dodecyl sulfate. Autoradiography of the resulting gels showed that the mutant had a surface iodination pattern identical with that of the wild type except for the absence of a major band that corresponded to a protein of apparent $M_r$ 22,000 (Kasper et al., 1982). Surface radioiodinated wild type-parasites were dissolved in detergent and then immunoprecipitated with monoclonal antibody G. When the precipitate was analyzed by acrylamide gel electrophoresis followed by autoradiography, the only iodinated protein recovered also had a $M_r$ of 22,000. We believe that the antigenic mutant does not have this protein on its surface. A less likely alternative is that the mutation that altered the relevant epitope also eliminated all potentially iodinatable sites. If, indeed, the $M_r$ 22,000 protein that is recognized by monoclonal antibody G is not represented on the surface of the mutant *T. gondii,* that protein cannot play a significant role in the growth of *T. gondii.* Additional antigenic mutants of *T. gondii* could be selected by using monoclonal antibodies directed to many other antigens on the surface of the parasite. The mutants may be useful for in vivo experiments to identify those antigens that play a critical role in immunity.

Drug-resistant mutants have been particularly valuable in studying DNA viruses such as the herpesviruses. The most useful mutants that we have obtained in *T. gondii* have been drug resistant. These mutants are particularly easy to isolate because the drug provides a positive selection pressure. Any drug that inhibits the growth of *T. gondii* can be used to select a resistant mutant provided only that the drug does not compromise the ability of the

host cell to support the growth of the parasite at least in short-term experiments and preferably in the week-long incubations needed for the resistant
mutants to make plaques. If a drug is particularly toxic to the host cell, it can
still be used to select a resistant parasite if a drug-resistant mutant host cell is
used. This mutant host cell, of course, cannot be resistant because of enhanced degradation of the drug or because of impermeability. We observed
that a mutant CHO cell with a temperature-sensitive lesion in DNA synthesis
supported normal growth of *T. gondii* at the restrictive temperature. Since
host cell DNA synthesis was apparently not required for the growth of the
parasite, we initially sought mutants resistant to drugs that block DNA
synthesis. Fluorodeoxyuridine (FUDR), is an analogue of deoxyuridine. The
active metabolite of FUDR is fluorodeoxyuridylic acid, a suicide inhibitor of
thymidylate synthetase. After optimal mutagenesis with nitrosoguanidine
about 0.01% of the surviving parasites made plaques in the presence of
FUDR and were 100-fold more resistant than was the wild type. We expected
the mutant parasites to lack deoxyuridine kinase but were unable to detect
this enzyme even in the wild-type parasite. It was also a surprise to find that
FUDR was not a specific inhibitor of DNA synthesis in *T. gondii* as it is in
mammalian cells. When infected Lesch-Nyhan cells were treated with FUDR
and labeled with [$^3$H]-hypoxanthine, DNA and RNA synthesis were inhibited
almost equally. Our first clue to understanding the metabolism of FUDR in
*T. gondii* came from tracing the incorporation of [$^3$H]-deoxyuridine into the
nucleic acids of *T. gondii*. Instead of being incorporated only into DNA,
[$^3$H]-deoxyuridine labeled both DNA and RNA of the parasite equally well.
Since this incorporation into RNA could only have happened if the deoxyri-
bose had been removed to yield the free base uracil, we turned our efforts to
the study of uracil metabolism in *T. gondii*. The parasite proved to incorporate uracil into its nucleotide pool and nucleic acids. We assayed the enzyme
responsible for the first step in this incorporation, uracil phosphoribosyltransferase and found that it was present in the parasite but not in the host cell.
This difference explained a previous observation that uninfected human
fibroblasts cannot incorporate [$^3$H]-uracil while infected cultures are heavily
labeled. Autoradiography of [$^3$H]-uracil-labeled cultures confirmed that only
the intracellular parasites were labeled.

Mutants of *T. gondii* resistant to FUDR had no detectable uracil phosphoribosyltransferase activity (Pfefferkorn, 1978). Since the same defect was
found in a number of independently isolated mutants, the lack of this enzyme
must be the basis for resistance. This enzymatic defect conferred resistance
not only to FUDR but also to fluorouracil and to fluorouridine. Consistent
with this pattern of resistance, mutant R-FUDR$^R$-1 was incapable of incor-

porating uracil, uridine, and deoxyuridine. The inability to incorporate deoxyuridine is illustrated in Figure 2 by the autoradiography of cells infected
wild-type and mutant parasites and labeled with [$^3$H]-deoxyuridine. This
precursor is incorporated specifically into DNA by the host cells and thus
only their nuclei are labeled. The wild-type parasites are also labeled by

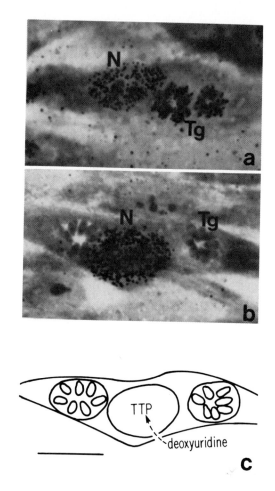

Fig. 2. Autoradiographic analysis of the incorporation of [$^3$H]-deoxyuridine by normal
human fibroblasts labeled 20–24 h after infection with *T. gondii*. **a:** Wild-type *T. gondii*. **b:**
FUDR-resistant *T. gondii*. **c:** Drawing based on the cell in b to emphasize that thymidine
triphosphate of the host cell is unavailable to the parasites, which remain unlabeled. Tg = *T.
gondii*; N = host cell nucleus. Bar is 10 µM (from Pfefferkorn and Pfefferkorn, 1977b).

[³H]-deoxyuridine, but the FUDR-resistant mutants are not. The lack of label in the mutant parasites shows that the thymidine triphosphate pool of the host cell is unavailable to the parasite.

We have also shown (unpublished) that wild-type *T. gondii* cannot incorporate cytosine, cytidine, or deoxycytidine. These observations together with the data described above can only be explained if *T. gondii* has the pattern of pyrimidine salvage shown in Figure 3. Note that all pyrimidine salvage is funneled through the base uracil. Since mutants incapable of pyrimidine salvage because they lack uracil phosphoribosyltransferase grow normally, this salvage is clearly not essential. This conclusion is consistent with the evidence that *T. gondii* is capable of de novo pyrimidine synthesis.

The second inhibitor of DNA synthesis that we used was adenine arabinoside, a drug that is used to treat certain herpes simplex viral infections. The active form of adenine arabinoside is the nucleoside triphosphate, which inhibits DNA polymerase. Adenine arabinoside proved to be a potent inhibitor of the growth of *T. gondii*. Analysis of the effect on the parasite-specific incorporation of [³H]-hypoxanthine in infected Lesch-Nyhan fibroblasts showed that DNA, but not RNA, synthesis was inhibited. After appropriate mutagenesis, we isolated a number of mutant parasites that were markedly resistant to adenine arabinoside. Further analysis of one of the resistant mutants showed a deficiency of the purine salvage enzyme adenosine kinase (Pfefferkorn and Pfefferkorn, 1978). The specific activity of this enzyme in extracts of wild-type parasites purified from fibroblast cultures was 940 pmol/mg protein/min. A similar extract prepared from the mutant R-AraA^R-1 had lower specific activity of only 36. However, this reduction in specific activity was not great enough to explain the increase in resistance to adenine arabinoside. A possible explanation for this discrepancy was that the mutant parasites had not been adequately purified. Since the human

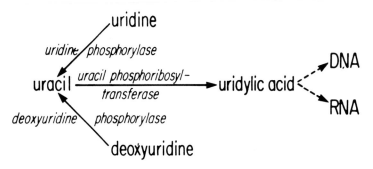

Fig. 3.   Salvage pathways for pyrimidines in *T. gondii*.

fibroblast host cells had a high specific activity of adenosine kinase, contamination with only 1.5% host of cell cytoplasm (by weight) would contribute all of the enzyme activity apparently present in the parasite mutant. Rather than attempting to improve our procedure for the purification of intracellular *T. gondii*, we chose the easier alternative of growing the mutant parasite in a mutant host cell that itself lacked adenosine kinase. In this case minor contamination of the purified parasites would not contribute artifactual enzyme activity. Mutant parasites prepared in this manner had no detectable adenosine kinase activity, a result consistent with their high degree of resistance to adenine arabinoside. The lack of adenosine kinase confers resistance to adenine arabinoside because this kinase also phosphorylates the drug, an essential first step in the formation of the active triphosphate form. As noted above, *T. gondii* is incapable of de novo purine synthesis and yet grows in cultures incubated in medium without purines. Thus the parasite must salvage purines from its host cell. The fact that the mutant resistant to adenine arabinoside grows normally means that the adenosine kinase of *T. gondii* cannot play an essential role in purine salvage by the parasite.

A variety of other drugs have also yielded mutants of *T. gondii*. The enzymatic bases of resistance to two other inhibitors of DNA synthesis have been tentatively identified: A mutant resistant to hydroxyurea has an altered ribonucleotide reductase (Kasper and Pfefferkorn, 1982), while a mutant resistant to aphidicolin (Pfefferkorn, 1984b) has an altered DNA polymerase. We have also isolated, but not yet characterized, mutants resistant to an inhibitor of protein synthesis (emetine), to an inhibitor of methylation (sinefungin), to each of the two drugs used clinically in toxoplasmosis (sulfadiazine and pyrimethamine), and to each of three anticoccidial drugs (monensin, emimycin, and arprinocid-N-oxide). One of the reasons for isolating these mutants was to use them in genetic crosses.

## GENETIC RECOMBINATION IN *TOXOPLASMA GONDII*

Before the introduction of gene cloning techniques, the only ways to study the arrangements of genes in viruses were recombination and complementation. We have used recombination to study the genetics of *T. gondii* even though the experiments are cumbersome. All of the previously described work was done in the convenient system offered by cultured cells. One of the major disadvantages of using *T. gondii* as a model obligate intracellular parasite is that only the rapidly growing asexual stage, the tachyzoite, can be reliably grown in cell culture. The establishment of a procedure for the in vitro sexual cycle of *T. gondii* would be a major advance. A variety of important questions ranging from the stimulus for sexual differentiation to the control of the synthesis of stage-specific antigens would be open to simple experiments. The greatest progress in establishing the sexual cycle of the parasite in vitro has been the development of methods to produce bradyzoites. Although the bradyzoite is not irreversibly committed to further sexual differentiation, it should be viewed as the obligatory first stage in this process. Encysted bradyzoites are readily produced in experimentally infected animals, but the stimulus that induces this response remains unknown. Although occasionally observed, the production of bradyzoites in vitro is incomplete and unpredictable. Thus the observation of Jones et al. (1986) that γ-interferon treatment of infected cultures appears to promote the differentiation of tachyzoites into bradyzoites is encouraging. In contrast to the hints of success with bradyzoites, the production of gametes and oocysts in vitro has never been accomplished. Gametogenesis in *T. gondii* is a complex process that involves several stages of schizogony as documented by Dubey and Frenkel's (1972) histological studies of infected kittens. The stimuli that drive bradyzoites to enter the schizogony pathway leading to gametogenesis are unknown. Since gametogenesis takes place only in members of the cat family, one of these stimuli must distinguish felines from other animals. Similarly, since gametes are produced only in intestinal epithelial cells, another stimulus must be something peculiar to these cells. We have attempted to supply both stimuli for gametogenesis by preparing monolayer cultures of intestinal epithelium and organ cultures of intestine from kittens. When either kind of culture was infected with bradyzoites prepared from the brains of chronically infected mice, it was eventually destroyed by overgrowth of tachyzoites. The only hint of success was the occasional production of what are probably, based on their time of appearance, first generation schizonts. The schizonts seen in cultured kitten intestinal epithelial cells are clearly morphologically distinct from the tachyzoites found in adjacent cells. Although our experiments have all met with failure, completing the sexual cycle of *T. gondii* in vitro may well be an achievable goal. It is encouraging to note that the complete sexual cycle of *Eimeria tenella*, a related intestinal parasite of chickens, has been accomplished in cultured cells.

Lacking an in vitro system, we have used chronically infected mice as the source of bradyzoites and weanling kittens as the definitive host. Before describing the use of this in vivo system for genetic crosses, it is essential to consider the question of sex determination in *T. gondii*. Two general models are possible. Sex could be genetically determined with each tachyzoite committed to differentiate either into a macrogamete or into a microgamete. Alternatively, the progeny of each tachyzoite could be capable of differentia-

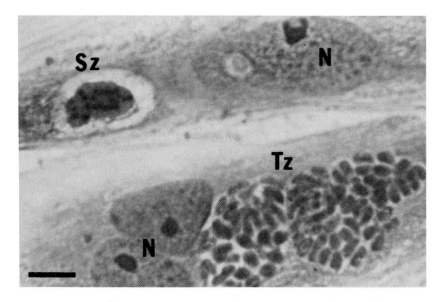

Fig. 4. A schizont (Sz) of *T. gondii* formed in cultured epithelial cells from the ileum of a newborn kitten. Confluent monolayer cultures were fixed and stained with Giemsa 2 days after injection with bradyzoites from the brain of a chronically infected mouse. In another cell, a brazyoite has differentiated into tachyzoites (Tz). N = host cell nucleus. Bar is 10 $\mu$M (L. Pfefferkorn, unpublished).

tion into both kinds of gametes with the individual decision as to sex determined by some local environmental stimulus in the feline intestine. To distinguish between these two alternatives, we prepared reliably cloned tachyzoites by three successive single plaque isolations. We used these clones to produce chronically infected mice and then fed the brains of these mice individually to kittens. The kittens produced the expected number of oocysts at the expected time (Pfefferkorn et al., 1977). These oocysts sporulated and were infectious. Thus tachyzoites are not genetically committed to making gametes of only one sex. This was a welcome conclusion because we had already isolated a good many mutants from a single cloned wild type. Having convinced ourselves that our mutants were capable of producing both gametes, we used them in genetic crosses. In our initial experiments we crossed a mutant resistant to FUDR with a mutant resistant to adenine arabinoside. We fed kittens the brains of mice that were chronically infected with these mutants and collected the resulting fecal oocysts. The sporozoites released from the oocysts were assayed by plaque formation in cultures with no drugs or with both FUDR and adenine arabinoside. The plaques in the latter cultures represented recombinants resistant to both drugs (Pfefferkorn and Pfefferkorn, 1980). We examined the biochemical basis of drug resistance in several recombinant clones. These recombinants totally lacked both uracil phosphoribosyltransferase and adenosine kinase. We concluded that this pattern of all-or-nothing inheritance is characteristic only of an organism that is haploid in all stages except for the zygote and its immediate product, the unsporulated oocyst (Pfefferkorn and Pfefferkorn, 1978). The process of sporulation must then include the meiosis that is essential to restore the haploid state. Our deduction as to the ploidy of *T. gondii* was subsequently proven correct by microspectrophotometry of individual Feulgen-stained cells (Cornelissen et al., 1984).

All of our crosses between FUDR- and adenine arabinoside-resistant mutants failed to show a frequency of doubly resistant recombinants greater than 12.1%. Having concluded that *T. gondii* was basically a haploid organism, we initially assumed that the maximal frequency of recombination for unlinked markers would be 25%. But this assumption did not take into account the bisexual nature of the parasite. Since the two parental types can each give rise to both male and female gametes in the cat intestine, not all matings take place between gametes that represent two different parents. When each parent contributes the same total number of gametes, half of the matings are self-crosses between gametes of the same genotype. The other half of the matings are out-crosses which do give rise to the expected 25% recombinants for unlinked markers. But the fraction of recombinants in the total progeny is reduced by a factor of two because half of the progeny are contributed by self-crosses.

The maximum of 12.5% recombinants is only achieved when the two parents supply an exactly equal number of gametes and becomes progressively lower as one parent comes to predominate. We devised a simple equation to predict the frequency of recombination between unlinked markers (Pfefferkorn and Kasper, 1983):

$$F = 1/2ab \qquad (1)$$

In this equation, $a$ and $b$ are the fractions of the total gamete population that are genotypes $a$ and $b$, respectively, and $F$ is the fraction of the total sporozoite progeny that is recombinant for genetic markers $a$ and $b$. When the genetic markers have a selective advantage, as with two drug-resistant genes, $F$ can be directly measured as the ratio between plaque titers in the presence of both drugs together and in drug-free medium. There is no direct way to measure the fraction of gametes that has a particular parental geno-type, but it can be measured indirectly. We assume that each gamete, regardless of genotype, has the same likelihood of forming a viable zygote. Thus the fraction of total gametes that have a particular parental gene should be the same as the fraction of the haploid sporozoite progeny that have that particular gene. This latter fraction can be measured as the ratio of plaque titers in medium containing one drug and in medium without drug. We confirmed the validity of Equation 1 through a series of binary crosses in which the ratio of the two parental types was systematically varied by roughly adjusting the number of encysted bradyzoites ingested. As shown in Figure 5, the observed recombination frequency conformed closely to that pre-dicted by Equation 1. The real value of this equation, however, becomes apparent when many drug-resistant mutants are available for use in crosses. Analysis of a large catalogue of mutants by binary crosses would require an unreasonably large number of kittens. We now carry out crosses with *T. gondii* by infecting the same kitten with a large number of mutants, a procedure that we term "orgy matings" by analogy with the original bacte-riophage procedure. Since each gamete operates independently in the kitten intestine and can mate with only one other gamete, Equation 1 can be used for analysis of the resulting data. The validity of this conclusion has been confirmed experimentally (Pfefferkorn and Kasper, 1983).

## INTERFERON AND *TOXOPLASMA GONDII*

Among the defense mechanisms that serve to contribute to recovery from viral infection is the production of interferons. Interferons block the growth of viruses through complex, biochemical mechanisms that are only partially understood. Two host cell enzymes induced by interferon have demonstrated antiviral activity through their ability to block virus-specific protein synthesis in infected cells. One antiviral enzyme is 2-5A synthetase, the product of which activates an endonuclease that hydrolyses mRNA. The other antiviral enzyme is a protein kinase that specifically phosphorylates initiation factor two and thereby inhibits protein synthesis. Although the analogy between *T. gondii* and viruses as obligate intracellular parasites is tenuous at best, we decided to examine the effect of interferons on the growth of *T. gondii*. We began by using a plaque-inhibition test in human fibroblasts that was based on the standard assay of the antiviral activity of interferon. This test showed that human recombinant γ-interferon had a potent antitoxoplasma activity while the corresponding α- and β-interferons had no such activity. Only 4 reference units/ml of gamma interferon were sufficient to block completely

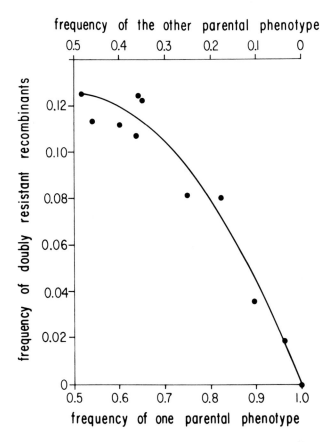

frequency of the other parental phenotype

frequency of doubly resistant recombinants

frequency of one parental phenotype

Fig. 5. The effect of altering the ratio of parental genotypes of two different drug-resistant mutants of *T. gondii* on the frequency of doubly resistant recombinants in the progeny sporozoites. The curve is the theoretical recombination frequency calculated from Equation 1 (from Pfefferkorn and Kasper, 1983).

the formation of plaques by *T. gondii*. One reference unit/ml is the minimal interferon concentration that shows significant antiviral activity. Thus the antiviral and antitoxoplasma potency of γ-interferon were roughly comparable.

The many studies of the antiviral activity of interferon suggested experiments that could be done with *T. gondii*. When cells are treated with interferon they do not immediately suppress viral growth. An interval of some hours is required to induce the synthesis of the antiviral enzymes and establish a virus-resistant state. Similarly, human fibroblasts treated with γ-interferon initially had no capacity to block the growth of *T. gondii*. One day later, the toxoplasma-resistant state was completely established. Once established in cells by suitable treatment with interferon, the virus-resistant state is no longer dependent upon the presence of interferon in medium. The toxoplasma-resistant state induced by γ-interferon was similarly independent of the continued presence of interferon in the medium. A classical experiment in animal virology showed that inhibition of host cell mRNA synthesis by actinomycin D blocked the antiviral activity of interferon. The inhibition of mRNA synthesis acted ultimately by preventing the host cell from synthesizing the antiviral enzymes. This experiment was possible because the nucleic acid synthesis of the RNA virus was dependent upon RNA-dependent RNA transcription and thus was resistant to actinomycin D. Since *T. gondii* is sensitive to actinomycin D, we could not exactly reproduce this experiment. We took advantage of an unpublished observation that mRNA synthesis of *T. gondii* is notably less sensitive to α-amanitin than is mRNA synthesis by the host cell. Thus we were able to block host cell mRNA synthesis without compromising the growth of the parasite. As shown in Table 2, when cultures were treated with α-amanitin at the time γ-interferon was added, the interferon did not inhibit the growth of *T. gondii* when the cultures were infected one day later. By itself α-amanitin had no effect. Our interpretation of these data followed the classical model for the antiviral action of interferon: we assumed that γ-interferon induced one or more "antitoxoplasma" enzymes and that without host cell mRNA synthesis these new enzymes could not be made.

Although the antiviral model of interferon action proved a good guide up to this point, it offered no clue as to the nature of the putative "antitoxoplasma" enzymes. A strong theoretical argument could be made against the two known antiviral enzymes. Since these enzymes are thought to be responsible for the antiviral activities of α-, β-, and γ-interferons, their role in the

**TABLE 2. Reversal of the Antitoxoplasma Effect of γ-Interferon by Treatment of Human Fibroblast Cultures With α-Amanitin[a]**

| | Percent of control parasite growth[b] | |
| --- | --- | --- |
| | Without α-amanitin | With α-amanitin[c] |
| Without γ-interferon | 100 | 97 |
| With γ-interferon, 16 units/ml | 17 | 93 |

[a]From Pfefferkorn (1986).
[b]Measured 1 day after infection.
[c]1 μg/ml added 1 day before infection.

antitoxoplasma effect would not explain the fact that only γ-interferon showed activity. Furthermore, it was difficult to imagine how the cytoplasmic antiviral enzymes could exert an effect upon the intracellular *T. gondii* since they would have to cross both the parasitophorous vacuolar membrane and the plasma membrane of the parasite itself. We stumbled upon the most important clue to the mechanism of the antitoxoplasma activity of γ-interferon while determining if this activity resembled the antiviral activity in being species specific. We observed that human γ-interferon was species specific and would not protect mouse cells against *T. gondii*. In the course of this work, a control culture showed that the antitoxoplasma effect of γ-interferon in fibroblast cultures was markedly dependent upon the tryptophan content of the medium (Pfefferkorn, 1984a). This dependence is illustrated in Figure 6, which shows the titration of γ-interferon in media that differed only in tryptophan content. As the tryptophan concentration in the medium was increased, more γ-interferon was required to suppress the growth of *T. gondii*.

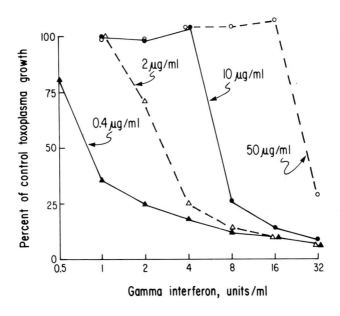

Gamma interferon, units/ml

Fig. 6. Titration of the antitoxoplasma activity of γ-interferon in media with four tryptophan concentrations. Human fibroblast cultures were treated with two-fold dilutions of interferon in four media that differed only in tryptophan content as noted in the figure. Twenty-four hours later all cultures were infected. Forty-eight hours after infection the parasite growth was measured by [³H]-uracil incorporation (from Pfefferkorn, 1984a, with the interferon concentrations converted to reference units).

We found that the antitoxoplasma activity of γ-interferon was closely dependent upon the tryptophan concentration of the medium because the interferon induced a potent indoleamine 2,3-dioxygenase, which degraded tryptophan (Pfefferkorn et al., 1986b). The activity of this enzyme was first detected by assays of the tryptophan content of the medium of cultures treated with various concentrations of γ-interferon. Figure 7 shows that the concentration of tryptophan fell in a manner that was dependent both upon time and upon the concentration of γ-interferon. The principal degradation products identified in the medium were N-formylkynurenine and kynurenine. Characterization of the indoleamine 2,3-dioxygenase induced by γ-interferon showed that it had a low $K_m$ of 3 μM for L-tryptophan. From the viewpoint of an intracellular parasite, the concentration of tryptophan in the medium is not as critical as the concentration within the host cell. Treatment of human fibroblast cultures with γ-interferon had a dramatic effect on the intracellular pool of tryptophan even when the medium contained an ample concentration of this amino acid (Pfefferkorn et al., 1986a). We measured the fate of [$^3$H]-tryptophan as it entered γ-interferon-treated human fibroblasts over intervals of a few seconds. As shown in Figure 8, the [$^3$H]-tryptophan was degraded to N-formylkynurenine and kynurenine as rapidly as it entered the cell. In

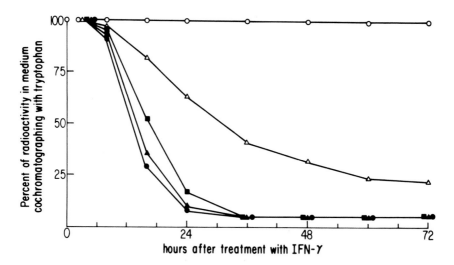

Fig. 7. The disappearance of tryptophan from the medium of confluent human fibroblast cultures treated with various concentrations of γ-interferon. Duplicate cultures were incubated with zero (O--O), 1 (△--△), 4 (■--■), 16 (▲--▲), and 64 (●--●) reference units/ml of γ-interferon medium that contained [$^{14}$C]-tryptophan. Samples of medium were obtained at intervals and analyzed by chromatography (from Pfefferkorn et al., 1986a).

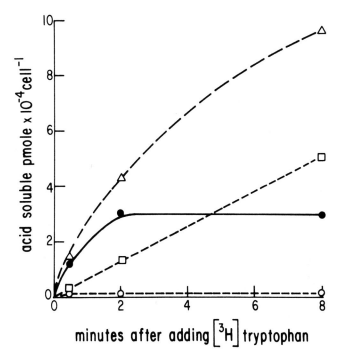

**minutes after adding [³H] tryptophan**

Fig. 8. The fate of [³H]-tryptophan in the soluble pools of control human fibroblasts and human fibroblasts previously treated for 12 h with 32 reference units/ml of γ-interferon. Tryptophan in control cells (●--●). Tryptophan (○--○), N-formylkynurenine (△--△), and kynurenine (□--□) in cells treated with interferon (from Pfefferkorn et al., 1986a).

contrast, no such degradation was observed in control cells, in which the intracellular [³H]-tryptophan rapidly rose to an equilibrium level.

We suspect that the growth of *T. gondii* is blocked in cultures treated with γ-interferon by the simple mechanism of starvation for tryptophan, an essential amino acid. An alternative hypothesis is that some product of tryptophan degradation is toxic to the parasite. This explanation can be ruled out for the two principal tryptophan metabolites, N-formylkynurenine and kynurenine. We measured the intracellular concentrations of these two substances and found that they were well below the toxic level. In addition, no toxic metabolites of tryptophan were found in the medium of human fibroblast cultures treated with γ-interferon. However, this observation does not rule out the toxicity of minor tryptophan metabolites that are either labile or exclusively intracellular. A theoretical argument against a role for toxic tryptophan metabolites is that they should play a more prominent role as the tryptophan concentration of the medium is increased. But we observed the opposite: $\gamma$-interferon was least effective in media with the highest concentration of tryptophan. This result is, however, consistent with our hypothesis that starvation for tryptophan explains the antitoxoplasma effect of $\gamma$-interferon. The in vivo significance of the antitoxoplasma effect of the indoleamine 2,3-dioxygenase induced by $\gamma$-interferon remains to be determined. $\gamma$-Interferon treatment of humans does, however, result in the depletion of tryptophan in the blood (Byrne et al., 1986).

## CONCLUSIONS

The paradigm of animal virology has served us well in suggesting experimental approaches to the study of *Toxoplasma gondii*. It may prove to be equally valuable for many currently unsolved problems, including the specificity of adsorption to host cells. Those who wish to pursue even further the analogy of animal viruses to obligate intracellular parasites might consider the immortalization of bovine lymphocytes by *Theileria parva*, a phenomenon reminiscent of the action of tumor viruses.

## ACKNOWLEDGMENTS

The personal research reviewed in this chapter was supported by grant AI-14151 from the National Institutes of Health.

## REFERENCES

Byrne GI, Lehmann LK, Kirschbaum JG, Borden EC, Caroll ML, Brown RR (1986): Induction of tryptophan degradation *in vitro* and *in vivo*: A $\gamma$-interferon-stimulated activity. J Interferon Res 6:389–396.
Cornelissen AWCA, Overdulve JP, van der Ploeg M (1984): Cytochemical studies on nuclear DNA of four eucoccidian parasites, *Isospora (Toxoplasma) gondii*, *Eimeria tenella*, *Sarcocystis cruzi* and *Plasmodium berghei*. Parasitology 88:13–25.
Dubey JP, Frenkel JK (1972): Cyst induced toxoplasmosis in cats. J Protozool 1911:155–177.
Hirsch JG, Jones TC, Len L (1974): Interactions in vitro between *Toxoplasma gondii* and mouse cells. In R. Porter and J. Knight (eds): Parasites in the Immune Host; Mechanisms of Survival, Ciba Foundation Symposium 25:205–220, Amsterdam: Elsevier.
Jones TC, Bienz KA, Erb P (1986): In vitro cultivation of *Toxoplasma gondii* cysts in astrocytes in the presence of gamma interferon. Infection Immunity 51:147–156.
Kasper LH, Pfefferkorn ER (1982): Hydroxyurea inhibition of growth and DNA synthesis in *Toxoplasma gondii*: Characterization of a resistant mutant. Mol Biochem Parasitol 6:141–150.

Kasper LH, Crabb JH, Pfefferkorn ER (1982): Isolation and characterization of a monoclonal antibody-resistant antigenic mutant of *Toxoplasma gondii*. J Immunol 129:1694–1699.

Pfefferkorn ER (1978): *Toxoplasma gondii*: The enzymic defect of a mutant resistant to 5-fluorodeoxyuridine. Exp Parasitol 44:26–35.

Pfefferkorn ER (1984a): Interferon-γ blocks the growth of *Toxoplasma gondii* in human fibroblasts by inducing the host cells to degrade tryptophan. Proc Natl Acad Sci USA 81:908–912.

Pfefferkorn ER (1984b): Characterization of a mutant of *Toxoplasma gondii* resistant to aphidicolin. J Protozool 31:306–310.

Pfefferkorn ER (1986): Interferon gamma and the growth of *Toxoplasma gondii* in fibroblasts. Ann Inst Pasteur/Microbiol 137A:348–352.

Pfefferkorn ER, Kasper LH (1983): *Toxoplasma gondii*: Genetic crosses reveal phenotypic suppression of hydroxyurea resistance by fluorodeoxyuridine resistance. Exp Parasitol 55:207–218.

Pfefferkorn ER, Pfefferkorn LC (1976): *Toxoplasma gondii*: Isolation and preliminary characterization of temperature-sensitive mutants. Exp Parasitol 39:365–376.

Pfefferkorn ER, Pfefferkorn LC (1977a): *Toxoplasma gondii*: Specific labeling of nucleic acids of intracellular parasites in Lesch-Nyhan cells. Exp Parasitol 41:95–104.

Pfefferkorn ER, Pfefferkorn LC (1977b): *Toxoplasma gondii*: Characterization of a mutant resistant to 5-fluorodeoxyuridine. Exp Parasitol 42:44–55.

Pfefferkorn ER, Pfefferkorn LC (1978): The biochemical basis for resistance to adenine arabinoside in a mutant of *Toxoplasma gondii*. J Parasitol 64:486–492.

Pfefferkorn ER, Pfefferkorn LC (1980): *Toxoplasma gondii*: Genetic recombination between drug resistant mutants. Exp Parasitol 50:305–316.

Pfefferkorn ER, Pfefferkorn LC, Colby ED (1977): Development of gametes and oocysts in cats fed cysts derived from cloned trophozoites of *Toxoplasma gondii*. J Parasitol 64:158–159.

Pfefferkorn ER, Eckel M, Rebhun S (1986a): Interferon-γ suppresses the growth of *Toxoplasma gondii* in human fibroblasts through starvation for tryptophan. Mol Biochem Parasitol 20:215–224.

Pfefferkorn ER, Rebhun S, Eckel M (1986b): Characterization of an indoleamine 2,3-dioxygenase induced by gamma-interferon in cultured human fibroblasts. J Interferon Res 6:267–279.

Schwartzman JD, Pfefferkorn ER (1981): Pyrimidine synthesis by intracellular *Toxoplasma gondii*. J Parasitol 67:150–158.

Schwartzman JD, Pfefferkorn ER (1982): *Toxoplasma gondii*: Purine synthesis and salvage in mutant host cells and parasites. Exp Parasitol 53:77–86.

Sibley LD, Weidner E, Krahenbuhl JL (1985): Phagosome acidification blocked by intracellular *Toxoplasma gondii*. Nature 315:416–419.

chapters on all aspects of *Caenorhabditis* biology, and includes a practical methods section. Results described below not specifically cited to the primary literature can be found in the appropriate chapters of this book.

## *CAENORHABDITIS* AS AN EXPERIMENTAL ORGANISM

*Caenorhabditis elegans* is a bacteria feeding soil nematode of worldwide distribution. It was recognized as an advantageous organism for experimental genetics by Dougherty and Calhoun (1949), but only began to be studied intensively in 1970 when Sydney Brenner improved the methods for genetic analysis and began a complete anatomical reconstruction of its nervous system by electron microscopy (Brenner, 1974). *Caenorhabditis* is grown in the laboratory on petri plates of agar seeded with *Escherichia coli*, much like a bacteriophage. The worm is normally a self-fertilizing, protandrous, hermaphrodite (karyotype 5AA,XX) so individual worms can be picked and "cloned" to separate petri plates or microtiter wells. The generation time is 3 days at room temperature with the worm passing through four juvenile stages (called larvae) separated by moults of the cuticle. A single hermaphrodite lays about 300 eggs. Since the adult worm is only a 1 mm long, up to 10,000 worms can be obtained on a single petri plate and millions can be grown synchronously in liquid culture on bacteria. The ease of handling a large number of worms makes it possible to find rare mutations and to obtain sufficient material for biochemical analysis. This contrasts with the difficulty of rearing most parasitic nematodes in the laboratory, although significant progress has been made in culturing *Ascaris* (Douvres and Urban, 1983).

The hermaphroditic mode of reproduction is one of *Caenorhabditis*'s main attractions for genetics; mutations can be induced by chemicals or radiation, two generations later 1/4 of the progeny will be homozygous for an induced mutation. In a natural population males arise by X chromosome nondisjunction (karyotype 5AA,X−) at a frequency of about 0.1%. These males mate readily to hermaphrodites, and since their sperm preferentially fertilizes the eggs male stocks can be easily maintained. Genetic crosses are performed by mixing males and hermaphrodites of different genotypes on a plate and following the phenotypes of their descendents so that mutations can be genetically mapped, tested for complementation to other mutations, and examined in various genetic backgrounds. Larval worms survive freezing in liquid nitrogen indefinitely so many stocks can be kept in the laboratory with no danger of loss or alteration.

*Caenorhabditis* is now being studied intensively in more than 50 laboratories worldwide, and a biennial worm meeting at Cold Spring Harbor in

May 1987 drew 304 participants. More than 685 genes have already been identified and genetically mapped to the six chromosomes of *Caenorhabditis*. These include mutants that affect nearly every recognizable step of development or behavior, including embryogenesis, muscle protein assembly, spermatogenesis, sensory or motor neuron function and development, sex determination, dosage compensation, cuticle formation, cell fate determination, etc. Mutant and wild-type strains are available from the *Caenorhabditis* Genetics Center, University of Missouri, Columbia, Missouri 65211. This Genetics Center also maintains the genetic map and publishes an informal newsletter twice a year which is available to anyone by subscription.

## MOLECULAR GENETICS

The small genome size of *Caenorhabditis* ($8 \times 10^7$ base pairs) combined with the available mutant collection has prompted intensive efforts to clone genes identified mutationally as well as genes encoding proteins that have been identified in other systems (Emmons, 1987). Among the novel results of characterization of some of these genes has been the discovery of the splicing of a leader sequence to the messenger RNA encoding the actin genes (Krause and Hirsh, 1987). Such transsplicing was first discovered in trypanosome parasites (Van der Ploeg et al., 1982); perhaps it occurs in parasitic nematodes as well.

The intensive molecular characterization of the *Caenorhabditis* genome has led to the development of a number of new methods for genomic analysis. Two methods of particular relevance to parasites are the demonstration of stable transformation of *Caenorhabditis* with injected cloned DNA (Stinchcomb et al., 1985; Fire, 1986) and the construction of a physical map of the genome by overlapping cosmid clones (Coulson et al., 1986). Because these methods could be applied to parasitic nematodes in which there is little hope of conventional mutant isolation and genetic characterization they will be described in some detail here.

### Physically Mapping the Genome

A genetic map consists of the ordering of genes, defined by mutations, on individual chromosomes (linkage groups) by measuring the frequency of recombination between mutations. A physical map consists of the ordering of cloned pieces of DNA by their overlapping sequences and assignment of the ordered regions to individual chromosomes by in situ hybridization. Since individual chromosomes are single lengths of DNA, the two maps should be colinear. The two maps can be connected when genetically mapped genes are identified in cloned DNA. Note that a physical map of the genome can be constructed in any organism whether or not mutants and a genetic map are available.

Coulson and Sulston have constructed a physical map of the genome using the method summarized in Figure 1. They first constructed libraries of genomic DNA in cosmid vectors which allow the cloning of 30–40 kilobase (kb) fragments of DNA, and grew each clone of DNA in microtiter wells. Since the genome of *Caenorhabditis* is $8 \times 10^4$ kb, it should be encompassed in about 2,500 cosmids if clones were perfectly representative of the genome. Clones are not representative, however, because some regions of DNA clone more readily than others so many more cosmids must be examined. DNA from individual cosmids was digested with the restriction enzyme HindIII; the fragments were end-labeled with $^{35}$S and recut with SauIIIA; and the resulting fragments, about 25 per cosmid, were fractionated on high resolution sequencing gels. The labeled fragments were identified by autoradiography and the fragment sizes found for each cosmid were entered into a computer data bank. The computer then compared the pattern of fragment sizes for each cosmid to every other cosmid which had been entered to identify cosmids that had at least 1/3 of the fragments identical, indicating that these cosmids must contain DNA that overlapped in the worm genome.

This method is called DNA "fingerprinting" because the pattern of fragments from each cosmid can be thought of as a fingerprint, uniquely identifying the worm DNA in the cosmid. As more cosmids are entered in the database, the overlaps grow and form larger and larger contiguous stretches of DNA called "contigs." The fingerprinting technique is sensitive enough that even DNA cloned into lambda vectors, which hold only about 15 kb of insert DNA, can be unambiguously assigned to cosmids.

The contigs were assigned to chromosomes either by identification of a cloned gene which had been mapped genetically within the contig, or by in situ hybridization of cosmid clones to chromosomal spreads (Albertson, 1985). As of May 1987 more than 16,000 cosmids had been "fingerprinted" and found to represent 90–95% of the genome. A total of 765 contigs had been identified, the largest being nearly 800 kb of DNA. About 20% of the contigs had been assigned to chromosomes either by in situ hybridization or identification of genetically mapped genes. The distribution of contig sizes suggested that the contigs were only separated by small gaps, and these are being joined by specific probing of libraries with DNA probes from the ends of the contigs.

Note that this entire physical map has been constructed in three years largely by only two people. Part of this time has been taken to develop the

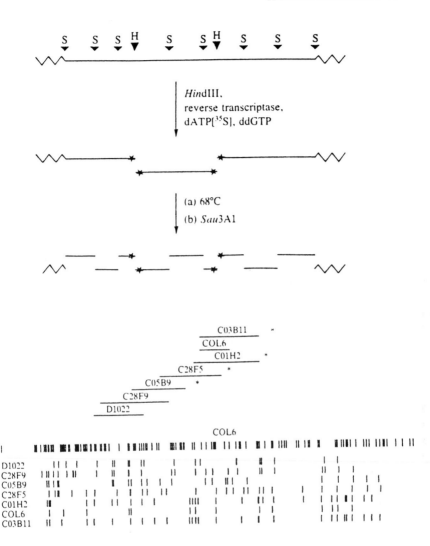

Fig. 1. **a:** Fingerprinting procedure. ddGTP, dideoxyGTP; S. *Sau*3A1 sites; H. *Hind*III sites. **b:** A contig displayed on the computer screen (**top**). Each clone is represented by a line of length proportional to the number of bands. Asterisk indicates the presence of hidden clones. Repeat of the name COL6 beneath the contig indicates location of this known gene; additional remarks can be added as required (**bottom**). Pattern of marker bands and clone bands, plotted from digitized data.

methods and write the computer programs for handling the data, which are written in Fortran and are available in VAX format from Coulson and Sulston. With the advent of more automated methods for reading autoradiograms, new yeast cloning vectors which can take larger inserts (Burke et al., 1987) and pulse field electrophoresis to separate chromosomes (Schwartz and Cantor, 1984), this kind of physical mapping will become easier in other organisms.

What can be learned from such physical maps? In *Caenorhabditis* the completion of the physical map and its more precise correlation with the genetic map will eventually mean that any gene which is genetically mapped can be assigned to a cosmid clone and then subcloned from that. Conversely, a cloned gene can be assigned to the physical map, then correlated with the genetic map. Preexisting mutations in that region of the genome can be examined to see if they affect the cloned gene, or the identified region can be specifically screened for new mutations either by selecting for mutations over a small deletion or selecting for mutations linked to a nearby morphological marker.

Already the physical map has been used to show that there is clustering of developmentally regulated genes in the genome. A multigene family of about 60 genes encodes the most abundant protein in the sperm, called the major sperm protein. This protein is the major structural element in the sperm pseudopod where it assembles into 2–3 nM filaments. Physical mapping of the major sperm protein (MSP) genes has shown that they are clustered into only three chromosomal locations (Ward et al., 1988). Several other sperm-specific genes are found in one of these locations. It is not yet known whether this clustering is necessary for coordinate regulation of these genes, or whether it has some other explanation, but such clustering would have been difficult to discover without the physical map. Additional examples of clustering are hinted at by genetic mapping and are being pursued with the aid of the physical map.

Functional clustering of genes may not be a common phenomena in eukaryotes, but when it does occur physical mapping should be useful for identifying related genes starting with a single cloned gene, even if no mutants are available, as is the situation with most parasites.

### Transformation With Cloned DNA

Surely one of the most important advances in molecular genetics has been the discovery that cloned DNA can be introduced into an organism and subsequently be stably inherited and properly expressed. The ability to create "transgenic" organisms enables a number of important experiments: identification of the essential functional and regulatory portions of a cloned gene; disruption of the normal copies of a gene either by integration into the gene, expression of antisense RNA, or misexpression of the protein gene product; and introduction of foreign and novel genes into an organism.

The first problem is to be able to get foreign DNA into an organism. This has been achieved for *Caenorhabditis* by immobilizing worms under oil and microinjecting DNA into the nuclei of oocytes (Stinchcomb et al., 1985; Fire, 1986). Equipment required includes facilities for pulling microelectrodes, micromanipulators, and a compound microscope for viewing the injection. An inverted compound microscope equipped with differential interference and fluorescence optics is helpful, but successful microinjection has been achieved on an ordinary upright microscope (Jeff Way, personal communication).

Fire has demonstrated that by injecting DNA into multiple oocyte nuclei, more than 60% of injected worms survive; of these survivors 10–50% will stably inherit the injected DNA either by integration into a chromosome or by formation of reasonably stable episomal DNA. He has developed a selection for transformation using a supressor tRNA gene to correct an amber mutation. Such a selection would not be applicable to parasites since the suppressor-sensitive mutations are not available. More general drug-resistance selections would be useful, and efforts are being made to develop these in several laboratories.

So far Fire and others have demonstrated successful correction of mutations in several muscle-protein-defective genes, by microinjection of the cloned wild-type gene. The injected genes appear to be properly regulated and are expressed only in the appropriate tissue. That is not true for injected genes which are not integrated into the chromosome, but are maintained as large tandem arrays (Stinchcomb et al., 1985). There is no evidence yet of homologous recombination of injected genes into their genomic counterparts. This would be particularly useful because it would allow disruption of the normal copies of genes which has been a powerful tool for yeast molecular biologists.

In addition to stable integration and inheritance of injected genes, injected genes are expressed transiently in a majority of injected worms, presumably directly from the injected DNA. Such transient expression may be particularly useful for studying the regulation of gene expression in parasitic helminths which cannot be carried through their complex life cycles after microinjection.

## DEVELOPMENT

The complete embryonic and postembryonic cell lineage of *Caenorhabditis* has been determined by direct observation of dividing cell nuclei (Sulston,

1988). This is the only complex metazoan for which this has been achieved. Together with electron microscopic reconstruction of the total cellular anatomy, the ontogeny and differentiated cell architecture of every cell is known. This information has been used to identify mutants altering cell lineage, to compare the role of cell ancestry and cell environment in the determination of cell fate, and to compare certain cell lineages with other nematodes to see how evolutionary changes could have occurred (Sternberg and Horvitz, 1982).

In addition to mutations which alter cell fate and identify genes controlling development, cell interactions and cell functions have also been investigated by ablating individual cell nuclei using a high-intensity laser focused with a compound microscope. By this method it was established that cells at the distal tip of the developing gonad control the proliferation of nearby germ cells (Kimble, 1981). Although many cell fates are established independent of their neighbors, laser ablations have shown that there are regulatory interactions between some developing neurons and between embryonic cells (Priess and Thomson, 1987). This technique could be easily applied to other helminth species both to study their development, and also to help establish the functions of identified cells.

A stage of postembryonic development which has been extensively studied is the formation of "dauer" larvae (Riddle, 1988). This larval stage is a modified form of the second-stage larvae which accumulates in crowded or starved cultures. It has a double thick cuticle and is resistant to desiccation and other insults so represents a diapauselike state specialized for dispersal and survival. This stage is of interest to parasitologists because it is similar to the infective stage of several parasitic nematodes. Many mutants have been obtained which alter the entry or exit from the dauer larvae and a genetic pathway of the sequence of gene actions has been constructed. A pheromone signal which triggers entry has been isolated and partially characterized. It is a stable, nonvolatile, fatty-acid-like compound. Mutants altering its production and altering response to the pheromone have been identified.

Another aspect of *Caenorhabditis* development which has been intensively studied is spermatogenesis and sperm motility (Kimble and Ward, 1988). As in other nematodes the sperm are amoeboid crawling cells and they appear to have a novel mechanism of motility by membrane flow. Unique proteins associated with this motility including the major sperm protein have been isolated and their genes sequenced in both *Caenorhabditis* and *Ascaris* (Burke and Ward, 1983; Klass et al., 1984; Bennett and Ward, 1988). Because nematode spermatozoa are so different from mammalian spermatozoa and the proteins identified in them so far have no known mammalian counterparts, they might be good targets for development of novel chemotherapeutic drugs which sterilize male nematodes and so disrupt the parasite's life cycle.

## NEUROANATOMY AND BEHAVIOR

The structure and function of the nervous system of *Caenorhabditis* has been reviewed by Chalfie and White (1988). The complete neuroanatomy of *Caenorhabditis*, including the interconnections of all neurons (the complete "wiring diagram"), has been reconstructed from serial section electron micrographs (White et al., 1986). This represents an anatomical tour de force not achieved in any other organism and is a fitting tribute to Richard Goldschmit's pioneering attempt to do this in *Ascaris* by light microscopy nearly 80 years ago. The hermaphrodite nervous system is composed of 302 neurons and 56 glial and supporting cells, whereas the male has 381 neurons and 92 glial and supporting cells. Comparison of the neurons in the hermaphrodite suggests that there are at most 118 different types of neurons. A simple functional classification into sensory, motor, and interneurons is not possible because some neurons have all three functions, such as an anterior mechanoreceptor that synapses onto both interneurons and directly onto head muscle processes. This may be a common property of simple invertebrate nervous systems and may represent an intermediate state in the evolution of specialized neural circuits.

*Caenorhabditis* is too small to allow intracellular recording from individual neurons, but physiological recording from *Ascaris* neurons, particularly the motor nervous system which has nearly identical anatomy to *Caenorhabditis*, has allowed inference of function of some of the neurons and plausible models for the control of body motion can be made (Stretton et al., 1985).

In addition to movement, the behaviors in *Caenorhabditis* that have been studied include chemotaxis, thermotaxis, osmotic avoidance, touch responses, egg laying, and male mating. Even a weak phototaxis has been observed (Burr, 1985). *Caenorhabditis* is attracted to cAMP, monovalent anions and cations, basic pH, some amino acids, and bacterial extracts. It is repelled by acid pH, D-tryptophan, and carbonate (Dusenbery, 1975). The behavioral response to gradients of attractants can be studied by countercurrent separation (Dusenbery, 1974) or more simply by monitoring the tracks worms leave on thin films of agar or the accumulation of worms in attractant gradients established in thin slurrys of Sephadex beads (Ward, 1973). By comparing the changes in concentration of attractant during the movement of the head, worms orient up a gradient and then they move directly up the gradient to accumulate at the gradient peak where they then habituate (Ward, 1973, 1978).

The sensory neurons mediating these responses are located in the amphids, sensilla located laterally in the tip of the head which open to the outside medium through the cuticle (Ward et al., 1975). Similar sense organs are found in all other nematodes (Wright, 1980). Each amphid contains eight ciliated sensory neurons exposed to the outside medium. The pattern of chemical response of individual neurons is not known, and the analysis of mutants that are defective in chemotaxis including some defective in amphidial neurons, does not give a simple correlation between anatomical and behavioral defect.

The sensory responses of parasitic nematodes and other parasitic helminths has been studied to some degree. It is clear that chemical senses are essential to locate hosts, to find tissues in hosts, and to locate mates. Interfering with such chemical responses would surely disrupt a parasite's life cycle. This would seem to offer an excellent opportunity for developing new classes of chemotherapeutic agents which would not have to penetrate the nematode cuticle in order to block the receptive portion of a sensory neuron. *Caenorhabditis* would be a useful model for screening for such agents because the behavioral assays are simple and economical so that they could be done in large numbers.

## NEUROCHEMISTRY

Several neurotransmitters have been identified in *Caenorhabditis* and investigation of their action, biosynthesis, and receptors is underway (Chalfie and White, 1988). Without direct neurophysiological recording, the evidence associating a transmitter with a particular synapse is necessarily indirect, but electrophysiological studies on *Ascaris* have verified the neuromuscular transmitters. Neurotransmitter studies are of particular significance to parasitic helminths, since most successful anthelmintics act against neurotransmission. The selection of mutants resistant to some of these anthelmintics has led to identification of genes responsible for resistance. When these genes are isolated it should be possible to study their protein products in more detail, some of which should be the direct target of the drug. This may make it possible to improve current anthelmintic drugs and to design new ones based on a better understanding of their action.

As has been established for *Ascaris*, acetylcholine (Ach) is the excitatory neuromuscular transmitter. Treatment of intact or cut worms with Ach agonists such as levamisole or Ach esterase inhibitors such as lannate (Aldicarb) causes muscular contraction and egg laying. By analysis of mutants resistant to lannate, the gene for choline acetyl transferase, *cha-1*, has been identified (Rand and Russell, 1984). It appears that the most common way for the worm to become resistant to inhibition of Ach degradation is to reduce the amount of Ach synthesized. There is a complex genetic interaction between mutations in the *cha-1* gene and a neighboring locus, *unc-17*. This may imply that the protein product of these genes is a single, large multidomain polypeptide with more than one function.

Three classes of Ach esterase have been identified enzymatically and shown to be the products of three distinct genes (Johnson and Russell, 1983; Kolson and Russell, 1985). Although mutations in individual genes alter the pattern of histochemical staining for esterase, they have no apparent behavioral phenotype implying that the genes are redundant. It is only when mutations in the two most abundant forms of the enzyme are combined that an uncoordinated and paralyzed phenotype is observed.

The Ach receptor has been identified by a binding assay with a levamisole analogue [$^3$H]-meta-aminolevamisole (Lewis et al., 1987). Twelve genes have been identified which can mutate to levamisole resistance; three of them appear to be necessary for binding, and another four alter the binding. It is not clear why so many genes affect the binding of a receptor. The receptor is likely to be complex multisubunit protein, thus these genes could encode different subunits or they could be involved in the processing or localization of the receptor complex. Cloning and identification of the gene products should clarify the structure of the receptor.

It is likely that the major inhibitory neuromuscular transmitter in nematodes is γ-aminobutyric acid (GABA). GABA-containing inhibitory motor neurons have been identified by biochemical and immunohistochemical studies in *Ascaris*. Application of GABA or muscimol, a GABA agonist, induces flaccid paralysis in *Caenorhabditis* (Chalfie and White, 1988). Mutants altered in the distribution of GABA have been identified, but their biochemical basis is not known. The response to GABA is of particular interest in nematodes, because the potent and successful drug avermectin appears to act at GABA synapses either by interacting with the receptor or an associated chloride channel (Kass et al., 1984). The possibility of isolating the genes responsible for drug resistance and identifying the target of avermectin may open new possibilities for development of novel anthelmintics.

The biogenic amines, dopamine, serotonin, and octopamine are all found in *Caenorhabditis*. Eight dopaminergic neurons have been identified by formaldehyde-induced fluorescence and mutants reducing or altering the distribution of fluorescence have been identified (Sulston et al., 1975). The only behavioral consequence of these mutations is a reduction in male mating efficiency. Serotonergic neurons have also been identified histochemically and are involved in controlling pharyngeal pumping and egg-laying, both of which are stimulated by application of serotonin to whole worms. Mutants with reduced serotonin and altered or eliminated staining of serotonergic neurons are found among the class of egg-laying defective mutants (Trent et al., 1983). Octopamine has an opposite effect to serotonin when applied exogenously, but neurons containing it have not been identified.

## CONCLUSIONS

From this brief summary, it is clear that research on *Caenorhabditis* can contribute to the study of parasitic nematodes in several ways. First, information learned about the genetics, anatomy, biochemistry, and development can be applied directly to other nematodes because of their similarity and common evolutionary origin. Second, the methods developed for studying *Caenorhabditis*, such as lineage analysis, laser ablation, behavioral analysis, and DNA transformation can be applied directly to other worms. Third, "reagents" developed in the study of *Caenorhabditis* such as drug-resistant strains, genomic and cDNA libraries, cloned genes, antibodies, and identified antigens can be used in the study of other nematodes. Fourth, *Caenorhabditis* and mutant strains can be used for initial screening for novel chemotherapeutic agents.

## ACKNOWLEDGMENTS

The work in the author's laboratory has been supported by NIH grant GM25243 and the MacArthur Foundation Consortium on the Biology of Parasitic Diseases, Johns Hopkins Program. The manuscript benefitted from comments by Diane Shakes, Dennis Dixon, and Jacob Varkey.

## REFERENCES

Albertson DG (1985): Mapping muscle protein genes by *in situ* hybridization using biotin-labeled probes. EMBO J 4:2493–2498.
Bennett KL, Ward S (1986): Neither a germ line-specific nor several somatically expressed genes are lost or rearranged during embryonic chromatin diminution in the nematode *Ascaris lumbricoides* var. *suum*. Dev Biol 118:141–147.
Brenner S (1974): The genetics of *Caenorhabditis elegans*. Genetics 77:71–94.
Burke DJ, Ward S (1983): Identification of a large multigene family encoding the major sperm protein of *Caenorhabditis elegans*. J Mol Biol 171:1–29.

Burke DT, Carle GF, Olson MV (1987): Cloning of large segments of exogenous DNA into yeast by means of artificial chromosome vectors. Science 236:806–812.

Burr AH (1985): The photomovement of *Caenorhabditis elegans*, a nematode which lacks ocelli. Proof that the response is to light not radiant heating. Photochem Photobiol 10:577–582.

Chalfie M, White JG (1988): The nervous system. The nematode *Caenorhabditis elegans*: Chapter 11 in Wood, 1988.

Coulson A, Sulston J, Brenner S, Karn, J. (1988): Towards a physical map of the genome of the nematode *Caenorhabditis elegans*. Proc Nat Acad Sci USA 83:7821–7825.

Dougherty EC, Calhoun HG (1949): Possible significance of free-living nematodes in genetic research. Nature 161:29.

Douvres FW, Urban JF (1983): Factors contributing to the in vitro development of *Ascaris suum* from second-stage larvae to mature adults. J Parasit 69:570–576.

Dusenbery DB (1974): Analysis of chemotaxis in the nematode *Caenorhabditis elegans* by countercurrent separation. J Exp Zool 188:41–48.

Dusenbery DB (1975): The avoidance of D-tryptophan by the nematode *Caenorhabditis elegans*. J Exp Zool 193:413–418.

Emmons S (1988): The genome. Chap 3 in Wood, 1988.

Fire A (1986): Integrative transformation of *Caenorhabditis elegans*. EMBO J 5:2673–2880.

Johnson CD and Russell RL (1983): Multiple molecular forms of acetylcholinesterase in the nematode *Caenorhabditis elegans*. J Neurochem 41:30–46.

Kass IR, Stretton AOW, Wang CC (1984): The effects of avermectin and drugs related to acetylcholine and 4-aminobutyric acid on neurotransmission in *Ascaris suum*. Mol Biochem Parasitol 13:213–225.

Kimble J (1981): Alterations in cell lineage following laser ablation of cells in the somatic gonad of *Caenorhabditis elegans*. Dev Biol 87:286–300.

Kimble J and Ward S (1988): Germ line development and fertilization. Chap 7 in Wood, 1988.

Klass MR, Kinsley S, Lopez LC (1984): Isolation and characterization of a sperm-specific gene family in the nematode *Caenorhabditis elegans*. Mol Cell Biol 4:529–537.

Kolson DL, Russell RL (1985): A novel class of acetylcholinesterase revealed by mutations in the nematode *Caenorhabditis elegans*. J Neurogenet 2:93–110.

Krause M, Hirsh D (1987): A *trans*-spliced leader sequence on actin mRNA in *C. elegans*. Cell 49:753–761.

Lewis JA, Fleming JT, McLafferty S, Murphy H, Wu C (1987): The levamisole receptor, a cholinergic receptor of the nematode *Caenorhabditis elegans*. Mol Pharmacol 31:185–193.

Office of Technology Assessment, U.S. Congress, (1985): Status of Biomedical Research and Related Technology for Tropical Diseases. Washington, DC: U.S. Govt. Printing Office.

Priess JR and Thomson JN (1987): Cellular interactions in early *C. elegans* embryos. Cell 48:241–250.

Rand JB, Russell RL (1984): Choline acetyltransferase-deficient mutants of the nematode *Caenorhabditis elegans*. Genetics 106:227–248.

Riddle D (1988): The dauer larvae. The nematode *Caenorhabditis elegans*. Chap 12 in Wood, 1988.

Schwartz DC, Cantor CR (1984): Separation of yeast chromosome-sized DNAs by pulsed field gradient gel electrophoresis. Cell 34:77–85.

Sternberg PW, Horvitz HR (1982): Postembryonic nongonadal cell lineages of the nematode *Panagrellus redivivus*: Description and comparison with those of *Caenorhabditis elegans*. Dev Biol 93:181-205.

Stinchcomb DT, Shaw JE, Carr SH, Hirsh D (1985): Extrachromosomal DNA transformation of *Caenorhabditis elegans*. Mol Cell Biol 5:3484-3496.

Stretton AOW, Davis RE, Angstadt JD, Donmoyer JE, Johnson CD (1985): Neural control of behavior in *Ascaris*. Trends Neurosci. 8:294-300.

Sulston JS (1988): Cell lineage. The nematode *Caenorhabditis elegans*. Chap 5 in Wood, 1988.

Sulston JE, Dew M, Brenner S (1975): Dopaminergic neurons in the nematode *Caenorhabditis elegans*. J Comp Neurol 163:215-226.

Trent C, Tsung N, Horvitz, HR (1983): Egg laying defective mutants of the nematode *Caenorhabditis elegans*. Genetics 104:619-647.

Van der Ploeg LHT, Bernards A, Rijsewijk FAM, Borst P (1982): Characterization of the DNA duplication-transposition that controls the expression of two genes for the variant glycoprotein in *Trypanosoma brucei*. Nucleic Acids Res 10:593-609.

Ward S (1973): Chemotaxis by the nematode *Caenorhabditis elegans*: Identification of attractants and analysis of the response by use of mutants. Proc Natl Acad Sci USA 70:817-821.

Ward S (1978): Nematode chemoreceptors and behavior. Taxis and Behavior. G. Hazelbauer (ed): London: Chapman and Hall, (Receptors and Recognition, Series B, Vol. 5.) pp. 143-168.

Ward S, Thomson N, White JG, Brenner S (1975): Electron microscopical reconstruction of the anterior sensory anatomy of the nematode *Caenorhabditis elegans*. J Comp Neurol 160:313-338.

Ward S, Burke DJ, Sulston JE, Coulson AR, Albertson DG, Ammons D, Klass M, Hogan E (1988): The genomic organization of major sperm protein genes and pseudogenes in the nematode *Caenorhabditis elegans*. J Mol Biol 199:1-13.

Warren KS, Bowers JZ (1983): Parasitology, a Global Perspective. New York: Springer-Verlag.

White JG, Southgate E, Thomson JN, Brenner S (1986): The structure of the nervous system of the nematode *Caenorhabditis elegans*. Philos Trans R Soc Lond [Biol]: in press.

Wood WB, ed. (1988): The Nematode *Caenorhabditis elegans*: Cold Spring Harbor NY: Cold Spring Harbor Laboratory, in press.

Wright KA (1980): Nematode sense organs. In B. Zuckerman (ed): Nematodes as Biological Models 2. New York: Academic Press, pp 237-295.

# APPENDIX

Participants in the Biology of Parasitism course, 1980: Sally Anderson, Sergio Arias-Negrete, Kenneth Boutte, Manuel Canla, Khalil Hafez, Barbara Johnson, Michael Johnston, Rena Jones, Perri Klass, Francis Klotz, Davy Koech, Eileen Lynch, Paul Mellen, Ian Rosenberg, Aloys Tumboh-Oeri, Dyann Wirth. Photographs of 1980-1987 course participants by Linda M. Golder.

Participants in the Biology of Parasitism course, 1981: Eva Avila, Ann Duerr, Pietro Liberti, Lawrence Lichtenstein, Kamini Mendis, Christian Ockenhouse, Francine Perler, Mario Philipp, Louis Safranek, Isabel Santos, Stuart Shapiro, Richard Sidner, Maggie So, Lindsey Unbekant, Carl Wahlgren, Ellen Winchell.

Participants in the Biology of Parasitism course, 1982: Boaz Avron, James Bangs, Robert Barker, Carl Boswell, Gregory Buck, Tecia Maria Ulisses de Carvalho, Marie-France Delauw, Kasturi Haldar, Michele Jungery, Pamela Langer, Pamela Moriearty, Martin Pammenter, Debra Rowse, Peter Tseng, Steven Zeichner, Dan Zilberstein.

Participants in the Biology of Parasitism course, 1983: Luis F. Anrya-Velazquez, Virendra Bhasin, Larry Chavez, Yara M. Traub Cseko, Dirk A.E. Dobbelaere, Galayanee Ekapanyakul, Alexandra Swiecicki Fairfield, Ana Flisser, Howard M. Goodman, Jessica Krakow, Amy J. Percy, Luis I. Rivas Lopez, Guillermo G. Romero, Ralph T. Schwarz, Laurence D. Sibley, Claire Wyman.

Participants in the Biology of Parasitism course, 1984: James R. Campbell, Dolores Correa Beltran, Paul R. Crocker, Gay Goodman, Max Grogl, Rabie Hussain, Halliday A. Idikio, Edwin G.E. Jahngen, Gale Wantanabe Jeffers, Jeffrey T. Joseph, Samuel I. Miller, Adolfo Martínez-Palomo, Mauricio Martins Rodrugues, William F. Pomputius III, Betty Kim Lee Sim, Nobuko Yoshida.

Participants in the Biology of Parasitism course, 1985: Norma Andrews, Carlos Arguello, Aldina Barral, Karen Bennett, James Burns, Jose Cordova, Tamara Doering, Maria Febbraio, Michael Foley, Rollin Johnson, Juan Lafaille, Moses Limo, Myriam Mogyoros, Babatunde Osotimehin, Gillian Woollett, Alejandro Zentella-Dehesa.

Participants in the Biology of Parasitism course, 1986: Pietro Alano, Birgitte Andersen, Parul Doshi, Josianne Eid, Sofia Haryana, George Lombardi, Richard Lucius, Subhash Morzaria, Jose Rosales, Nicholas Samaras, Opeolu Shonekan, Photini Sinnis, Pamela Stucky, Patricia Talamas, William Weidanz, Ronald Zimmerman.

Participants in the Biology of Parasitism course, 1987: Enrique Acosta-Gio, Raquel Alvarez, Lena Aslund, Wendy Barry, Carla Cerami, Dalia Gordon, Barbara Herwaldt, Marc Karam, Mo Klinkert, Sridhar Mani, Reginaldo Prioli, Cesare Rossi, David Sherman, Anders Sjolander, Andrew Slater, Miriam Tendler.

# Index